The Miocene Stratigraphy Of California Revisited

Also Pliocene Biostratigraphy Of California

AAPG Studies in Geology No. 11

The Miocene Stratigraphy of California Revisited by R. M. Kleinpell

Also

Pliocene Biostratigraphy of California by C. R. Haller

with special sections by:

G. Hornaday	illustrations by Mary E. Taylor
A. D. Warren	extra plates by Margaret Hanna
A. Tipton	

Published by
The American Association of Petroleum Geologists
Tulsa, Oklahoma, 74101, U.S.A.

Published December 1980
Library of Congress Catalog Card Number: 80-69781
ISBN: 0-89181-015-3

AAPG Staff members who have contributed to this project include:

Peggy Rice, E. M. Tidwell, Deborah Zikmund, Gary Howell, Grace Hower, Ronald Hart, Douglas White, Sally Hunt, Nancy Wise, Regina Gill, Peggy Pendergast, Laura Denson, June McFarland, Carol Short, Pat Calvert, Amy Brown, A. A. Meyerhoff, E. A. Beaumont, and Robert H. Dott.

The Association gratefully acknowledges the partial financial assistance provided by the Pacific Section of The American Association of Petroleum Geologists in preparation of this volume.

Printed by
Edwards Brothers, Inc.
Ann Arbor, Michigan, U.S.A.

Table of Contents

Miocene Stratigraphy of California: Introduction

Background and Acknowledgments

In the autumn of 1970 Kingsley Nash, president of the Pacific Section of the Society of Economic Paleontologists and Mineralogists, invited the senior writer (R.M. Kleinpell) to update his 1938 book, *Miocene Stratigraphy of California*. Two problems were immediately evident. First, should the book be supplemented or the entire subject updated? This question was readily resolved because updating the subject as a whole was too comprehensive a task to be undertaken at the time. With the aim of this work thus qualified as a supplement to the 1938 book and its 1955 reprint, the reader will in consequence wish to have at hand a copy of one of the earlier works, since references thereto will be made in order to avoid unnecessary repetition. Second, funds were required to support preparation of such an updated supplement. Early in 1971, the Pacific Sections of The American Association of Petroleum Geologists and the Society of Economic Paleontologists and Mineralogists agreed to furnish the necessary funds.

In addition, the Pacific Sections conducted field trips to critical areas of Miocene outcrop in California and prepared a program for their March 1972 joint meetings in Bakersfield, all with a view to eliciting data pertinent to an updating effort. Many geologists, paleontologists, and others contributed to these preliminary efforts. An entire afternoon of the Bakersfield meetings was given to a general discussion, paneled by Orville L. Bandy, R. Stanley Beck, John L. Browning, Clifford C. Church, Manley Natland, and R.M. Kleinpell, with E.H. Stinemeyer and R.C. Blaisdell as moderators. The 21 papers presented at the Bakersfield meetings were subsequently published as "The Proceedings of the Pacific Coast Biostratigraphic Symposium Presented at the Forty-Seventh Annual Pacific Section, S.E.P.M. Convention, March 9-10, 1972, Bakersfield, California" (Stinemeyer, 1972.) Additional information has been acquired through field trips by the joint writers, assisted by Weldon Rau, Charles Fulmer, and others. Further data have been published in the guidebooks for more formal field trips and symposiums.

The writers express their thanks and appreciation to all who contributed to these preliminary and related efforts. Too numerous to mention here, many contributors will be noted in subsequent pages. The writers particularly wish to acknowledge, however, Kingsley Nash and John Curran and their associates and successors in office in the Pacific sections of AAPG and SEPM and A.A. Meyerhoff, former publications manager of AAPG, and his associates and successors.

In turn, the writers express appreciation to their many colleagues in foraminiferology and geology for help through long years of direct or indirect association in joint professional efforts; appreciation is also due many former students at the University of California, Berkeley and Santa Barbara, whose arduous legwork and sometimes tedious research contributed so much. To professors S.W. Muller of Stanford University, J.D. Barksdale, V.S. Mallory, and Howard Coombs of the University of Washington, A.O. Woodford of Pomona College, and to colleagues of former days at University of California at Berkeley, the senior writer gratefully acknowledges his indebtedness for many fruitful discussions, and to professors Zach Arnold, W.B.N. Berry, and Gordon Hornaday of the Museum of Paleontology at University of California at Berkeley for their assistance in problems of stratigraphic paleontology and foraminiferology. A.A. Almgren, M.N. Bramlette, C.C. Church, J.M. Hamill, W.P. Woodring, and W.W. Wornardt graciously loaned or gave critically pertinent paleontologic materials. Herschel L. Driver provided a welcome and timely financial contribution. At the risk of inadvertently underestimating the help of many others, the writers note with particular appreciation the discussions and critical commentaries of R. Stanley Beck, Clifford Church, Joseph C. Clark, Charles Fulmer, Margaret Hanna, W.H. Holman, Kenneth Lohman, Weldon Rau, and Walter Wornardt, as well as the staff of the Department of Geological Sciences at the University of California at Santa Barbara who contributed both scientific commentary and technical assistance.

The writers also thank Mary E. Taylor for her care and patience in illustrating the fossil foraminifers, Margaret Moore Hanna for figuring certain of the siphogenerinids and valvulinerids, and Jay Phillips for his help in preparing the list of references. Secretarial assistance has been most efficiently provided by Evelyn Gordon, Diane Mondragon, and the secretarial staff of Anderson, Warren and Associates, Inc.

Scope

What are the important areas for updating in a supplement such as the present work? First, there should unquestionably be illustrations of foraminiferal species that previously were widely scattered and commonly inaccessible to workers. Second, the criteria should be clarified for interpretive conclusions and updatings in stratigraphic paleontology with special emphasis on the principles of chronology, chorology, and phylogeny. These disciplines constitute the basis for evaluating data and arriving at the scientific conclusions basic to this work. Some of these criteria were presented in the original work in foreign languages. These should be translated because many workers are unacquainted with the German and French scripts. Third, some history of the stratigraphic paleontology of the West Coast Tertiary is appropriate, particularly an elucidating of the Oligocene age of the so-called lower Miocene and a clarification of a variety of problematic boundaries and misusages of terms.

Many pertinent data are now available to supplement the original work, including references to papers published since 1938 and a checklist of species and their distribution in the type section of the Luisian Stage. Some additional chorological data concerning matters of facies are provided, because the 1938 book offered only the earliest attempts to arrive at paleoecologic interpretations (in the form, mainly, of compilation of data on living foraminifers). Furthermore, several interpretations in the original work need to be corrected.

A vital phase of the updating is the presentation of foraminiferal lineages within a superpositional context and discipline, where these have become demonstrable. In this connection, several significant new genera and species are described. Next, the extensions in the stratigraphic ranges of several species are demonstrated wherever adequate disciplines on such species range extensions exist. Revisions and updatings of chronologic classifications are vital, primarily in correlations in age of zonal refinement. As indicated previously, certain correlations emphasized in the original work need correcting because problems of facies, not understood at the time, contributed to miscorrelations. Some of these corrections already have been published, to which publications reference will be made.

As an updating of the earlier book, the present work will be subject in due time to further corrections and clarifications of interpretation. In this context, the writers wish to note that any misinterpretations have been made in good faith and have been based on the disciplines of the data and the demonstrated principles available at this time. The position taken by the writers in this matter is succinctly stated in the following passage from A.M. Davies (1934, p. 48).

> The practical value of fossils lies in the use that can be made of them in geologic mapping and correlation. If, however, fossils are used by rule of thumb alone, there are pitfalls and traps that cannot be avoided. Some understanding of the conditions which control life and evolution of organisms is essential to the intelligent judgment of the evidence that fossils offer.

Thus, *facies,* both paleontologic and lithologic, enter the establishment of a geochronology which is accomplished, as pointed out by Davies (1934, p. 56), by a "slow process of trial and error." However, through this very process we come to recognize the various paleontologic *facies* which are contemporaneous.

This brings us back to the need for clarifying the principles of chronology, chorology, and phylogeny, beyond what had been possible in 1938. Although updated coverage of these three groups of principles that are vital in stratigraphic paleontology has been accomplished by the writer, this portion of the updated manuscript, even when reduced to its bare minimum, proved quite extensive and has been considered too cumbersome to include in the present supplement. Through the cooperation of the Pacific Sections of AAPG and SEPM with the national AAPG organization in Tulsa, it has been possible to arrange for the publication of such a section, devoted to Criteria in Correlation, as a separate publication under that title, which is to be made available through the Pacific Sections.

In essence, this separate publication will serve to update and expand upon, as needed, that portion of the 1938 book between pages 11-19, 79-84, and 87-99. Included will be pertinent commentary on the passage presented on pages 90-91 in French (from Munier-Chalmas and de Lapparent, 1893), and translations of the passages originally presented in German (from Oppel, 1856-58, and Diener, 1925), on pages 92-95 and 96-97.

For the purposes of the present updating, the disciplines in reference above may all be grouped under the heading of *principles.* As mentioned, these principles constitute the basis for the interpretations and conclusions arrived at in the present supplement, (i.e., those that are considered as being valid in the light of currently available data). Suffice it to present here, therefore, a basic definition of the term *general principle* as herein employed. It is taken from the work of J.S. Mill, as given in Webster's 1934 unabridged dictionary:

> A principle ascertained by experience is more than a mere summing up of what has been specifically observed in the individual cases which have been examined; it is a generalization grounded on those cases.

History of Stratigraphic Paleontology of West Coast Tertiary

R.M. KLEINPELL

Beginnings

The beginnings of intensive micropaleontologic studies on the Pacific Coast coincide rather closely with the first big search for petroleum following World War I. From the turn of the century only a few stratigraphically isolated foraminiferal faunules had been described in print. Beyond general references, accounts published by Chapman (1896, 1900) and Bagg (1905) were the only local foraminiferal records available. As it happened, Chapman's assemblage was not accessible and, with a few local exceptions, the species identified by Bagg were represented by illustrations of European specimens. Bagg's genuine Henry Ranch material was also unavailable for comparison, until J.P. Smith located some of it at Stanford University. The material had been left there by J.C. Branner, who had collected it, along with a note as to the locality. Meanwhile, Joseph Cushman had re-collected on the Henry Ranch for Marland Oil. He decided that the three new California siphogenerine species described and named by Bagg had probably been illustrated in the same stylized manner as Bagg's other "species" and that probably they would be conspecific with some siphogenerines he, Cushman, had named from the collections made by W.D. Kleinpell in San Luis Obispo County. Thus, a mistaken synonymy appeared in Cushman's 1926 monograph on the genus *Siphogenerina*. Bagg's illustrations of *Siphogenerina branneri* proved meticulously accurate, however. Unfortunately, the consequent semantic confusion in the taxonomy of Pacific Coast siphogenerinds has persisted (see Kleinpell, 1938, p. 21-22, 183, 299-300; 1972, p. 89-90). With the growing interest in geologic exploration for petroleum and its micropaleontologic adjunct, the taxonomy of a number of microfaunules, such as those recorded by Chapman and Bagg, was soon being formally recorded (see Kleinpell, 1972, p. 2).

Old Epoch Boundary Problems

The West Coast superjacent stratigraphic column previously had been blocked out on the basis of marine megafossils, with certain mammalian and paleobotanical correlates. This classification had followed the recognition on the West Coast of the major Tertiary subdivisions of Lyell, including the work of Deshayes, Bonelli, Desnoyers, deBasterot, Prevost, and their successors (see Lyell's *Principles,* v. 3, 1833, p. xii-xiii, 12-17, 18-22, 30-34, 41-44, 49-52, 53-61, 62-68, 155-182, 202-216, 241-256, 275-316, 351, and the tables). Of these series-epochs, the Miocene was at first only vaguely recognized, much more tentatively than the Eocene, Pliocene, or Pleistocene.

Series boundary problems in the West Coast ranges became substantial. During early railroad surveys, fossil mollusks in a float-boulder picked up by William P. Blake in the vicinity of Tejon Pass were recognized by Timothy A. Conrad as Lyell's Eocene. Lyell's Pliocene was recognized in the Merced beds along Seven Mile Beach on the San Francisco Peninsula as a result of work by the California Geological Survey (1856-66) and by William Gabb of the Whitney Survey.

It was long thought that ammonites had persisted into the Tertiary in the West Coast ranges. The controversy and correlation problems that resulted seem to have stemmed primarily from the conclusions of Gabb. These conclusions were sustained by John B. Trask. The controversy was resolved, at least as to age, by studies of the giant pelecypod "*Venericardia planicosta*" conducted by Angelo Heilprin of the Philadephia Academy. As a result of Heilprin's work, this pelecypod was recognized as the worldwide "fingerpost of the Eocene." Thus, although distinct species of this genus were eventually established in California's lower Tertiary rocks, the Eocene age of these controversial strata was confirmed as Conrad had concluded.

The Tertiary ammonite controversy nevertheless persisted. In spite of Lyell's criteria for subdividing the Old World Tertiary, many early workers could not conceive that a species could persist from one geologic epoch to another, let alone cross the boundary of a period or an era. With respect to both period and stock of the "Tertiary ammonites," this incredulity, however invalid in general, in this particular instance proved correct. Hand-lens inspection of foraminifers finally revealed, without benefit of megafossil evidence, that the great "Cretaceous" Chico Formation behind the Stanford campus, although ammonite-bearing at base, ranged high into the Eocene (see Kleinpell 1972, p 3). Other controversial localities were revisited and their fossils restudied. Giant venericards and ammonites were not found in association and, as J.P. Smith often noted to his students, the last of the supposed West Coast Tertiary ammonites turned out to be poorly preserved and misidentified sea urchins.

Two other aspects of the original controversy persisted well through the first quarter of the century, however. First, many structural geologists could not accept that the structural strain of the Laramide disturbance, so well recorded in the continental interior, would be barely discernible on the West Coast. Stratigraphers and geologists could hardly avoid noticing the great lithologic similarity between the Late Cretaceous and early Tertiary formation of West Coast ranges and, in an orogenic sense, the feeble structural reflection of the stress of the Laramide disturbance. Hence, the tendency was to correlate by lithology.

Second, concordant formations exhibiting essentially the same lithologies were considered single time-rock units and were dated by fossil species or fauna found within the limits of a formation. Inherent in this view is the supposition that a fossil species or faunal aggregate cannot survive geologic changes in the environment and that, furthermore, formations cannot transgress time.

It should be noted, however, that, for example, the Cretaceous-Tertiary boundary along the west side of the San Joaquin Valley is generally a shale-on-shale contact; a thin glauconite bed is practically the only mapping clue. The time significance of the sequence was eventually demonstrated largely through careful collecting and foraminiferal studies by C.C. Church, R.W. Burger, and others.

The Pliocene-Pleistocene boundary was soon clarified also, at least to the extent that this boundary has ever been clear in the type terrain of Asti in northern Italy versus that in the far more southerly Ischia-Val de Noto area. Arnold's (1906) monographic studies of the pectens and of the San Pedro Formation contributed substantially to resolving this series boundary problem locally. As usually drawn, the boundary was previously set between a lower, warmer water and an upper, colder water molluscan fauna as found in the typical Santa Barbara Formation. This reflected a climatic trend that had been obscured and confused in the original studies of the San Pedro, where float from the upper (subsequently termed Palos Verdes) beds cropping out at the top of a cliff had intermingled with fossils from beds cropping out at the cliff's base.

Following U.S. Geological Survey usage since about 1952, the local Pliocene-Pleistocene boundary generally has been placed at the base of the Santa Barbara Formation and its age-equivalents—thus at the base of the "mud pit" shale of the uppermost Pico mudstone as found along the Ventura River. An upper limit of the Pleistocene, involving a post-Ice Age Holocene series and epoch, has only recently been recognized by the U.S. Geological Survey.

Of all the West Coast Tertiary series-epoch boundaries, however, the upper and lower boundaries of the Miocene and subsequently of the Oligocene were the least readily drawn.

The Miocene-Pliocene boundary controversy in California stems partly from the relative vagueness of this upper Miocene boundary in the Old World. In terms of continuous marine deposition, an unfossiliferous interval exists between the type Miocene and the type Pliocene of northern Italy. Furthermore, different terminologies have been used for marine, brackish, and nonmarine beds in this interval. Traditionally, French workers and mammal specialists have included the transitional interval in the Pliocene; German students and many marine invertebrate specialists include the same transitional interval in the Miocene. Today, the Zanclean Stage commonly is designated as the basal Pliocene, instead of the overlying Plaisancian of past usage. The brackish-water Sarmatian beds and the Pontian deposits, rich in *Hipparion* horse faunas, occupy this interval in southern and parts of central and eastern Europe and are thus difficult to relate to the normal marine sequence elsewhere. *Hipparion* is known to have appeared in Europe as early as latest Tortonian times (Charles Reppening, personal commun.). Along the West Coast of North America, this same interval occurs within the foraminiferal sequences designated as upper Mohnian, Delmontian, and Repettian; the molluscan sequences are commonly designated as Briones, Cierbo-Neroly, Jacalitos, or the more recently recognized Cerrotejonian-Montediablan (i.e., Clarendonian) mammalian sequence. In view of the above, the designation "Mio-Pliocene" is often used for this transitional interval.

The question of the precise position of the Oligocene-Miocene boundary in the Old World (i.e., the lower boundary of the typical Miocene) is closely tied to faunal and lithologic facies problems associated with the Chattian-Aquitainian interval of France and Germany. It has yet to be resolved in many places even in the Old World.

Von Beyrich established his Oligocene Series by designating beds that were age-equivalents of the upper beds of a stratal sequence selected by Lyell as type for his Eocene. In addition, von Beyrich included beds found in the German lowlands previously referred to the Miocene by earlier workers and by Lyell himself.

Included in the Oligocene were strata "which are intermediate in age between the Miocene formations of the *Faluns jaunes* of Bordeaux, the Touraine, . . . and the older Eocene formations of the 'Calcaire grossier' of Paris and the lower Tertiary deposits of England to, and including those of the Barton clay. . . if one restricts the name Miocene to only those deposits of the age of the Mussel beds (Faluns) of Touraine or the yellow sands of Bordeaux."

The recognition of the Oligocene Series in the West Coast ranges is symptomatic of the history of the Eocene-Miocene boundary problem on the Pacific Coast. Certain fossiliferous strata in the Caribbean were pronounced Oligocene equivalents after von Beyrich had distinguished and defined his Oligocene. Next, with these newly recognized New World Oligocene beds, William Dall then correlated certain fossiliferous strata in and around Astoria, Oregon. Principally on the basis of their biogenetic affinities with the Astoria fossils, Arnold next designated as Oligocene the fossils found stratigraphically high in the typical San Lorenzo Formation of California's Santa Cruz Mountains. The latter fossils were discovered at horizons transitional to the overlying Vaqueros Formation, in which, at still higher levels, truly Vaqueros fossils had been found.

Thus recognition of the Oligocene Series had arrived in California. However, the Caribbean beds in reference were, in fact, Miocene, and Dall's correlation between Oregon and the Caribbean proved correct at the series-epoch level. The San Lorenzo mollusks, however, turned out not to be chronologically equivalent to those around Astoria, though they are biogenetically and apparently ancestrally related. In view of this tangle of miscorrelation, it is not surprising that Eocene-Oligocene and Oligocene-Miocene boundaries in both the Pacific Coast ranges and in Europe have continued to be controversial. Nor is it surprising that the term "Oligocene" was ignored by early workers on the West Coast. F.M. Anderson insisted vehemently that typologically there was no such epoch as Oligocene, only Eocene. Mrs. Hubert G. Schenck, whose husband was considered an authority on the Oligocene, remarked that it was a somewhat dubious honor to be the wife of a scientist who was an esteemed authority on something the very existence of which was in doubt. Even as recently as 1962, Eames et al stated in reference to the Americas "that, in the whole region we have considered (and even as far north as the State of Washington), there are no published records of stratigraphical sections of fossiliferous marine beds which can be dated as Oligocene. Almost all those beds previously dated as Oligocene, and even some previously referred to the Upper Eocene, we regard as being undoubtedly Miocene (mainly Lower Miocene)." It should be noted that Woodring (1970) has clarified the subsequent change of mind by Eames et al (1968).

In California, the base of the Miocene has been placed by some workers as low as the lowest beds of the Vaqueros Sandstone, now recognized on the basis of foraminiferal studies by Kleinpell, Fulmer, Rau, and others as a Blakeley Oligocene age equivalent. As Schenck (1935) pointed out, Timothy Conrad had simply noted that the fossils brought to him from Salinas Valley strata named Vaqueros by Hamlin (1904) and Fairbanks (1904) more closely resembled those of Miocene than Eocene age. Neglecting Conrad's qualification that these strata were not necessarily contemporaneous with European Miocene, workers were soon considering the Vaqueros to be of early Miocene age. The threefold sequence of Tertiary formations in the Salinas Valley eventually became the classic sequence for the California Miocene.

Above the Vaqueros formation lay some unique siliceous shales, termed "Monterey" by Blake (1856), and then the sugary white sandstones that Antisell of the Pacific Railroad Survey camped on in 1856 and which were formally named Santa Margarita Sandstone by Fairbanks (1904). A rare scattering of arc shells and other facies megafossils in the Monterey caused it to be considered Miocene, largely on the basis of superposition. The Santa Margarita, once thought to be correlative with the Pliocene Etchegoin Formation of the San Joaquin Valley because of the abundance of the cockle *Pseudocardium* ("*Mulinia*") *densatum*, was, on the basis of scallops and sand dollars, correlated with the late Miocene San Pablo of the San Francisco Bay area, which contains ±23% living species (Clark, 1915). Reed (1925) dispelled the earlier notion of a widespread great unconformity between the Monterey and Santa Margarita, further consolidating the threefold sequence as the standard local Miocene sequence.

Meanwhile, Anderson (1905, 1908, 1911; Anderson and Martin, 1914) had described as the Temblor Formation certain megafossiliferous sandstones interbedded with siltstones and mudstones of the San Joaquin Valley and adjacent San Juan district of San Luis Obispo County. The discovery that stringers of these sandstones interfingered locally with Monterey shales promoted the term "Temblor-Monterey" as part of California's standard Miocene usage (Smith, 1919). It was eventually realized, largely through the work of Richards (1936), that the higher Santa Margarita sandstones also interfingered with, and even graded laterally into, Monterey shales. Yet the presence of the Monterey Shale, with its sparse and in many places undiagnostic megafauna, has contributed to the persistence on the West Coast of middle Tertiary boundary problems.

Formational Approach to Series Subdivisions in California

In California the practice gradually arose of referring to California subdivisions of Lyell's series-epochs in terms of local formations and their names. This practice was distinctly at variance with even the earliest U.S. Geological Survey policy, which stated that formations were mappable entities with geologic ages possibly varying from place to place and that two or more formations in different areas might be essentially the same age and might even interfinger locally. The need for a dual terminology in such matters, inherent in the designations employed by the writer (Kleinpell, 1938), was summarized by Schenck and Muller (1941), and has been further emphasized by Weaver and Tipton (1972, p. 52-62), particularly in reference to lithogenetic as distinguished from time-rock units. The system, series, and stage boundaries are also involved (Weaver and Tipton, 1972; Kleinpell, 1948, 1964, 1972; D.W. Weaver, 1969).

However valid the use of formations as supposed age symbols may have been in some places, they are entirely inadequate for stage and zone level correlations. Many times these divisions were defined by the mere presence at some formational horizon of a fauna or species thought to be an "index fossil." Formational terminology was all too often transferred directly into time terminology. This practice persisted in spite of ample clarification by Schenck and Muller (1941).

European Tertiary Stages and Zones

A good look at the classic type localities of the Tertiary stages and zones of Europe reveals that they have not been established on the basis of the principle of Oppel's zonation, as at least implied by Kleinpell (1938, p. 90-91). In contrast, the Tertiary "letter classification" of Umbgrove (Leupold and van der Vlerk, 1931) in the former Dutch East Indies *is* an Oppelian time-rock classification (based on large foraminifers in carbonate rocks). The European Tertiary stages are for most part simply sequences of exposed strata with fossils present at one or more zones within such a sequence. A few of them have the merit of avoiding a single name for a stage and formation (e.g., Lutetian and Calcaire Grossierre) such as followed in older California usage, and as criticized by Schenck and Muller (1941); but in only a few instances, not in many. Thus, no wonder that early workers in the California Tertiary, industrial or otherwise, found it difficult to correlate their fossiliferous strata with the stages of Europe to any degree of refinement beyond that of the guide-fossil, guide-assemblage, or Lyellian percentage-based correlations which permit age determinations to the system, series, or subseries level of refinement at best. Moreover subseries (e.g. "middle Miocene, etc.), seldom if ever have type localities (Weaver and Tipton 1972, p. 55) as standards of reference in correlation. Thus theoretical "evolutionary" age correlations were arrived at on the basis of things such as "stages of evolution" or "guide-fossil" taxa, without regard to general superpositional discipline. In more than one place these correlations have come in upside down, or at best, as classifications of geologic time, have changed fairly radically from week to week.

Refined correlation by fossils is difficult enough at best and is subject to revisions of interpretation from time to time as the number of relevant facts increases. Two facts, in any event, should be emphasized. First is the invalidity of the "stratotype" approach to the recognition of standards of reference in such refined correlations. Implied in this approach is the validity of purely lithologic correlation as a basis for time and time-rock classification and correlation based on such formal standards of reference; whereas such an inference is of course inherently invalid for correlating the top or bottom of any such time or time-rock unit. Moreover, it should be remembered that even Lyell, in defining his "periods" of the Tertiary—the series and epochs of the present hierarchy—did not set up limits to his series (see Weaver, 1969); nor did "stages" even exist at the time he first subdivided the Tertiary.

Second, Kleinpell's (1938, p. 90-91) deduction, that since Munier-Chalmas and de Lapparent set up a classification of European stages which included Tertiary stages, the same was possible in California, was *not* justified as he implied, in the sense that an Oppelian classification was being attempted for the Tertiary of California. Clearly, the criteria employed by Munier-Chalmas and de Lapparent for classifiying "stages" in the Tertiary of Europe do *not* lead to a stage-and-zone classification of the California Tertiary based on the principles Albert Oppel demonstrated in his subdivision of the Jurassic.

Another mistake inadvertently made by the writer in 1938 was the use of capital letters for the subdivisions of time-rock terms such as series and stages. "Where is the type section of the Middle Miocene?" professor H.G. Schenck would thunder, and quite significantly so, for *there is none*. "That is why stage subdivisions have been recognized, where possible," he would continue. An example, wherein lower case letters for vague and untyped time-rock units should have been used, may be cited from the Eocene. The lower part of the West Coast "Upper Eocene" correlates in age with the upper part of the Gulf Coast "Middle Eocene," with consequences that will be obvious to the reader (see Phillips 1972b).

Infra-Series Subdivisions in California

Because some sort of series-subdivision time-rock terms for such entities found in the West Coast ranges were needed in local correlation, zones based on the seemingly consistent stratigraphic ranges of certain species came into use. In early work, the scallops, sand dollars, and turritellid gastropods were the most commonly used. These finer subdivisions, based essentially on fossil species, were given zonal terms. Many of these turned out to be more biostratigraphic than biochronologic and as variable in time significance as formations. Some are actually teil-zones, as for example, J.C. Merriam's (1904) two early mid-Tertiary *Turritella* zones (*"hoffmani"* and *ocoyana*). Though still of value, these teilzones were shown by Loel and Corey (1932) to overlap in time-stratigraphic range. The Paleogene turritellid phylogenies of Merriam (1941) and the Neogene pecten phylogenies of Arnold (1906) continued to be the most useful West Coast Teritary age indicators, but these fossils were restricted to a relatively shallow-water facies and in many places general superpositional discipline was lacking. In short, correlations more refined in magnitude than an epoch or a vague and informal subepoch, however valuable in reconnaissance work, were merely local and particular synchroneities.

Vaqueros

At the base of the Salinas Valley "Miocene" sequence, the Vaqueros Sandstone at its type locality (Hamlin, 1904) acquired an age connotation on the basis of its fairly rich fossil mollusk and echinoid faunas. These faunas occur stratigraphically only through several hundred feet of section near the top of the mappable formation (which is over 4,000 ft thick; Dorrance, *in* Kleinpell, 1938, p. 7). The local megafossiliferous beds of Hamlin's type Vaqueros were correlated widely and thoroughly by Loel and Corey (1932, p. 31-410). It should nevertheless be noted that their "zones" are not the kind used in the present paper nor in Kleinpell (1938). Instead, Loel and Corey's zones are the kind described by Diener (1925) as "believige Schichte oder Bank in einem Lokal profile, in der ein bestimmtes Fossil haufig vorkommt"—that is, a "fossil bed" of the American geologic vernacular or a "biostrome" in more formal usage. Diener's discussion of the distinction of such zones from others (such as those of Oppel, 1856-58) was presented in the original German by Kleinpell (1938, p. 96-97).

Other aspects of the mistaken assumption that the type Vaqueros carried an age connotation were already noted in the 1930s (Kleinpell, 1934, 1938; Schenck, 1935). Not only was it indicated that the megafossiliferous beds were probably largely of Oligocene rather than early Miocene age (Kleinpell, 1938, Fig. 14, p. 168-172, and especially p. 181), but also it was noted that the type Vaqueros included intervals of finer grained sedimentary strata bearing foraminifers as old as Eocene (Kleinpell, 1938, p. 7 footnote). A few years later, Thorup (1941, 1943) reclassified the Vaqueros at the type locality. Using fossils lithogenetically, he established the Vaqueros (restricted) and, in the lower and unmegafossiliferous portion, four distinct members: in ascending order, the Junipero Sandstone, Lucia Shale, The Rocks Sandstone, and Berry Conglomerate. In the subsequent decade, W.C. Wardle mapped the area and studied the pre-Oligocene foraminifers in the Lucia Shale Member (Wardle 1957).

Other pertinent facts regarding the Vaqueros are: (1) in the Santa Cruz Mountains, some distance north of the typical Vaqueros area, a well-developed extension of the marine Vaqueros locally grades laterally into a formation of terrestrial origin, the Zayante (Clark, 1966b, p. 1184-b); (2) the most northerly known occurrence of its characteristic megafossils is in the Skooner Gulch Formation of the Mendocino County coast (Addicott, 1967); and (3) at its type locality, the top of the Vaqueros falls very close to the Oligocene-Miocene Series boundary.

Monterey Formation and Related Facies Problems

Inherent in the "Temblor-Monterey" usage for the supra-Vaqueros part of the Salinas Valley standard Miocene sequence was a facies factor, including both lithofacies and biofacies. The lower part of the type Temblor Formation of Anderson (1905) could actually be demonstrated as equivalent in age to the upper part of the type Vaqueros through both foraminiferal faunas (Kleinpell, 1930, 1938 Fig. 14) and megafossils (Arnold and Anderson, 1910). Students of megafossils were loathe to admit this, however, until Clark and Clark (1934) pointed it out (L.M. Clark *in* Schenck, 1935; *in* Kleinpell, 1938, p. 39).

That a facies factor was involved in Vaqueros-Temblor-Monterey relations was noted early in the century by Louderback (1913), but with singular consequences. The writer will never forget his experience in professor J.P. Smith's "Tertiary Conchology" class. Given a Miocene bibliography to peruse with no mention of the Louderback monograph on the Monterey, the monograph was spotted in the geologic library and taken to Smith for comment. It seemed to cover the entire subject assigned for study that week. Down to the tip of his nose went "J.P.'s" spectacles as he viewed the find with a heavy sigh. There followed the crushing comment, "Oh, my goodness. I wish you hadn't found that paper so soon. It's not a scientific paper, you know; it's a legal brief." And "J.P." resumed his whistling and returned to the "peeling back" of ammonites in which as usual he had been engaged. Smith was, of course, quite correct. Louderback had accurately called attention to the facies relations that existed between the Temblor Formation (which was recognizable from the San Joaquin Valley as far west as the Highland homocline east of Paso Robles) and the Vaqueros Sandstone of the more nearly coastal terrain in the western Salinas Valley. He had then, however, attempted to solve the age-relationship predicament by erecting a "Monterey Series" that included the Vaqueros, Temblor, and Monterey formations as facies. Such a "series" might have been better termed simply the Miocene Series of California. Nevertheless, the newly corrupted usage of "Monterey" was adopted by the U.S. Geological Survey and appeared in its San Francisco Folio, Professional Paper 193 (Lawson et al, 1914, p. 123). About 20 years passed before this legalistic and nonscientific confusion of terms was qualitatively clarified and formally dropped in favor of the original and typologically formational usage of "Monterey Shale." Meanwhile the unnecessary term "Salinas Formation" had been coined. It appeared in print as a local expression of the nearby type Monterey Shale—under the prevailing circumstances, quite unavoidably so (English, 1918; see Kleinpell 1938, p. 1, footnote 1).

Another facies aspect of the West Coast Miocene became apparent when foraminiferal faunas of the California "lower Miocene" (i.e., upper type Vaqueros, lower type Temblor formations) were shown to be age equivalents of the foraminiferal faunas of the Oligocene Blakeley Formation of Washington (Kleinpell, 1938, p. 77, Table XVII, and Fig. 14), despite marked differences in nearshore molluscan faunas of the two formations. The same facies problem came to include faunas of the San Lorenzo, upper Pleito, San Ramon, and other related West Coast formations. Much was clarified and brought into focus by the work of Fulmer (1975), which was begun before 1950, and of Rau (1948, 1951, 1958, 1964, 1966, 1967, 1975).

Santa Margarita

Late recognition of the age equivalence of much of the upper Monterey Shale and the Santa Margarita Formation perhaps was due in part to the noticeable lithogenetic differences between the sandstones of the Santa Margarita and those of the San Pablo, Modelo, and Puente Formations. Above the interbedded Monterey shales and local sandstones of the San Francisco Bay area, the San Pablo is essentially a Santa Margarita age equivalent, as shown by studies of San Pablo fossils (Weaver, 1909; Clark, 1915; Trask, 1922). Hudson and Craig (1929) demonstrated that the sandstone members of the typical Modelo were partial Temblor equivalents below and Santa Margarita equivalents above. Megafossils are too scarce in the Puente sandstone for age determinations of a refined sort, though in many places the shales carry rich foraminiferal faunas. The Monterey and Santa Margarita of the Salinas Valley-Monterey area are in many places age equivalents, although one or the other is absent in still other areas. This was not clear, however, until demonstrated by Richards (1935a, b, 1936). Previously, geologists customarily correlated the Monterey-Santa Margarita contact in the Highland School District or in the Santa Margarita type area with that in Reliz Canyon. Yet the basal type Santa

Margarita is of Briones of even somewhat pre-Briones middle Miocene age equivalence. In the type Monterey area, the basal Santa Margarita is earliest Pliocene or perhaps even younger.

Micropaleontologic Biostratigraphy

Since microfossils are abundant on the West Coast in the offshore and generally finer grained age-equivalents of the mollusk-bearing strata, micropaleontology came to play the leading role in the clarification and redefinition of stages and zones.

Early Problems and Applications

When "Brick" Elliott perfected the core barrel for use in rotary drilling, the significance of micropaleontology for refined correlations in oil geology was considerably enhanced. The vehicle used was detailed biostratigraphy. This took the form of precise stratigraphic allocation of fossil assemblages in biostratigraphic continua (see Reed, 1933, p. 134, on the nature of continua), modeled after the procedures of the 19th century paleontologist August Quenstedt and conducted in both surface and subsurface sequences of fine-grained clastics. California micropaleontologists pursued this approach to refined correlation almost from the outset; it was especially characteristic of the monthly meetings of the SEPM Pacific Section from the 1920s to the 1940s (see Kleinpell, 1971).

Most of these studies began in subsurface Pliocene sequences in southern California. Nomenclature varied greatly at first, and a certain amount of scrambled formational and age names persisted. Even within this context, a committee of 13 geologists (chaired by micropaleontologist George H. Doane) representing the SEPM Pacific Section set up a classification of seven Pliocene "faunal zones" (see Reed, 1933, p. 229-231). These faunal zones were based on many detailed biostratigraphic sequences in the more fine-grained Pliocene rocks of the Los Angeles and Ventura basins.

Although company and consultant files were replete with pertinent data, they remained unpublished. The first published detailed record of such a biostratigraphic continuum was prepared by Driver (1928) and sponsored by W.S.W. Kew. The manuscript covered the Pliocene surface sequence in Adams Canyon and relied on an informal taxonomy for natural phyletic realities. The same year, O.C. ("Jimmy") Wheeler demonstrated (see Kleinpell, 1972) the notable consistency along strike of biostratigraphic groupings in the same Pliocene sequence. Presently Hoots (1931) included in a U.S.G.S. Professional Paper a record by Wilbur Rankin of detailed foraminiferal biostratigraphy from a nearby late Miocene sequence, employing a partly formal taxonomy. An extensive diatom sequence by Kenneth Lohman was also included by Hoots. By 1931 Cushman and Laiming had presented another detailed biostratigraphic record with completely formal foraminiferal taxonomy for an even older middle Tertiary sequence nearby.

Yet biostratigraphy, left in a purely empirical state, contains hazards over and above the factor of error inherent in any scientific deduction. In a field where time is often money, too many earlier micropaleontologists were sufficiently intoxicated with the "new" as to suppose that empiricism, basic as it is in all research, was enough. Fortunately, especially as a result of the classic "San Diego controversy," both geologists and megapaleontologists became wary of the sweeping decisions of many of the new "magic men." Based on careful contact-snooping in the field, Thomas L. Bailey's criticisms (in his 1935 paper on lateral change in faunas) were particularly telling. Yet in the closing remarks of that classic paper it is clear that Bailey was still thinking biochronologically in terms of William Smith's principle. Because of the problem of facies, this principle is irrelevant below system-period levels in geologic time-refinement based on correlation by fossils.

In 1933, Natland had demonstrated the parallels that exist between five living bathymetrically controlled community types on the sea floor and comparable biostratigraphic groupings in the nearby Pliocene sequence. (Admittedly the parallels did not extend to whole true communities of faunules, nor did biochronologic classification appear until later.) Through Natland's work the significance of chorology and facies in biostratigraphic chronology based on foraminifers was clearly documented. Since then, the work of O.L. Bandy, R.D. Cifelli, and a few others has incorporated the role of facies in stratigraphic micropaleontology with the relationship between individual organismal form and bathymetry and with thermal and latitudinal gradients, current phenomena, and floating levels in plankton.

Meanwhile, the late Ralph Reed (active field geologist, oil company executive, and voracious reader in the field of earth history generally) suggested that seemingly Albert Oppel (Quenstedt's student) had circumvented the stumbling blocks of local ecologic as well as lithologic facies by segregating them from biostratigraphic data for purposes of refined biostratigraphic chronology. The suggestion was appreciated by many of his associates, especially W.D. Kleinpell, D.D. Hughes, J.M. Hamill, Boris Laiming, by J.P. Smith, H.G. Schenck, and S.W. Muller at Stanford, Chester Stock and Ian Campbell at Cal Tech, and also W.P. Woodring. The germ of the consequences appeared in Cushman and Laiming's (1931) paper. Through numerous subsequent micropaleontologic works of this sort, some 15 benthonic foraminiferal stages and more than twice as many zones were recognized within the Pacific Coast provinces of the Tertiary.

In 1926, the writer had his first experience with detailed biostratigraphy. He was employed by Roy Fergusson to sample extensive Pliocene cores (temporarily stored in the basements of San Francisco buildings) lest they be thrown out altogether in the course of moving oil company headquarters to Los Angeles. Correlation of Marland Oil Company foraminiferal surface sections across the Coast Ranges followed, involving primarily age relationships between the typical Vaqueros and Temblor formations and the overlying lower Salinas and lower Maricopa shales.

Following the discovery of major middle Tertiary production in the Elwood oil field by Frank Morgan of the Rio Grande in cooperation with Barnsdall's Dick Sherman, the writer, while engaged as a field geologist for Richfield Oil Company, ran head on into the problems of refined correlation in the Monterey Shale, in some areas 9,000 ft (2,743 m) thick without trace of a significant megafossil. Extensive sampling was subsequently supplemented by H.G. Schenck, W.E. Dunlap, and other oil company staffs, Pacific Section AAPG and SEPM discussion, further private collecting, and the aid of W.P. Woodring, M.N. Bramlette, and K. Lohman of the U.S. Geological Survey.

All this, with Ralph Reed's suggestion about Oppelain zonation still in mind, led to the writer's proposal (Kleinpell, 1933) of a half dozen stages and more than twice as many zones based on the stratigraphic distribution of benthonic foraminifers. All were seemingly demonstrable within the stratigraphic and geologic limits of the Monterey formation *sensu lato*. The number of stages (six, the same proposed by Oppel in his work on the Jurassic) disturbed Reed, however. In 1936, Schenck and Kleinpell suggested a seventh and subjacent stage, the Refugian, a proposal acceptable to Reed, and which eased his mind (Kleinpell 1972, p. 97-98, 101, 110).

Mid-Tertiary Stages

Of the six middle Tertiary stages first proposed in 1933, the major portions of four were apparently deposited at medium marine depths over wide areas of low relief west of the Sierra Nevada. At that time, these mountains began a progressive rise along an eastern fault scarp, which interrupted previous drainage that brought clastic deposits from farther east. Thenceforth, such sediments apparently deflected more to the north (San Pablo) and south (the more sparingly megafossiliferous and deeper water Modelo and Puente), or were dumped mainly east of the scarp as the newly extensive terrestrial deposits of the Rosamond Group.

Marked orogenies were widely associated with deposition of the lowest (Zemorrian) and highest (Delmontian) of the six stages and to some extent with the superjacent portions of the succeeding (Saucesian) stage and the immediately subjacent portions of the preceding (Mohnian) stage. Widespread marine transgressions occurred especially during the middle of the sixfold stage sequence. In particular, throughout Relizian time marine waters flooded extensive areas in the Transverse Ranges (the same general region that in the early Eocene received California's only widely transgressive tropical limestone, the Sierra Blanca). During the Mohnian the Los Angeles basin proper was inundated for the first time and the great marine trough of Pliocene times, from Waltham Valley to the Santa Cruz coast, was probably initiated.

In the middlemost of the six stages a slight discordance in places separated the Relizian from the overlying Luisian, but shifting of major marine deposition was notable in more than one area. The Relizian transgression followed the initiation of much block-faulting, breccia formation (San Onofre), and volcanism associated with Dibblee's Lompocan orogeny. A marked foraminiferally zoogeographic provincialism set in through much of the West Coast ranges.

A sharp late Luisian orogenic disturbance which has been termed the "Zuman orogeny" (Kleinpell 1972, p. 98, 101) shifted isotherms equatorward. Although the stress was widely reflected in epeirogenic strain, it produced sharp folding strains mainly in a few relatively narrow geosynclinal belts. A relative dearth of strandline deposits persisted, as during the Relizian and Luisian. It is interesting to note that three of the foraminiferal stages (Saucesian, Relizian, Luisian) occur within the stratigraphic limits of the *Turritella ocoyana* teilzone, though after the Zuman orogeny a modified megafossil fauna ancestral to that of the later Mohnian Briones Formation appears at a few localities. Toward the close of Mohnian time, however, tectonic instability increased more generally, again with several local intraformational breccias and brea conglomerates. This reflected Dibblee's Rafaelan orogeny at the opening of the Delmontian Age and was seemingly associated with a cold nip in the waters.

Thenceforth, at least beneath surface waters, relatively sheltered shelf-sea conditions prevailed from the Huasna region north to Monterey and the Berkeley Hills, with coarse clastic megafossiliferous inshore deposition more extensively developed. The southern California basins deepened notably, as they continued to do in the early Pliocene, until Dibblee's Zacan orogeny again greatly altered Coast Range paleogeography and even the deepest basins began to fill, as Pliocene benthonic foraminifers clearly reveal. In the intermediate area, Delmontian shelf-sea deposition (more truly open ocean than farther north) seems to have been widespread, as in the Santa Maria region and south of Coalinga (see Woodring and Bramlette, 1950). With this increasing regional surface relief of Mio-Pliocene times, associated facies problems rendered benthonic foraminiferal zonation of the upper Delmontian precarious from the outset. However, the possibilities for marine megafossil and terrestrial mammal zonations were enhanced, as was zonation based on diatoms which were particularly rich throughout this interval.

During Zemorrian and early Saucesian times, bathymetry and high-relief surfaces (causing complex facies problems) were, if anything, more extreme than during the Mio-Pliocene. Surface temperatures were then on the tropical side, but with marine zoogeographic affinities eventually extending in all directions. The marked fragmentation of marine provinces had held over from the preceding age (late Refugian) well into the early Zemorrian, however. A great foredeep had developed in the southwestern San Joaquin Valley—the "deep dark hole" of local micropaleontologic jargon.

Wrestling with these bathymetric facies problems, the writer (Kleinpell, 1938) still placed the top of the Zemorrian far too high, for no better reason than that a number of forms common in the deeper water facies of the Zemorrian elsewhere carried upward at least to the contact between Pack's Vaqueros and the base of the type Maricopa. Diagnostic Zemorrian congregations found in the type Zemorrian section, however, do not include all the deep-water forms that subsequently proved to be principally younger, though still deep-water, Saucesian assemblages (Kleinpell and Weaver, 1963). Moreover, deep-water deposition continued in this area through much of Miocene time. This led to the designation "pseudo-Saucesian" (see Beck, 1952; Rudel, 1968) for these local and exceptional deep-water, post-Saucesian, mid-Miocene foraminifer aggregates that in bathymetric facies contrast markedly with the almost uniformly widespread medium-depth, community-type assemblages of the Relizian, Luisian, and early Mohnian.

Subsurface biostratigraphic sequences in the southwestern San Joaquin Valley reveal some of the best late Eocene-Oligocene-early Miocene benthonic foraminifer continua in the Coast Ranges. Experience with these subsurface sequences had in fact encouraged the writer to select as the type section for the Zemorrian Stage the best of the nearby outcropping sections. It is invariably difficult to pass through marine Oligocene biostratigraphic continua on the surface in California without encountering coarse clastic beds that are either barren or contain shallow-water facies.

A comparable problem faced Schenck and Kleinpell in selecting a surface type for the seventh middle Tertiary stage, the Refugian. In the Pacific Northwest excellent sequences were known to pass through this stage directly from the subjacent late Eocene stage (Schenck and Kleinpell's "Unnamed Stage," which became the A-group of Laiming's Eocene zones and the Narizian Stage of Mallory). These sequences did not pass on up through the superjacent Zemorrian, however (e.g., at Bassendorf Beach and Pittsburg Bluff in Orgeon). In spite of the general overburden of vegetation in the Pacific Northwest, subsequent studies (especially by Fulmer, 1954, 1975; Jeletzky, 1973) have yielded the sequences looked for earlier in California. Even so, shallow-water foraminiferal assemblages tend to characterize the surface expressions of the upper Refugian Stage in its so-called Lincoln facies. To recognize a stage, however, more than one or two normal facies of the time-stratigraphic interval are necessary. Meanwhile Frizzell and Blackwelder (1933) had found a shelf-sea assemblage associated with some of the most short-ranged species diagnostic of the Lincoln-zone molluscan congregation. The same molluscan congregation, though in stratigraphically isolated outcrops, had also been found in the Pacific Northwest superpositionally in association with deeper water uvigerines such as were recorded later by Cushman and Simonson in California.

Eventually, a mapping assignment with North American Consolidated Oil in the Santa Barbara embayment yielded the writer the medium-depth foraminifers known from the upwardly incomplete surface sequences of Oregon, Washington, and the San Joaquin Valley. These foraminifers occurred stratigraphically below beds with Lincoln molluscan correlates, and these in turn occurred stratigraphically below what were presumably early Zemorrian shelf-sea foraminifers, and certainly hundreds of feet below late Zemorrian congregations. The whole sequence was as near to a biostratigraphic continuum through this interval as had been found in surface outcrop in California. Employing the molluscan congregation (i.e., the shallow-water facies) as intermediary in this long-sought biostratigraphic continuum, the type section for the seventh foraminiferal stage was named and defined in 1936 in the old Spanish land grant of Nuestra Senora del Refugio. Formal zonation of the new stage was not attempted, however, pending discovery of better surface discipline.

W.J. Classen, G.R. Hornaday, Herlyn, and the writer have carried a continuous biostratigraphic sequence of the entire Paleogene through the heart of the western Santa Ynez Mountains into the type Refugian, and Eugene Wilson has carried the Refugian eastward from the type area across the Horseshoe fault. Hugh P. Smith has recorded a zonally complete Refugian sequence in Harry Johnson's Wagonwheel Formation, between the type Narizian below and De Witt Taylor's upper Zemorrian sequence in Van Couvering and Allen's Hannah Formation; if present, the lower Zemorrian is unfossiliferous; however, 40 ft (12 m) of glauconitic siltstone intervenes concordantly here but is overlapped immediately to the north. L. Forrest, F.R. Sullivan, Earl Brabb and associates, W.W. Fairchild, P.R. Wesendunk, D.W. Weaver, Roberta Smith, and others have found the Refugian Stage in the Santa Cruz Mountains in both medium- and shallow-water facies biostratigraphically between late Narizian foraminiferal strata below and early Zemorrian above. Comparable subsurface sequences are known from the Santa Barbara Embayment.

The data most suitable for zonation of the Refugian Stage still occur in subsurface sequences rather than in outcropping sections of the sort most appropriate for type zonal standards of reference. Tipton et al, (1973) examined and correlated four such sections in the San Joaquin basin with an aim toward constructing a sequence of foraminiferal faunas from latest Eocene to earliest Miocene times. In addition, previous zonations of the Zemorrian Stage were revised. The pertinent data were placed under adequate discipline both surface and subsurface, with two zones, a lower *Uvigerina gesteri* Zone and an upper *Uvigerinella sparsicostata* Zone.

Some Sequels

Thus biostratigraphic disciplines for the recognition of foraminiferal stages, especially from the late Zemorrian to the early Delmontian, had seemed adequate for publication by 1933-34 in spite of the "pseudo-Saucesian deep" and the Zuman orogeny. In turn, Ralph Reed's worries about a sixfold Oppelian classification had been assuaged before further publication in 1936. His concern about "symmetries" had been well founded, however. All known superpositional evidence to the contrary, by 1944 the majority of a committee still placed the Blakeley of Washington as an age equivalent of the Refugian and the Briones of California as age equivalent of the Luisian. Could some abstract concept that somehow age sequences based on organic evolution in different stocks should be rationalistically and neatly "symmetrical" right across the line have been involved?

The geologic exploration in search of petroleum that led to the discovery of foraminifers in the biostratigraphic continuum selected subsequently as type for the Refugian Stage was part of a private paleontologic consulting practice conducted by the writer. He was assisted at times by R.R. Wilson and L.J. Simon, in association with geologist and producer Lowell Saunders, geologist and petroleum engineer Warren Ten Eyck, and W.D. Kleinpell. In the course of this practice, many problems in biofacies, lithofacies, and time correlations were encountered in the central California Coast and especially the San Joaquin Valley, and for help from these associates the writer is grateful.

It has long been known by industrial micropaleontologists that middle Tertiary sandstones on both sides of the valley were more or less lenticular. There are many local "climbings," "droppings down," and lensings-out of sand bodies, and through studies of the distribution of benthonic foraminifers therein, recognition of such "shale-outs" became important in locating combination structural-and-stratigraphic trap oil accumulations, especially on the east side. Thus facies changes recognized within the context of stage and zone correlations based on benthonic foraminifers were often involved.

Lowell Saunders, incorporating, blossomed into the Intex Petroleum Corporation and clients Al Jergins and Harry Campbell of A.T. Jergins Trust made extensive discoveries especially in the Edison area. Seaboard (formerly Milham Exploration) and Kern County Land Company prospered also. By the late 1930s, the writer, through Richfield Oil, had acquired valuable assistance also from R.S. Beck, L. Forrest, S. Carlson, M.B. Payne, Rod Cross, R.T. White, M.L. Natland, W. Rothwell, P.H. Dudley, J.W. Sheller, J. LeConte, and Frank Tolman among others, under the aegis of H.W. Hoots, Rollin Eckis, Mason Hill, and T.W. Dibblee, Jr., and a close relationship with General Petroleum (now Mobil Oil) through E.C. Edwards, H.D. Hobson, P.P. Goudkoff, Frank Carter, and Carrol

Wagner. Some of the resulting correlations were eventually incorporated into Alex Diepenbrock's California Division of Oil and Gas reports on east side oil fields, the U.S.G.S. Professional Paper on the Kettleman Hills, and M.N. Bramlette's (1946) classic monograph on the origin of the siliceous Monterey Shale.

Two of the more curious sequels to Ralph Reed's suggestion about the possible role of Oppelian zonation in the geologic exploration for petroleum are sufficiently interesting to merit mention. The first involved an article in a trade journal that suggested we had been using the "wrong fossils" (benthonic rather than planktonic foraminifers) in this sort of work, thereby causing time errors that are costly in contemporary petroleum exploration. With all due regard for the improvements and refinements in micropaleontology that have developed since the 1940s, this seemed a singular evaluation. One was led to wonder whether all the petroleum discovered with the aid of these "wrong kinds of fossils" should be returned to the ground (Kleinpell, 1972, p. 102).

The other instance occurred at a fairly recent meeting devoted to the exploration and economics of the petroleum industry. A university professor discussed oil and gas in three dimensions and informed us that in this connection the geologic ages of petroleum reservoirs are unimportant. Just how this viewpoint helps us locate "shale-out" traps is not clear. In this regard, some fairly recent experiences in the Four Corners area come to mind. There, to be sure, fossils played little if any role; only local synchroneities in the strictest empirical sense were involved. Yet the fourth dimension—time or geologic age—proved critical. After much miscorrelation based on otherwise helpful but time-wise "paleoelectrically" inappropriate electric logs, great tracts of land were quit-claimed as unproductive. In the westward feathering out of the Cretaceous shoreline deposits there several ash beds occur, and some geologists noted that these ash beds could be distinguished petrologically and superpositionally as of different ages. The subsurface sequences in the abandoned dry holes were recorrelated, with most productive results.

Many papers on middle Tertiary benthonic foraminifers appeared in the earlier decades of active micropaleontology. Some of these formally recorded geologically significant occurrences of foraminifers; others expressed incipiently or even fully the biostratigraphic Oppelian traditions (for a list, see Kleinpell, 1972, p. 15-16). By 1959 Mallory had presented an Oppelian classification of West Coast Paleogene.

In 1952-53, M.L. Natland proposed three post-Miocene foraminiferal stages in the West Coast Pliocene (a fourth involved the entire Quaternary). This followed some singular sequels to his earlier work (Natland, 1933). The Doane committee's Pico Formation had been subdivided into the "Pico brown" and the "Pico blue," and a lower time-unit was still designated the "Repetto" after the informal term given to richly foraminiferal mudstones of the Repetto Hills. A "Repetto-Pico transition zone" 400 ft (122 m) thick was placed in varied positions by workers from different laboratories. Wissler (1943, p. 210) summarized the entire local Tertiary column and the Neogene foraminiferal column of southern California in great detail. Following Natland's 1933 facies studies, however, geologists became alarmed at the interpretation of abyssal depths accorded the "Repetto" by Natland, since mud cracks, supposedly indicative of subaerial deposition, were found in the interbeds between strata bearing the abyssal foraminifers. Woodring's studies of the few "Repetto" mollusks nevertheless corroborated Natland's bathymetric interpretations. Not until oceanographic studies showed that such features are found at great oceanic depth was this apparently conflicting evidence resolved.

Another controversy revolved around *Bolivinita* ("*angelina*") *quadrulata,* an invading species from the South Pacific Neogene. This species commonly occupies a very restricted interval of about 13 ft (4 m) in the "Repetto," an interval that as a whole attains thicknesses of 3,000 to 6,000 ft (400 to 1,800 m). Why then, some asked, could the "Repetto" not be consistently subdivided into 13-ft intervals?

Meanwhile, the new ecologic rather than chronologic significance of biostratigraphic foraminiferal subdivisions of the Pliocene persisted. Furthermore, Natland's living samples from the Santa Barbara Channel came from no greater depth than between 1,000 and 1,500 (abyssal) fathoms and his stratigraphic column went from shallow depths at the top progressively toward abyssal depths downward in the column toward the base. Consequently, some maintained that his dredgings of the living samples had not reached the "Miocene" simply because the channel was nowhere deep enough.

In 1952-53 and again in 1957, Natland proposed stage names for the three Pliocene subdivisions and designated type sections for them. It is true that although diverse lithofacies are well known, nearshore biofacies of these stages were not extensively involved and no diagnostic species congregations were designated. Nevertheless it had long been known that certain species have particular relationships with the Repettian Stage of the California province. For example, *Plectofrondicularia californica* makes its last appearance here, *Uvigerina peregrina* occurs for the first time, and *Bolivinita quadrulata* is restricted to this stage. Thus Natland rescued the term "Repetto," in the context of its original time-stratigraphic meaning, from the use as a formational designation (see Haller, 1967, and this volume).

This time-diagnostic congregation of the Repetto holds true from as far north of the Repetto Hills as the Humboldt basin, as substantiated by the work of R.E. and K.C. Stewart, D.D. Hughes, Doug Crawford, C.R. Haller, and others (see Haller, 1967). It is the same with the appearance of *Bulimina subacuminata* higher in the column and *Uvigerina juncea* ("*U.* cf. *tenuistriata*") still higher. Furthermore, zonally diagnostic congregations are also present within these benthonic foraminiferal stages of the Pliocene.

Factors such as faunal floods in the quantity (often meticulously quantified and even graphed), and coiling ratios in planktonic foraminifers, are frequently employed for time-correlation purposes (e.g., in the California Pliocene). In terms of transgressing time because of ecologic facies factors, this usage seems to be inconsistent. Two distinct relations of the fossil, or fossils, are here telescoped into one. Even those correlations based on either unique phyletic taxonomy or stage of evolution stem not from the interpreted taxonomy itself but from the morphologic organismal facts behind the evolutionary interpretations. Synthesis is accomplished through comparing the facts behind the diverse relationships interpreted, not through piling interpretations upon interpretations.

The periodic buildup of published emendations can engender a certain disillusionment with worldwide planktonic zonations. Many so-called paleontologic "sequences" turn out to be "evolutionary" rather than actually superpositional columns. Sometimes, however, so-called stratotypes are resorted to, by relying on classic type areas or sections where fossils that are stratigraphically restricted at certain horizons *within* a formation have served as basis for a supposed time-rock unit of lesser magnitude than a series, whereas rock stratigraphy as well as time stratigraphy are intrinsically involved. Microfossils are then meticulously collected through the entire formational sequence to render it a "stratotype." Thus a sort of medieval nominalism is established, to become a standard of reference in further correlation, whereas the actual stratal sequence involved is usually far greater than the biostratigraphic interval or intervals in reference, and rock stratigraphy and time stratigraphy become telescoped into an unscrambleable omelette (see Schenck and Muller, 1941, for the consequence of such procedures).

Finally, just as *series* boundaries plagued the earlier workers in West Coast stratigraphic paleontology because recognition of such boundaries required correlations based on time-stratigraphic units of *stage* magnitude, so today *stage* boundaries appear to plague contemporary workers. The reason is that these boundaries call for correlations based on time-stratigraphic units of no broader than *zonal* magnitude.

In any event, the suggestion made in the 1920s and 1930s by Ralph Reed concerning the potential of Oppelian zonation for correlations in the West Coast Tertiary has borne fruit (Kleinpell, 1972, p. 110) in the fields of structural and historical geology and paleogeography as well as in stratigraphic paleontology and micropaleontology. There has been concomitant success, of course, in their application to geologic exploration for oil. As data accumulate and with accompanying extensions in species ranges, as emphasized by Davies (1934, p. 52, 56) in his passages on trial-and-error procedure and on the long ranges of all invertebrate species, the generalizations derived through the procedures of Oppelian zonation become progressively better disciplined. This is especially so at progressively more refined stratigraphic levels in geologic time correlations.

Zones and Stages of Middle Tertiary of California

Correlation of Stage Boundaries

The California middle Tertiary stages in reference in the present work are, in principle, stages such as those recognized by Albert Oppel more than a century ago, although based on the organic evolution and stratigraphic distribution of benthonic small foraminifers rather than nektonic ammonites. These stages thus correspond, both qualitatively and in magnitude, to the "Zonengruppen" of Oppel (1856-58). That is, each stage is a group of two or more of Oppel's zones, grouped together and distinguished from each other, almost, though not entirely, as a matter of convenience in handling and communication; yet both the zones and the stages are recognized on the basis of the same principles.

Often data may not be adequate for the recognition of so refined a time-rock unit as a zone, though adequate for the recognition of a broader grouping of two or more of them. Hence, instead of employing the larger (or stage) units as the basic time-rock units in his classification, and in order to avoid the confusing usage of such terms, as "lower lower," "middle lower," "lower middle," "lower upper," "upper upper," etc., Oppel used separate terms in a twofold hierarchy of time-rock units. He employed d'Orbigny's term "stage" (Etages) for his "Zonengruppen" or larger units, and coined in a technical sense the term "zone" for his more refined and fundamental time-rock units. Thus, time-rock units of stage magnitude can be correlated with type sections of stages set up as standards of reference in a stage classification; yet the boundaries between stages—usually selected because of some apparent greater significance than the boundaries between the zones within a stage—are nevertheless still boundaries between two zones.

The correlation of such stage boundaries to any degree of precision automatically involves recognition of prehistoric time intervals not of stage-age magnitude but of zonal magnitude in refinement. Weaver (1969) presented an analogous situation in correlating the boundaries of Lyellian series-epochs, which requires recognition of time distinctions that are no longer of series-epoch but of stage-age magnitude in refinement.

A succinct summary of the essentials in the stage and zone classification of the California middle Tertiary was presented by Wornardt (1972, p. 284-333). This summary included the zonation of the Mohnian Stage of Pierce (1956), plus the equivalent time units in the megafossiliferous sequence (based mainly on fossil echinoids as recognized by the late George L. Richards, Jr., and on mollusks as previously recognized by many workers for many years), and the more essentially descriptive rock units (formations) and biostratigraphic units involved. Therefore, a further review of this classification, beyond an updating of its component zonal congregations (Berry, 1964, p. 70), seems superfluous here. More recently a succinct, though general, summary of the Oppelian stages in the California Tertiary as a whole has been presented (Berry, 1974). In turn, the historical context in which the middle Tertiary classification here under review was first developed and proposed has been summarized by the writer (Kleinpell, 1971, 1972).

To the best of the writer's knowledge the California middle Tertiary stages, insofar as the sequence of their type sections as selected is concerned, can still be recognized as such. Concerning their component zones, however (and correlations with them as presented in 1938 and 1955), a great deal has been subsequently learned, involving especially chorologic facies phenomena and the extensions of species ranges as known then and now, as well as a better understanding of the phyletic relations of at least some of the stocks involved.

The review which follows, then, will be presented stage by stage. Special emphasis will be placed on their component zones as they are now seemingly recognizable and upon the zonally diagnostic congregations (Berry, 1964, p. 70) which, through the additional data that have accumulated since 1938, permits their recognition as of now. Collectively, the congregations diagnostic of the zones within a stage may then be taken automatically to constitute the congregations of the "Zonengruppen" which are the stages into which these zones have been grouped. Additional remarks on superpositional discipline, on the basis for previous errors in particular correlations as published, and on diverse interpretations of age, chorology, and phyletic taxonomy that clearly still persist, will be included wherever seemingly pertinent and possible within the space permitted.

Refugian Stage

The Refugian Stage, first proposed and described in 1936 (Schenck and Kleinpell, 1936, p. 215-225), directly underlies the lowest and oldest of the six stages here under review. It has been suggested that the stage was originally based on a molluscan sequence rather than on a sequence of benthonic foraminifers (Addicott, 1972, p. 5), an interpretation which, however, is only partly true. The difficulties involved if the recognition of the Refugian Stage were actually to be based on both mollusks and foraminifers, as suggested by Addicott, are apparent both at Zemorra Creek (Addicott, 1972, p. 8-9) and at the type Alsea Formation in Oregon (Snavely et al, 1975, p. F16). In both sections the top of the "Refugian Stage" on the basis of mollusks occurs somewhat higher in the section than that based on foraminifers, exemplifying the biological principle that the times of greatest evolutionary change in one group of organisms do not necessarily coincide with major evolutionary changes in another. A stage, to be recognizable as such, must include some examples of at least its major diverse facies within its province. In the case of the Refugian Stage, the employment of molluscan data was supplementary to that of foraminiferal data, because at the time of its proposal difficulty was encountered in selecting a depositionally unbroken and continuously fossiliferous surface-outcrop sequence as a type section. The foraminiferal age diagnosis and the superpositional relations of the stage to both subjacent and superjacent stages were well known from many subsurface sections at the time of its proposal, but it has been inherent in the problems of an Oligocene sequence that continuity of fossiliferous strata of medium-depth origin are not readily found in surface outcrop. Part of the responsibility for this situation involves the closing phases of Dibble's Ynezan orogeny. At least two desirable surface sequences have been subsequently found in California: on Monocline Ridge in Fresno County (Phillips et al, 1975) and at Arroyo el Bulito, Santa Barbara County (Tipton, 1976a, b; Warren and Newell, 1976a, b).

Refugian foraminifer faunas when first discovered in the Coast Ranges (Cushman and Schenck, 1928) were thought to be of Oligocene age, but subsequent studies have indicated that the stage as a whole more probably encompasses both the uppermost Eocene below and lowermost Oligocene above. Suffice it to say that the *Uvigerina vickburgensis* Zone as tentatively proposed by Kleinpell and Weaver (1963, p. 32-33) and based on the subsurface sequence of Cushman and Simonson (1944) is the age equivalent of the mollusk-bearing surface beds in the upper part of the stratigraphic sequence selected originally as the Refugian type section; and that the highest occurrence of such species as *U. vickburgensis, U. cocoaensis, U. atwilli, Plectofrondicularia packardi robusta,* and *Cibicides haydoni* mark the highest beds of the Refugian Stage. (see also H.P. Smith, 1956, and Waters, 1970).

Further discussion of this stage is beyond the scope of this supplement. The Refugian Stage has been given intensive study by Ann Tipton Donnelly (Tipton, 1976a, b) in which a three fold zonation appears to be demonstrable. The age equivalents of the stage in the Pacific Northwest have been, and still are, under extensive study by Weldon Rau (1948, 1951, 1958, 1964, 1966, 1967, 1975).

Zemorrian Stage

Superposition

Superpositional discipline, in terms of its overlying the Refugian Stage, is lacking at the type locality of the Zemorrian Stage, though observable at the type locality of the Refugian Stage. Primarily however, it was considered demonstrable through subsurface biostratigraphic sequences available for study during a period of active drilling and coring by the petroleum industry in the San Joaquin Valley during the 1930s. At that time these data constituted confidential information, and because they were subsurface sequences, were unsuitable as type sections for general reference. Hence, the outcrop section nearest to these subsurface sequences and bearing the complete supra-Refugian sequence of foraminiferal assemblages was selected for the type and standard of reference, in spite of the unfossiliferous 400 ft (120 m) or more of beds that underlay the selected type section, and the prevalence of glauconite and a possible though indefinite disconformity at the base (i.e., at the "Tejon"-Temblor contact) on Zemorra Creek.

Inclusion of the Salt Creek shale, the lowest member of the typical Temblor Formation, within the type Zemorrian depended entirely on the presence of *Buliminella curta* at sample-locality CM-70 (Kleinpell, 1938, Table VI, p. 40, 45) along strike in the bed of Chico Martinez Creek. This left a total of 450 ft (137 m) of strata barren of age-diagnostic species between the horizon of *Buliminella curta,* a post-Refugian species, and the next-lowest foraminifer-bearing beds, which are of late Eocene (Narizian) age (Kleinpell, 1938, p. 106-108).

The mollusk-bearing "Lower Temblor Sandstone Member" (or "*Phacoides* Reef" as it is usually referred to) has been left out of a "Vaqueros" designation in a recent paper (Addicott, 1972, p. 3), but more usually it has been considered an offshore facies of the lower Vaqueros with an offshore Vaqueros fauna, especially because the Agua Sandstone, taken as the uppermost interval of the type Zemorrian, carries a characteristically upper Vaqueros fauna farther north along strike (see L.M. Clark and Alex Clark in Schenck, 1935, p. 523). This more northerly upper Vaqueros fauna occurs in an old and only subsequently rediscovered fossil locality recorded by Ralph Arnold. This fossil locality had led Arnold and Anderson (1910) to designate these beds and their sandstone continuation farther north as "Vaqueros Formation." There is no known evidence as yet for referring to the Salt Creek Member of the Temblor Formation as "Refugian" (Addicott, 1972, p. 3; see also Tipton, et al, 1973, p. 27-28), granting, to be sure, that the lower 50 ft (15 m) of the Salt Creek Member is barren of age-diagnostic fossils. Unhappily, the rediscovery of Arnold's old locality led, at the time, to raising the top of the Vaqueros Formation several hundred feet higher (i.e., top of the Agua Sandstone) in the sequence serving as typical of the Temblor Formation of F.M. Anderson directly along strike. This unfortunate procedure was due to the fashion of those times of considering a formation as having a definitely fixed time dimension (see Schenck and Muller, 1941, for a clarification of the principle involved).

The superpositional relations between Refugian and Zemorrian at the type section of the Refugian Stage have led to comparable controversies, although for different reasons. In 1938 excellent Zemorrian foraminifers were found at L.S.J.U. locality 1436, then considered to occur stratigraphically in a shale member of the Alegria Formation, a local marine lateral equivalent of the terrestrial redbed Sespe Formation (Kleinpell, 1938, p. 33, 108, 111). The shale or mudstone unit in reference subsequently was designated unit F of the Alegria Formation when Dibblee (1950) gave the name Alegria to the beds formerly referred to as the "marine Sespe" (Dibblee, 1950, p. 30-31). Kleinpell and Weaver (1963, Fig. 4, p. 33) listed the same locality as B-6915 with the same assemblage, and added another Zemorrian foraminiferal assemblage as coming from unit D (locality B-6917) somewhat lower in the Alegria. The stratigraphic position of the latter as given turned out clearly to be mistaken: a reverse dip between the two localities, at first taken to be an overturned attitude, proved instead to reflect a normal reverse dip on a plunging structural nose, and the two foraminiferal localities, if not from the identical horizon, were clearly from the same shale unit, apparently unit F insofar as the writer could trace this shale (or greenish mudstone unit) westward from the area of the structural nose to Dibblee's type section of the Alegria.

The Zemorrian foraminifers at L.S.J.U. locality 1436 (B-6915) at least served to place fossiliferous Zemorrian—thought at the time to be lower Zemorrian—superpositionally above the type Refugian. The latter consisted of both the lower and upper foraminifer-bearing Gaviota mudstones and middle sandstone lens plus the overlying mollusk-bearing lower Alegria sequence (units A through C, and subsequently extended upward by Weaver to include unit E) and the upper Zemorrian, represented as well by the local supra-Alegria, Vaqueros, and Rincon (lower member) formations. Controversy has continued over the stratigraphic position of the foraminiferal locality L.S.J.U. 1436, and subsequent studies of the zones in the Zemorrian Stage have not shown its assemblage to bear diagnostically lower Zemorrian species. The question thus has arisen as to whether a lower Zemorrian zone may be missing in this sequence and a hiatus of zonal magnitude be present in its stead. Since, superpositionally, a zonal continuum from upper Refugian to lower Zemorrian has subsequently been found elsewhere and published upon, the controversy over L.S.J.U. 1436 has become essentially academic; yet for purposes of the local geology involved, and for the possible offshore significance of its assemblage, the controversy can appropriately be pursued as to its present status. It seems to have at least three or four facets.

From the west side of Canada de Santa Anita and on westward, an angular unconformity—less than 5 degrees at its first appearance—becomes discernible between the Vaqueros and Alegria formations (Weaver and Kleinpell, 1972), and the angle of its discordance increases in magnitude farther along strike to the west as the Vaqueros progressively overlaps older and older beds. At Cojo Canyon the bentonite bed (at the Rincon-Monterey contact) has graded westward into an agglomerate before turning into the true Tranquillon Volcanics still farther west (Dibblee, 1950). A few miles west of Cojo Canyon these volcanic rocks lie directly on top of the lower Gaviota mudstone in subsurface, having overlapped all the intervening formations. Still farther northwest, south of Lompoc, the overlapping strata are equivalent in age to the *Siphogenerina hughesi* Zone and those bearing *Pecten lompocensis* (the local "Vaqueros" of Arnold et al, 1907, and the "basal Monterey" calcarenites of Dibblee, 1950). This situation is characteristic of the Santa Maria basin as a whole and of the area as far east as the so-called "weatherdeck" in the wilderness area between the upper Santa Ynez and upper Sisquoc Rivers (Keenan, 1932; Reed and Hollister, 1936; Dibblee, 1969).

However, on the south slopes of the Santa Ynez Mountains, from west of Canada de Santa Anita to the ridge east of Canada del Agua Caliente (L.S.J.U. 1436): (1) correlation in age by unconformity from one such locality to another cannot be made; and though there are pebble beds along the contact throughout this area it is necessary to bear in mind the admonition of Reed (1933, p. 134) that pebble beds *per se* are not evidence for unconformity (see also Blackwelder, 1909). (2) The stratigraphic allocation of locality 1436, whether Vaqueros, Rincon, or Alegria, has been brought up as a potential criterion; but since formations can vary in age from place to place this only begs the question. Moreover, the foraminiferal assemblage in reference is a shelf-sea faunule which, lacking zonally diagnostic species, could occur either above or below any sandstone body of shallower water origin. At Camp Comfort on San Antonio Creek, miles to the northeast (south of Ojai), a considerable sandstone body lithogenetically indistinguishable from those in the Alegria Formation (and there bearing *Pecten magnolia*) occurs stratigraphically hundreds of feet above the Vaqueros-Rincon contact and well within the Rincon Shale, Still farther northeast, though still present in lower Lion Canyon, the Rincon has disappeared from the column in the isoclinal syncline of Thompson Ridge (where a thin fine-pebble bed occurs as basal Monterey), only to reappear much farther north, across the Santa Ynez Range, in the upper Sespe Creek column. Finally, even on Los Sauces Creek, the lower 300 ft (91 m) of the Rincon Shale is barren of fossils (Cushman and Laiming, 1931, Figs. 2, 3, 5). (3) Vaqueros megafossils have been reported stratigraphically below the foraminiferal locality, but again this only begs the question. The *Turritella variata* megafossil fauna is celebrated for its joint occurrences of "Tejon" and "Vaqueros" megafossils. *Pecten sespeensis* (a former "Vaqueros index fossil") has already been figured from the middle Gaviota sandstone (E.J. Wilson, 1954) stratigraphically well below the highest local occurrence of Eocene giant venericards, adding, along with *Pecten ynezianum,* to the problem cited by Davies (1934, p. 159, footnote) about the magnitude of Neogene as distinguished from Paleogene scallops. (4) An abstract (Weaver and Frantz, 1967) stemming from local mapping in the Alegria Canyon-Canada del Agua Caliente area has shed further light on the controversy. However, the controversy is not solved, because mapping by Weaver and Frantz was carried out without benefit of bed-tracing from any unbroken sequence of stratal units in a superpositionally well-defined column of mappable strata. What is needed, if it can be managed, is a study in areal geology that traces the nine units of Dibblee's (1950, p. 30-31) type section of the Alegria Formation eastward to the east side of the ridge between Canada del Agua Caliente and Gaviota Canyon, a procedure which, if successful, would serve to place L.S.J.U. locality 1436 in its stratigraphic relationship to this Alegria sequence. It was such an effort that the writer attempted and which led him to place this locality in Dibblee's unit F (as Dibblee, 1950, p. 31, himself had done), although there is no assurance that this attempt on the part of the writer is infallible and could not be improved on, modified, or corrected by a more careful and thorough job of bed- and contract-tracing in this area. The need for this more detailed contact-mapping was emphasized by the fact that, in the early mapping, only scattered *Cyclammina* were to be found in unit F farther west, though the Alegria appears to be of most deep and truly marine origin just in the most foraminiferally critical area of locality 1436 where it disappears eastward under the Horseshoe fault, and by the fact that Dibblee's Alegria units A and B when traced eastward from the type Alegria section were clearly lenticular along strike.

Zonation and Congregations

The Zemorrian Stage has received more than an ordinary amount of attention in terms of attempting to subdivide it into zones, perhaps largely because of the great lateral variations in lithology and in thicknesses of formations that obviously occur from place to place, and of biofacies as well as lithofacies problems that its component strata present.

In 1938 a twofold subdivision seemed the only possible classification that would in any way accommodate the available evidence; yet a threefold subdivision, used by industrial micropaleontologists in the San Joaquin Valley, was temptingly attractive because of the conspicuous floods of *Atwillina* ("*Siphogenerina*") *pseudococoaensis* that seemed to characterize the middle of the Zemorrian in subsurface sequences over a wide area there—the so-called "Hub fauna" of those days. But the latter was simply a faunule, and the "flooding" as illusory a phenomenon as it was widespread (see Davies, 1934, p. 52 for the false age significance).

In 1963 a fourfold subdivision was tentatively proposed, but this, too, has not held up. The most extensive and best disciplined studies have been those of Ann Tipton Donnelly, whose classification returns to the original twofold zonation (see Tipton et al, 1973) but with a different and greatly improved and superpositionally better disciplined pair of zonal congregations, especially for the lower of the two zones. Biostratigraphically the type sections for the stage and for the two zones remain the same as the 1938 classification. To follow the development of disciplined zonation that has resulted in Donnelly's classification, the reader is referred to the following publications: Barbat and von Estorff (1933); Fulmer (1954, 1975); Sullivan (1962); Kleinpell and Weaver (1963); Fairchild et al (1969); R.K. Smith (1971); L. N. Edwards (1972); Hornaday (1972); Patet (1972); Weaver and Tipton (1972); and Tipton et al (1973; 1974, p. 132-150, Pl. 1). Some of the species most diagnostic of both the Zemorrian Stage and of its two zones appear graphically placed as to range in the phyletic charts on uvigerines and siphogenerines and on valvulinerids (Figures 6, 7).

Species restricted to the *Uvigerina gallowayi* Zone are *Uvigerina gesteri*, and the cryptogenetic and rate *Epistomina ramonensis*. Species making their first appearance in the zone are *Atwillina pseudococoaensis, Bolivina marginata marginata, Bulimina carnerosensis, carnerosensis, B. rinconensis, Buliminella subfusiformis, Cassidulina crassipunctata, Cibicides americanus americanus, Nodosaria estorffi, Nonion ynezianum, Planulina cushmani, Siphogenerina nodifera, S. transversa, Uvigerina obesa obesa, U. obesa impolita, Uvigerinella californica,* (?)*Cibicides hodgei wilsoni*, and (?)*Valvulineria casitasensis subcasitasensis*. Species making their last appearance in the zone are *Cibicides hodgei hodgei, Gyroidina condoni, Nodosaria holserica, Nonion affinis, Plectofrondicularia barbati*, and *Pullenia octoloba*.

Species restricted to the *Uvigerina sparsicostata* Zone, or upper zone of the Zemorrian Stage, are *Eponides pseudoaffinis* ("*affinis*"), *Laimingina smithi*, and *Uvigerina sparsicostata*. Species making their first appearance in this zone include *Anomalina glabrata, Baggina californica, B. robusta robusta, Bulimina carnerosensis mahoneyi, Estorffina mayi, Marginulina dubia, Nonion incisum kernesis*, "*Nonion incisum* (Cushman) var." of Cushman and LeRoy, and *Siphogenerina cymricensis;* and the following species make their last stand in this zone: *Atwillina pseudococoaensis, Guttulina hughesi, Eponides frizzelli, E. kleinpelli, Nonion ynezianum, Sigmomorphina reedi, Siphogenerina nodifera*, and *Uvigerina kernensis*. A few cryptogenetic species of "*Bolivina*" also occur only in the Zemorrian Stage, apparently, but are too rare to be of stratigraphic significance.

Miscorrelations Published

In closing a discussion of the Zemorrian Stage as updated, mention should be made of three previously published miscorrelations that have long since been corrected but not clarified in print. Several species originally thought to be diagnostic of the Zemorrian Stage have turned out to be deep-water species of much longer stratigraphic range than supposed, and have led to miscorrelations to be found in Tables XII, XV, and XVI of Kleinpell (1938, p. 57-61, 74-75, 76-77). In Table XII only the two lowest samples appear to be of Zemorrian Age. Because most beds involved are of Saucesian Age, they will be mentioned again in somewhat greater detail under the discussion of the Saucesian Stage which follows.

Saucesian Stage

Superposition

The type section of the Saucesian Stage conformably overlies the type section of the upper zone of the Zemorrian Stage on Los Sauces Creek, Ventura County (Kleinpell, 1938, p. 112-113). If the type section of the Zemorrian Stage were to be extended into a "type area," that would include beds traceable as far north as Carneros Creek (type section for the Temblor Formation; see Anderson, 1905; Curran, 1943). A conformable relationship may also be observed on the west side of the San Joaquin Valley between the highest beds of the type Zemorrian Stage (the local Agua sandstone member) and the lowest beds (i.e., the upper Santos Shale, or "C Shale" of John Mahoney) of the local Saucesian Stage (see Kleinpell, 1938, p. 54-55, Table XI), which interfinger with the lowermost sands of the Carneros Sandstone Member of the Temblor Formation at its type locality or area at Carneros Creek.

In the Coalinga area and northward on the west side of the San Joaquin Valley, deposition of the Temblor Formation appears to have shoaled perceptibly. In the Seaboard 1 Welsh well the Temblor uncomformably overlies beds of Refugian age (Cushman and Simonson, 1944). Two shallow-water foraminiferal faunas consisting of 38 species and varieties—three of the species and one of the varieties being new—have been recorded and figured by Garrison (1959) from a Temblor sequence about 1,000 ft (305 m) thick. The faunas in this subsurface Temblor sequence straddle, in age, the Zemorrian-Saucesian boundary, from late Zemorrian below into early Saucesian above. Many miles farther north but still on the east side of the San Andreas fault, Saucesian foraminifers have been found (Robert Rose, personal commun.) in isolated outcrops of Rincon-like mudstone in the otherwise topographically nearly flat area in the vicinity of the city of San Jose.

Finally, with reference to Zemorrian and Saucesian benthonic foraminifer faunas, mention should be made of the taxonomic paper by Bandy and Arnall (1957) in which three of the type Rincon Shale foraminifer species and one variety of Cushman and Laiming (1931), from Ventura County in southern California, have been reidentified or renamed, along with the renaming of one variety (of

Kleinpell, 1938) of comparable age. In the same paper five new species and one new variety have been described, named, and figured, of which one, *Rotalia becki,* is important among the deep-water faunas of the California Miocene, the facies-fauna aptly termed "pseudo-Saucesian" by Beck (1952) and referred to elsewhere in this supplement (also see Rudel, 1968).

Zonation

Three zones may be recognized within the Saucesian Stage, so that it is now possible to speak of a lower, a middle, and an upper Saucesian, instead of simply a lower and an upper substage as previously. They occur conformably one above the other in the Saucesian type section on Los Sauces Creek. Lower Saucesian also succeeds upper Zemorrian conformably upward in several other columns, as in the shallow-water sequence described by Garrison (1959), and including other subsurface sections generally of deeper water facies farther south in the San Joaquin Valley (Tipton et al, 1973).

Siphogenerina transversa zone—The congregation diagnostic of this zone includes the restricted occurrence of *Eponides nanus,* (a descendant of *E. pseudo-affinis, "E. affinis"* of Cushman and Laiming, perhaps), *Uvigerina mexicana,* and *Siphogenerina tenua;* the earliest occurrences of *Bolivina adelaidana, B. advena advena, B. floridana floridana, B. floridana striatella, Bulimina alligata, Nonion costiferum costiferum, Nonionella miocenica, Plectofrondicularia miocenica laimingi, Uvigerinella californica ornata, Valvulineria depressa,* and *Virgulina californiensis californiensis;* and the highest known occurrences of *Bulimina carnerosensis mahoneyi, Cancris sagra, Estorffina ("Siphogenerina") mayi, Planulina cushmani,* and *Uvigerina becarrii.*

Plectofrondicularia miocenica Zone—Cushman and Laiming (1931) and Snedden (1932) do not draw the limits of the zone they use under this name at precisely the same horizon, Snedden's being 173 ft (52.7 m) stratigraphically higher (Snedden, 1932, p. 43) in the column; but this seems to be based on horizons of increased or decreased abundances of certain species. Neither do they draw their Vaqueros-Rincon contact at exactly the same horizon in terms of stratigraphic feet below the base of the bentonite bed which constitutes the Rincon-Monterey ("Modelo shale") contact (Snedden, 1932, p. 41-42). Cushman and Laiming's zone, having date priority, is taken as type for the zone, especially since Snedden (1932, p. 41) specifically stated that "The present paper is primarily an attempt to show the age relationships of the 650 feet of the 'Modelo' shale which rests upon the Rincon Shale."

There still remains the problem of whether or not to include the *"Haplophragmoides trullissata* Zone" of Cushman and Laiming (1931, Fig. 5), which is a zonule rather than a zone (Kleinpell, 1938, p. 116); but since the only species in the zonule that does not range both below and above the zonule (i.e., *Gaudryina* cf. *flintii*) ranges only upward, the zonule is arbitrarily included within the *Plectofrondicularia miocenica* Zone in the Los Sauces Creek type section. Nor, in the form of recorded occurrences, is the congregation complete in the type section, though *Siphogenerina kleinpelli* has been found therein (Carlton M. Carson, personal commun.) as well as higher in the same section (Kleinpell, 1938, p. 302). The congregation of the *P. miocenica* Zone is characterized by the first appearance of *Bolivina adelaidana,* and of *Siphogenerina kleinpelli* which has evolved apparently from *S. transversa* at about this point (see Blacut and Kleinpell, 1969; Kleinpell, 1938, p. 116), together with the highest known specimens of *Bulimina carnerosensis carnerosensis, Bulimina rinconensis, Cibicides americanus crassiseptus, Cibicides elmaensis, Uvigerina gallowayi,* and *Valvulineria casitasensis casitasensis;* and perhaps also the highest known *Siphogenerina cymricensis* and *Beckina* sp. (*"Siphogenerina* sp." of Tipton et al (1973, Fig. 11; 1974, p. 132-151, Pl. 1).

Uvigerina obesa Zone—The lowest 425 ft (130 m) of the Sandholdt Formation (its lowest member) in Reliz Canyon was taken as the type section of the *U. obesa* Zone. At the type section of the Saucesian Stage on Los Sauces Creek it is represented by the upper member of the typical Rincon Shale (below the bentonite bed) plus the 300 ft (91 m) of Monterey ("Modelo") Shale overlying the bentonite bed there (Kleinpell, 1938, p. 113, 116-117). The rare and cryptogenetic *Cibicides relizensis* may be restricted to this zone, but the congregation is characterized especially by a great influx of bolivinid species (*brevior, californica, conica, perrini, salinasensis,* and *tumida*), by the first appearances also of *Bulimina pseudoaffinis, B. "subcalva"* (of Snedden), of cassidulinids (*panzana,* and perhaps *pulchella*), more than one *Epistominella ("Pulvinulinella": relizensis, subperuviana subperuviana,* and *subperuviana minuta*), and of *Eponides keenani, Planularia luciana, Plectofrondicularia miocenica directa, Pullenia miltiloba, Robulus hughesi, Robulus reedi, Siphogenerina branneri,* and *Valvulineria williami;* and by the highest occurrences of *Anomalina* cf. *A. patella, Bulimina alligata, Cibicides floridanus, Epistominella parva, Marginulina subbullata, Robulus mayi, Robulus warmani, Siphogenerina transversa,* and *Uvigerina obesa impolita.*

Snedden's *"Siphogenerina hughesi"* of his *"Siphogenerina hughesi* Cushman zone" (1932, p. 44) are only this species in part (Kleinpell, 1938, p. 301); the stratigraphically lower ones are worn specimens of either *S. transversa* or, more commonly, *S. branneri*. However, *S. hughesi* is present in his sample from 300 ft (91 m) above the bentonite bed and again at 450 ft (137 m) above the bentonite bed (Snedden, 1932, p. 44). These are all stratigraphically above the *Uvigerina obesa* Zone here, and immediately above the top of the type Saucesian Stage. The identity of his *"Bulimina subcalva"* was not clear to the writer when Snedden kindly showed his collections in the mid-1930s (see Kleinpell, 1938, p. 259), but the following species were present in the *Uvigerina obesa* Zone above the bentonite bed on Los Sauces Creek though listed under various names, some tentative (see Snedden, 1932, p. 43-46): *Baggina robusta robusta, Bulimina ovula, B. pseudoaffinis, Buliminella subfusiformis, Nonion costiferum, Plectofrondicularia miocenica miocenica, Uvigerinella californica, Uvigerina obesa, Siphogenerina branneri, Siphogenerina transversa,* and *Valvulineria depressa*. Both the formal and the tentative taxonomic terms used by Snedden (1932) are synonymized under the designations in the above list, in the Systematic Catalogue in Kleinpell (1938, p. 182-356). Snedden carried the Los Sauces Creek column on above the horizons in reference here. The bentonite bed occurs 2,260 ft (689 m) above the Vaqueros-Rincon contact as he placed this contact, and he records one sample stratigraphically lower than any in Cushman and Laiming's column though he lists no species from it in his 1932 paper. He listed species from as high as 650 ft (198 m) above the bentonite bed, which are also synonymized by Kleinpell, (1938, p. 182-356). The interval goes through beds of Relizian age up into the lower Luisian Stage before structure breaks the continuity of his stratigraphic sequence.

Remarks

Other *U. obesa* Zone faunas are also found above the bentonite bed in still other localities in nearby areas, as in the "lower" part of the coastal "Naples Bluffs" sequence between the mouths of Las Varas and Dos Pueblos Creeks (as check listed in Table 4 and graphically shown in Fig. 6 of Kleinpell, 1938), and in the road cuts on the inside of the turn where Highway 101 turns sharply from a westward direction into a northward approach to Gaviota Gorge. Geologists are prone to assume that paleontologic changes of age significance directly accompany changes in the lithologic sequence, but this is an environmental approach to organic evolution. A time-lag in a faunal change of evolutionary and age significance is usual, and the greater the turnover in species at a sharply drawn horizon in a concordant biostratigraphic sequence the more likely it is to be due to change in local ecology rather than to age. Changes in zonal congregations seldom involve more than 5% of the species involved. As a matter of convenience, in correlating between oil wells the age line is usually drawn at the nearest horizon of lithologic change in the cores or at the nearest change between zonules. However, this has generally been understood by geologists and paleontologist alike, and allowance has automatically been made for the small factor of time error in a cored sequence within a restricted area. By the more mechanically and inorganically minded, this "lag" has even been attributed to the "reworking" of older foraminifers, occasionally a very real phenomena but usually only when the diameters of the clasts in the sediments attain the diameters of the microfossils—a situation originating from the same causes as an approach to a similarity in size between float-boulders of silicate and of limestone rock observed in ascending a stream bed.

With relation to the bentonite bed at the Rincon-Monterey contact in the western Transverse Ranges, radiometric dating of some beds within the Tertiary has revealed another facet of stratigraphy with a potential of sufficient significance to call for a few additional words in passing. Turner (1970, p. 111) has been able to date the Zemorrian-Saucesian contact at 22.5 m.y. and the Saucesian-Relizian boundary at 15.3 m.y. Turner's paper gives all the pertinent details. The item in special reference here however has to do with the relative ages of the bentonite bed and its near equivalent somewhat farther north, the Obispo Tuff. There the dates at the base of the two pyroclastic sequences are very nearly the same, yet whereas several hundred feet of uppermost *U. obesa* Zone Saucesian overlies the bentonite bed, the Obispo Tuff (a much thicker stratigraphic unit) is directly and conformably overlain by the Point Sal Formation and/or other beds of earliest Relizian age. Radiometric dating independently corroborates this very minor age difference. Thus Turner (1970, p. 103-105) placed the Saucesian-Relizian boundary at 15.3 ± 0.5 m.y. and, as he stated, "The tuff bed, its bentonite equivalent, and the Tranquillon Basalt are correlative with the middle part of the Obispo Tuff . . ." and that it "is of interest to note the close agreement of the paleontologic and radiometric data here, each indicating that the Santa Barbara County volcanics are slightly older than the top of the Obispo Tuff. The fact that the

Tranquillon Volcanics are overlain conformably by uppermost Saucesian, rather than by Relizian beds, supports the previous assignment of a 15.3 m.y. age to the Saucesian/Relizian boundary."

Before closing a discussion of the Saucesian Stage it should be pointed out that some, though not all, of the species that apparently become extinct at or near the close of the Saucesian Stage—especially those whose morphotypes are indicative of a deep-water environmental optimum—may owe their disappearance in most of California to the immediate widespread reduction of areas of truly deep water, as also happened to those species, which once were thought to range no higher than the top of the Zemorrian Stage but where the local upward limits of which proved to be under bathymetric-facies rather than age control, and which led to the miscorrelations in Tables XII, XV, and XVI (Kleinpell, 1938) discussed under the Zemorrian Stage. This probability is emphasized by the situation in the so-called "deep dark hole" (a Standard Oil Company term) at the southwestern end of the San Joaquin Valley. Both Relizian and Luisian Stages are characterized by widespread deposits of medium-depth origin. The medium-depth ocean apparently was a consequence of Dibblee's Lompocan orogeny, which appears to have been epeirogenic rather than orogenic, complete with block-faulting and associated depsitional phenomena both clastic (San Onofre Breccia) and volcanic (pyroclastics). On the other hand, in the area of the "deep dark hole," deep waters of the abyssal and lower bathyal districts persisted through the Miocene, complete with foraminiferal faunules typical of such depths, and known by Beck's appropriate term "pseudo-Saucesian" (Beck 1952; see also Rudel, 1968; Foss and Blaisdell, 1969, p. 92). In this "deep," species such as *Anomalina californiensis, Bulimina* aff. *inflata, Cibicides* cf. *mcKannai, Cibicides pseudoungerianus evolutus, Pullenia moorei, Rotalia becki,* and others, including many of the long-range uniserial lagenids and their parallels, persisted long after disappearing from the widespread Miocene faunas of California characteristic of medium water depths. *Plectofrondicularia californica,* for example is present from late Eocene (Narizian) time until it flourished and spread again more widely in the extensive abyssal marine troughs that developed during the Repettian (early Pliocene) Age in southern California and in Humboldt County in the West Coast ranges.

Relizian Stage

Superposition and Zonation

The Relizian Stage (Kleinpell, 1938, p. 117-121) typified by the middle and upper members of Thorup's Sandholdt Formation in Reliz Canyon where it lies conformably between uppermost Saucesian beds below and lowermost Luisian beds above, seems to require a minimum of updating. It is clearly of Miocene age, and post-Oligocene, as is much if not all of the underlying Saucesian Stage, although as a stage the Saucesian involves in part at least the Aquitanian problem of the Old World.

Perhaps worth reemphasizing are the extensive marine transgressions indicated by the Relizian distribution, and its superpositional relations to a great diversity of underlying rocks; the essentially calcareous nature of most of its component strata; and the minimum of clastic deposits, other than locally, from all but its lower zone. In many places these lower Relizian beds are as calcareous as they are silty or fine-grained sandy. In southern California they include not only *Pecten (Amusium) lompocensis*-bearing quasi-clastic limestones (see Kinney et al, 1954; Truex, 1976, p. 84-86), but also huge blocks of San Onofre Breccia, and along the Santa Barbara coastal bluffs the lower of two conspicuous Tertiary intervals of lenticular intraformational breccia (Reed, 1935a, b).

Finally, it should also be reemphasized that by Relizian time, and perhaps even during latest Saucesian time and certainly in association with Dibblee's Lompocan orogeny, the zoogeographic connections with the marine Caribbean area that had characterized most of Zemorrian and Saucesian time in the West Coast ranges area were severed (Blacut and Kleinpell, 1969); and that an independent marine province had developed in the California area. Inshore, the Astoria megafossil fauna of the Pacific Northwest, much of which extends as far south and southeast as the Sobrante of the Berkeley Hills (Lutz, 1951) and to the Vallecitos area still farther southeast (Kleinpell, 1938, p. 119, 124, 258), and the Temblor megafossil fauna of most of California farther southwest and south, have much in common. However, the Astoria fauna lacks the cones, lyropectens, and turritellids of the Temblor. Offshore, however, the Relizian benthonic small-foraminifer faunas are remarkably uniform from southern California to Alaska. Through Luisian time a barrier may have existed between the two molluscan subprovincial areas, as reflected by the subsurface "Zilch Zone" of Barbat (1943, in Calif. Div. Mines Bull. 118, p. 483, 588).

As to congregations in the two zones of the Relizian Stage, it should be noted that *Baggina californica* appears much earlier in the column than previously supposed; that *Valvulineria californica obesa* appears to have evolved at essentially the same time as did *Valvulineria californica appressa* though morphologically in a different direction and apparently with less significant evolutionary consequences, and is thus as much a part of the *Siphogenerina hughesi* Zone congregation as of that of the higher *Siphogenerina branneri* Zone; that speciation proceeded significantly in the genus *Siphogenerina* as well as among the valvulinerids, as may be seen in the phyletic charts of these two stocks presented in Figures 6 and 7; and that the lagenids noted (Kleinpell, 1938, p. 121), though certainly cryptogenetic, exhibit restricted and characteristic stratigraphic ranges within the West Coast Province during early and middle Miocene time. Perhaps as rich, diverse, and typical a Relizian foraminiferal faunule, in its uppermost zonal and climax community phase, as can be found anywhere in the West Coast range, is that collected by Branner, studied by Bagg, and misinterpreted by Cushman as to stratigraphic position. The true nature and stratigraphic position of this sample was mentioned by Kleinpell (1938, p. 21-22; 1972, p. 89-90), and a check list of the West Coast species represented therein can be compiled by any interested reader through pursuing the references noted in the index under the heading "Henry Ranch" (Kleinpell, 1938, p. 420).

Luisian Stage

General

The Luisian Stage lies conformably on the uppermost strata of the Relizian Stage at the type localities of both the Relizian Stage and the Luisian Stage, as well as at many other localities. In fact the only nonconformable relationship of which the writer is aware is the possible discordance in the general area behind Laguna Beach (Kleinpell, 1938, p. 123), observed by both the late R.W. Burger and by H.D. Hobson in the course of private industrial studies there. On the whole, the lower Luisian in most places is more noticeably clastic than the middle part of the stage, as is in fact the upper part also in some areas. Also, the Luisian overlaps the Relizian in some areas to lie directly on older rocks, whereas, in other areas of extensive marine Relizian occurrence, marine deposition ceases either at the base or within the Luisian.

The type section of the Luisian Stage has been taken as the richly foraminiferal 900-ft-thick (275 m) sequence of Monterey Shale exposed conformably between the Temblor Formation below and the Santa Margarita Formation above, in a gulley tributary to the canyon that has come to be known as "Edwards Canyon" (Kleinpell, 1938, p. 121-123, Fig. 11). These formational contacts were mapped by Anderson and Martin (1914). The Edwards Canyon sequence is but one of several such richly foraminiferal Monterey Shale sequences exposed on the northern flank of the granodioritic La Panza Range, structurally on the northward dipping so-called Highland homocline (named for its location in the old Highland School district, in San Luis Obispo County).

Lateral change in lithofacies is a conspicuous feature of the Temblor–Monterey–Santa Margarita sequence here on the Highland homocline. Along "Edwards Canyon," the most westerly of the three local Monterey columns and the one selected as the type Luisian, the Monterey consists mostly of soft punky shales, which are directly overlain by Santa Margarita sandstones that compare very well with those of the type section of the Santa Margarita somewhat farther west (Richards, 1935a, b, 1946). Locally to the east, the Monterey becomes progressively more of a siliceous shale lithologically more typical of much of the Luisian Stage elsewhere. Indeed, regionally this is generally true, for a chert bed is conspicuous at the base of the Luisian from Santa Barbara to San Luis Obispo and Monterey Counties, where in the vicinity of Monterey the Luisian corresponds to the lowest member of the type Monterey Shale. In turn, the local overlying lower Santa Margarita becomes fine-grained and diatomaceous, and even shaly on Indian Creek east of "Edwards Canyon Creek" east of "Edwards Canyon," rendering mapping of a Monterey–Santa Margarita contact difficult were it not for the persistent beds of chert concretions, ash, and sporbo. Locally, in the underlying Temblor along the Highland homocline, more and more argillaceous shaly mudstone appears eastward between the sandstone strata of the Temblor Formation, much as in the type section of that formation on the west side of the San Joaquin Valley. In fact, the Temblor Formation column of the Highland homocline area reflects a depositional history much more like that of the Temblor Range and west side of the San Joaquin Valley than like the areas of typical Vaqueros deposition farther north and northwest in the Salinas Valley and farther north in the Coalinga–Kettleman Hills area of the San Joaquin Valley.

On the northwest beyond the type Santa Margarita area, in the Adelaida quadrangle on the west side of the southern Salinas Valley, the middle Tertiary, including the Relizian and Luisian Stages and the type locality of *Siphogenerina hughesi* near Chimney Rock (Cushman, 1925a), acquires the geosynclinal character, with many folds, that persists northward into the San Antonio Hills and the Arroyo Seco drainage area, where extensive exposures are present in the Vaqueros Valley and Reliz Canyon. Many of the Relizian and Luisian foraminiferal faunas of the Adelaida quadrangle have been studied and identified by Robert R. Wilson (Kleinpell, 1938, Table VIII, p. 48-49).

In the opposite direction from the Highland homocline, several miles southeast but still west of the San Andreas fault, this middle Tertiary column becomes more and more like the sequence at the type locality of the Temblor Formation together with the overlying Monterey ("Maricopa") Shale, though of a somewhat shallower water marine origin as on Caliente Mountain and the adjacent Cuyama Valley toward the east end of which it grades laterally into terrestrial beds. Whatever the magnitude of lateral slip along the San Andreas fault, the trend of the fault appears to have occupied a topographically high area during much if not most of Tertiary time, especially during the middle Tertiary. This area apparently was highest on the western side (see Oakeshott, 1964).

Mapping of Caliente Mountain has been essentially of a reconnanissance nature (English, 1916; Hill et al, 1958) and many puzzling stratigraphic anomalies have appeared on the various maps of the areal geology there. Recently, with the aid of carefully collected fossil foraminifers and some equally careful contact tracing, Phillips (1972a) greatly clarified the stratigraphic sequence there. Phillips' paper is in the nature of a preliminary report, the published sequel of which, complete with areal and stratigraphic and paleontologic details, is awaited with great interest. From a half-dozen localities or more on the north side of Caliente Mountain, Turner (1970, p. 111) gave a radiometric date of somewhere between 13.7 and 14.5 m.y. for the base of the Luisian and "less than 13 m.y." for the top of the Luisian Stage. The stratigraphic interrelations among foraminifers, mollusks (especially those of the unique Branch Canyon Formation; see Kleinpell, 1938, p. 73), and mammals on the north side of Caliente Mountain have been discussed with comprehension and in considerable detail by Stirton (1960), complete with pertinent bibliographic references and especially a summary chart (p. 362).

It may also be added in passing that, on the basis of diatoms (W.W. Wornardt, personal commun.), not only the *Merychippus* beds of the Coalinga district but also the well known "bone bed" of the celebrated Sharkstooth Hill locality and sequence near Bakersfield in the San Joaquin Valley, fall definitely within the limits of the Luisian Stage; and that the littoral-living "monster" discovered some years ago on the Stanford University campus and being studied by Repenning is stratigraphically associated with fossil foraminifers of Luisian age (J.J. Graham, personal commun.). As noted previously (Kleinpell, 1938, p. 154-155), truly pelagic mammals, including cetaceans and especially the desmostylids so common in the Temblor-Monterey sequence and which Roy Reinhart has shown to constitute a distinct order (and not the true sea cows, *Sirenia*), also have long been known from the middle Tertiary of California (Vanderhoof, 1936, 1937, 1941, 1942; Domning, 1972; Wilson, 1973). For a review of these and the occurrences of nonmarine Tertiary West Coast mammals, see Savage and Barnes (1972).

Superpositional relations between foraminifers and larger invertebrates are essentially those recorded in 1938 (Kleinpell, 1938, p. 136, 152-157). Addicott (1970) recorded a number of additional megafossil occurrences in the Temblor Formation of the east side of the San Joaquin Valley, notably between the "Temblor A" and "Temblor B Zones" of Anderson (1911). Since the still higher "C Zone" is foraminiferally of Luisian age and was not expanded upon in 1938 (Kleinpell, p. 155, also footnote 154), it will be well to reemphsize here, in passing, that this "Temblor C Zone" megafossil fauna, as originally recorded, included many "Temblor B Zone" (Barker's Ranch fauna) elements. The inclusion of these fossils was due to the fact that in the critical fossiliferous area, a normal fault on the north side of the Kern River had thrown "B Zone" beds against "C Zone" strata, unbeknown to the early reconnaissance geologists and fossil collectors. Interested readers can find the structural features involved in Fox (1929). From a foraminiferal standpoint, it is significant that extraction of these "B Zone" elements in reference still leaves such critical species as *Turritella ocoyana* and *Pecten andersoni* in the "C Zone" assemblage, so that these two stratigraphically significant species of Mollusca are here within the limits of the foraminiferal Luisian Stage.

Updating of the Luisian Stage requires chiefly the presentation of a check list of species and their distribution in the type section of the stage on the Highland homocline in San Luis Obispo County, a discussion of the stratigraphic ranges of certain species as now better understood, and additional illustrations of both valvulinerids and siphogenerinids. The need for further illustrations of valvulinerids is due to the fact that Cushman's original figures of the types of the genus *Valvulineria,* coming early as they did in the history of Coast Range micropaleontology, were limited to line drawings. The need for additional illustrations of siphogenerinids, on the other hand, is due not only to a shortage of adequate pictures, but also is due to the lack of clarity in distinguishing those heritable traits that serve to distinguish one species from another within the genus *Siphogenerina* (especially in the Luisian), and to the confusion inadvertently brought about by Cushman's mistaken taxonomy as presented in his monograph on the genus (Cushman, 1926d). To some extent it is also due to the unawareness of the role that typology plays in stratigraphic paleontology and chronology.

Additional illustrations of "siphogenerines" have been presented here in the hope of clarifying the phylogeny of the various stocks involved. In the first place, as the generic name *Siphogenerina* has been used, it covers a group of species of polyphyletic origins (see systematics section of Tipton et al, 1973): species of various uvigerines that have attained to a siphogenerine stage, or evolutionary grade, in plans of growth (after all, the Cretaceous genus *Siphogeneroides* for long masqueraded under the name "*Siphogenerina*" for comparable reasons). In short, these "siphogenerinids" include *grades* rather than all being a single *clade* (see Huxley, 1958). These several lineages have herein been distinguished, described, and named (see the section herein devoted to Taxonomy, by Kleinpell and Tipton).

Most of this polyphyletic confusion involves primarily pre-Luisian species. Much confusion has resulted, nevertheless, even in species ranging upward into the Luisian, from failure to distinguish ontogenetic stages one from the other, and in recognizing metagenetic dimorphism as a nonheritable trait. The role played by the latter *lapsus* in relation to the so-called genus "*Rectuvigerina*" has been mentioned elsewhere (Tipton et al, 1973, p. 57), but the indiscriminate lack of attention to ontogenetic stages and metagenetic dimorphism dates back to Bagg (1905) who accurately described *Siphogenerina branneri* but "discerned" and named three separate species in a cluster that actually represent but one. Comparable practice seems to have carried on ever since. Workers look for a pointed or a well-rounded intitial end of the test, or they look for a blunt growth-end in a "short" specimen as distinguished from a well-rounded neatly apertured growth-end in a "longer" specimen, but these are not heritable traits. To the best of the writer's knowledge the morphologic traits of phyletic significance in *Siphogenerina* are primarily the nature of the ornamentation: relatively few costae with wide intercoastal areas, or the reverse; thick, broad, or so-called "lamellate" costae, in some species sharp, but in others broader and coarser, or so low as to seem to be merely striae rather than true costae; and the degree to which the costae are affected by the sutures, either hardly if at all affected as they pass across sutures from one chamber to a later chamber and so on, or rather strongly affected by the sutures and dropping to a notable extent into the sutural depressions between the chambers, the latter usually being still somewhat inflated in the ancestral uvigerine sense even in the uniserial stage.

The valvulinerids seem to fall into well-defined species and subspecies, though, as in the case of *Siphogenerina*, the writer confesses his early confusion with both of these stocks until Cushman sent him type specimens, "in the flesh" as it were, for actual comparison. Here again the recognition and distinction of ontogenetic stages is important, and the writer blushingly admits that his "*Valvulineria californica* var. *minuta*" are nothing other than immature specimens of the *Valvulineria californica* stock. The distinction and use of the term was due to inexperience, and also to the hopeful caution that leads stratigraphic paleontologists to "oversplit" to begin with, as can be said of certain other "varieties" (termed "*parva*" or its equivalent "*minuta,*" etc, in Kleinpell, 1938). Among the valvulinerids the nature of the dorsal and ventral sutures, the chamber heights, the nature of the periphery, and the degree of symmetry or asymmetry of the coiling-plane, and whether umbilicate or not, seem to play the most important roles in the speciations within *Valvulineria*. Two additional and historically related items of interest seem to be (1) that this mid-Tertiary genus in California appears to be an anomalinid and not a true rotalid, so that the "*Valvulineria*" of Eocene age are parallels rather than being cladally related; and (2) that prior to Cushman's description and naming of the genus *Valvulineria* and his designation of its genotype, these forms were known to industrial workers in California as "*Anomalina*" and the belts of their stratigraphic occurrence were known as the "*Anomalina* Zone." Phyletic charts, as now thought to be understood for both the siphogenerines and the valvulinerids, are presented here in Figures 6 and 7.

Zonation

Three zones still seem to be recognizable in the Luisian Stage. Nor do the congregations as now recognized differ greatly (Kleinpell 1938, p. 125-126), though quantitative factors (as mentioned with reference to *Anomalina salinasensis* and *Bolivina advena ornata,* and a few others in the *S. collomi* Zone) have no place in zonal diagnosis and should not have been mentioned. Furthermore, *Siphogenerina collomi* is not known from the *S. nuciformis* Zone as stated (Kleinpell, 1938, p. 125). These two species might be treated as subspecies, yet this is a possibility only where there happens to be an extreme wealth of data. It is also doubtful that *Dentalina barnesi* has the restricted stratigraphic range implied. Also, *Epistominella ("Pulvinulinella") gyroidinaformis* is certainly known to range throughout the Luisian Stage and, in fact, well up through the upper Miocene or higher, wherever the water was deep enough or even where the clasts in the containing beds have become large enough to permit its local preservation in "reworked" form. *Epistominella capitanensis* and *Eponides rosaformis* both perhaps appear as early as the *S. reedi* zone. Finally, there seems to be a possibility of recognizing two subzones in the *S. reedi* Zone, though it is still far from demonstrable on the basis of present data. The only known occurrences of *Siphogenerina obispoensis* are from the upper part of this zone, which stratigraphic interval also carries a few individuals that begin to approach, though do not attain, *Siphogenerina nuciformis* (e.g., see Weaver et al, 1969, p. 76, 79, 81, 83, Pl. 24, Fig. 11). Such individuals are perhaps to be considered "sports" unless populations of this form are actually found. The form in reference here is the "*Siphogenerina* sp. (?) aff. *S. nuciformis*" of the phyletic chart.

The needed check list of species in the type section of the Luisian Stage is presented on Table 3, along with check lists for two neighboring columns on Quailwater Creek (Table 2) and on Indian Creek (Table 1), both somewhat farther east along strike. This area is where W.D. Kleinpell, working for R.D. Reed and Roy Collom, collected most of the Monterey Shale species first described, figured, and named by Cushman (1925, 1926) mostly in the earliest volumes of his "Contributions."

A curious historical incident should be noted before closing this discussion of the Luisian Stage, for the light that it throws on the role of typology in stratigraphic chronology. It came about probably through the writer's failure to include a check list for the type section of the Luisian Stage in his 1938 work. Two undergraduate students were assigned to fill this gap and to present their results at a meeting of the Pacific Sections of AAPG-SEPM, and they were good enough to send their results to the writer for his comment prior to their presentation at the meeting. Disaster loomed, with the following additional history behind it. Industrial paleontologists working on the Monterey ("Maricopa") sequence that overlies the typical Temblor Formation on the west side of the San Joaquin Valley, had long recognized a so-called "flood zone" of foraminifers that occurs from about 1,254 to 1,329 ft (382 to 405 m) above the Temblor-Monterey contact. This so-called "flood zone" had been called the "*Anomalina* Zone", then the "*Valvulineria zone,*" and then the "*Valvulineria californica zone" in common parlance, and it is underlain by on the order of 1,000 ft (300 m) of Siphogenerina reedi* Zone beds that overlie the Gould Shale Member (upper Relizian) there. When a Luisian Stage was named (Kleinpell, 1930, 1933a, b, 1934, 1938) it was assumed by many workers to be the equivalent of this "flood zone," the foraminifers of which constitute an excellent example of the *Siphogenerina nuciformis* Zone. Thus, when the latter zone was found by the two students stratigraphically high in the type Luisian section on the Highland homocline, they decided that "the base of the Luisian Stage occurs half way up in the column which is the type section for the Luisian Stage." So much for usages, names, and typology in "new scientific discovery." One hesitates to belabor the incident, but it can serve as an object lesson in how conscientious effort and accurate particular correlation can be undone by nothing more than (1) a lack of understanding of the parts played by a concrete type, and (2) by sheer semantics, in the desired objectivity of scientific interpretation. The intricately complex problems presented by the facts of nature are themselves sufficiently difficult to place objectively in their true relations with each other, without the added burden of terminological confusion in the field of communication.

Another problem emphasized by the writer's failure to include a check list of species in the type section of the Luisian Stage in his 1938 work is the entire matter of species lists, pictures of species, and actual specimens of fossil foraminifer species. This problem is of more than ordinary importance since speciation within a genus is at the very root of refined biostratigraphic chronology and correlation. A few examples may serve to bring this situaion into bold relief, while the type Luisian check list follows them, along with five other check lists from Luisian sequences in superpositionally critical areas (Tables 1-6).

In the bit of history recounted above, only three species of *Siphogenerina* were recorded in the type Luisian check list presented by the students, whereas six well-defined species are present in that sequence. Especially "*S. collomi*" is recorded wherever siphogenerines with heavy costae of any sort (other than those of *S. transversa*) are found. One might wish that comparisons with the actual specimens in this type Luisian sequence could be made in such instances of recording *S. collomi*; yet the particular recording in the check list in reference was precisely in this very sequence involving the very same specimens in what was at least Cushman's topotypical area for this species, and thus overlumping heaven knows all of what. However, the writer is not without sympathy for the students in this matter; until Cushman's actual types and a topotypical sequence were examined the writer also had troubles. Thus, in Table V (Kleinpell, 1938, p. 34-37) he over-recorded *S. collomi,* indicating their presence in samples 4, 5, 7, 8, and 9. Subsequently he has found (other than in those additional samples noted on p. 300 of the 1938 work) *S. collomi* in only two samples of Galliher's member 5 of the type Monterey Shale: 7 (L.S.J.U. locality 333) and 8 (sample CV-5). The others (Kleinpell, 1938, nos. 4, 5 and 9 of Table V, p. 35) are clearly within the limits of Kleinpell's *S. nuciformis* which Galliher also recorded. This *lapsus* has not been corrected, even in the face of the fact that the 1938 illustrations (Kleinpell, 1938, Figs. 10, 11, and 12 of Plate XV) were at least adequate and an improvement over those that were possible more than a decade earlier.

Siphogenerina collomi was again recorded from the Santa Maria basin by Woodring and Bramlette (1950, p. 22), and although the writer examined their specimens of the new species and varieties in the form of either topotypes, "metatypes," or "idiotypes," no specimens of *Siphogenerina collomi* actually were included. However, it may very well occur in the Monterey Shale there, even at the locality recorded, for the Monterey sequence seems to be a contiuum there. What actually is involved is the absence of pictures of the foraminiferal species recorded, in hypotypical form, although one is grateful indeed for the two foraminiferal plates, 22 and 23, which serve to clarify a great deal with reference to the species figured as distinguished from those simply listed.

At least one other comparable instance can be mentioned, hopefully to some profit. For the California State Division of Mines, a special report (Richmond, 1952) was prepared on the geology of Burruel Ridge, northwestern Santa Ana Mountains. Accompanying this report are an excellent map of the local Tertiary formations, a thorough bibliography worthy of pursuit, and lists (prepared by W.T. Rothwell and M.L. Natland) of Miocene Mohnian foraminifers (principally from the lower, middle, and lowest-upper Puente Shales) from about a half-dozen localities (Richmond, 1952, p. 8, 9), though with no pictures. Within a decade this publication was followed by a foraminiferal paper by P.B. Smith (1960) with a few illustrations (Pl. 57-59) and four correlated columns (Fig. 157), and also many lists of foraminiferal species together with a section on systematics. Smith brought her materials to the writer for discussion, and during three of four such visits much in the way of diverse interpretations, both phyletic and chronologic, were ironed out in a partial meeting of the minds, though several disagreements—as would not be unnatural—remained. The three plates of illustrations are a fortunate addition to the literature. Although, in general, there was much agreement, an example of disagreement that can be pointed to with assurance is *"Bulimina carnerosensis"* (Smith, 1960, Figs. 3, 4 of Pl. 58) which clearly is not conspecific with the species of that name from the Temblor Range, nor is it of Luisian age. It is more probably a Mohanian *Galliherina* subspecies not as yet described at that time.

Mohnian Stage

General

At its type locality, the Mohnian Stage does not lie conformably on the Luisian Stage, although it does lie conformably on the youngest zone of the Luisian at several other localities. The type locality, between Mohn Spring and the site of the old community of Girard (type locality of the lower Modelo Formation of Hoots (1931), was selected for more than one reason. Principally, it was chosen because of the diversity and richness of its foraminiferal fauna, in a biostratigraphic continuum already on record (Rankin, *in* Hoots, 1931, p. 113, Pl. 27, A, C), which required only the completion of the formalization of some of the tentative taxonomy used and was soon accomplished (Rankin, *in* Cushman and Kleinpell, 1934). Though, as at the base of the type section, a biostratigraphic "hiatus" (in the form of some unfossiliferous beds, mainly sandstones) also exists at the top, this type Mohnian section is conformably overlain by a foraminiferal sequence which was thought to correlate with the uppermost two cherty to diatomaceous members of Galliher's stratigraphic sequence of the Monterey Shale at its type locality (Galliher 1931, 1932, Fig. 3, p. 22). It was also thought that the richness of the type Mohnian foraminiferal faunas would facilitate its correlative comparisons with other upper Miocene sequences of California, as for example that of the Palos Verdes Hills. Ironically, these hopes of having selected a standard of reference for correlative purposes have, instead, eventuated in a controversy that has not only muddied the waters of upper Miocene correlations in California but which has encompassed a personal tragedy of fatal proportions. The controversy in reference, however, has to do primarily with the age of the type Delmontian Stage rather than with that of the type Mohnian Stage, and further discussion of the controversy is in the section devoted to the Delmontian.

To update the status and zonation of the Mohnian, it may be appropriate to restate at the outset that the Mohnian Stage, like the Relizian, is one of the two most widely transgressive marine stratal units of the California Miocene. The unconformity at the base of the Mohnian type section—altering from place to place from a concordant disconformity to an angular nonconformity up to 43 degrees in magnitude—is a function of an orogeny. This orogeny was termed the Zuman orogeny by Kleinpell (1972) for lack of any previous or more distinctive designation and because (1) of its marked effects on the geosynclinal pre-Mohnian rocks of the central Santa Monica Mountains, though it is reflected elsewhere in the Coast Ranges as an epeirogeny involving either positive or negative movements; (2) there are many areas in which marine deposition was continuous from latest Luisian to earliest Mohnian time though with a notable concentration of highly phosphatic shales in the lower Mohnian; and (3) faunally the change from Luisian to Mohnian times reflects a fairly sudden and conspicuously extensive southward shift of an isotherm which affected the Coast Range area as far south as the Mexican border or beyond, though apparently not so markedly in the offshore oceanic region west of much of the southern part of southern California.

Some Mohnian Sequences

One of the most well-known and extensively studied biostratigraphic sequences of uninterruptedly deposited marine beds of the Mohnian Stage overlying upper Luisian beds, is in the Newport Lagoon area near the southern California coast behind the town of Newport. The recently published check list of foraminifers found throughout the exposed portion of the Miocene in the area (Warren, 1972, 1973) is presented here as Figure 1. Warren lists the species along a vertical coordinate in sequence from top to bottom on the chart according to the stratigraphic horizon at which each species first appears locally in the Miocene column here, just as Cushman and Laiming (1931) organized their check list for their older sequence along Los Sauces Creek in Ventura County. At the bottom of Warren's check list and chart the writer has presented an alternative interpretation of the boundaries of a stage and some of the zones as they would occur if his stage and zone boundaries were applied on an Oppelian basis. Included, in the form of footnotes, are a few alternative interpretations of phyletic relations among certain species represented. Phyletically and taxonomically affected are one of the principal *Nonion* lineages in the West Coast Tertiary, several generic names, and the four distinct taxa of the *Galliherina* lineage, heretofore overlumped and previously assigned by the writer to the genus *Bulimina*. The origins of the *Galliherina* lineage are still not clear, but seem perhaps closer to *Uvigerinella californica* than to *Bulimina sensu strictu* among the Buliminidae. Their occurrences in Warren's Newport Lagoon column appear to reflect very closely the respective stratigraphic ranges of these four galliherine taxa. Finally, the first appearance in the column of *Bolivina rankini* and of *Galliherina ("Bulimina") delreyensis* indicates to the writer that deposition here proceeded from late Mohnian time into the Delmontian.

Northward from Newport on the southern California coast another stratigraphic continuum from the Luisian to the Mohnian is found in the Altamira Shale Member of the Monterey Shale in the Palos Verdes Hills (Woodring et al, 1936; Woodring et al, 1946). Here the Tertiary column begins with transgressive volcanic flows and strata which contain foraminifers of the upper Relizian *Siphogenerina branneri* Zone (Zone A of Woodring et al, 1936, p. 133) and continues unbroken upward through the remainder of the Miocene into the Repettian (West Coast lower Pliocene) Stage before continuity is interrupted by faulting and/or erosion due to the affects of Dibblee's Zacan orogeny there. The foraminiferal continuum requires much offsetting of traverse lines and careful contact-snooping, so that the result is a composite column. One item of special interest here is that the highest known occurence of San Onofre Breccia is in the lower Mohnian at Point Fermin (see Woodring et al, 1936, p. 139-140). Another item of interest is the presence of pelagic mammals of late Mohnian age in the Valmonte Diatomite Member of the Palos Verdes Hills (Wilson, 1973), an occurrence which invites comparison with the presence of pelagic mammals of possible Mohnian age in the Santa Cruz Mountains (Vanderhoof, 1941).

East of the Palos Verdes Hills the Luisian also passes upward into the Mohnian Stage in parts of the Los Angeles basin subsurface sections, although only in a few small patches around the edges of the basin do pre-Mohnian marine Tertiary beds underlie the Mohnian Stage. Prior to Mohnian time the Los Angeles basin seems hardly to have been a Tertiary depositional basin at all, for widely throughout this area and on eastward for some distance the highly phosphatic shales of the lower Mohnian Puente Formation lie with depositional contact on the pre-Tertiary schist complex, reflecting a widespread early Mohnian marine transgression over this area. Wissler (1943, p. 210, Table 1) has graphically summarized the entire stratigraphic column here from the Jurassic to the Holocene, with a discussion of details and of correlations based on the foraminiferal faunas of beds ranging in age from Oligocene through Miocene to Repettian (Wissler, 1943). Wissler's subsurface Divisions E and D correspond to the lower and upper Mohnian Stage, respectively, and to the more northwesterly surface sequence of Hoots (1931) from units 1 through 16 and possibly 17. Wissler (1943, p. 218) also gave a summary of the number of upper Miocene zonules ("subzones") that may be recognized for local subsurface correlations within the oil fields of the basin—60 within Divisions A, B, C and the upper part of D, coring operations on down through the remainder of D and Division E having been too few to permit comparable subdivisions.

Geologically, it is also of more than special interest that the Fernando-like deposition of the Sycamore Canyon Formation (Krueger, 1936) had already begun before the close of Mohnian time in the general Whittier area. An analogous situation, though in reverse, is to be found in the Pico Canyon–Santa Susana Mountains area where "Modelo-like" (Monterey shale with coarse clastic lenses) deposition persisted locally into Repettian time, reflecting the diverse local effects of Dibblee's Rafaelan orogeny in southern California as tectonic instability increased there in late Miocene time.

In his paper on Burruel Ridge, which is north some distance from Newport Lagoon and on the eastern edge of Los Angeles basin, Richmond (1952, p. 9-11) included a list of foraminiferal species from six or more Miocene localities and from one Repettian locality which shows the Puente to be of Mohnian age in most, if not all, of its three local members. The stratigraphic relations of the Puente Formation to the locally underlying Topanga and the overlying Sycamore Canyon Formation and Fernando (Pliocene) Group are also depicted and discussed by Richmond (1952). In the San Jose Hills father north (Cushman, 1926b) and also farther northeast (Stewart and Stewart, 1930b), Puente deposition continued into Pliocene time.

A more extensive area, and many more foraminiferal assemblages, are reported on in the same general Santa Ana Mountains vicinity, in the paper by Smith (1960). Here too the foraminifers listed are mostly of Luisian and Mohnian age. However, in addition to the misidentification of a pictured species previously noted, another should be added: *Uvigerina joaquinensis* (Smith, 1960, Figs. 1, 2 of Pl. 57). Although its stratigraphic occurrences do not seem too unusual, many inconsistencies as to age and correlation will be apparent to the reader. Since Smith and the writer went over her material together several times it seems worthwhile, in reviewing Luisian-Mohnian relations, to discuss if not to attempt to clarify some of the more apparent inconsistencies in reference.

Perhaps faulting, stratigraphic isolation of samples (Smith, 1960, p. 468, 470, 472, 475), and other structural complications, of which there are many in the area, may have had something to do with these discrepancies. Perhaps even zoogeography played some role, as it certainly has offshore nearby. Subdividing the Mohnian three ways instead of four need not cause serious trouble, and encountering an extension in the stratigraphic ranges of species is not an altogether unusual experience. At the time of the conferences in reference, however, that which seemed the most telling basis for differences in interpretation was a statement that appeared in print later (Smith, 1960, p. 485) as follows: "No primary types of *S. nuciformis* were examined, but plesiotypes assigned by Kleinpell to this species appear to be within the range of variation of *S. collomi*. Therefore, *S. nuciformis* is questionably included in the synonymy of *S. collomi*." However, on an earlier page (p. 466) there seems nothing overly puzzling in the related statement: "Type specimens of these two *Siphogenerina* species were examined, and no satisfactory distinction could be made between them (p. 485)." This problem has already been discussed, and though subspecies rather than species may be involved, distinct taxa with distinct stratigraphic ranges are definitely involved. Another even more extreme example of "over-lumping" occurred in the case of *"Bolivina obliqua,"* listed by Smith (1960, p. 484) as from "Luisian(?) to upper Mohnian." This example was especially confusing to the writer because he saw none of Barbat and Johnson's species among the supposed representatives of *Bolivina obliqua* which were shown to him and Smith's specimens remained unfigured. Thus it came to appear, on study of the materials, that certain species had become "lumped" to the point of whole lineages being recorded as single species: for example, the four *Galliherina,* the *"Bolivina" seminuda* line (coming to include *"B. rankini"* etc.), and some of the *Epistominella* and *Nonion*. Open nomenclature had been prudently employed up to a half-dozen times or so in most if not all the tables, but many other identifications must be taken on faith by the reader owing to lack of illustrations. The writer is constrained to express disagreement with many of these identifications, such as on Table 3 *(Bolivina decurtata, B. obliqua),* Table 4*(B. obliqua, Robulus smileyi),* Table 8*(Bolivina sinuata alisoensis, B. subadvena),* Table 9*(B. subadvena),* and Table 10 (ranges of forms such as the aforementioned, plus *Discorbinella valmonteensis, Pullenia miocenica, "Uvigerinella californica,"* and a few other minor extensions in range that seem equivocal).

Singularly, after all these ambiguities and *non sequitur,* most of the conclusions as to age come out convincingly well though all too often not seeming to stem from the evidence presented. Also, the Pliocene and Miocene depth facies analogies presented on pages 478-479 do not seem to stem from evidence in the case of the Miocene, as an examination of the pseudo-Saucesian deep-water faunas of the southwestern San Joaquin Valley will reveal. Otherwise, the discussion of "Conclusions" is on the whole very valuable.

Across the Los Angeles basin to the north, north of Beverly Glen, a foraminiferal section collected by Thomas A. Baldwin served Pierce (1956) in modifying the zonation of the upper Mohnian (the former *"Bolivina hughesi* Zone"), although that stratigraphic section has no well exposed foraminifer-bearing bottom or top beds. A few miles farther west on the north slope of the Santa Monica Mountains, in a section from which beds are not areally traceable back to Pierce's section across a northward extension of crystalline and other older rocks toward Encino Park, the sequence selected for the type section of the Mohnian Stage is exposed. The distribution of foraminiferal species in this sequence from Mohn Springs to the former "Girard" was recorded by W.D. Ranking (*in* Hoots, 1931, p. 113), the species therein only informally identified were named, described, and figured by Rankin (*in* Cushman and Kleinpell, 1934) and the sequence was again highlighted by the writer in six out of seven critical assemblages from six of seven listed samples therefrom (Kleinpell, 1938, p. 46-47, Table VII).

The very basal Mohnian beds, where they lie with depositional contact on the schist in Brown Canyon, carry a megafossil fauna. The *Astrodapsis* and *Pecten* of this fauna have their closest affinities with those of the Briones Sandstone of the East Bay area far to the north. Woodring (*in* Hoots, 1931, p. 110-111, 123-124) considered the basal lower Modelo beds that contain *Astrodapsis* and *Pectens* as "older than the Briones sandstone of the San Francisco Bay region" (also see Kleinpell, 1938, p. 155).

Somewhat farther north, along Oak Ridge, the Mohnian lies conformably on the Luisian Stage once more, but as deep-water facies (David Berry, 1971) rather than the shoaling characteristic of this upward sequence in many other places. Along the ocean farther northwest, the same conformable Luisian-Mohnian sequence is discernible all along the Santa Barbara coastal bluffs, with the same concentration of phosphatic shales in the lowermost Mohnian. The upper part of this Naples Bluff foraminiferal sequence east of the mouth of Dos Pueblos Creek was presented by Kleinpell (1938, Table IV). Here the entire sequence from the mouth of Las Varas Creek (west of which there is a comparable sequence extending to the gravel cover at the mouth of Canada del Capitan) east to the Naples bluffs is check listed in Tables 4 and 5 as one of the most nearly complete Miocene foraminiferal sequences in the Coast Ranges. Only the upper Relizian sequence is interrupted here at the mouth of Dos Pueblos Creek. A complete sequence through this interval occurs at the type section of the Relizian Stage far to the north, in Reliz Canyon where, however, the Mohnian (i.e., supra-Luisian) interval carries no well-preserved fossil foraminifers, as it does in this Naples Bluff section (Table 5).

From Tables 4 and 5 it can be seen that, in age, the strata comprising this foraminiferal Rincon-Monterey-Sisquoc column extend from the middle Saucesian *Plecto miocenica* Zone below to the upper Delmontian *Bolivina obliqua* Zone above. Thus, essentially the entire Miocene proper, as found in the West Coast ranges, is here represented. Isolated outcrops of calcareous Monterey Shale in lower Dos Pueblos Creek, in gulleys and in the railroad cut on the northeast barranca of the creek, bear upper Relizian *(S. branneri* Zone) foraminifers and their diatomaceous age equivalents (W.W. Wornardt, personal commun.), helping to fill in the one missing zone in the one major unexposed portion of the biostratigraphic continuum.

Foraminifers are scarce in the argillaceous and calcareous to nodularly clastic Rincon shale beds below the bentonite bed, and in the lower Delmontian brown Sisquoc shales near the top. Although all but one of the Miocene zones as well as all the stages are represented, there are also other minor drawbacks to the column in its entirety as a standard of reference. This is ture especially in terms of the local thicknesses of the zones, particularly in the lower part of the column (Table 4) where more than one minor structural feature disturbs the otherwise perfect continuity of the sequence. For example, a structural flexure, its axis broken, occurs in the Monterey Shale and thin ash beds between samples

SBC-16 and SBC-17. Although the zone represented on both flanks of the flexure is the same, measurement of stratal thickness here is made difficult as to precise detail, as is true also of the interval of locally contorted shale at SBC-3, between SBC-9 and SBC-11, and between SBC-25 and SBC-26. Sample SBC-29 is at the base of a conspicuous intraformational breccia (stratigraphically the lower of the two referred to in Reed, 1935a, b) about 100 ft (30 m) or more thick. The breccia is almost a conglomerate at the top, with about 6 ft (2 m) of buff, fairly coarse, friable and ashy sandstone directly above it. This sandstone is succeeded upward by thin limestone ledges and punky laminated bituminous shales typical of the Monterey. Samples SBC-34 and SBC-35 are from clasts in the breccia; their assemblages are zonally indistinguishable from those in the darker colored matrix of the breccia.

Another stratigraphically isolated sample from the lower part of the Naples Bluff column, collected by Jay Phillips in November, 1971, is significant insofar as *Siphogenerina hughesi* itself, not found in the well-exposed sequence in the bluffs and along the beach, occurs here. Phillips collected the sample "about one-third the distance from the Bentonite Bed to the mouth of Dos Pueblos Creek, just east (about 20 feet) of a small stream that flows down the face of the cliff, and about 10 feet above beach level." It contained the following species: *Bolivina advena advena, B. californica californica, B. floridana floridana, B. floridana striatella, Buliminella curta, Globigerina* spp., *Nodogenerina advena advena, Siphogenerina hughesi, S. kleinpelli, Uvigerinella californica californica, Valvulineria depressa,* and *V. williami*. With reference to the lower part of this column, it may also be added that small 20-ribbed mud-pectens were found at the horizons of samples SBC-27 and SBC-30.

The upper part of the column (Table 5) exhibits no notable minor flexures in the Monterey Shales. A minor discrepancy in measurement of thickness is, of course, doubtless involved across the mouth of Dos Pueblos Creek, where thickness has been determined by averaging the attitudes in the sea bluffs on either side of the creek mouth. The top of the 10-ft (3 m) thick chert bed at the east side of the creek beds constitutes a better stratigraphic datum plane than its base because the underlying beds exposed here are also cherty, although with partings of softer organic shales. Fresh surfaces of bedding-planes at sample localities SBC-53 (5 ft or 1.5 m below SBC-54) and SBC-55 reveal a fragile cycloid foraminifer that does not survive sample-washing techniques. It apparently represents *Cassidulinella renulinaformis* Natland.

Finally, it should be pointed out that sample SBC-64 carries no diagnostic species, and that the Mohnian-Delmontian boundary has been drawn here on the basis of a notable lithologic feature under local foraminiferal discipline as to age at several other localities in the sea bluffs along this Santa Barbara coast. From a horizon 15 ft (4.6 m) below the horizon of sample SBC-64, the underlying diatomite gives way to chocolate-colored shales and mudstones (with a few flinty stringers) that become notably clastic within a few feet stratigraphically higher and especially a few feet still higher where they are almost graywacke. Conspicuously, above these notably coarse clastic beds, the brown "shales" become massive, and also more or less finely silty, with conchoidal fracture. They are much jarosite stained. The strata of this entire brown shale and mudstone interval are lithologically characteristic of the Sisquoc Formation as mapped by Dibblee (1950) from this general area westward along the coast, above the Monterey Shale. In turn, field relations indicate that the interval of more or less clastic beds here corresponds to the interval of lenticular, coarse, lowermost Sisquoc brea conglomerates at the mouth and in the general area on either side of Gaviota Creek, and to the interval of conspicuous intraformational shale breccia in the sea bluffs a short distance west of the beach entrance to Hope Ranch in the area east of Naples (the uppermost of the two breccias referred to in Reed, 1935a, b). Wherever along this Santa Barbara Coast these coarse clastic beds interrupt the sequence from more shaly beds below to more chocolate-colored massive beds above, and are under local foraminiferal discipline, they mark the change from upper Mohnian below to Delmontian above. The coarse clastics, as previously mentioned, appear to reflect Dibblee's Rafaelan orogeny (see also Reed, 1926; Reed and Hollister, 1935; Dibblee, 1950). In closing the discussion of this Mohnian sequence near Naples, it should be noted that, for purposes of comparing the updated check list. (Tables 4 and 5) of the Naples Bluffs column with the original (Kleinpell, 1938, Table IV, p. 29-31), sample SBC-44 is taken from the same stratum as was the original sample N-30. The stratigraphically higher horizons in Table 5 can be calibrated commensurately.

Across the Transverse Ranges to the north, in the Santa Maria basin, the Mohnian also apparently lies conformably on the Luisian, within the Monterey Shale of that region. No continuous foraminiferal sequence is on record within this area, unless it be in subsurface columns, but the basin has been mapped in considerable detail and the foraminiferal localities plotted and their species listed, so that ages can be given for the sequence of members in the local Monterey Shale. Canfield (1939) placed much of the local Miocene subsurface sequence on record though his foraminiferal zonation remained local, but Wissler and Dreyer (1943, p. 235-238) summarized the stratigraphy of the area as a whole and for the various oil fields of that time. Miocene foraminiferal zones were numbered, from zone 7 (the Point Sal Formation which, in the west especially, overlies the terrestrial variegated Lospe Formation) of Relizian age, upward through zones 6 and 5, both of Luisian age, and zones 4 and 3 which are lower Mohnian, to zones 2 and 1 and an overlying "Santa Margarita" unit which together represent the upper Mohnian. Their still higher Sisquoc Formation is Delmontian below and "lower Pliocene" above, with an "Upper Middle Pliocene to Recent" sequence above the Sisquoc. Several valuable plates figuring Miocene species of foraminifers, among others (which range in age from Cretaceous to Pleistocene), are also included by Wissler and Dreyer (1943), but these come from widely separated geographic areas. Included in this group are *Bolivina obliqua* and *"Nonion* sp." Barbat and Johnson, and seven other species from the Reef Ridge shale of the western San Joaquin Valley, two species from the Temblor, and one species from the Monterey (Wissler and Dreyer, 1943, Fig. 67. On Figure 98 (in connection with Glenn C. Fergusson's summary of the San Joaquin Valley east side Tertiary, p. 239-246), *"Bolivina* sp. of Barbat and Johnson, *B. marginata, Epistominella ("Pulvinulinella") gyroidinaformis, Valvulineria californica, Baggina ("Valvulineria") robusta, Uvigerina ("Uvigerinella") obesa impolita, U. sparsicostata,* and *Laimingina ("Siphogenerina") smithi* are shown. Also Figure 67 illustrates some species of younger occurrences but which are also known from the Miocene, and Figure 101 shows an unidentified Zemorrian *Pseudoglandulina*.

Seven years after the Santa Maraia publication by Wissler and Dreyer, the U.S. Geological Survey publication (Woodring and Bramlette, 1950) contained much foraminiferal data, mostly Miocene and Pliocene, and two plates of figured foraminifers (mostly but not all from the Santa Maria basin), seven foraminiferal check lists, and discussions of new forms encountered (p. 58-61) and of local age correlations (p. 100-106). A few foraminiferal species or zones are mentioned in the section devoted to a review of the local oil fields. Fossil localities are described (many of them referring to the foraminifers recorded in the check lists) from pages 136 to 142 (Woodring and Bramlette, 1950).

Though most of the foraminiferal assemblages of Woodring and Bramlette from the Santa Maria basin come from stratigraphically isolated occurrences, their check lists give ages of the stratigraphic units from the Point Sal Formation at the base of the column through the Monterey, Sisquoc, and Foxen formations. At five localities (Woodring and Bramlette, 1950, p. 17-18, 100, 136) the Point Sal Formation yields a typical lower Relizian *Siphogenerina hughesi* Zone fauna. Included in this fauna are *"Valvulineria* aff. *V. casitasensis"* (perhaps a form close to *V. depressa?*) which most people might not identify, and a very faintly striate variety of *Siphogenerina hughesi* which is typical of the forms included in most well-preserved populations of that species. From the lower member of the overlying Monterey (p. 21-22, 100, 137) come seven assemblages, again not stratigraphically allocated but, with only one exception (10b, "not plotted"), showing conflicting evidence. These assemblages range from upper Relizian through the entire interval of the Luisian Stage. A single sample from the middle member of the Monterey yields (when a few instances of the open nomenclature used are allowed for) an assemblage that is typical of the lower Mohnian Stage except for one of the species recorded: *"Bulimina delreyensis."* Without an illustration it is impossible to corroborate the identity of this species, but one has so often encountered the heretofore unnamed *Galliherina uvigerinaformis warreni* masquerading under this specific name that one is inclined to doubt that the species *Bulimina delreyensis* is present at this stratigraphically low horizon. The next seven localities from the upper member of the local Monterey yielded a typical upper Mohnian (old *"Bolivina hughesi* Zone") fauna (Woodring and Bramlette, 1950, p. 25-26, 137). These are apparently isolated samples from near Redrock Mountain in the Purisima Hills. Open nomenclature of *"Cassidulina* cf. *williami"* avoids conflicting evidence from rearing its head, but two records here and one later (Woodring and Bramlette, 1950, p. 58) are of special interest and significance: (1) the statement (p. 60) that Bramlette's new species, *"Ellipsoglandulina" fragilis,* with a generic assignment that is tentative only, is "The species listed by Wissler [1943] as '*Bulimina* sp. (large, crushed)' in his zones B and C of the Delmontian of the Los Angeles Basin" (further stratigraphic occurrences of this species will be discussed on a later page here); and (2) the presence here, and stratigraphically higher also, of *Epistominella ("Pulvinulinella") purisima* Bramlette, n. sp., a situation which, given illustrations of this species (Woodring and Bramlette, 1950, Pl. 23), implies from its appearance and its occurrences that it may very well be a descendant of *E. ("P.") capitanensis* Cushman and Kleinpell (see Kleinpell, 1938, p. 147, 327), with a possible overlapping but on the whole higher stratigraphic range; and (3) the possible conspecificity of *Virgulinella pertusa* and *V. miocenica* (Woodring and Bramlette, 1950,

p. 58). The meager shallow-water assemblage recorded "from Tinaquaic sandstone member of Sisquoc formation at locality 52″ (Woodring and Bramlette, 1950, p. 34), containing *Nonion belridgensis* as it does, makes one think this facies-faunule may also be of Miocene age. This is also true of the stratigraphically higher assemblages listed later.

Northward, beyond the Santa Maria basin proper, in the Huasna syncline and adjacent areas, the Mohnian is represented by the *"Pecten discus* shales" and some of the overlying Santa Margarita and Pismo sandstones. The top of the Mohnian is at a bed whose equivalence in terms of foraminiferal sequence is still undemonstrable; it may be as high as the beds that bear *Astrodapsis cierboensis*. Over the pass still farther north, and into the southern most Salinas Valley, the Mohnian is apparently represented by the lower and middle type Santa Margarita Sandstone. In the lower beds small, poorly preserved, low-petalled and perhaps even immature specimens of an *Astrodapsis* closest to *A. brewerianus* are present (George L. Richard, personal commun.).

Along the west side of the Salinas Valley as far north as the San Antonio Hills are folded geosynclinal deposits of hard, platy, brown siliceous shales that bear few signs of microfossils other than scattered imprints of foraminifers suggesting nonionids. On the more cratonic eastern side of the Salinas Valley, the lowermost lithologically McLure-like portion of Reed's Pancho Rico Formation (Woodring and Bramlette, 1950, p. 106, correlation chart) is probably also part of the Mohnian Stage (see Kleinpell, 1972, p. 94-95). Still farther east and southeast, as at the type section of the Luisian State along Edwards Canyon and even eastward to Quailwater Creek, the Mohnian is represented by a sequence of Santa Margarita sandstones much like those of the type Santa Margarita. The contact seems conformable, even gradational, but, as previously mentioned, a bed of phosphatic sporbo; a horizon of chert concretions in which bedding-planes pass from the adjacent laminae of siliceous shales directly across the edge of, into, through, and out of the opposite side of the concretion back into laminated siliceous shale; and a conspicuous bed of volcanic ash characterize the thin transitional stratigraphic interval at the contact. At a locality not too many mile distant Max Steinecke collected, from an "unnamed brown sandstone formation below white *Ostrea titan*-bearing sandstone equivalent to the Santa Margarita of the type locality and above fossiliferous Temblor sandstone" (Barbat and Johnson, 1934, p. 7), a megafossil assemblage which "is correlated on paleontological grounds with the Briones formation of Contra Costa and Alameda counties." Among the 15 species recorded are *Pecten crassicardo, P. c. nomlandi, P. estrellanus, P. raymondi, P. vickeryi,* and *"Astrodapsis* sp. indet; with raised petals, but found below Santa Margarita." All of these localities further corroborate the conclusion that through most of this general area the Mohnian Stage is represented largely by sandstones lying conformably above uppermost Luisian Monterey Shale. On the east side of the Carrizo Plain there are some anomalous and relatively unstudied Santa Margarita sandstones, and in the Cuyama Valley, post-Luisian as well as Luisian strata pass eastward into terrestrial deposits, some of which may be of Mohnian age.

Passing eastward across the San Andreas fault into the southern San Joaquin Valley, on the east side of that valley the Mohnian may be represented by the *Bulimina*-bearing beds of Comanche Point. Certainly the subsurface "Santa Margarita" sandstones and much of the more or less associated Fruitvale Shales are of early Mohnian age (Ferguson, 1943, p. 240-241, Figs. 96A, B). Though overlapped by younger formations on the surface at the south end of the valley, much of the immensely thick, so-called "brown shale" of the subsurface San Joaquin Valley is probably of Mohnian age, though it carries only few and sporadically distributed poorly preserved fossil foraminifers, mostly bolivinids (one of them, oddly enough, close to the Pliocene *B. interjuncta*). One thin bed, high in the "brown shale" column but apparently below equivalents of the subsurface Stevens sands and the Chico Martinez Cherts of the west side outcrop sections (see Barbat and Weymouth, 1931, on the occurrence of *"Borophagus littoralis"*), is composed almost entirely of *Uvigerina subperegrina*. Whether this bed is just below (in Mohnian) or just above (in Delmontian) is still not demonstrable. The Chico Martinez Cherts of the Temblor Range and the mid-Santa Margarita sandstone-and-conglomerate reefs farther west that are especially rich in *Ostrea titan* and commonly carry *Astrodapsis cierboensis* as well as *Pecten crassicardo* and *P. estrellanus* are also close to the Mohnian-Delmontian boundary.

Before moving on to the Mohnian outcrops of the west side of the San Joaquin Valley and the more northerly central California area, it may be well to correct an error of long standing that appeared in Kleinpell (1938) and which was found to be mistaken almost at the time the book went to press: the stratigraphic position of the lowest and oldest *Hipparion* horse teeth in the West Coast foraminiferal Neogene sequence (see Kleinpell, 1938, p. 70-71, 155-157, 173-176, 180-181). The data at the time were inadequate as discipline for any firm conclusion, and despite all the qualifications with which they were surrounded and the extensive dialectic devoted to alternative possibilities, the decisions eventually made as to the greatest likelihoods indicated, were erroneous. To begin with, the statement at the bottom of page 71 about the foraminifers being "a great deal like those of the uppermost Miocene of the Los Angeles basin" was a conclusion revised by Donald D. Hughes in 1939 (personal commun.) by which time he had decided that they were most like those at about the middle of the Mohnian Stage (complete with the *Uvigerina hootsi* mentioned). This conclusion alone was enough to drop the proposed horizon of correlation in reference, as given on page 157, by a stage-and-a-half: ". . . the *Hipparion* cf. *mohavense* fauna of the Mint Canyon Formation and its correlatives is here included within the Delmontian Stage, following the local stratigraphic allocation by Stirton [1933] and others." It was soon learned and pointed out (Repenning, personal commun.) that *Hipparion* had appeared in Europe before the close of the Tortonian Stage (the most likely European age equivalent of the West Coast range Mohnian Stage). By 1960 Stirton had placed the Barstovian *(Merychippus*-bearing)-Clarendonian *(Hipparion*-bearing) boundary at a mid-Mohnian horizon—with a dashed line to be sure, but still commensurate with the pertinent discipline. Thus, with Bode's *Merychippus* faunas of early Luisian age at the oldest, it is only the short interval of time between latest Luisian and earliest Mohnian from which we lack fossil mammals in the Coast Ranges. If *Hipparion* was derived in the New World from *Merychippus* we must look not to the extensive land bridges of Miocene-Pliocene times but to the paleogeographic affects of the Zuman orogeny for the first migratory movements of *Hipparion* from the New World to the Old. Finally, the celebrated piece of diatomaceous float bearing foraminifers and the tooth of a *Hipparion* cf. *mohavense* still has not been traced to its outcrop of origin somewhere in the Puente Hills and its full significance still eludes us. To the best of our knowledge the diatoms still have not been studied and the foraminifer is *Galliherina uvigerinaformis warreni,* which is at least not out of keeping with the other known data pertinent to this problem.

To return to the San Joaquin Valley west side: there, in the Zemorra Creek–Chico Martinez Creek–Carneros Creek area, several hundreds of feet stratigraphically above the top of the type Temblor Formation, the Mohnian again conformably overlies the Luisian Stage. Above the so-called "*Valvulineria* Flood Zone," which is the *Siphogenerina nuciformis* Zone of this column (Kleinpell, 1938, Table VI) and on the order of 200 ft (61 m) thick, the *S. collomi* Zone is many times thicker, and though it contains no specimens of *S. collomi* (a siphogenerine in the uppermost sample of this zone appears to be *S. kleinpelli)* it carries 39 Luisian species, of which three seem especially significant: *Bolivina parva, Uvigerina joaquinensis,* and *Valvulineria joaquinensis*. The higher beds here bear no foraminifers but when traced northward they do, and the evidence in this more southerly area is that the change in deposition from uppermost Luisian to Mohnian equivalent took place in deep water. Such was also the case in subsurface sequences at the southern end of the valley where the *S. collomi* Zone is also exceptionally thick, and carries *S. collomi* at several intermittent horizons. At the southeastern end of the valley a sandstone body, known locally as the Wicker sand, occurs at or near the top of the *S. collomi* Zone, and persists southwestward and westward in subsurface. Its equivalent, about 10 ft (3 m) thick and unlithified and therefore inconspicuous unless the column is studied in detail, occurs in outcrop at about the same stratigraphic horizon on Chico Martinez Creek.

Northward along the west side of the San Joaquin Valley, in the McDonald Ranch–Shale Point area, this surface equivalent of the Wicker sand suddenly increases tremendously in thickness (see Fig. 2) in the form of the Twisselman sand, which straddles the Luisian-Mohnian boundary here and eventually overlaps the entire sequence of Temblor-Gould-Devilwater beds and lies directly on rocks as old as Cretaceous before disappearing beneath the alluvium of the Antelope Valley. Heikkila and MacLeod (1951) have published cross sections, columns, and a map of this somewhat structurally complicated area, but the excellent paleontologic studies carried on in conjunction with them by Carlos Key, under the supervision of S.W. Muller of Stanford University, has never been published. Through the good offices of Muller and Key it has been possible to publish a check list of foraminifer species from Key's most critical localities and samples and to present this check list, with locality descriptions, as Table 6 here.

The stratigraphic and geologic significance of this Shale Point area can hardly be overemphasized. Here the deep-water ocean bottom of the south has suddenly shoaled, and lower Mohnian overlap has taken place northward across an old high area. The unfossiliferous Antelope Shale Member of the more southerly area persists northward as the essentially unfossiliferous McLure Shale Member. At its base, and locally at least, unconformably on the remaining sandstones of the west side Temblor

Formation, a very thin unit of phosphatic siltstone, the remnant of the Mohnian portion of the more southerly McDonald Shale Member of the Monterey, is present.

A few foraminiferal assemblages have been found and reported from this thin basal McLure phosphatic and glauconitic siltstone, pointing up its age equivalence with the basal McDonald shale of the Shale Point area: one by Kleinpell (*in* Woodring et al, 1940, p. 127-128, Pl. 50), and one from a nearby locality, with better preserved Foraminifera, by an old friend C.C. Church who sent the writer a slide of his assemblage which included more species than did the writer's earlier collection. The foraminifers from the latter locality have been listed by Church (1972, p. 75-76). The writer is in full agreement with most of Church's identifications, but the same cannot be said of some of Church's age and correlation discussion. Also, some of the species listed by Church could not be found on his slide (*Cibicides* sp., *Glandulina* sp., and *Uvigerinella* sp., none of them of significance in this discussion) whereas others not noted by Church, some significant and some not, *were* found: *Baggina subinaequalis* Kleinpell (C), *Bolivina* sp. (perhaps the *B.* cf. *cuneiformis)* (C), *Bolivina modeloensis* Cushman and Kleinpell (two specimens only; see Fig. 16 of Pl. VII), *Buliminella curta* Cushman (F), *Cassidulina cushmani* listed) (F), and *Uvigerina hootsi* Rankin (one specimen only). Not a great difference, but combined with some of the species listed previously (especially *Cibicides* cf. *C. illingi),* the presence of *Baggina subinaequalis* and *Bolivina modeloensis* make the basal McLure look even more like lower Mohnian, and seem to leave little if any basis for the statement that in ". . . the present study of this more complete faunal assemblage, its stratigraphic position . . . now appears to be much higher in the section than was originally postulated . . ." (Church, p. 76). Except for a postulated unconformity Church's subsequent discussion of the various valley columns is beyond cavil, especially that of the Tidewater Reward 101-A upper Mohnian "brown shale" sequence. Westward, far into the Waltham Valley, where the basal McLure lies unconformably on steeply dipping Temblor sandstones, the basal sandy McLure carries *Astrodapsis* with raised petals in an area where Mohnian overlap has proceeded much farther than at the edge of the San Joaquin Valley. However, in the final column of the article, the discrepancy is accentuated by the comparison of the basal McLure assemblage with that of a sample from the Woodland Hills area on the north slope of the Santa Monica Mountains far to the southwest (a slide of the assemblage was provided by A.A. Almgren). Certainly the two assemblages have in common *Bolivina vaughani, Buliminella subfusiformis,* a *Cassidulina, Planulina* cf. *ariminensis,* and *Uvigerina hootsi,* but the Woodland Hills sample contains at least 25 species of which only two or three are discussed in the comparison mentioned: "a large, heavily costate *Uvigerina* similar to *U. subperegrina* and a *Bolivina,* which at one extreme might be identified as *B. woodringi* while in its smaller less inflated form is very similar to the *Bolivina* in the lower McLure shale, identified as *B. cuneiformis* . . . ." Examination of the *Uvigerina* in reference reveals that it more closely approaches *U. peregrina,* a Pliocene species, than *U. subperegrina.* An immature specimen of *Bolivina woodringi* is present, and the other *Bolivina* (presumably the "*B. cuneiformis*" referred to, but questionably identified as such Kleinpell, 1938, p. 270) is much as described in the subsequent passages, except that those from the McLure are completely smooth and may involve two species (or at least subspecies), whereas those from the Woodland Hills sample distinctly bear costae, especially a prominent one along the early median line. The most abundant species in the Woodland Hills sample is *Bolivina subadvena spissa* Cushman, first described from the Pliocene and occurring in the "Pliocene" of the San Jose Hills, though it also is recorded from horizons as low as upper Mohnian. Finally, the stratigraphic horizon of the Woodland Hills sample, at the corner of Murietta and Roblar Streets north of Valle Vista but south of Ventura Blvd. is well up in Hoots' upper Modelo diatomite, and on stratigraphic grounds alone may be Pliocene, or at least high in the local Delmontian, and can hardly be Mohnian in any sense of geologic age. The writer does not profess to have the answer to any of the problems raised, but thinks that the interpretations currently on record in published form, both phyletic and chronologic, should be pursued further than they have been, and certainly with other possible alternatives in mind. With this in mind, some of the stratigraphically critical species from both the basal McLure of Reef Ridge and from the supra-type *Bolivina obliqua* Zone upper Modelo diatomite of Woodland Hills, are figured in the accompanying plates.

Passing westward from Shale Point and Reef Ridge to the coastal area north of the crystalline Sierra Salinas, a comparable overlap of the early Tertiary and many middle Tertiary formations of the Arroyo Seco-Vaqueros Valley area by the Mohnian as well as by the Luisian occurred in the Carmel–Monterey Bay area. Here the Luisian (Monterey member 5 of Galliher, 1932) transgresses widely over granite and scattered small pockets of sandstone, that may or may not be pre-Luisian in age, and is conformably overlain by the lower Mohnian (Galliher's Monterey member 4). In Coyote Gulch, however, on the north side of the lower Carmel River Valley, this lower Mohnian has overlapped the Luisian to lie with depositional contact on an old granite ridge, as noted by Kleinpell (1938). Upward in the Monterey the lower Mohnian is succeeded by Galliher's member 3, which bears an upper Mohnian foraminiferal fauna (Kleinpell, 1938, Table V). Here the Mohnian is generally in neritic facies, in both senses of the word, and an excellent lineage of Nonionidae has evolved, apparently from the Luisian *N. incisum* through *N. montereyanum carmeloensis* and its subspecies *N. m. montereyanum* to the Delmontian *N. schencki*. *N. multicameratum,* apparently a descendant of *N. kernensis* Kleinpell and its relatives, appeared soon after the opening of Mohnian time, as it did elsewhere.

At the latest, *N. multicameratum* appears in the first zone above the lowermost Mohnian in the Rodeo Shale Member of the Monterey northeastward across San Francisco Bay (Kleinpell, 1938, Table XIII, p. 56, 62-66) and at Selby Smelter, where *N. montereyanum carmeloensis,* the more essentially mid-Mohnian subspecies, also occurs at the highest Monterey foraminiferal locality. Here, however, in contrast with the Monterey Bay area, the late Mohnian and higher Miocene strata have been deposited in progressively shallower water. The Hercules Shale finger (still Mohnian) apparently is a shelf sea deposit, within the more generally and more nearly littoral megafossil-bearing Briones Formation (Trask, 1922).

It should also be mentioned that the Tice Shale, at its fossil-rich locality at the west end of San Pablo Dam, during extensive post-1938 collecting, has yielded in its lowest beds several *Valvulineria californica californica* and even more numerous *Elphidium,* so that it may be compared with the Twisselman-McDonald sequence at Shale Point, where again the change from Luisian to Mohnian occurred in shallow water. More than 100 ft (30 m) of Tice Shale was exposed for many years, in road cuts west of the west end of San Pablo Dam. The Tice here is mostly black, argillaceous shale with soft yellow calcareous lentils or yellowish shale. However, a few feet of cherty shales of the normal Monterey type are present at the base of the exposed section. A new highway has been built at a slightly higher topographic level, exposing sandstones that unconformably overlie the Tice here and that have been mapped as Briones. In order that students might have the opportunity to examine the poorly preserved sort of fossil foraminifers they would often be faced with in well cores after they left school, this exposure was visited annually before the new highway was constructed. Some laminae also yielded many well-preserved foraminifers, including *Valvulineria californica californica* and *V. c. obesa* (the longer-ranging and more generalized subspecies of the stock) in about the lower half of the exposed column. *Elphidium* was present both stratigraphically higher and lower in the same sequence. Nor was it ever possible to be sure, in any detail, as to the stratigraphic allocation of Lawson's CC-1 sample (Kleinpell, 1938, Table XIII) in the sequence. This sample carried unidentifiable crushed specimens of four or five more species than those recorded in 1938 and the impression was left that the location "west of the San Pablo dam" (Kleinpell, 1938, p. 65) might indeed was accurate placing the sample no higher than the middle of the exposed Tice sequence. The conclusions reached were; (1) that the Luisian Stage passed upward to the Mohnian Stage there stratigraphically *within* the Tice; (2) that it reflected fairly shallow depositional depths in keeping with the arc-shell faunas of the immediately underlying Oursan Sandstone Member and the overlying Hambre Sandstone Member of the Monterey there; and (3) that the beds from which sample SPD-4 were collected represented the deepest water deposits laid down in that vicinity during the transition from Luisian to Mohnian time.

Still farther northwest in California, the Mohnian Stage is involved in sequences that include the overlying Delmontian Stage in diverse local facies. Mention of these sequences is therefore included in the review of the Delmontian Stage.

Zonation

At the beginning of Mohnian time, an isotherm swept southward over the California faunal province accompanied by the invasion of the province by new elements of northern origin, while many of the more truly subtropical elements that had persisted through the Luisian in California either moved farther south or, as species, became extinct. At the same time, other warm-water elements with a somewhat greater thermal tolerance continued to live side by side with the new cold-water elements through at least a part of the Mohnian Age. Such southward movements of isotherms continued in California until a cold peak in mid-Pleistocene time (Smith, 1919). Though apparent fluctuations in this progressive southerly sweep of isotherms may have occurred at one horizon in the Repettian (early Pliocene), such fluctuations are of very minor magnitude in this overall cooling trend. By mid-Pliocene time temperatures on the West Coast appear to have corresponded to those pertaining at comparable latitudes today, although terrestrial climates at that time appear to have been more dry. Because of such progressive changes in the latitudinal gradient of the thermal environment, it is apparent that difficulties also would arise in correlating from north to south in the California area where the most critical of these temperature changes occurred from Miocene until mid-Pleistocene time when a single cold-water province seems to have existed from far to the north to a latitude south of the Mexican border.

The greatest difficulties in correlating from north to south from late Miocene time onward were among the provincial inshore megafossil faunas, the environmental niches of which were most sensitive to these climatic changes. The writer and D.W. Weaver were told by Ralph Arnold of the difficulties that he and his contemporaries had in the early days in correlating from north to south in the Coast Ranges, even though employing the most age-diagnostic late Neogene mollusks. Thus *Pecten terminus* was named because it was the last species of lyropecten — a group common in California since Oligocene time — remaining in the Tertiary column of central California north of the Transverse Ranges. However, it soon was learned that *Pecten cerronsensis,* the Lyropecten presently living off the coast of lower California, lived through most of the Pliocene of southern California. Although the offshore benthonic foraminifers would not be expected to reflect this difficulty in correlation as much as the inshore mollusks such as *Pecten caurinus,* which was living far north as early as Miocene time although it was not present in Santa Barbara until deposition of the upper, or so-called San Pedro, zone of the late Pliocene–early Pleistocene Santa Barbara Formation. Still, the effect on upward species ranges of the cold nips that clearly inaugurated the openings of both the Mohnian and Delmontian Ages in the West Coast ranges must be borne in mind. We must be prepared to recognize zoogeographic facies as well as ecologic facies—if indeed the boundaries of fossil zoogeographic provinces can eventually be found—for Oppelian stages and zones are based on the stratigraphic interrelations of the teilzones of species (*i.e.,* the stratigraphic ranges of species *within a faunal province).* Although such stages and zones cannot be recognized as such until the ecologic facies factor has first been culled out, intrinsically Oppelian zonation does *not* circumvent the problems of zoogeographic facies (Kleinpell, 1964).

Another feature of significance from earliest Mohnian time onward is that progressively more living species enter the biostratigraphic column. With this in mind, it will be seen that from earliest Mohnian time on the role of species extinctions in the recognition of a congregation will progressively diminish whereas the horizon of the appearance of a new species (the horizon that has come to be called a "datum plane") becomes progressively more important in the recognition of zones. However, it must also be borne in mind that Oppelian stages and zones are zoogeographically provincial phenomena, so that provincial species extinctions, where demonstrable, are still important in the recognition of zones for a particular province.

Thus, the California foraminiferal faunal province, which had begun to take form about the time of the Lompoc orogeny and which had lasted on the West Coast through Relizian time, and, at least in California, on through the Luisian Age, apparently broke up at the time of the Zuman orogeny. Therefore, at the beginning of Mohnian time, the dominant elements of the California foraminiferal province either became extinct or, with the zoogeographic essentials of its composition, moved south.

In the early days of applied micropaleontology in the California petroleum industry, especially in subsurface work, the biostratigrphic sequence that in 1934 came to be designated as Mohnian was generally referred to in its lower portion as the "*Baggina californica* Zone" and in its upper portion as the "*Bolivina hughesi* Zone." In southern California, nearby surface standards of reference available were those of the Modelo Formation cropping out on the north slope of the Santa Monica Mountains (Rankin *in* Hoots, 1931, p. 113), the sequence from the Altamira Shale Member of the Monterey to the Repettian of the Palos Verdes Hills (Woodring et al, 1934), and the numerous subsurface sequences in the Los Angeles basin as summarized by Wissler (1943, Table 1, p. 210, 222). Presently it became apparent that two subdivisions of the "*Baggina california* Zone" (lower Mohnian) could be recognized over a fairly wide area, and for these the writer (Kleinpell, 1938) coined formal names: the *Bolivina modeloensis* Zone below and the "*Bulimina*" *uvigerinaformis* Zone above, with types in the north slope Santa Monica Mountain section (Rankin *in* Hoots, 1931) and in the Naples Bluff Monterey Shale section of the Santa Barbara coast, respectively. An attempt also was made, although only tentatively, to subdivide the previously recognized and higher "*Bolivina hughesi* Zone" into at least two subzones (Kleinpell, 1938, p. 130-131). Subsequently, Pierce (1956) formally subdivided the old "*Bolivina hughesi* Zone" into two zones, based on the biostratigraphic sequence in the Modelo Formation as sampled by T.A. Baldwin in the Benedict Canyon area, which is several miles east of the type Mohnian sequence that had been check listed by Rankin *(in* Hoots, 1931) between Mohn Springs and Woodland Hills (old "Girard"). In spite of Pierce's scrambling of the local bolivinid taxonomy (see the section herein devoted to Systematics) and his subsequent (1970) supposition of an age equivalence between the type section of the Mohnian and the Delmontian Stages (see the following pages for an attempt at clarifying this theoretical supposition), Pierce's twofold subdivision of the old "*Bolivina hughesi* Zone" was a marked improvement over any earlier attempt, and it was formally adopted (Kleinpell *in* Stirton, 1960, and others), though not its twofold terminology, since, in organic evolution, different stocks are not necessarily affected to the same selective extent by the same environmental changes, and a comparably discrete zonal terminology is therefore requisite in any time-rock classification of fossiliferous strata.

Unhappily, scrambled taxonomic terminology has direct effect even on any single zonal terminology based on foraminifers such as in Pierce's zonation. Moreover, the name *Bulimina uvigerinaformis* Zone also has certain disadvantages: (1) this zone was typed in isolation, as it were, in the Naples Bluff section on the Santa Barbara coast; and (2) the "*Bulimina*" involved in the zonal terminology probably represents a distinct genus (see the accompanying notes on Systematics), *Galliherina,* the lineage of which includes not only one but several species and subspecies with differing stratigraphic ranges, however, much these ranges overlap. This situation has led to many zonal miscorrelations based on the supposition that *Galliherina* represents a single species and that its genotype *("Bulimina uvigerinaformis senu stricto)* and its unrecognized subspecies or related species are restricted in stratigraphic range to the zone named after it, a demonstrably mistaken supposition. Thus it has seemed desirable, herein, to coin new names for the four zones recognizable between the top of the *Bolivina modeloensis* Zone (lowermost Mohnian) and the base of the *Bolivina obliqua* Zone (upper Delmontian), and to type them in the same biostratigraphic sequence in which the two zones mentioned also have been typed. It is hoped thereby that this may clarify much past confusion and especially may help clarify the controversy that has recently developed over the age relations of the Mohnian and Delmontian Stages.

Bolivina modeloensis Zone—This lowermost Mohnian zone with Hoots' units 1 through 5 as type, still seems to be characterized by a congregation in which *B. marginata gracillima, Eponides mansfieldi* (as a notably ornate variant of the species), *E. rosaformis,* and *Nonion mediocostatum* make their last local stand, and several species in addition to the name-giving species *Bolivina modeloensis,* come into the province for the first time: *Bolivina bramlettei, Baggina subinaequalis, Cassidulina monicana, C. modeloensis, Cibicides illingi, Galliherina uvigerina-formis uvigerina formis, Gyroidina soldanii rotundimargo, Epistominella capitanensis, Nonion montereyanum carmeloensis, N. multicameratum, Uvigerina hootsi, U. segundoensis, "Valvulineria" araucana,* and *Valvulineria grandis.*

Bolivina barbarana Zone.—This zone is, at least in part, essentially equivalent to the old "*Bulimina*" *uvigerinaformis* Zone, but now is more specifically typed within the same sections as the subjacent and superjacent zones of the Mohanian. The type for the zone is within Hoots' unit 6, as faunally characterized by Rankin's samples 113, 114, and 115, and particularly by sample T-4 of Kleinpell (1938, Table VII, p. 46-47) which is stratigraphically almost immediately below Rankin's sample 115. The zone appears to be characterized by the lowest known occurrences of *Bolivina barbarana, B. decurtata, B. pseudospissa, B. sinuata alisoensis, B. woodringi, Eponides healdi, Galliherina uvigerinaformis warreni, Nonion montereyanum montereyanum, Robulus mohnensis* (apparently restricted to this zone), and *Uvigerina modeloensis,* and by the highest known occurrences of the single apparently restricted species mentioned and notably the highest know occurrences of *Bolivina modeloensis, Nonion costiferum sensu lato, N. multicameratum, Pullenia moorei,* and *Uvigerina carmeloensis.*

Bolivina wissleri Zone—This zone has for its type the sequence in Hoots' unit 9 from the horizons of Rankin's samples numbers 119 through 136. Hoots' units 7 and 8 with Rankin's sample number 117 may also be this young although the foraminiferal assemblages seem zonally undiagnostic. The congregation of this zone seems to combine the lowest occurrences of *Anomalina hughesi, Bolivina girardensis, B. granti, B. hughesi, B. wissleri, B. seminuda, B. sinuata sinuata, Buliminella semihispida, Discorbinella valmonteensis, Galliherina uvigerinaformis doanei, Hopkinsina magnifica,* and *Virgulina delmonteensis,* with the highest known occurrences of *Baggina californica, Bolivina advena advena, Cibicides illingi, Epistominella capitanensis, E. gyroidinaformis, Galliherina uvigerinaformis uvigerinaformis, G. uvigerinaformis warreni, Uvigerina modeloensis,* and *Virgulinella miocenica* (a form possibly conspecific with *V. pertusa).*

Bolivina goudkoffi Zone—The type for this zone is taken as the upper part of Hoots' unit 9 from Rankin's sample 137 upward through Hoots' unit 13 with Rankin's sample 152. A congregation in this zone appears to combine the restricted known occurrence of *Cassidulinella renulinaformis* (a very fragile form), the lowest known occurrences of *Bolivina goudkoffi* (perhaps restriced to this zone), *B. hootsi, B. seminuda foraminata, Cassidulina quadrata,* and *Uvigerina hannai,* with the highest known occurrences of *Anomalina hughesi, Bolivina bramlettei*(?), *B. Californica, B. decurtata* (?)*B. girardensis* (?), *B. granti, B. hughesi*(?) *Buliminella californica, B. semihispida*(?), *B. granti, B. hughesi*(?), *Buliminella californica, B. semihispida*(?), *Cassidulina modeloensis, Discorbinella valmonteensis* (?), *Nonion montereyanum carmeloensis,* and *Planularia cushmani.*

In view of the relatively negative evidence from the superjacent lower zone of the Delmontian Stage it should be borne in mind that some species previously supposed to have their highest occurrences in this zone and questioned in the above list, may very well range into this higher stage. Especially, *Bolivina bramlettei, B. decurtata, B. girardensis, B. goudkoffi, B. hughesi, B. woodringi,* and *Discorbinella valmonteensis* seem to have upward extended ranges, as indicated by their persistence in the Newport Lagoon column (see Fig. 1) along with lower or upper Delmontian species, or both, such as *Bolivina rankini* and *Galliherina delreyensis*. Data from the Santa Maria basin (Woodring and Bramlette, 1950) further corroborate such upward extension in range.

Delmontian Stage

"The Miocene Series in California was divided into six benthonic foraminiferal stages by Kleinpell (1938); the upper Miocene was separated into the Mohnian (lower) and Delmontian (upper) Stages. Bandy (1972b) and Ingle (1967), however, have suggested that these stages are time transgressive in part when referenced to radiolarian and planktonic foraminiferal biostratigraphy.

"Kleinpell's type Delmontian Stage near Monterey, California, contains benthonic foraminifers and diatoms that elsewhere in California are restricted to strata correlated with the lower and middle parts of the Mohnian Stage (Pierce, 1972; Ruth, 1972). Consequently, Bandy (1972b), Ingle (1967), Pierce (1972) and others have challenged the validity of the 'Delmontian' as a stage. The term 'Delmontian,' however, has been retained in this paper to refer to rocks dated as younger than Mohnian and older than Pliocene" (Barron, 1975, p. 619).

In many ways the two preceding paragraphs state, perhaps as succinctly as possible, the nature of a controversy that has come to center on the age relations between the type Mohnian and type Delmontian sequences. Although the first paragraph has little bearing on the controversy in reference, the statement embodied therein has, as a general postulate, helped lead to the particular phase of the controversy that is summed in the second paragraph. The statement in reference in the first paragraph reflects a bugaboo of some decades standing, and needs to be clarified at the outset of any discussion of the very real and particular controversy embodied in the statements of the second paragraph lest it be brought in speciously, however inadvertently so, as a criterion having some bearing on the particular controversy in reference—which bearing it does not have.

The Controversy

First phase—For some years specialists engaged in correlation on the basis of planktonic foraminifers have felt that the widespread geographic dispersals of planktonic species circumvented the facies problems faced by specialists working with benthonic foraminifers. This illusion has been further fostered by the fact that planktonic specialists have operated on a basis of correlation by index fossil, a practice well adapted to reconnaissance work. However, Davies (1934, p. 52, 56) has already called attention to the hazards, in principle, to correlations based on individual species in this manner as well as on whole assemblages of species *per se,* Another related headache long suffered by West Coast micropaleontologists is that many of the so-called "biostratigraphic columns" of planktonic workers elsewhere have turned out to be theoretical columns based on "evolutionary bioseries" of planktonic species, without any superpositional discipline over the stratigraphic ranges of such species. One hopes this phase is a thing of the past. To this hope one can add another, the realization at long last by some planktonic specialists that facies phenomena, both thermal and even "bathymetric" (insofar as certain plankton floats at certain levels and other plankton at other levels, as found in the Okhotsk Sea, for example) do affect the distribution of planktonic as well as benthonic species of foraminifers. Stratigraphic ranges of planktonic species also need to be brought under superpositional discipline adequate to give them chronologic significance. To this problem has been added that of iterative evolution and parallelism in the evolution of planktonic foraminifers in Tertiary time as demonstrated by Cifelli (1969).

So then, to return to the ecologic facies facet of the "time-transgressive" bugaboo. Among the first planktonic specialists to recognize that facies factors played a role in the distribution of planktonic foraminifers were J.C. Ingle and Orville Bandy, yet they continued to employ ecologic criteria (e.g., coiling ratios as criteria for paleoecologic temperature evaluations) for purposes of chronologic correlation based on planktonic foraminifers from the Miocene-Pliocene through the Pliocene and Pleistocene portions of the Cenozoic column. Everyone can be lucky for a while, or, as the saying goes, a stopped clock is correct at least twice a day, but in principle one cannot continue to have it both ways—that is, ecologic facies cannot bring about "time-transgression" in one case and precise stage and zone correlations in another. Even principles have their limits; for not even principles and the interpretive deductions they soundly lead to can be metamorphosed into facts.

In reality, there is inherent in any correlation of Oppelian stages and zones, whatever dominant taxa they are based on, a certain element of time-transgression, since any particular biostratigraphic sequence is in fact a sequence of fossilized communities (devoid of its unfossilizable elements but including a certain number of allochthonous elements as well as the autochthones of the once living community itself). The type for an Oppelian stage or zone is simply such a biostratigraphic sequence, selected from one of many throughout a province, in which the congregations diagnostic for a time-rock unit of such magnitude seems best in evidence. Even in the selected type section for such a time-rock unit the boundaries remain to some extent fuzzy (see Weaver, 1969) owing to causes that are inherent, philosophically much the same as in a group of organisms constituting a species, from which group a single organism is selected as morphologically the most typical. Hence, the "sports" that defy classification require open nomenclature, because they are the organismal "links" between the "lumps in the soup" which are the species in a lineage. It is simply that an Oppelian zone is the smallest unit to which an objectively single designation as time-rock unit can be applied, other than those zonules (in the sense of Fenton and Fenton, 1928) which can locally be employed for short-distance correlation in many oil fields on a basis much the same as a local lithologic unit can be employed for such purposes. In short, not even the surface of a zone is as smooth as a billiard table (Kleinpell, 1972, p. 108-109).

Here an item of historical rather than technical significance may be inserted in passing, in the hope it may help clarify further that which aims at being a scientific discussion. In a textbook in which the Kleinpell (1938) book came under discussion (Glaessner, 1945), it was stated in effect that Kleinpell had divided his zones into zonules, as though one could divide a bushel of oranges into so many smaller bulk units of apples. The experience was repeated in an analogous manner in an industrial problem here on the West Coast: a young industrial successor of Kleinpell expressed disgust, with the zonules formally designated in the Reliz Canyon column. By way of response, and in sheer curiosity Kleinpell asked, "What was wrong with them?" "A guy cannot pick them up in any other canyon or in any set of wildcat cores from the area, and use them for correlation in any case," was the response. This shifted the degree of impatience to the other end of the dialogue, "Well, who the devil ever said or wrote that they could?" Unhappily the young man continued innocent of the fact that a zonule, in the Fenton definition (Fenton and Fenton 1928), could even be recurrent in the self-same column, not to mention its different age in, or even its disappearance from, a column elsewhere, even nearby. Orville Bandy exhibited more of a sense of humor in an analogous dialogue prior to the recognition of facies in planktonic foraminifer distribution. Without reference either to unknown species ranges or facies problems, Bandy was, as volubly as neatly, correlating "planktonic zones" from the California Tertiary to the classic terrains of Europe, so that the writer was constrained to comment as politely as possible, paraphrasing Galsworthy, that "idealism seems to vary in direct proportion to the distance from the facts." Bandy was not irritated but, delighted, took it in stride. It was no lack of a sense of humor that brought Bandy into the controversy here under discussion, but instead an apparent inability to inhibit an impulse to publish voluminously, in the days of "publish or perish" in academic circles.

Second Phase—The problem referred to in Barron's second paragraph is one whose solution depends entirely on evidence and relevant principle. Discussion of this, the major and particular phase of the controversy, is in itself a sad undertaking, since it involves a critique of a promising young micropaleontologist who became so emotionally involved in the controversy that he literally worked himself to death over it, a sad loss to the profession and his family. Were it any comfort to anyone, one can only add that both errors and attempted corrections of errors in the fields of science can be made in equally good faith, as is true throughout the instances in reference. Until 1964, Richard Pierce was in good spirits with a deep and genuine interest in all the ramifications of his profession. A few years later he became engrossed in the problem of the age relations between the type sections of the Mohnian and Delmontian Stages. In the many conversations the writer had with Pierce between that time and his tragically premature death in 1972 (see Hickernell, 1972), it became progressively apparent that, for all the details of great interest and great value that came up, a meeting of the minds was growing farther apart. This, insofar as the writer could grasp, was due not so much to the particulars involved but instead to the context in which they were being viewed. There seemed to be no common meeting ground as to an understanding of (1) the difference between a zonule (descriptive basically, though interpretable enough although not for a time-significance) and a zone (interpretive basically, though soundly interpretive only through the application of principle relevant to the describable and diverse facts at hand); (2) the role of typology in biostratigraphic chronology and the criteria, in the form of principles, requisite in age correlations with that type; and (3) the unavoidable trial-and-error method required in building up a refined prehistoric time scale of any sort as emphasized by Davies (1934, p. 56), and which unavoidably encounters the need to extend the stratigraphic ranges of species from time to time in order to accommodate the evidence, based on definite stratigraphic occurrences in relation to those of other species, as this evidence comes in. For a part of these difficulties the writer feels responsible, since his so-called "Range Chart" (Kleinpell, 1938, Table XVIII, p. 137-151) is not really so much a "range chart" as simply a "known occurrence chart." It should be noted, however, that neither Eocene nor Pliocene occurrences of any of the species listed in Table XVIII were given; nor, through ignorance as much as anything else, were they mentioned in noting the species considered diagnostic of stages and zones in the section on Correlation and Age (p. 99-135). Sadly, the writer could not get these contextual realities across to Pierce in his conversations with him.

Furthermore, the writer must add in passing that he has no assurance that his own views on the controversy are of necessity in any way final. Like anything in science they must stand the test of evidence through time. The Delmontian Stage as first conceived was a troublesome interval to deal with, at best. Even the original zonation, partial and tentative as it was, was already "updatable" many years ago (Woodring and Bramlette, 1950; Stirton, 1960). One criticism of many of the present references to the controversy I am constrained to make is their uncritical and tacit acceptance of the rejection of the Delmontian Stage as though this were already a demonstrated fact, as for example, in the second paragraph of Barron's opening remarks quoted at the outset of this section on the Delmontian. The summaries of Ingle (1972) and Bandy (1972a), simply repeat the conclusions of Richard Pierce, presented *post mortem,* in order that these views could appear in the Bakersfield Symposium of 1972. As for pertinent diatom data and other planktonic data (Ingle 1967; Ruth), more on subsequent pages where these elements, as related in time to other sequences of benthonics, can be discussed at greater length.

Additional remarks concerning the controversy–Another curious aspect of the controversy is the inherent relatively negative nature of the Delmontian in terms of sparcity of benthonic foraminifers. Wherever encountered, the so-called Delmontian benthonic foraminifers are spottily distributed and few specimens are found. Fossil echinoids and scallops, especially, are well developed in the shallow-water sandstone sequence that roughly corresponds to the Delmontian, as are fossil diatoms in the planktonic facies; but what stratigraphic relations these sequences may have in relation to the sequence of benthonic foraminifers disciplining the Delmontian Stage are only clear in part. For these relations in continuum the diatoms seem to offer the best hope. Perhaps they can serve to translate the spottier occurrences of the others, if not into a single language, at least into a set of disciplined time-rock relations.

It is also interesting to note that elsewhere in the world, at about some such horizon, or horizons, in the Tertiary column, similarly "negative" or highly localized faunal sequence or sequences occur. Thus, in the Indo-Pacific Province, there is the Tertiary "g" of Umbgrove (1938) and Leupold and van der Vlerck (1931) where both lepiodocyclines and miogypsines suddenly became extinct and a more clastic type of deposit succeeded a preponderantly carbonate sequence. Another example is in Europe where a sequence including diatomite ("Sahelian," though not type Sahelian) below, sulfur-bearing terrestrial and unfossiliferous beds (the Gessoso-Solfifero of Italy or "Messinian" in the strictest sense) in the middle, and an upper marl rich in brachiopods and foraminifers and certain plankton (the Zanclean) and in places a sandstone facies equivalent (Tabianian), intervenes between the lower beds from which Lyell took his highest typical Miocene fossils and the overyling bed where his lowest typical Pliocene fossils occur. This is the so-called "Mio-Pliocene" interval of many authors, which variously has been included by some with either the Miocene or with the Pliocene or has been handled by still others with a Solomonean judgment.

The "negative" stratigraphic gap in reference appears to reflect the effects of an orogeny that has greatly increased surface relief and thereby created a great diversity of bathymetric facies from place to place, rendering correlations at or around this interval commensurately difficult. A comparable sequence of geological events at about the same time has also been recorded from New Zealand (Kennett, 1965, 1968). In California the diversity of bathymetric facies during this interval seems to be as great as anywhere else, and faunally seems to have been accompanied by a sudden cold nip, that is, the southward shifting of an isotherm such as also brought Luisian time to a close and inaugurated the Mohnian Age at a horizon one stage earlier in the West Coast column. Bathymetric depths (and commensurate facies phenomena) ranged from abyssal through medium depths to the littoral, whereas open-ocean (pelagic) biofacies passed laterally into "neritic" biofacies. In any evaluation of the Delmontian, such great variations in facies must be borne in mind.

It is, moreover, of interest and significance to note that most West Coast micropaleontologists (including Barron, 1975) recognize such an "interval" between the latest Miocene and earliest Pliocene, whatever these may locally be, but others seem to think the type Delmontian Stage does not occupy this interval. Therefore, the controversy actually boils down to the question of whether the type Delmontian occupies this interval or whether, instead, it correlates with the Mohnian. This question was in all probability most emphatically brought to a head in a paper presented by Richard Pierce at an SEPM meeting at Newport Beach in March, 1970. In the review of the Delmontian Stage in its original sense, some of the ideas from this presentation will be referred to from time to time, as the questions of data, evidence, and principled evaluation thereof come up for discussion.

Review of Delmontian Stage in California

At the time of its original proposal the type section of the Delmontian Stage was the sequence of Galliher's members 2 and 1 (Galliher 1932, p. 22, Fig. 3, Pl. 2) as exposed from the ridge at the head of Canyon Segundo northeastward to the contact with the overlying Santa Margarita Sandstone north of Benchmark 226, (i.e., the Guidoti Ranch are of Kleinpell 1938, p. 131-133, Fig. 13). This sequence was selected because (1) a foraminiferal sequence was as well developed within Galliher's members 2 and 1 of the type Monterey Shale; (2) it conformably overlay the thinly bedded soft organic shale of Galliher's member 3 with its relatively rich foraminiferal fauna which, together with the fauna from his member 4 opal shales, appeared to correlate with the faunas of the type section of the Mohnian Stage in southern California (which in turn was overlain there by foraminiferal diatomite that seemingly correlated with the foraminiferal faunas of this type Delmontian); (3) it was conformably overlain by sandstone beds that bore a megafauna usually considered to be of Pliocene age; and (4) whatever the facies of parts of such sequences at other California localities, this entire sequence seemed to correlate in age with several other entire sequences elsewhere in California.

Extremes in local bathymetric and other facies diversities had seemingly developed in the West Coast range area in the wake of growing tectonic instability of late Mohnian times. All of this, like the thermal cold nip that characterizes passage from Mohnian to Delmontian time generally, appears to have been related to, or was a function of, the Rafaelan orogeny (Dibblee, 1950). When Bowen (1965, 1969) subsequently mapped the Monterey area, he was unable to employ Galliher's chert member 2 as a mappable unit, a not surprising situation in view of the fact that the type Monterey Shale area is highly folded structurally (Galliher, 1932, Pl. II) and that diatomite readily turns into cherty shale and bedded chert locally under tectonic stress or even the mere pressures of overload (Bramlette, 1946), as does for example the so-called indicator bed of upper Relizian diatomite which locally turns to chert across the plunging axis of the Coalinga nose. However, the fact that the chert member 2 is not traceable in no way changed the stratigraphic position of the type Delmontian Stage in the type Monterey Shale area. East of upper Canyon Segundo some miles, a sequence of diatomite is again well exposed in roadcuts along the Los Laureles Grade. Still farther east it grades laterally into a white fine-grained sandstone resembling, and mappable as, Santa Margarita Sandstone, though it is not so coarse grained as the Santa Margarita that overlies the type Delmontian sequence. This sandstone if coarse grained enough to permit the hazards of reworked older foraminifers (which there are of clast-size) to enter into any interpretaion of the sedimentary origin and age of these more easterly sandy equivalents of the upper Monterey diatomite. Moreover, though not so fortunate as Kenneth Gow in finding identifiable fossiliferous float apparently from the coquina-like uppermost beds of the Santa Margarita where it is folded into a little syncline and anticline before disappearing unconformably beneath the terrestrial Paso Robles Formation, Bowen (1965, 1969) was able to find megafossils in place in the lowermost of these Santa Margarita beds among which was *Astrodapsis salinasensis* Richards. Thus, this stratigraphic sequence exactly repeats that found in many other localities in the Salinas Valley and as far south as the Huasna syncline where Merriam (1941) found a *Nonion schencki*-bearing Delmontian foraminiferal fauna 200 ft (61 m) stratigraphically below the upper Santa Margarita Sandstone which there carries *Astrodapsis salinasensis* at its base and *Astrodapsis* cf. *jacalitosensis* and other Pliocene megafossils at its top. In the Huasna area, however, and unlike the downward stratigraphic sequence in the Monterey area, the upper Monterey is separated from the uppermost lower or "*Pecten discus* shales" of the Monterey by a lower Santa Margarita sandstone which bears *Astrodapsis tumidus* in it upper portion and *A. cierboensis* in its coarse lower beds (Richards, 1935a, b, 1936; Kleinpell, 1938, Table II, p.24-25, p.165-168, and footnotes on p.166-167).

In the essentially terrestrial Paso Robles Formation, which unconformably overlies the marine Santa Margarita in both the northern and southern Salinas Valley, no marine fossils have been found in the north, though many miles north of the type Monterey area marine Pliocene reappears in the Sargent, Moody Gulch, and Purisima outcrop areas, as an extension from the geosynclinal Kettleman Hills–Coalinga area via the Waltham and Priest Valleys. In the south, however, a megafossil correlate of the Pliocene fauna in the Etchegoin Formation, with *Dendraster gibbsi, Pecten healeyi,* and *Dosinia,* is to be found in lenses of Monterey-Shale-pebble-conglomerate within the otherwise terrestrial lower Paso Robles Formation in the area east-southeast of Paso Robles. The presence of marine fossils in these beds would seem to have some bearing on the reversal of major drainage direction of the Salinas River that apparently has occurred since the time of Dibblee's post-Miocene Zacan orogeny.

North of the type Delmontian, *Astrodapsis salinasensis* (misidentified as "*A. antiselli*" in the U.S. Geol. Survey Santa Cruz Folio, but rectified by Richards, 1935b) occurs in the Santa Margarita Sandstone of the Scott Valley which underlies an upper ophiuroid-bearing Monterey Shale. This shale was traced by Joseph C. Clark into Monterey Shale shown to extend along the coast northwest of Ben Lomond Mountain (Santa Cruz Folio) and to carry a Delmontian foraminiferal fauna. U.S. Geological Survey Map MF-577 shows the more southerly type-Delmontian locality which carries the fossils originally designated as of Delmontian age (see Clark and Rietman, 1973, for many details in the Santa Cruz–Monterey area Tertiary, and Clark et al, 1974). Still farther north, the isolated upper Miocene and Santa Margarita-like Wimer (or Wymer) beds of Del Norte County (Watkins, 1974) may be entirely or in part of Delmontian age. As previously mentioned in discussing the Mohnian Stage, northeastward from Monterey across San Francisco Bay, the Mohnian Briones Sandstone passes upward into the San Pablo proper (lower Cierbo and upper Neroly Members) and higher pre-Pliocene marine beds (which appear to be partly of Delmontian age), and terrestrial equivalents along strike there.

In connection with these facies problems, the entire type Monterey Shale presents data of possible paleogeographic interest. In the Table IV check lists of Kleinpell (1938, p. 34-37) only the very lowest samples from each of Galliher's Monterey members 5, 4, and 3 carry rare specimens of the planktonic foraminifer *Globigerina,* and by Delmontian time only a rich diatomaceous plankton is listed. Although marine throughout, it is as though, at the first appearances of the Luisian, lower Mohnian, and upper Mohnian benthonic faunas, an open-ocean current momentarily entered an otherwise current-sheltered area which in Delmontian time was reached by a colder open-ocean plankton from the north and which remained there for at least as long as the Delmontian. The Luisian and lower Mohnian faunas seem to be of shelf-sea "neritic" type, the upper Mohnian fauna approaches the bathyal, after which progressive shoaling apparently set in through Delmontian time in the Monterey area.

Farther south, in the Salinas Valley, cores from the Standard Oil 1 Piedmont (Kleinpell, 1938, Table IX, p. 51) carry a few rare globigerines at Delmontian depths, but none in the Mohnian cores. What is chronologically even more important, *Bolivina obliqua* appears at what would appear to be a Delmontian depth (1,995 to 2,000 ft; 608 to 609 m). Thus, although this species is lacking in the type Delmontian diatomite, it is clearly represented in the Salinas Valley biostratigraphically at the same horizons. Also present at apparently Delmontian horizons in these well cores are *Nonion belridgensis, Nonionella miocenia, Virgulina californiensis,* and "*Bolivina* sp?" Barbat and Johnson (variously between depths of 1,854 and 2,000 ft; 565 and 609 m). These species are known also from the Reef Ridge Shale fauna of the San Joaquin Valley from which *Bolivina obliqua* was first described. Lower in the Piedmont well (2,215 to 2,218 ft; 675 to 676 m) both *Nonion montereyanum montereyanum* and *N. m. carmeloensis* (conspicuous species in the type Monterey Mohnian immediately below the type Delmontian) are abundant along with *Bolivina goudkoffi,* another upper Mohnian form. Again, the Delmontian strata higher in the well (1,854 to 2,000 ft; 565 to 609 m) which carry *Bolivina obliqua* low in this interval, carry more than a half-dozen species known from the type Delmontian diatomite, including *Bolivina parva, B. seminuda,* and *Virgulina delmonteensis,* as well as the longer ranging *Buliminella curta, B. subfusiformis, Nonionella* aff. *miocenica, Virgulina californiensis,* and *V. c. grandis.*

The Reef Ridge Shale in the Reef Ridge area of the San Joaquin Valley was reviewed by Church (1972, p. 64), with a view toward updating its foraminiferal fauna and its stratigraphic relations with underlying and overlying formations. In listing Barbat and Johnson's originally recorded species from listing the Reef Ridge Shale, the *"Bulimina pseudotorta"* listed is *B. montereyana* (Kleinpell, 1938, p. 254, 258) and not the ancestral Relizian-Luisian species. Church also has added the six species collected in 1931 by Bramlette, Lohman, and Kleinpell from just north of the Big Tar Canyon Road: *Bolivina foraminata, B. vaughani, Bulimina* cf. *ovata, B.* aff. *montereyana, Buliminella curta,* and *Virgulina californiensis* (see Woodring et al, 1940). Church also noted that the McLure–Reef Ridge contact appears to be gradational rather than unconformable as reported from farther north (Barbat and Johnson, 1934, p. 5-6). From a nearby subsurface section of Reef Ridge Shale that carries *Bolivina obliqua, Nonion belridgensis,* and much of the surface Reef Ridge Shale fauna, Church added the *Bulminella elegantissima, Eponides exigua,* and *Virgulina subplana* of the original records (Barbat and Johnson, 1934, p. 10).

The Reef Ridge Shale was first named by Barbat and Johnson (1934), and in that publication *Bolivina obliqua* was named, figured, and described for the first time. The underlying McLure Shale Member, barren except at the very base, is the northern extension of the barren Antelope Shale (brown shale) Member of the more southerly Monterey of the west side. The type Reef Ridge Shale underlies the type Jacalitos Formation which is rich in early Pliocene megafossils. Farther south in the east side of the valley, the lower part of the terrestrial Chanac Formation (Ferguson, 1943, p. 240, Fig. 96a) seems to be of Delmontian age, lying as it does above the marine Mohnian Fruitvale Shale— Santa Margarita subsurface sequence there. The upper Chanac terrestrial lens, there, is of early Pliocene age, lying below the transgressive Etchegoin "marine finger" or "*Mulinia* beds" of the east side subsurface. Also to the south, on the west side, in the Chico Martinez Creek area, the Reef Ridge Shale equivalent is represented by the lower part of the "punky diatomite" of what is known,as the Belridge Diatomite Member of the Monterey. From the upper part of the Belridge diatomite, *Pecten terminus,* a lower Pliocene lyropecten of the Jacalitos Formation, has been collected from the Buried Hills just east of the mouth of Chico Martinez Canyon.

The Delmontian Stage is probably best represented in the Santa Maria basin. From the basin facies of the Sisquoc Formation, exclusive of the Todos Santos Claystone Member, Woodring and Bramlette (1950, p. 35) have check listed a fauna of 35 species including *Bolivina obliqua,* plus another variety thereof, also *B. rankini,* from 37 localities; from 10 localities in the Todos Santos Member of the Sisquoc (p. 36) 19 species which include *B. obliqua, B. rankini, B. foraminata, Nonion belridgensis* and other species in a fauna that looks very much like that of the Reef Ridge Shale; and from the marginal Tinaquaic sandy facies of the Sisquoc (p. 34) five shallow-water species which include *Nonion belridgensis.* From the stratigraphically higher Pliocene Foxen mudstone, Woodring and Bramlette (1950, p. 39-40) listed a large foraminiferal fauna that includes, among other forms, *Bolivina obliqua* and a variety thereof, indicating that both of these range upward into beds of Pliocene age, as does the *B. seminuda* lineage as exemplified by *Bolivina* aff. *B. rankini* here. The stratigraphic significance, especially of the foraminiferal assemblages from the Santa Maria basin, is extensively discussed in excellent summaries (p. 58-61, 100-106) which include far more information than can be itemized here, beyond quoting the following significant passage (p. 100): "The large fauna from the upper [Monterey] member in the eastern Purisima Hills . . . represents the upper part of the *Bolivina hughesi* zone, of late Mohnian age, and also in part a younger faunal division found in about 1000 feet of strata between the *Bolivina hughesi* zone proper and the *Bolivina obliqua* zone in the their type regions on the north slope of the Santa Monica Mountains. The intervening strata were thought to be virtually barren of Foraminifera when Kleinpell [1938, p. 130, 134] defined those zones. *'Ellipsoglandulina' fragilis, Hopkinsina magnifica,* and *Pulvinulinella purisima* are characteristic of this intermediate faunal division in the Santa Maria district and in other areas, including the north slope of the Santa Monica Mountains. Kleinpell assigned the *Bolivina obliqua* zone to the lower part of his Delmontian Stage. Assignment, however of the intermediate faunal division to the lower Delmontian and the *Bolivina obliqua* zone to the upper Delmontian appears to be preferable. The intermediate faunal division is presented at locality 21, in the Casmalia Hills, in strata mapped with the Todos Santos claystone member of the Sisquoc formation."

This proposed lower Delmontian, with underlying upper Mohnian, is also represented in the upper member of Monterey Shale as shown in isolated samples check listed from the Purisima Hills (Woodring and Bramlette, 1950, p. 25-26, 137).

With the preceding zonal revision, the writer is in complete agreement. Woodring and Bramlette's discussion of the Sisquoc Formation, which follows the quoted passage, is of further significance and worthy of careful attention, especially with reference to the possibly lower Sisquoc age equivalence of the "massive silty more or less diatomaceous mudstone" in the trough of a syncline a mile north of Point Conception" (see Dibblee, 1950) and "in the syncline near Lompoc" where it "overlies the 1,000-foot diatomite and diatomaceous mudstone of the Monterey shale exposed in the well-known Lompoc quarries. The diatomite and diatomaceous shale, also barren of Foraminifera, are probably of early Delmontian age. They overlie a 1,000-foot section of alternating units of porcellaneous shale and diatomaceous shale containing late Mohnian Foraminifera of the *Bolivina hughesi* zone, which occur also in the upper part of the underlying main body of porcellaneous and cherty shale . . .," as pointed out by Bramlette (1946, p. 7). Again, (Woodring and Bramlette, 1950, p. 101) "The diatoms may afford a basis for a determination of the age relations." With reference to the barren 1,600 ft (488 m) of upper Monterey Shale mudstone in Reliz Canyon, Monterey County, and "the characteristic Sisquoc species *Yoldia gala* and also *Anadara* cf. *A. trilineata,"* this barren upper Monterey porcellaneous mudstone overlies porcellaneous shale containing a few Foraminifera assigned by Kleinpell to his lower Delmontian Stage. Though the range of *Yoldia gala* outside the Santa Maria district is unknown, its occurence and the stratigraphic position of the mudstone suggest correlation with the Sisquoc formation" (Woodring and Bramlette, 1950, p. 101). A valuable discussion of the Salinas Valley east side Pancho Rico Formation of Reed, and of the Pliocene, follows, with an excellent summary chart (p. 106). Plates 22 and 23 illustrate several of the more important foraminifers, including both the Delmontian and Pliocene *Bolivina obliqua*. The so-called *Uvigerina foxenensis* n. sp. (Pl. 23, Fig. 16) is probably a synonym of *U. peregrina*. Previously (on p. 60), we significantly learned that a species listed by Wissler (1943) as *Bulimina* sp. (large, crushed), in his zones B and C of the Delmontian of the Los Angeles basin, is identified by Woodring and Bramlette as *"E." fragilis* (tentatively assigned to *Ellipsoglandulina)* and figured on Plate 22 (Figs. 4, 8-10).

South of the Santa Maria basin, around the western end of the Santa Ynez Mountains, Dibblee (1950) has referred to the Sisquoc Formation the massive whitish silty mudstones of the northwest-southeast trending synclinal trough immediately north of Point Concepcion. Mapping eastward along the Santa Barbara County coast, Dibblee continued to apply the name "Sisquoc" to the lowermost massive brown silty mudstones of this unit (formerly often miscalled "Santa Margarita shale") that overlie the hard platy upper Monterey shales as far east as Gaviota Beach. Farther east these platy shales hold up much of the bluffs along this coastal area, and when stratigraphically higher beds appear again eastward they tend to be punky diatomaceous shales and even diatomite. Dibblee (1950, p. 34-37) discussed the Monterey sequence above the Tranquillon Volcanics of the westerly area here and its easterly Bentonite Reef equivalent in terms of units B through F. He presented a tabular description of these for the area between Cojo Canyon and Gaviota Canyon (p. 36), showing them to be Relizian and uppermost Saucesian (B), Luisian (C), and lower Mohnian (D, uppermost of his "lower Monterey"), upper Mohnian (E), and Delmontian (F, his highest "upper Monterey)" noting that the latter "carries lenses of chert conglomerate east of Cuarta Canyon" and that the overlying Sisquoc shale has "basal sand and chert conglomerate east of Sacate Canyon." The contact between Monterey and Sisquoc is not a precise stratigraphic horizon, though in general it is very nearly so, but most significant are the intermittent lenses of brea conglomerate that occur at about the same stratigraphic interval along this whole coast and generally separate upper Mohnian beds below from more sparsely foraminiferal Delmontian beds above. One such tar-soaked lens of unsorted pebbles, of both chert and of crystalline rock, is particularly conspicuous just west of the mouth of Gaviota Creek and commented on by Dibblee (1950, p. 37). Other lenses occur at about the same horizon both on the west and the east, until in the highly diatomaceous upper Monterey Shales at Hope Ranch a lens of truly intraforational breccia appears at this same horizon. This horizon of brea conglomerates and intraformational breccias is the highest of two such horizons discussed by Reed (1935a) for their tectonic, historic-geological, and stratigraphic significance (see also Kleinpell, 1938, p. 131, with footnote). This highest of two horizons of intraformational breccia lenses separating Mohnian from Delmontian strata appears to reflect Dibblee's Rafaelan orogeny.

Upper Monterey diatomite in the Naples Bluffs (section Table 5; also Kleinpell, 1938, Table IV, p. 29-31) carries Delmontian foraminifers, including *Bolivina obliqua* and *B. malagaensis,* a little over 180 ft (55 m) stratigraphically above the highest upper Mohnian foraminiferal sample found there. The age of the upper Monterey diatomites at Coal Oil Point and Goleta Point is not clear. In the vicinity of Rincon Mountain, the Monterey Shale on the upward (northern) side of the huge thrust fault there strikes inland, but just west, at Rincon Point, the uppermost shales exposed carry a Repettian foraminiferal fauna, as do the highest beds on the south side of Red Mountain (where the Los Sauces Creek section has graded eastward into a thin Oak Ridge-like Miocene column) but still north of the surface expression of the thrust fault.

East of Ventura River, on the west end of Sulphur Mountain, the horizon of highest intraformational breccia and brea conglomerate lenses has passed laterally into a locally tar-soaked sandstone body into which the Union Oil Company once tunnelled almost horizontally for heavy-oil recovery. At the very top of the platy Monterey Shale that underlies this sandstone — the westernmost exposures of what may be termed the upper Modelo sandstone — an uppermost Mohnian foraminiferal fauna which carries *Galliherina uvigerinaformis doanei* has been collected in Fresno Canyon. Over Sulphur Mountain to the east-southeast the platy Monterey that holds up the heights of the mountain has been thrust southward into contorted overturns in recumbent folds which are thrust-faulted to some extent wherever, prior to thrusting, a dome existed in the anticlinorium that then existed. In the "swales" between such previous and now-faulted domes, some continuous stratigraphic sections can be found, as in the vicinity of Coche Canyon (Evans, 1928). The type locality for *Bolivina sinuata alisoensis* Cushman and Adams (1935) is in the laminated shales below the upper Modelo sandstones. Here, in turn above the overturned upper Modelo sandstone, are once more relatively massive and poorly bedded brown conchoidally fracturing silty and more or less siliceous mudstones that carry spottily distributed Delmontian foraminifers. These upper Modelo shales are here gradationally succeeded upward by the lowest local beds of Repettian age within the Fernando Group which here bear, among other species, *Bolivina cochei* and *B. subadvena sulphurensis* Cushman and Adams above the basal Pliocene sandstone of Cartwright. Here, the lower Repettian beds and many of the higher Fernando ("Pico Brown") beds consist chiefly of heavily bedded lenticular conglomerates. Eastward, however, the Repettian beds grade laterally into the thin-interbedded and in many places greenish siltstones and argillaceous shales which disappear still farther northward and northeastward beneath the ramifications of the San Cayetano fault.

Farther east and south of the Santa Clara River valley, on the north slopes of the Santa Monica Mountains, the Miocene column is very different although there are sufficient sandstone bodies interbedded with the more or less siliceous Monterey-like shales for the term "Modelo Formation" to be retained for them in preference to the term "Monterey" which is more appropriate for the organic shales of the coast. Hoots (1931) has divided the Modelo into lower and upper members. The lower has many more or less lenticular bodies of sandstone, and the upper member is primarily diatomite and diatomaceous shale. Both members are further subdivided into lithologically distinctive units of even smaller magnitude. In the section from Mohn Spring on the south to Woodland Hills (old "Girard") in the north, Hoots' units 1 to 16 inclusive were selected as the type section for the Mohnian Stage. A sample from a few feet below the top of the gently dipping and highly diatomaceous unit 18, just south of old Ventura Boulevard, carries a rich foraminiferal fauna of late Delmontian age, including abundant *Bolivina obliqua,* common *B. Rankini,* and rare *Cassidulina delicata* (Kleinpell, 1938, Table VII, p. 46-47). As Wissler (1943, p. 218, 223) was careful to note, the Mohnian-Delmontian contact in this section was arbitrarily placed at the top of Hoots' unit 16 because of the relatively unfossiliferous nature of units 10 through 17 (Wissler, 1943, p. 223), and, as he added, "Unit 18 carries definite lower Delmontian foraminifers . . .," the *Bolivina obliqua* Zone then still being considered "lower" rather than upper Delmontian. It was also of these intervening strata, which were thought to be virtually barren of Foraminifera at the time when the arbitrary stage boundary was designated (Kleinpell, 1938), that Bramlette wrote (Woodring and Bramlette 1950, p. 100) that the upper member of his Monterey Shale included not only the age-equivalents of the old *"Bolivina hughesi"* zone but also part of a younger faunal division found in about 1,000 ft of strata between the *Bolivina hughesi* Zone proper and the *Bolivina obliqua* Zone in their type regions on the north slope of the Santa Monica Mountains (Kleinpell, 1938, p. 130, 134). Moreover, the inclusion of units 17 and 18 as type for the *Bolivina obliqua* Zone was also arbitrarily taken (Kleinpell, 1938, p. 134), the zone being based faunally on the sample almost at the top of unit 18, plus the widespread distribution of the "joint occurrences" therein and the consistent stratigraphic position thereof. It was this supposedly unfossiliferous interval of which Bramlette (1946) wrote " *'Ellipsoglandulina' fragilis, Hopkinsina magnifica,* and *Pulvinulinella purisima* are characteristic . . . in the Santa Maria district and in other areas including the north slope of the Santa Monica Mountains." This situation led him to believe that "Assignment . . . of the intermediate faunal division of the lower Delmontian and the *Bolivina obliqua* zone to the upper Delmontian appears to be preferable." There is nothing at the type section of the Delmontian Stage that is not in keeping with such a zonation; the assemblages from Galliher's member 2 were originally thought to be too much of a a facies fauna to be zonally distinctive. In fact, the *Bolivina obliqua* Zone was, as early as 1960 (Stirton, 1960, p. 362), formally considered as *not* representing the lowest part of the Delmontian Stage.

W. C. Holman has spent a great deal of the restricted spare time of his active professional life carefully studying the fossiliferous sequence from the Miocene to the Pliocene, including the Woodland Hills (old "Girard") area and its general vicinity (Holman 1958). Holman is of the studied opinion (written commun.) that Hoots' unit 19 embraces an unconformity and hiatus within its stratal sequence, and that the lower part is still Miocene and presumably Delmontian (Stirton, 1960, p. 362, left room for two possible zones within the Delmontian Stage, one post-*Bolivina obliqua* Zone and one pre-*Bolivina obliqua* Zone), whereas the upper part, mainly sandstone, carries Pliocene fossils, including megafossils, and is so young as to warrant being called part of the Pico Formation.

The Mohnian fauna collected by T. A. Baldwin and check listed by Pierce (1956) is from a sequence in the Beverly Glen area not many miles east of the type Mohnian sequence, though in a separate column. The sequence has no faunally distinctive top nor bottom, and is at a horizon well above the highest locality of Pierce. Almgren's sample, discussed by Church (1972) with reference to foraminifers and age of beds at the base of the McLure Shale Member in the San Joaquin Valley, is from the same area. That this assemblage—well up in Hoots' upper Modelo diatomite—is Mohnian, however, seems highly unlikely. Nor is it clear as to whether it is Delmontian in the sense discussed by Bramlette, or even lower Pliocene in some medium-depth facies.

On the north slope of the Santa Monica Mountains, the Delmontian, which overlies the Mohnian there, is represented by a shelf-sea facies. Eastward and southeastward, in and around the edges of the Los Angeles basin, its age-equivalent Puente Formation which includes all the normal bathymetric facies from littoral to abyssal. Especially toward the east of the basin's edge, it is very difficult to distinguish "uppermost Miocene" from "lowermost Pliocene." The so-called "upper Puente shale" includes locally lenticular clastic bodies that range from sandstones to coarse conglomerates, much as does the "upper Modelo shale" on the north side of the Santa Clara River Valley somewhat farther west. Wissler (1943, p. 209-234, Table 1, p. 210, Figs. 90-94) has summarized the basin subsurface sequence, in which his divisions C, B, and A, in ascending order, represent the Delmontian and constitute beds deposited at abyssal to bathyal depths. His divisions C and B of the Puente Formation which in subsurface are the beds with "*Bulimina* sp. (large, crushed)" referred to by Bramlette as probably lower Delmontian, and his uppermost Puente division A, characterized by *"Rotalia" garveyensis* (a species of fairly long range and deep-water optimum) are assigned to the upper Delmontian

Wissler (1943, p. 222-223) has correlated the basin subsurface sequence of the upper Miocene with neighboring outcrop sections. Those on the north slope of the Santa Monica Mountains have already been mentioned. Accurately allowing for the arbitrary placing of the Mohnian-Delmontian boundary at the type section of the Mohnian Stage, Wissler correlated Hoots' units 1 through 5 with the lower half of his subsurface division E, those of units 6 through 8 with its upper half, unit 9 with the lower half of division D, units 10 through 16 and possibly 17 probably with the upper part of subsurface Puente division D, and unit 18 as definitely Delmontian. He noted that in the city of Los Angeles and in the Repetto Hills farther east, Miocene division A (with *"Rotalia" garveyensis*) "conformably underlies the lower Repetto," and that Eldridge and Arnold (1907, p. 144) included the upper 2,500 ft (762 m) of the city Miocene within the Fernando Group (Pliocene), a reconnaissance procedure analogous to that of English (1926, p. 40, Pl. 1) also, in the Whittier Hills near the northwestern end of the Puente Hills. Krueger (1936) pointed out that these early workers had included a maximum of 3,800 ft (1,160 m) of upper Miocene shales, sandstones, and conglomerates in the Fernando of supposed Pliocene age. Krueger named the mappable unit the Sycamore Canyon Formation, which except for its very lowermost beds represents the Delmontian Stage. This takes on added interest when it is realized that, apparently as a consequence of Dibblee's Rafaelan orogeny, surface relief was increased tremendously and, in a geographic sense, differentially. Monterey-type deposition, with increasing amounts of clastic material, persisted locally into early Pliocene time in parts of the Santa Susana Mountains and the San Jose Hills (north of Puente), but with a minimum of clastic material in the Santa Maria basin and the southwestern San Joaquin Valley. Clastic Fernando-type deposition already had begun in Delmontian time in parts of the Whittier Hills (e.g., Sycamore Canyon Formation). Progressively fine clastic to subsequent coarse clastic material was deposited across the Miocene-Pliocene (Delmontian Reef Ridge Shale—Pliocene Jacalitos Sandstone) boundary in the Coalinga region, and terrestrial deposition occurred in areas such as the Salinas Valley and the east side (Chanac Formation) of the San Joaquin Valley.

Wissler (1943, p. 223) showed that in the San Jose Hills both zones of the lower Mohnian are represented by the lower Puente shale and that on the basis of stratigraphic position the middle Puente sandstone member would represent the lower part of subsurface division D (i.e., the *"Bolivina benedictensis"* Zone of Pierce). In the area north of La Habra, where the Sycamore Canyon Formation unconformably overlies the upper Puente shale, foraminifers are relatively scarce, but in a nearby well both divisions C and D equivalents (lower Delmontian to upper Mohnian) were found. Crumpled upper Puente shale along the Hacienda Cayon road yielded well-preserved upper Mohnian forminifers. These data summarized by Wissler serve to emphasize some of the marked lithofacies changes along strike that occur in this area in beds deposited in late Mohnian, Delmontian, and Repettian times.

Wissler (1943, p. 222) also compared his subsurface foraminiferal column as developed in more westerly fields such as Torrance, Wilmington, El Segundo, and Playa del Rey with the sequence cropping out in the nearby Palos Verdes Hills. He emphasized a correlation between his subsurface division C and U.S. Geological Survey Foraminifera locality 24 in the Malaga Mudstone Member of the Monterey, which is of Delmontian age (Woodring et al, 1936, p. 147).

The Malaga Member which overlies the upper Mohnian Valmonte Diatomite Member of the Palos Verdes Hills, lithogenetically resembles much of the Sisquoc Formation of the Santa Maria basin and the Santa Barbara County coast, and the massive conchoidally fractured jarosite-stained so-called "Santa Margarita shale" of early workers in the foothills north of the Santa Clara River Valley of Ventura County. On the basis of locality 24 Foraminifera and two other meager assemblages the Malaga appears to be of the same Delmontian age (Woodring et al, 1936, p. 147 and check list). Woodring et al (1936, p. 147-148) also pointed out that, "Mudstone characterized by the same lithologic features occurs at the top of the Miocene section along Newport Lagoon at the west end of the Laguna Hills, 30 miles southeast of the Palos Verdes Hills, and also at the east end of the Laguna Hills, where it forms part of Woodford's Capistrano formation." They also stated that "the assemblages from localities 24 and 25 may be correlated with the fauna of Hoots' lithologic unit 18 of the upper member of the Modelo formation of the Santa Monica Mountains, as listed by Rankin from exposures near Girard, and with the upper shale member of the Modelo formation in the type region in Modelo Canyon" and "that the Malaga mudstone at locality 24 was deposited at abyssal depths, about 500 fathoms or more."

The Newport Lagoon column in reference has been shown, complete with a check list of foraminifers recorded by A. D. Warren from the lowest to the highest beds exposed in continuous sequence, in Figure 1. To the writer it has seemed that beds of Delmontian age, with *Bolivina rankini* and *Galliherina delreyensis,* are represented in the uppermost beds of this column, and that the distribution of the species of the *Galliherina* lineage, as they have been recorded as to stratigraphic occurrence there, is not only chronologically significant in particular but in general, with *G. u. doanei* ranging on up into the *Bolivina foraminata* Zone (lower Delmontian) in more than one column elsewhere.

The more southerly Capistrano Formation has been studied by White (1956, p. 237-260, Pl. 27-32). Lithogenetically it is reminiscent of the Sisquoc Formation farther north and appears to be essentially of the same age. A darker colored lower 1,500 ft (457 m) lies conformably on the Monterey Shale (Woodford, 1925, p. 216). Although largely unfossiliferous, it carries in a 15-ft (4.6 m) thick diatomaceous shale unit (U.S.C. locality 150), apparently above the middle of this lower part of the formation, a foraminiferal assemblage of long-ranging species, mostly known from both Mohnian and Delmontian horizons, which includes *Bolivina bramlettei* and *B. decurtata,* both thus far recorded from no higher than Mohnian horizons. These two species indicate a horizon "in the uppermost part of the Mohnian" (though how high these two species range in the Tertiary, especially this far south, is still under limited discipline). However, as White (1956, p. 241) also pointed out, "there is no denying that the fauna is closely related to the younger stage." White continued "the evidence of upper Mohnian at this horizon does not exclude the possibility that somewhat older Mohnian assemblages could be present within the unit. Also it is possible that typical Delmontian may be present higher in the section, although no such faunal evidence was found." He went on to say that there was a possible correlative unit at Malaga Cove—the Malaga Mudstone Member of the Monterey, which is lithologically identical to parts of the lower Capistrano Formation. However, the only recorded Foraminifera from this unit were found at localities 23, 24, and 25 of Woodring et al (1946; also Woodring et al, 1936, p. 147) and these Foraminifera are more typically Delmontian. White proceeded, "the possibility that these two units are equivalent at least in part is worthy of consideration, since at this time several hundred feet of upper Miocene sediments above U.S.C. Locality 150 have as yet not yielded Foraminifera, and could conceivably contain assemblages with a younger aspect than those thus far encountered." The overlying 300 to 400 ft (91 to 122 m) of the upper Capistrano member are lighter in color and, beginning about 600 ft (183 m) stratigraphically above the Miocene foraminiferal horizons, carry foraminiferal assemblages that "are comparable to faunas found in the upper part of the Repetto formation in the central part of the Los Angeles Basin" (White, 1956, p. 242), including *Bolivina subadvana sulphurensis, Bulimina rostrata, B. subcalva, Anomalina ("Nonion") pompilioides,* and *Plectofrondicularia californica.* This is a deep-water assemblage, lower bathyal or abyssal, of early Pliocene Repettian age, an age assignment for the Capistrano Formation in keeping, as White (1956, p. 239) noted, with earlier assignments by Reed (1933, p. 238), Reed and Hollister (1936, p. 125), Kleinpell (1938, p. 168), and Driver (1948, p. 116). For a summary of the Pliocene stratigraphy and megafossils of this general area, the terminology in use, and a valuable bibliography, the reader is referred to Vedder (1972).

Zonation

Little can be added to previous understanding of zonation within the Delmontian Stage beyond emphasizing the recommendations of Woodring and Bramlette (1950) that the type *Bolivina obliqua* Zone be considered to represent an upper Delmontian zone, and that the barren stratigraphic interval beneath it be recognized as a lower Delmontian zone. This lower zone conformably separates the upper from the uppermost fossiliferous type Mohnian, and was found in more than one area to characteristically bear *"Ellipsoglandulina" fragilis, Hopkinsina magnifica,* and *Epistominella ("Pulvinulinella") purisima.*

In keeping with the conclusions of Woodring and Bramlette (1950), Wissler (1943), and further corroboratory data made available to the writer by W. H. Holman (personal commun.), Hoots' units 15, 16, 17, and perhaps the lower part of unit 18 (that part below Rankin's sample number 172) may be taken to typify such a lower Delmontian zone. Woodring and Bramlette (1950) have written that " *'Ellipsoglandulina' fragilis, Hopkinsina magnifica,* and *Pulvinulinella purisima* are characteristic . . . in the Santa Maria district and in other areas, including the north slope of the Santa Monica Mountains." They synonymize their *"E." fragilis* with Wissler's *Bulimina* sp. (large, crushed) from the Los Angeles basin subsurface. Of the three species, mentioned, the first two seem, from the data presented, to have appeared in upper Mohnian horizons. The *Epistominella ("Pulvinulinella") purisima* appears to be a Delmontian descendant of *Epistominella capitanensis.* Negative evidence, however, recommends that this zone and its congregation receive additional study in order to determine diagnostic elements. For the present, the name *Bolivina foraminata* Zone is tentatively proposed because of the common and widespread occurrences of this species at lower Delmontian horizons even though it occurs both lower and higher in the column.

From the data presented from the type Monterey Shale, the Standard 1 Piedmont well in the Salinas Valley (material furnished by A. M. Hubbell), the Santa Maria basin, the Newport Lagoon column, and elsewhere (W. W. Wornhardt, personal commun.), the following suppositions may be made.

1. *Galliherina delreyensis* and *Nonion schenki* have their first occurrence in the *Bolivina foraminata* Zone and their last occurrence in the *Bolivina obliqua* Zone.

2. The three "characteristic" lower Delmontian species of Woodring and Bramlette, along with *Bolivina decurtata, B. girardensis, B. granti, B. hughesi, Buliminella semihispida*(?), *Discorbinella valmonteensis, Galliherina uvigerinaformis doanei,* and *Nonion montereyanum montereyanum,* persist upward from Mohnian horizons at least into the *Bolivina foraminata* Zone.

3. *Bolivina barbarana* (which is very close to the lower Pliocene *B. cochei* Cushman and Adams), *B. bramlettei, B. goudkoffi, B. parva, B. woodringi, Buliminella curta,* and *Uvigerina hannahi* persist up into the *Bolivina obliqua* Zone.

4. *Bolivina obliqua* makes its first appearance in the upper Delmontian zone, and persists, along with a new subspecies thereof, into the lower Pliocene.

5. The *Bulimina* sp. (large, crushed) subsurface interval of Wissler (1943), also Hoots' units 15, 16, 17, and perhaps the lower part of his unit 18 (Hoots, 1931), also the assemblage from near the base of the Malaga Mudstone Member (locality 24 of Woodring et al, 1934), also the proposed lower

Delmontian assemblages of Woodring and Bramlette (1950) from the Santa Maria basin, and the subsurface interval between about 2,000 and 2,200 ft (610 and 671 m) in Standard Oil 1 Piedmont in the Salinas Valley, are age correlates of the foraminiferal assemblages in Galliher's type Monterey Shale member 2 and its lateral diatomite equivalents in the lower part of the type section of the Delmontian Stage.

Thus, the *Bolivina obliqua* Zone represents the upper, and *not* the lower, portion of the Delmontian Stage, as already pointed out by Woodring and Bramlette (1950) on the basis of data from both the Santa Maria basin and the Santa Monica Mountains.

Concerning key species found in the Santa Maria basin by Woodring and Bramlette(1950, p. 36) the cited locality 21 from the Todos Santos Claystone Member of the Sisquoc Formation in the Casmalia Hills appears to lack the *Hopkinsina,* but does carry *Bolivina foraminata, Buliminella subfusiformis, Cassidulinoides californiensis, Globigerina bulloides* (rare), and *Virgulinella pertusa* (?"*V. miocenica*"). In the stratigraphically isolated samples in the upper Monterey member from around Redrock Mountain in the Purisima Hills (Wooodring and Bramlette, 1950, p. 25026, 100, 137), sample 14 carries the *Hopkinsina* though not the other two "characteristic" species in reference (although "*Baggina* aff. *B. californica,*" a bolivinid, and *Bulimina* cf. *B. galliheri* are listed). Samples 15, 16, 19, and 20 bear both the characteristic "*E.*" *fragilis* and the *Hopkinsina* and samples 16 and 19 have *Epistominella purisima*. In addition to other species, *Bolivina subadvena spissa* (in 16), *Gyroidina multicameratus* (in 18), and *Suggrunda kleinpelli* (in 19) occur in these samples. These samples are exclusive of the clearly upper Mohnian localities check listed. Thus, either the three "characteristic" lower Delmontian species already appeared in the latest Mohnian, or that which is implied and much more likely, these other species in reference above lived on at least into early Delmontian time. With this in mind it is likely enough that *Bolivina bramlettei, B. decurtata, B. girardensis,* and *B. woodringi* may have done likewise, as Pierce (1970) told us orally regarding occurrences in beds that correlated with the barren Capistrano interval of White (1956) and which on superpositional grounds had been considered possibly Delmontian. Other than noting *B. decurtata* in Hoots' Delmontian unit 18, no actual localities for these occurrences were presented by Pierce. That *Bolivina obliqua* occurs in White's Mohnian and that *Bolivina obliqua* and *B. rankini* form peak zones in the type Mohnian, as reported by Pierce, also have not been corroboratable as to such actual occurrences, but on the contrary, range from upper Delmontian into the Pliocene.

Whether a third and uppermost zone also may be recognized in the Delmontian is also not altogether clear. Pierce (1970) mentioned a diatom flora like that in the type Delmontian that occurs in beds above those from which Joseph C. Clark reported *Astrodapsis salinasensis,* which Pierce considered Pliocene. Above these are beds carrying *A. jacalitosensis* in a Pliocene megafauna, overlain by an early Sisquoc diatom assemblage considered to be early Pliocene; this is in keeping with all previously known evidence. Perhaps Pierce was unaware of the work of Richards (1935a,b; 1936) on *Astrodapsis*. In any event, this only begs the question of a possible third Delmontian zone.

Thus, to sum up our knowledge, or relative paucity of knowledge, of Delmontian zones: (1) we are in agreement with Bramlette's conclusions (Woodring and Bramlette, 1950) that the *Bolivina obliqua* Zone probably represents an upper rather than a lower zone of the Delmontian Stage; (2) that, if not the congregation for the *Bolivina obliqua* Zone proper, the species so listed (Kleinpell, 1938, p. 134-135) are at least a congregation diagnostic of the Delmontian Stage; (3) that whether a third and still higher zone can be recognized in the Delmontian, as possibly represented in the upper Monterey Shale of the Scott Valley and the coast north and northwest of Ben Lomond Mountain in the Santa Cruz Mountains, or whether these are post-Delmontian beds, is not clear; (4) that localities shown on U. S. Geol. Survey Map MF-577 in the general area in reference are of Delmontian age and the type Delmontian shown thereon lies below upper Santa Margarita Sandstones that carry *Astrodapsis salinasensis* (the former "*A. antiselli*") at base and *A. jacalitosensis* at top. The latter occurrences are stratigraphically the same as in the Huasna syncline area where, however, the Delmontian foraminifers occur 1,800 ft (549 m) above the uppermost local lower Santa Margarita bearing *A. tumidus,* whereas at the type Delmontian a sequence of Mohnian shales—Galliher's members 3 and 4 of the type Monterey Shale—lies conformably between the lowest Delmontian beds and the basal type Monterey Shales which are of Luisian age.

Repettian Stage

Beds of Delmontian age are succeeded upward by foraminifer-bearing beds representing Natland's Repettian Stage (Natland, 1952, 1953, 1957) in deep-water facies at a number of localities: (1) in the Palos Verdes Hills (Woodring et al, 1934; Woodring et al, 1946); (2) in the Los Angeles basin subsurface and at the type Repettian (Wissler, 1943); (3) at Rincon Point on the Los Angeles–Ventura County line; (4) in shelf-sea facies throughout much of the Santa Maria basin (Woodring and Bramlette, 1950), along the Santa Barbara County coast (Dibblee, 1950), and scattered along the locally faulted southern scarp of Sulphur Mountain in Ventura County where continuous depositional sequences are exposed; and (5) in a sequence from the shelf-sea assemblages of the type Reef Ridge Shale below to the overlying littoral-facies mollusk- and echinoid-bearing Jacalitos Formation of the Coalinga area.

Holman's (1958) Miocene-Pliocene contact in Hoots' upper Modelo unit 19 involves a disconformity and possible slight hiatus, probably reflecting the early effects of Dibblee's Zacan orogeny on the north slope of the Santa Monica Mountains. The age of the lower part of unit 18, below the disconformity, is uncertain in terms of zonal detail, as is the age of the Union Oil Co. locality well up in the upper Modelo diatomite above Pierce's highest Mohnian in the Beverly Glen section several miles east. However, the lower part of the unit 18 is distinctly post-Mohnian. A Repettian foraminiferal assemblage has been described by Stewart and Stewart (1930b) from the uppermost Puente shale of the eastern end of the Puente Hills, apparently above beds of Delmontian age. Repettian foraminifers also are recorded from beds above the highest strata in the Newport Lagoon section west and across the lagoon from the Miocene beds. All these Repettian foraminiferal assemblages, whatever their various bathymetric facies, are of early Pliocene age in West Coast range Tertiary terminology. They are beyond the purview of the present supplement except for serving as a sort of stratigraphic "ceiling" and superpositionally disciplined upper age limit for the stratigraphic sequence of benthonic foraminiferal faunas here under updated review.

Diatoms and California "Mio-Pliocene"

Fossil diatoms occur in the California Coast Ranges in varying degrees of concentration, from Upper Cretaceous (Moreno Shale) to Holocene. Many Miocene horizons are significantly diatomaceous and can be widely correlated (Lohman, 1974). One of the stratigraphic intervals in which they are especially abundant, commonly in the form of real diatomites over extensive areas, is the interval that embraces the Delmontian and Repettian Stages. These two ages began more or less at the peak of Dibblee's Rafaelan orogeny, closed with the inception of his Zacan orogeny (Dibblee, 1950), and exhibit a maximum of diversity in both bathymetric biofacies and lithofacies, and also to some extent faunally, in thermal facies. Thus, a cold-nip inaugurated (or accompanied) the beginning of Delmontian time, as well as a diversity in lithofacies which is reflected in the local persistence of Monterey-type deposition in one area at the same time that Fernando-type deposition began in another area, in many places nearby. Although Delmontian strata tend to bear coolish water faunules even at shelf-sea depths, those of the Repettian carried, even in their abyssal foraminiferal faunas, a warm-water element in *Bolivinita "angelina"* (conspecific with *B. quadrulata*) for a very short interval of time. This species is apparently an invader from the Indo-Pacific Province, where it ranges throughout the Neogene into the Holocene, whereas in California it occurs (in place) through only a maximum of about 13 ft (4 m) of strata in the local Repettian. Most important, however, is the fact that age correlations are made difficult by facies diversities that characterize these two stages in the West Coast ranges—littoral to "protected" shelf-sea to open-ocean shelf-sea to bathyal to abyssal biofacies. Offshore faunules, especially of benthonic foraminifers, are commonly very spotty in distribution which compounds the problems of time-rock classification and correlation; therefore, the widespread abundance of diatoms throughout the "Mio-Pliocene" interval becomes significant.

In an unpublished PhD dissertation, Wornardt (1963), whose research was supervised by G Dallas Hanna among others, has summarized the stratigraphic distribution of diatoms in this general "Mio-Pliocene" interval, carrying on his work in a few carefully selected and critical sections. The taxonomy of the diatoms he studied has appeared in print (Wornardt, 1967), but not their biostratigraphy as he summarized it in detail. Since, to a greater extent than any other stock of organisms, the stratigraphic distribution of these diatoms serves to translate the age significance of "Mio-Pliocene" benthonic foraminifers, planktonic foraminifers, larger invertebrates (mainly inshore echinoid and molluscan megafossils), and even vertebrates (both marine and terrestrial) into a single "language," a short summary of some of Wornardt's most critical and stratigraphically significant early conclusions may shed further light, by way of independent evidence, on the age relations of the more spottily distributed benthonic foraminifers under special scrutiny in this supplement.

Wornardt began his diatom studies on the distribution of the species of this stock in the Sisquoc

Formation as exposed in the roadcuts along Harris grade in the western Purisima Hills some distance north of Lompoc and south of Santa Maria, and in the upper diatomite member of the type Monterey Shale as exposed in the diatomite quarry near Del Monte. The latter is the type sequence of the Delmontian Stage which is based on benthonic foraminifers, and much the same sequence is exposed somewhat farther east on the Los Laureles grade, where the chert that is Galliher's type Monterey member 4 has graded laterally into diatomite between the two type Monterey areas. Along the Harris grade, north of Lompoc, the diatomaceous Sisquoc Formation is conformably overlain by the Foxen Formation of middle Pliocene age, carries at or near its base a conspicuous stratum, often locally referred to as a "Foraminite," consisting of little more than foraminifers, *Uvigerina ("foxenensis") peregrina* in particular being abundant. The lowest beds in the Sisquoc of Harris grade are those exposed in the axis of an anticline at its southernmost end.

The two diatom floras—the one from the Harris grade Sisquoc and the one from type Delmotian upper type Monterey Shale—were found by Wornardt to be significantly different: one early Pliocene, the other late Miocene. Further studies permitted him to realize (1) that the Harris grade Sisquoc flora overlay superpositionally a southern equivalent of the upper type Monterey flora in the upper Monterey Shale of the area of the commercial Lompoc diatomite quarries south of Harris grade, although they did so in an unconformable sequence there; but (2) that a comparable sequence, complete with a flora from the interval missing at the quarry, occurred along Sweeney Road in a conformable sequence somewhat farther east.

Through Wornardt's studies it is thus possible to state that diatom floras characterisitic of the type Delmontian Stage in the uppermost strata of the type Monterey can be correlated with others elsewhere (in superpositional relations with other diatom floras both older and younger) such as the lower flora in the Monterey Shale on Sweeney Road, the lower flora in the Malaga diatomaceous mudstone, the flora associated with Delmontian foraminifers in the Naples Bluff sequence on the Santa Barbara County coast, Delmontian foraminifer-bearing diatomaceous beds at Newport Beach (see Fig. 1) that have also been considered "Mohnian" (i.e., above WNPB 13, those from WNPB 12 down being Mohnian and correlating with the upper Mohnian Valmonte diatomite flora; (see Barron 1975, Fig. 6, p. 625-626), and with the diatomite of Hoots' unit 18 at locality T-1 (Kleinpell, 1938, Table VII) well above the top of the highest type Mohnian. The upper Malaga flora and the Newport flora from N-14 upward are Pliocene (Repettian) and correlate in age with the Harris grade Sisquoc flora, the upper flora at the Lompoc diatomite quarry, and the upper flora on Sweeney Road.

Planktonic and Large Foraminifera in Middle Tertiary of California (Zemorrian Through Delmontian)

GORDON HORNADAY

Planktonic Foraminifera

General

It has long been known that this group of Foraminifera as found in the post-middle Eocene Tertiary of California exhibits a much lower level of specific diversity than is found in tropical regions, or in other middle- or high-latitude regions. The specimen abundance of those species which do occur in the California mid-Tertiary (in appropriate environments) is, relative to benthonic forms, also much lower in general, although there are scattered floods of specimens in thin stratal layers or individual laminae. These, however, usually consist of one or two overwhelmingly dominant species with perhaps a few additional rare species. A good demonstration of this is found in the Tecolote Tunnel faunal lists (Zemorrian-Mohnian) of Bandy and Kolpack (1963). Bandy and Kolpack listed only six species of planktonic Foraminifera for the entire interval, and in the sample with the highest percentage (25%) of planktonic foraminifers only two species are present.

For California in General

From the Saucesian through the Luisian, two species—*Globigerina concinna* Reuss and *G. bulloides* d'Orbigny—make up the bulk of the planktonic fauna, with *G. concinna* predominant in the lower portion of the interval and *G. bulloides* increasing in proportion upward, until it displaces *G. concinna* completely somewhere in the upper Luisian. *G. bulloides* continues on through the Mohnian, where it is joined by *G. pachyderma* (Ehrenberg) as a dominant form. Both continue to be dominant to the Holocene. Conspicuous by their rarity are the globorotalids and turborotalids, and by their virtually complete absence the keeled globorotalids, morphotypes (in the sense of Cifelli, 1969) so prominent and important in zonal classifications in tropical areas. Thus the California middle Tertiary planktonic foraminiferal faunas consist largely of globigerinids and a relatively small number of species at that.

That this situation is largely due to conditions of temperature in surficial waters and that these conditions in turn are the result of the south-flowing cool-water California Current (see Ingle, 1967, p. 282) can hardly be doubted, for the evidence from both benthonic Foraminifera and Mollusca indicates tropical to subtropical conditions from Zemorrian through Luisian time. The benthonic Foraminifera indicate an abrupt climatic cooling with the beginning of the Mohnian, and again, though perhaps to a lesser extent, at the beginning of the Delmontian. Among the plankton, the presence of *G. pachyderma* and the holdover *G. bulloides* is ample evidence of a source of cool water, but warmer water elements (including small globorotalids) are also present, resulting in a mixed and moderately diverse transition fauna in the lower Mohnian. Lipps (1964, p. 112) recorded a drop in specific diversity from the Luisian to the Mohnian at Newport Bay, but, on the whole, the upper Miocene planktonic faunas are probably more diverse than those of the "lower" and middle Miocene in spite of the climatic deterioration that had taken place. Apparently the zone wherein mixing between the cool California Current and warm subtropical waters occurred was far enough north, during at least the warmer intervals of the late Miocene, to allow some warmer water species into the southern California area. Considerable climatic fluctuation throughout the late Miocene is revealed by the plankton. Much work analyzing this situation has been based on dextral-sinistral coiling ratios (of *G. pachyderma* in particular). Bandy, Ingle, their collaborators, Lipps, and others have utilized coiling ratios as well as the occurrence of particular species and genera to determine the various warm and cool intervals. Unfortunately they also display a propensity to attempt to correlate time horizons on the basis of these purely ecologic phenomena.

Until the 1950s, when various zonal classifications based on tropical planktonic foraminifers began to appear, little interest had been accorded the California Miocene planktonic foraminifers but the publication of Bolli's zonation for the Tertiary of the Caribbean (Bolli, 1957a,b,c) stimulated a worldwide interest in the subject. Others already at work in the Caribbean and elsewhere were quick to follow with their own zonal schemes. Blow (1959,1969), Eames et al (1962), Banner and Blow

(1965) with their now widely referred to N-zones, Postuma (1971), and others all presented zonal classifications based on tropical faunas. Jenkins (1966, 1971a,b) devised a planktonic zonal classification for the cooler climate of the New Zealand area. However, Bolli's 1957 work, supplemented in 1966, remains pivotal to most of that which has followed, although much has been modified, added to, or subtracted from it. In any event, several workers began trying to correlate strata in California by use of the various planktonic zones. It quickly became apparent that the zones themselves were not recognizable as such in California. This does not preclude the possibility of making correlative ties of individual samples to some of the zones within certain limits of error. What is not possible is division of a continuous fossiliferous section into a sequence of planktonic zones with established boundaries as can be done with the benthonic-based zones and stages in California.

California Miocene Planktonic Foraminifera

In the last decade and a half, many papers have been written on, or include comment on, California Miocene planktonic Foraminifera. It is not the purpose here to discuss or list them all. The bibliographies of those discussed will provide ample references for those interested in pursuing the subject further.

The subjacent Refugian is perhaps of all the California stages the one most deficient in planktonic Foraminifera. This is in keeping with the comments of Cifelli (1969, p. 159) on the late Eocene and Oligocene reduction of specific diversity among the planktonic Foraminifera. Lipps (1967a) reported a *Globigerapsis semiinvoluta* Zone fauna from near the top of the Refugian Stage which would make the Refugian Eocene, but the samples cited by him as containing this fauna are actually lower in the Refugian than Lipps believed (see Blacut and Kleinpell, 1969, p. 10). They are lower Refugian and hence not immediately subjacent to the interval of concern here. McKeel and Lipps (1972) discussed very briefly a sparse planktonic foraminiferal fauna from the Refugian "Siltstone of Alsea" of Oregon which they concluded is "uppermost Eocene or basal Oligocene in age." McKeel and Lipps (1975) mentioned a sparse planktonic foraminiferal fauna from the Bastendorff (Refugian in its upper portion), also an Oregon formation.

Eames et al (1962) had implied that the Oligocene was missing in California as well as in much of the rest of the world. Lipps (1965) refuted this by demonstrating the presence of Oligocene plankton in the upper Zemorrian of Los Sauces Creek. The species list (22) seems unnaturally long for, as has been pointed out, the Oligocene was a time of reduced specific diversity for the planktonic Foraminifera, and indeed Lipps stated that he regarded some of the species of Eames et al as synonyms of others.

Lipps (1967a) correlated the Saucesian with the *Globigerina ciperoensis* (part), *Turborotalia kugleri*, and *Catapsydrax dissimilis* (part), planktonic zones; he correlated the Relizian with the *C. dissimilis* (part), *Globorotaloides stainforthi*, and *Globigerinitella insueta* Zones. Lipps (1967b) subsequently reiterated the same correlation. However, Bandy et al (1969) reported the presence of the *Catapsydrax stainforthi* Zone in the upper part of the lower Saucesian of Reliz Canyon (and in this connection it should be noted that the generic designation of several of the zonal name-bearers changes from author to author). Subsequently Tipton (1972, p. 116) reported *C. stainforthi* from the subsurface lower Media Shale ("upper" lower Saucesian of Kleinpell, 1938, equals the middle Saucesian in this work) of the San Joaquin Valley. This *Plectofrondicularia miocenica* Zone occurrence was more fully documented in Tipton et al (1973). Meanwhile Bandy and Ingle (1970) had sharply criticized Lipps' correlations, and equated the *Globorotalia kugleri* Zone with the upper Zemorrian, the *Catapsydrax dissimilis, C. stainforthi,* and *Globigerinatella insueta* (part) Zone with the Saucesian, stating at that time that Lipps' correlations "are about one stage too high." Part of the problem seems to lie in Lipps' acceptance of a supposed downward extension of the Relizian into the type Saucesian at Los Sauces Creek as implied by Redwine et al (1952) and by Carson (1965; see Lipps, 1967b, p. 60). There has been no demonstration that there is any basis for so amending the Saucesian and Relizian Stages. Lipps and Kalisky (1972, p. 241) acknowledged this and accepted Kleinpell's original Saucesian-Relizian boundary "pending full documentation of the revisions." Other references to Saucesian plankton are to be found in Bandy (1972a, b) in which he reiterated the views expressed in Bandy and Ingle (1970).

Eames et al (1962, p. 35, 91) reported that two samples from the lower Relizian of the Salinas Valley represented the *Globigerina ciperoensis ciperoensis* Zone, based on the presence of that taxon, and they correlated the Relizian to the Chickasawhay Formation of the Gulf Coast, thus proving to their satisfaction the Miocene age of the Chickasawhay. Refutation came quickly as Parker (1964, p. 620) examined the same two samples and declared that the identification of *G. ciperoensis ciperoensis* Bolli was in error. She called the species in question *G. dubia* Egger and, based on other species present in the samples, gave a broad possible age range of not older than the *Globorotalia kugleri* Zone nor younger than the *Globorotalia fohsi barisanensis* Zone. Bandy and Ingle (1970) also disagreed with the *G. ciperoensis* identification. They called the species in question *G. concinna* Reuss. They also recorded (p. 159) *Globoratalia scitula* Brady, the first appearance of which is in the *G. peripheroronda* Zone, and stated that "its occurrence here is consistent with this" (i.e., a *G. peripheroronda* Zone age) whereas Parker had recorded *G. scitula praescitula* Blow which ranges as low as the *G. kugleri* Zone. That the single commonest species in these two samples could be given three different names by three different authors or pairs of authors, all of whom are experts on planktonic Foraminifera, gives some food for thought. Nonetheless, from the evidence itself it is clear that the Relizian does not correlate with the *G. ciperoensis ciperoensis* Zone.

The erroneous correlation of the Oligocene Chickasawhay Formation of the Gulf Coast with the Miocene Relizian Stage of California, which supposedly identified the former as Aquitanian and lent credence to the claim of there being no marine Oligocene on the Gulf Coast, makes the following passage from Woodring (1970, p. 307) seem appropriate:

> As part of the sweeping allegation that no marine Oligocene is known in America except in the Tampico area of Mexico and Cuba, the Bohio and Camito formations have been alleged to be of Aquitanian (early Miocene) age (Eames and others, 1962, p. 38-37). The sweeping allegation has been tacitly repudiated (Eames and others, 1968, p. 292-295).

Lipps (1967b) discussed planktonic Foraminifera and calcareous nannoplankton from Reliz Canyon, the type area of the Relizian Stage. His misconception of the lower limits of the Relizian has already been discussed. He also mentioned briefly faunas from the Luisian and Mohnian. Bandy et al (1971) described a new subspecies, *Globorotaloides suteri relizensis,* from the Relizian of the Salinas Valley to which they attached some stratigraphic importance.

The Monterey Shale at Newport Bay in southern California has been the site of many investigations of diverse types of microfossils and provides one of the best sources of planktonic foraminifers in California. Lipps (1964) studied the sequence at Newport Bay and recorded 17 species from the upper Luisian and 9 species from the lower Mohnian. Ingle (1967) studied the planktonic faunas of Mohnian, Delmontian, and Pliocene strata from several localities in southern California including Newport Bay. Bandy and Ingle (1970) discussed the occurrence of several supposedly diagnostic species in the Luisian, Mohnian, and Delmontian at Newport and elsewhere in southern California, as well as in the overlying Pliocene. Bandy (1972b) reiterated some of the conclusions of the previous paper and concerned himself with correlations from southern California to the Mohole, international correlations, temperature fluctuations, magnetic reversals, and radiometric dates for the Luisian-Mohnian-Delmontian and overlying Pliocene interval of southern California. Ingle (1972, 1973a,b) included comments about the late Miocene to early Pleistocene plankton of southern California.

The several attempts to correlate the Mohole (Experimental Mohole) cored interval to California are interesting from more than one standpoint. In these attempts the Newport Bay section has been the principal California point of reference. Parker (1964) was the first to attempt foraminiferal correlations between the Mohole, located near Guadalupe Island off Baja California at lat. 28° 59'N, and Newport Bay at lat. 33°38'N, a distance of about 320 mi (515 km). Utilizing planktonic Foraminifera almost exclusively, she assigned her Mohole samples (1 to 12) to the Luisian and Mohnian Stages (sample 13 was considered Pliocene) on the basis of plankton observed in those stages at Newport Bay. She recorded the presence of a very few benthonic species supposedly diagnostic of the Mohnian but of these only *Bolivina girardensis* Rankin, listed as very rare from one sample, would seem to be of possible significance. Ingle (1967) discussed Parker's Miocene correlations only very briefly. Bandy and Ingle (1970) assigned the California stage names Relizian, Luisian, lower Mohnian, upper Mohnian, and Delmontian to the Mohole cores entirely on the basis of planktonic species, none of which ever played a role in the definition of those stages. From the reports that have been issued so far it would appear that the diagnostic benthonic species of the California stages are lacking in the Mohole with the one possible exception noted above. Bandy (1971) claimed to have recognized N-zone 18 in the Mohole and then (Bandy, 1972a), on no better basis than ecologically, though not chronologically, significant *Globigerina pachyderma* coiling ratios, correlated back to the southern California Mohnian and Delmontian which he had suddenly and on purely theoretical grounds come to regard as synonymous, leading him to propose suppressing the Delmontian Stage. In the same year Bandy (1972b) freely used California stage names for cored intervals of the Mohole which contain no Foraminifera diagnostic of those stages. Included in the text were several radiometric dates tied to his

Mohole "stages" which he then "transported" back to the mainland to give radiometric ages to the California stage, a truly remarkable performance. The point here is that as long as the various planktonic zones (or the ranges of the various planktonic species) cannot be recognized in their entirety (i.e., with upper and lower limits in California stratal sequences relatable to the California benthonic zonal succession), the type of correlation that has been made to date between California and the Mohole is not only meaningless, but in terms of actually relevant evidence is altogether specious and thoroughly spurious.

Here employing "Pliocene" in the classic California sense of "post-Delmontian," the superjacent Pliocene planktonic faunas are almost entirely modern in character. During the warmer periods of the Pliocene a few tropical forms such as "*Sphaeroidinella dehiscens*" (Parker and Jones) and *Pulleniatina obliquiloculata primalis* Banner and Blow appeared in southern California (see Bandy and Ingle, 1970, p. 161-163). Overall, the Pliocene was a time of fluctuating warmer and colder periods which is reflected in the planktonic faunas. Taken as a whole the Pliocene planktonic foraminiferal fauna is a mixture of boreal, temperate, and tropic elements, much like the present transitional fauna off the California coast. Papers dealing with the Pliocene include Ingle (1967), Bandy and Ingle (1970), Bandy (1972b), and Ingle (1972, 1973a,b).

The majority of references to California Miocene planktonic faunas given above have dealt with the coastal regions of southern California. It would appear that throughout the Miocene a certain degree of restriction from the major oceanic currents prevailed for the San Joaquin Valley area. Bandy and Arnal (1960, 1964, 1968, 1969) have discussed the general pattern of planktonic distribution in the Miocene of the San Joaquin Valley and have concluded that the major oceanic connection was west across the San Andreas fault from the southern end of the San Joaquin basin. They further postulated that the prevailing current pattern in the basin was counter-clockwise as shown by planktonic-benthonic foraminiferal ratios for each of the stages studied by them. As elsewhere in California, specific diversity is low among the Miocene plankton although large populations did occur at times, particularly in the Relizian and again at certain times in the Mohnian. These large populations are generally dominated by one or two species. After Mohnian times rapid shoaling and pronounced and rapid restriction of oceanic connections took place in the San Joaquin Valley, and in central California generally, with the virtual elimination of planktonic Foraminifera there by the end of the Delmontian.

Interregional Correlations

With the recent great interest in and proliferation of publications on the planktonic Foraminifera has come an equally great proliferation of worldwide correlations (often presented in chart form), the more so for the contributions made in the last few years by workers on calcareous nannoplankton, diatoms, silicoflagellates, and Radiolaria. These correlative schemes are too numerous and too rapidly amended by their own authors to review in any detail here. Since the correlational framework for them is basically tied to planktonic foraminiferal zonation (including the positioning of the various nonforaminiferal "zones" such as Martini's NP and NN zones) and since these charts generally contain the almost obligatory radiometric time scale, which for some important Miocene dates is based in part on California rocks, it seems pertinent to discuss the subject very briefly.

For years, one of the chief goals of many paleontologists has been to equate their local stratigraphic sections to the European "standard" stages and to delineate the series boundaries with precision. The use first of planktonic organisms and second of radiometric dates was supposed to accomplish this. However, the European Tertiary stages are extremely poorly defined and do not represent a continuous, nonoverlapping or even nongapping sequence. Consequently "stratotypes" have been erected by various interested parties for the more familiar stage names. These have been based in large part on planktonic zones (first established elsewhere in the world) recognizable in the stratotypes or supposedly correlatable to them. Since the Tertiary series as established by Lyell and added to by von Beyrich and Schimper never had any discrete boundaries to begin with, and since in delineating the boundaries of biostratigraphic units the level of refinement called for is that of the next smaller unit in the hierarchy of a prehistoric time scale, a stage level of refinement is required in order to recognize a series boundary. Thus the Aquitanian "stratotype" became (by recommendation of the Committee on Mediterrean Neogene Stratigraphy) the lowermost Miocene, with the Chattian below it the upper Oligocene, and the boundary between them the series boundary. The base of the Aquitanian was defined as being the "*Globigerinoides*" datum which in turn was supposed to be the base of zone N-4 (the *Globorotalia kugleri* Zone). A radiometric date of 22.5 m.y.B.P. had been obtained for the

Zemorrian-Saucesian contact in California (Turner, 1970). The Saucesian had been tentatively equated to the Aquitanian as long ago as 1938 by Kleinpell; Lipps (1967) had placed the *G. kugleri* Zone in the Saucesian (although not at the very base), therefore the Oligocene-Miocene boundary is 22.5 m.y. old. This date seems to have been carried around the world despite the fact that the *G. kugleri* Zone more likely correlates, in part at least, with the upper Zemorrian (Bandy and Ingle, 1970; Hornaday, 1972).

Turner (1970) provided additional radiometric dates of: Saucesian-Relizian boundary approximately 15.3 m.y.; Relizian-Luisian boundary, 13.7-14.5 m.y.; Luisian-Mohnian, less than 13 m.y. Dymond (1966) obtained some K-Ar dates from glass shards in Mohole cores. Two of these have been assigned to the Luisian (12.3 **c** 0.4 m.y.) and the Mohnian (11.4 **c** 0.6 m.y.) as per Parker's (1964) interpretation, the potential fallacy of which has already been discussed.

Berggren (1969a, b, 1971, 1972) compiled a series of correlation charts accompanied by extensive texts by incorporating data based on several kinds of microfossils, radiometric dates, the paleomagnetic scale, and magnetic anomalies. Using this web of data, supplemented by deep-sea sedimentation rates which he used to arrive at age estimates for planktonic zones not otherwise datable (even though sedimentation rates of any sort long ago were shown to be inherently unreliable as criteria for the classificatory measurements of prehistoric time), Berggren has covered the Tertiary waterfront. The latest of these at hand as of this writing is Berggren and Van Couvering (1974) and covers only the late Neogene (the last 15 m.y.). His most recent complete Tertiary correlation scheme was Berggren (1972) and in his words, "version 4."

An examination of these charts reveals much ebb and flow of the dates of important datum planes and boundaries. However, as chart succeeds chart the amount of modification and alteration appears to be diminishing slightly and a degree of stability seems to be forthcoming. The interested reader is referred to Berggren's works in the area of interregional correlations as they are by far the most comprehensive available and his extensive lists of references provide entry to the more specialized aspects of geochronology.

Large Foraminifera

General

The literature records the presence of only two species of the larger Foraminifera in the California "Miocene" and each of these is recorded at only one locality. *Miogypsina panamensis* (Cushman) is known from a single locality in San Mateo County in northern California and *Lepidocyclina californica* Schenck and Childs is recorded from a single locality in San Luis Obispo County in central California. The *Miogypsina* locality is in a gritty sandstone lens from within the San Lorenzo Formation as mapped by Cummings et al (1962) and also reported by Hornaday (1972). Subsequently, Brabb (1972, personal commun.) has mapped the shale interval containing the sandstone lens as the undifferentiated San Lorenzo–Lambert Formations. The locality is of late Zemorrian age. The *Lepidocyclina californica* locality is from a conglomerate within the Vaqueros Sandstone and is Saucesian in age according to Schenck and Childs (1942).

It is puzzling that only these two localities of large Foraminifera are known from California. While it is easy to point out that general paleoecologic conditions in the California Oligocene and Miocene (general scarcity of carbonate sediments, presence of terrigenous sediments in the Oligocene and siliceous shales in the lower Miocene) were not conducive to the growth of large foraminifers, it is difficult to conceive of each of these two thriving colonies (specimens are numerous in both; *M. panamensis* constitutes a considerable proportion of the rock in which it is found—perhaps as much as 25%) living totally isolated from their nearest relatives approximately 1,500 mi (2,415 km) away on the Gulf Coast. It would seem likely that other localities, at least at synchronous horizons, remain to be discovered. However, it also seems likely that neither taxon was in California for any appreciable duration of time and each probably was a chronologically short-term invader of the province. Wide dispersal of nepionic ontogenetic stages plus the turbid waters that also apparently prevented the growth of reef corals during the Oligocene of California may have played a part in a sporadic occurrence here of large foraminifers. James Perrin Smith compared the paleogeography of California during deposition of the Vaqueros, Rincon, and type Temblor formations with that of the Marquesas Islands of today, where high surface relief and muddy waters prevent the growth ofotherwise available reef-coral larvae. Most micropaleontologists, in any event, do not prospect the type of coarse clastic sediments involved in each occurrence and this doubtless also has played a role in the paucity of recorded occurrences. Interestingly enough, each locality was discovered by a field geologist.

Miogypsina

The *Miogypsina* occurrence was first described by Graham and Drooger (1952) from material collected by R. N. Hacker. It was identified by them as *M. (Miolepidocyclina) ecuadorensis* Tan, a species Cole (1967) considered to be a synonym of *M. panamensis* (Cushman). Cole also considered *Miolepidocyclina* to be a synonym of *Miogypsina*. Graham and Drooger considered it to be of middle or early-late Oligocene age, Cole placed it in the late Oligocene and more specifically in the *Globorotalia kugleri* Zone of Bolli. The writer is in basic agreement with this age assignment (Hornaday, 1972, p. 39-41) and this correlation then bears heavily on the controversy over the position of the *G. kugleri* (approximately = N4 zone) and the succeeding *Catapsydrax dissimilis, C. stainforthi,* and *Globigerinatella insueta* Zones *vis à vis* the Zemorrian and Saucesian stages. The *Miogypsina* locality is clearly of late Zemorrian Age on the basis of the associated small benthonic Foraminifera. It would seem Bandy and Ingle (1970) were correct in placing the *C. dissimilis, C. stainforthi,* and *G. insueta* (in part) Zones in the Saucesian and the *Globorotalia kugleri* Zone in the Zemorrian. Further complications nevertheless then arise, for if the *G. kugleri* Zone is correlative to at least a part of the Oligocene Zemorrian Stage, and if the *G. kugleri* Zone and the *Globigerinoides* datum supposedly defining its base are Aquitanian, then the Aquitanian (based on the stratotype erected for it) is in part Oligocene and older than the 22.5 m.y. Zemorrian-Saucesian boundary. Such complications will be left to others to resolve.

Lepidocyclina

The *Lepidocyclina* locality from the Vaqueros was discovered by N. L. Taliaferro, who turned the specimens over to H. G. Schenck for study. Designated at first simply as *Lepidocyclina* sp. in several abstracts and a brief article and mentioned as such by Kleinpell (1938, p. 356), the taxon was formally described as *Lepidocyclina (Lepidocyclina) californica* by Schenck and Childs (1942). Schenck and Childs placed this locality in the Saucesian, which they regarded as Oligocene. Little help in interregional correlation can be expected from a species known from only one locality. Although Schenck and Childs stated that *Lepidocyclina, sensu stricto* ranges no higher than Oligocene, it is now known to range from the late Eocene into the early Miocene in the Western Hemisphere. They compared *L. californica* to several other *Lepidocyclina* from the Gulf Coast and Central America which they considered as closely related and which they considered to be of Oligocene age. Stratigraphically the ranges of some of the closely related species are more extensive than they recognized. Furthermore, Cole (1957, p. 42; 1961, p. 373) has synonomyzed first *L. californica* with *L. L. waylandvaughani* Cole and then the latter with *L. canelli* Lemoine and R. Douville. The validity or lack of it in this procedure cannot be resolved here. Both the taxonomic situation and the biostratigraphic sequence of *Lepidocyclina* in the Western Hemisphere seem to be more complex than realized by Schenck and Childs. Suffice it to say that the California *Lepicocyclina* from the Vaqueros Sandstone, on the basis of associated benthonic small foraminifers, is early or early-middle Saucesian in age.

Calcareous Nannoplankton Biostratigraphy of Cenozoic Marine Stages in California

A. D. WARREN[1]

Foreword

R. M. Kleinpell inquired about the possibility of this writer preparing a chapter on the relation of the calcareous nannoplankton zonations to some of the Cenozoic marine stages of California. Ultimately it was decided to present a brief review of published data and interpretations plus observations made by the writer and his associates regarding all of the Cenozoic marine stages now recognized in California. Time and space limitations render it impractical to attempt to document fully all interpretations presented in succeeding paragraphs; therefore the reader is cautioned that this chapter is intended to serve only as a working hypothesis. Some of the zonal assignments discussed are tentative and will remain so until more data are available on the vertical distribution of nannofossil index species. In the interim, it is hoped that the zonal assignments presented may serve as a framework which other nannoplankton investigators may substantiate or disprove on the basis of their further studies.

Introduction

Calcareous nannofossils are the individual microscopic calcite skeletal elements produced largely by marine, planktonic, single-celled, golden-brown algae. They range between 1 and 50 microns in size and may be found in marine strata of Early Jurassic to Holocene age.

In 1954, Bramlette and Riedel of Scripps Institution of Oceanography published a landmark paper in which they called attention to the value of calcareous nannoplankton in biostratigraphic applications. Reaction to this paper was slow and difficult to discern, although a few major oil companies and universities commenced programs to study calcareous nannoplankton on a research basis before the end of that decade. However, between 1961 and 1971, there was a remarkable acceleration in refinement of the calcareous nannoplankton biostratigraphic discipline largely as a result of the interest generated by, and material made available from, the JOIDES Program (Deep-Sea Drilling Project). During that same time interval, some major oil companies began using nannofossil biostratigraphy as an integral part of their exploration and development programs. Many workers have contributed data, so that by utilizing such information as well as their own observations, Martini (1971) and Bukry (1973a, 1975) each proposed a calcareous nannoplankton zonation for the marine Cenozoic. The two zonation schemes are very similar, although some different taxa are used to define zones of essentially equivalent age. At any rate, it is possible to correlate these two widely used zonations one with the other. The purpose of this chapter is to show the relation of the nannoplankton zonations of Martini (1971) and Bukry (1973a, 1975) to the marine stages of the Cenozoic in California as defined by Schenck and Kleinpell (1936), Kleinpell (1938), Goudkoff (1945), Natland (1952), and Mallory (1959) to the degree that these relationships are now understood by the writer. At the same time, an integration of the nannoplankton zones with the planktonic foraminiferal zones of Banner and Blow (1965), Blow (1969), and Berggren (1971) will be indicated. Both Martini (1971) and Berggren (1972) have put forth their opinions as to this particular interdisciplinary integration and there are some areas of conflict between the two. The integration implied here is based on the ideas of Martini and Berggren as well as on the observations of the writer working with material from the Gulf Coast, West Coast, and Caribbean areas and from the Far East. Again, the reader is advised to regard this integration as a tentative correlation between the nannoplankton and planktonic foraminiferal zonations until such time as these relations can be documented in detail.

[1]Anderson, Warren & Associates, Inc., 11526 Sorrento Valley Road, Suite G, San Diego, California 92121.

Before proceeding with a discussion of nannofossil biostratigraphy and the California Cenozoic stages it would be well to consider some of the problems encountered by workers in that area. Failure to recognize or to solve these problems has resulted in some confusion and the comparatively slow development in use of the nannoplankton discipline for biostratigraphic analysis on the West Coast.

Reworking and Redeposition

More often than not, marine sediments of the West Coast Cenozoic contain some degree of redeposited older species of nannofossils in addition to their indigenous nannoflora. Early workers everywhere were plagued by this situation, because the lack of exact knowledge of the stratigraphic range of a particular species made it difficult or impossible to recognize it as reworked. This problem has been largely alleviated by the development of more precise knowledge on the ranges of the key nannoplankton taxa; data from the Deep Sea Drilling Project contributed greatly in this respect. Now it is relatively easy for the nannoplankton investigator to recognize the presence of older species in incompatible association with a younger nannoflora. However, a problem remains where older species are reworked into younger strata that contain no indigenous nannofossils. Should this situation arise in a continuum under study, it is usually apparent, but if a single spot sample is observed, there is no way to prevent a misinterpretation without resorting to one of the other paleontologic disciplines.

Specimen Dilution

Most marine sediments of the West Coast Cenozoic contain diagnostic calcareous nannoplankton. However, in general, nannofossils in West Coast strata are diluted in the sedimentary matrix to a much greater degree than they are in other areas where workers have utilized them in biostratigraphy. This results in the interpreter not being able to find a sufficient number of specimens on a conventionally prepared smear slide to make a reliable analysis. A cubic millimeter of nannoplankton ooze from deep sea cores can contain more than 20 million specimens of calcareous nannofossils, whereas a cubic meter of marine formation from the West Coast Cenozoic commonly contains less than that number of specimens. It is apparent that one drop of slurry prepared from 1 cu cm or less of that formation (a standard nannoplankton slide procedure) might contain no specimens. Obviously, a method of concentrating and isolating the nannofossils from a slurry of macerated rock and distilled water is needed. The extremely minute size of calcereous nannofossils makes it most difficult to find a satisfactory, yet practical, technique that will work for all sediments that might come under scrutiny. Perhaps a short centrifugation technique to remove particle sizes below 5 microns and above 30 microns is the most practical conceived to date in order to concentrate nannoplankton from macerated formation slurries.

Diversity Versus Latitude

As is common in other paleontologic disciplines, species diversity in the calcareous nannoplankton tends to decrease proportionately with increasing degree of latitude. This results in some loss of biostratigraphic resolution where sediments deposited in higher latitudes or in colder waters do not contain all of the diagnostic taxa necessary to subdivide a continuum as finely as is possible with sediments laid down in lower latitudes and warmer water where species diversity approaches a maximum.

Species Concepts

Apparently, differences in species concepts between individual investigators working on nannoplankton sequences from identical sections have resulted in conflicts of interpretation. This phenomenon is common during early phases of development of all paleontologic disciplines. It is the writer's position that the *sensu stricto* species concept ("splitting") is nowhere more essential than in the study of the biostratigraphic distribution of calcareous nannoplankton.

California Stages Versus Standard Nannoplankton Zones

In correlating the standard nannoplankton zones to the Cenozoic marine stages of California, it is essential that the nannoplankton worker has available material from biostratigraphically continuous sections, preferably type sections, which contain both Foraminifera and nannofossils. In addition to a nannoplankton analysis, the materials must also be analyzed by a paleontologist proficient in diagnosis of the California stages primarily on the basis of smaller benthonic Foraminifera. If this last condition is not met, then any conclusions drawn as to precise relation of nannoplankton zone to California stages must remain speculative. Consequently, all new data on correlations of the standard nannoplankton zones to the California marine stages discussed in paragraphs to follow have been preceded by foraminiferal interpretation of the sections involved.

Other Problems

Since 1960, a number of workers have published data that indicate that some of the benthonic foraminiferal indices used to define certain of the California stages may be time transgressive (see Bandy, 1960b, 1967, 1972; Bandy and Ingle, 1970; Gibson and Steineck, 1972; Ingle, 1967; Steineck and Gibson, 1971, 1972). The writer has intentionally avoided this issue and limited conclusions to evidence from the type sections of the Cenozoic marine stages of California. Data derived on nannoplankton distribution within type or nearby sections may be of considerable importance in proving or disproving the temporal integrity of the benthonic Foraminifera in question.

Cenozoic Marine Stages of California as Related to Standard Nannoplankton Zonation

The Cenozoic marine stages of California were erected and more or less formally defined between 1936 and 1959 by the previously mentioned authors. Altogether there are 17 Cenozoic stages currently recognized in California, ranging in age from early Paleocene to late Pleistocene, and they all have been defined largely on the basis of the overlapping stratigraphic ranges of species of benthonic Foraminifera, the "congregations" of Berry (1964, 1966, 1968). In paragraphs to follow, these stages and their relations to the nannoplankton zonation proposed by Bukry (1973a, 1975) will be discussed. Relations of the California stages to the nannoplankton zonation of Martini (1971) and integration of both nannoplankton zonations with the planktonic foraminiferal zonation of Banner and Blow (1965), Blow (1969), and Berggren (1971) are illustrated (Figs. 1, 2), although not discussed in detail. Each of the Cenozoic stages is considered in order from oldest to youngest beginning with the Paleocene Cheneyan Stage and ending with the late Pleistocene Hallian Stage. Previously published data and opinion reported by various workers are cited and new data are submitted to substantiate opinions proffered herein.

Cheneyan Stage

Originally defined and considered to be probably Paleocene in age by Goudkoff (1945), this stage subsequently was shown to be equivalent to Danian strata elsewhere by Loeblich (1958) and then by Martin (1964) on the basis of planktonic Foraminifera. No data on the nannofossil associations of the Cheneyan Stage have been published, nor has the writer had an opportunity to study strata from this age rock from California.

Ynezian Stage

This stage was originally defined on the basis of Foraminifera and considered to be Paleocene in age by Mallory (1959).

The following workers have referred either directly or indirectly to the Ynezian Stage and its calcareous nannofossil associations: Bramlette and Sullivan (1961), Bukry (*in* Gibson, 1976), Bukry et al (1973), Gibson (1974), Hay and Mohler (1967), Martini (1971), Mohler and Hay (1967), Schmidt (1970, 1975), Sullivan (1964, 1965), and Sullivan (*in* Weaver et al, 1969).

Data from the papers cited above indicate that the Ynezian Stage might include all or parts of the following calcareous nannoplankton zones defined by Bukry (1973a) and listed from oldest to youngest: *Fasciculithus tympaniformis* Zone, *Heliolithus kleinpelli* Zone, *Discoaster mohleri* Zone, *Discoaster nobilis* Zone, *Discoaster multiradiatus* Zone, *Discoaster diastypus* Zone.

The writer has not examined enough Ynezian material from surface sections in California to offer any different interpretation from that shown. This is not to imply that nannoplankton zones older than the *Fasciculithus tympaniformis* Zone may not occur in strata of Ynezian age.

Bulitian Stage

This stage was originally defined on the basis of Foraminifera and considered to be of Paleocene age by Mallory (1959).

The following workers have referred either directly or indirectly to the Bulitian Stage and its calcareous nannofossil association: Bramlette and Riedel (1954), Bramlette and Sullivan (1961), Bukry (*in* Gibson, 1976), Bukry et al (1973), Gibson (1974), Hay and Mohler (1967), Martini (1971), Mohler and Hay (1967), Schmidt (1970, 1975), Sullivan (1964, 1965), and Sullivan (*in* Weaver et al, 1969).

Data from the papers cited above indicate that the Bulitian Stage might include all or parts of the following nannoplankton zones defined by Bukry (1973a) and listed from oldest to youngest: *Discoaster multiradiatus* Zone, *Discoaster diastypus* Zone, *Tribrachiatus orthostylus* Zone, *Discoaster lodoensis* Zone.

The writer has examined continua at or near Lodo Gulch (see Bramlette and Sullivan, 1961; Martin, 1943), Devils Den (see Cook, 1950), the Aqueduct section (just south of center of Sec. 34, T25S, R18E, M.D.B.& M. between Devils Den and Mabury Hills), Mabury Hills (see Cifelli, 1951), and several sections at Media Agua Creek (see Mallory, 1970); all of these areas contain strata readily assignable to the Bulitian, Penutian, and Ulatisian Stages. Indeed, Media Agua Creek is the type area for the zones of the Ynezian and Bulitian Stages as well as the type locality of the Penutian Stage as defined by Mallory (1959).

After completing examination of sections from those five areas, the writer came to the conclusion that all occurrences of *Tribrachiatus orthostylus* above the Bulitian, and possibly the lowermost part of the Penutian, represent reworking. Rare occurrences of *Tribrachiatus orthostylus* have been observed in all of the Cenozoic marine stages of California from within the Penutian through the Hallian and even in Holocene bottom samples off southern California. Bukry (1973a, p. 689) cited occurrences of that species in the *Nannotetrina quadrata* Zone (early Narizian) of Arroyo el Bulito. All such occurrences apparently represent reworking and redeposition.

Hay and Mohler (1967, p. 1506) commented on the propensity of calcareous nannofossils to endure the rigors of erosion and redeposition with little effect on the condition of the reworked specimens. *Discoaster lodoensis* is another species, similarly large and robust as *Tribrachiatus orthostylus,* that is also common in the reworked state. However, the net effect of the high degree of reworking of *Tribrachiatus orthostylus* in California on correlations is most disconcerting. Data on the true stratigraphic ranges and compatible associations of calcareous nannofossils that have been made available by the Deep Sea Drilling Project are beginning to relieve somewhat the problems presented by reworking of diagnostic nannoplankton species. In most sections from each of the five areas mentioned earlier in this paragraph, reworked *Tribrachiatus orthostylus* above the Bulitian and lower Penutian Stages are present but frequency of specimens is from very rare to few in number, whereas presumably indigenous occurrences within the Bulitian and lower Penutian Stages are consistent and frequency of specimens is common to abundant. Bukry (1973a) has indicated by implication that the last typical occurrence of *Tribrachiatus orthostylus* is approximately coincident with the earliest occurrence of *Coccolithus crassus*. In the light of all these considerations, examination of the check list of calcareous nannofossils prepared by Bramlette and Sullivan (1961) on the Lodo Gulch section leads to an interesting reinterpretation of that particular continuum. The writer would place the top of the *Tribrachiatus orthostylus* Zone of Bukry (1973a) between samples 51 and 52. This point represents the last indigenous occurrence of common to abundant *Tribrachiatus orthostylus* (=*Marthasterites tribrachiatus*) and the approximate earliest occurrence of *Coccolithus crassus;* all occurrences of *Tribrachiatus orthostylus* from samples 52 to 110 are considered to be reworked into the *Discoaster lodoensis* Zone of Bukry (1973a).

In the opinion of the writer, the Bulitian Stage, at least in the zonal type area at Media Agua Creek and in areas north to Lodo Gulch, includes all or parts of the *Discoaster diastypus* Zone and *Tribrachiatus orthostylus* Zone, nannoplankton zones of Bukry (1973a).

Penutian Stage

This stage was originally defined on the basis of Foraminifera and considered by Mallory (1959) to be of early Eocene age.

The following workers have referred either directly or indirectly to the Penutian Stage and its calcareous nannofossil association: Bramlette and Riedel (1954), Bramlette and Sullivan (1961), Bukry (1973a; *in* Gibson, 1976), Bukry et al (1973), Gibson (1974), Hay and Mohler (1967), Sullivan (1964, 1965), and Sullivan (*in* Weaver et al, 1969).

Data or implications from papers cited above suggest that the Penutian Stage may include all or parts of the following nannoplankton zones defined by Bukry (1973a) and listed from oldest to youngest: *Discoaster nobilis* Zone, *Discoaster multiradiatus* Zone, *Discoaster diastypus* Zone, *Tribrachiatus orthostylus* Zone, *Discoaster lodoensis* Zone.

The writer has examined samples from the type section of the Penutian Stage at Media Agua Creek as well as from other nearby continua at Mabury Hills, the Aqueduct section, and Devils Den. At Media Agua Creek, the last common occurrence of *Tribrachiatus orthostylus* is in the early Penutian although the species continues to occur in rare numbers considerably higher than that. As mentioned, these rare occurrences above the early Penutian are considered to be reworked specimens. However, at the other three sections examined, *Tribrachiatus orthostylus* is either absent or extremely rare and sporadic in occurrence above the early Penutian. The top of the Penutian at the type and at each of the other three continua mentioned above occurs within the *Discoaster lodoensis* Zone of Bukry (1973a).

Therefore, it is the writer's interpretation that the Penutian Stage at and within the vicinity of the type section contains only parts of the nannoplankton zones defined by Bukry (1973a) from oldest to youngest: *Tribrachiatus orthostylus* Zone and *Discoaster lodoensis* Zone.

Ulatisian Stage

This stage was originally defined on the basis of Foraminifera and considered to be of middle Eocene age by Mallory (1959).

The following workers have referred either directly or indirectly to the Ulatisian Stage and its calcareous nannofossil associations: Bramlette and Sullivan (1961), Bukry (*in* Gibson, 1976), Bukry et al (1973), Hay (1967), Hay and Mohler (1967), Martini (1971), Mohler and Hay (1967), Schmidt (1970, 1975), Steineck and Gibson (1972), Sullivan (1964, 1965), and Sullivan (*in* Weaver et al, 1969).

Data or implications from papers cited above suggest that the Ulatisian Stage may include all or parts of the following nannoplankton zones of Bukry (1973a), from oldest to youngest: *Discoaster diastypus* Zone, *Tribrachiatus orthostylus* Zone, *Discoaster lodoensis* Zone, *Discoaster sublodoensis* Zone.

Based on data from Sullivan (1965), it appears that the type Ulatisian, in the Vacaville shale, Solano County, California, contains at least parts of the *Discoaster lodoensis* and *Discoaster sublodoensis* Zones of Bukry (1973a). This is in keeping with other data from Bramlette and Sullivan (1961) as reinterpreted here, Sullivan (1965), and with new data developed by the writer from Garza Creek, Devils Den, the Aqueduct section, and Media Agua Creek; all of these other and new data indicate that the base of the Ulatisian Stage lies within the *Discoaster lodoensis* Zone, whereas the top of the Ulatisian is coincident with, or very nearly coincident with, the top of the *Discoaster sublodoensis* Zone of Bukry (1973a). Consequently, it is the opinion of the writer that the Ulatisian Stage as manifested in its type area and in five other continua from Lodo Gulch to Media Agua Creek on the west side of the San Joaquin Valley contains all or parts of the nannoplankton zones of Bukry (1973a), from oldest to youngest: *Discoaster lodoensis* Zone, *Discoaster sublodoensis* Zone.

Narizian Stage

The Narizian was originally defined on the basis of Foraminifera and considered to be of late Eocene age by Mallory (1959).

The following workers have referred either directly or indirectly to the Narizian Stage and its calcareous nannofossil associations: Bramlette and Sullivan (1961), Bukry (1973a, 1975), Bukry (*in* Gibson, 1976), Bukry et al (1973), Gibson (1974), Hay (1967), Martini (1971), Schmidt (1970, 1975), Sullivan (1965), Sullivan (*in* Weaver et al, 1969), Tipton (1976a,b), and Warren and Newell (1976a,b).

Data or implications from papers cited above suggest that the Narizian Stage may include all or parts of the following nannoplankton zones of Bukry (1973a, 1975), listed from oldest to youngest: *Rhabdosphaera inflata* Subzone of the *Discoaster sublodoensis* zone, *Nannotetrina quadrata* Zone, *Reticulofenestra umbilica* Zone, *Chiasmolithus oamaruensis* Subzone of the *Discoaster barbadiensis* Zone.

Data on nannoplankton from the type area of the Narizian Stage in the upper Gredal, Point of Rocks, and Kreyenhagen formations of the western San Joaquin Valley, Kern Country, California are limited to the early Narizian only, as manifested mainly in the upper Gredal and Point of Rocks Formations. Unfortunately, the Welcome Member of the Kreyenhagen Shale, which largely constitutes the type upper Narizian, appears to be devoid of calcareous microfossils at the surface. It should be pointed out that the base of the Point of Rocks Formation is time-transgressive so that at Media Agua Creek it is considerably older than the base of the Point of Rocks Formation at the Aqueduct section; thus a portion of the upper Gredal Formation at the Aqueduct section is a time equivalent of parts of the lower and middle Point of Rocks Formation at Media Agua Creek.

As will be discussed, most knowledge of late Narizian nannoplankton associations has been developed from studies of continua of the Sacate Formation in the type area of the Refugian Stage at Arroyo el Bulito, Santa Barbara Country, and of the Butano and San Lorenzo formations in the type area of the San Lorenzo Formation, Santa Cruz County, California.

Most previously published data and new data derived from lower Narizian strata at Devils Den, the Aqueduct section, and Media Agua Creek suggest that the base of the Narizian Stage in its type area is coincident with, or very nearly coincident with, the base of the *Nannotetrina quadrata* Zone of Bukry (1973a). Studies conducted on Narizian strata of the Sacate Formation in Arroyo el Bulito and of the Butano and San Lorenzo formations in the type area of the San Lorenzo Formation by Warren and

Newell (1976) indicate presence of the *Reticulofenestra umbilica* Zone and, in the latest Narizian at both localities, the *Chiasmolithus oamaruensis* Subzone of the *Discoaster barbadiensis* Zone of Bukry (1975). Therefore, it is the writer's interpretation that the Narizian Stage contains all of the following nannoplankton zones or subzones of Bukry (1973a, 1975), from oldest to youngest: *Nannotetrina quadrata* Zone, *Reticulofenestra umbilica* Zone, *Chiasmolithus oamaruensis* Subzone of the *Discoaster barbadiensis* Zone.

Refugian Stage

This stage was originally defined on the basis of Foraminifera and mollusks and was considered to be of late Eocene or early Oligocene age by Schenck and Kleinpell (1936).

The following workers have referred directly to the Refugian Stage and its calcareous nannofossil associations: Brabb et al (1971), Bukry et al (1973), Lipps and Kalisky (1972), Tipton (1976a, b) and Warren and Newell (1976a, b).

All data from papers cited above suggest that the Refugian Stage includes only the *Isthmolithus recurvus* Subzone of the *Discoaster barbadiensis* Zone of Bukry (1975).

Zemorrian Stage

The Zemorrian was originally defined mainly on the basis of Foraminifera and considered to be Oligocene (Rupelian equivalent) by Kleinpell (1938, Fig. 14).

The following workers have referred directly or indirectly to the Zemorrian Stage and its calcareous nannofossil associations: Bandy and Ingle (1970), Bukry et al (1973), Lipps (1967, 1968), Lipps and Kalisky (1972), and Warren and Newell (1976a, b).

Data or implications from authors cited above suggest that the Zemorrian Stage may include all or parts of the following nannoplankton zones or subzones of Bukry (1973a, 1975): *Helicopontosphaera reticulata* Zone, *Sphenolithus predistentus* Zone, *Sphenolithus distentus* Zone, *Sphenolithus ciperoensis* Zone, *Cyclicargolithus abisectus* Subzone of the *Triquetrorhabdus carinatus* Zone.

The writer concurs with these previously published data, although none of these data were derived from study of the type area of the Zemorrian Stage.

Saucesian Stage

The Saucesian was originally defined mainly on the basis of Foraminifera and was considered to be approximately equivalent in age to the Aquitainian (and in part possibly Chattian) Stage by Kleinpell (1938, Fig. 14).

The following workers have referred directly or indirectly to the Saucesian Stage and its calcareous nannofossil associations: Bandy and Ingle (1970), Bramlette and Wilcoxon (1967), Bukry et al (1973), Lipps (1967a, b 1968), Lipps and Kalisky (1972), and Warren and Newell (1976b).

Data or implications from authors cited above suggest that the Saucesian Stage may include all or parts of the following nannoplankton zones of Bukry (1973a): *Triquetrorhabdulus carinatus* Zone, *Sphenolithus belemnos* Zone, *Helicoptosphaera ampliaperta* Zone.

The writer has examined a continuous section from the type area of the Saucesian Stage in the middle and upper Rincon Shale and lowermost Monterey Shale of Los Sauces Creek, Ventura County, California. At this locality the Rincon and Monterey Shales are separated by a 2-m thick bentonite bed, often correlated with the Tranquillon Formation, which here lies approximately 610 m above the top of the Vaqueros Sandstone. This bentonite bed serves as a convenient reference point in discussing the type Saucesian as it separates the middle Saucesian from the late Saucesian as well as separating the uppermost Rincon Shale from the lowermost Monterey. In this study, the *Discoaster deflandrei* Subzone of the *Triquetrorhabdulus carinatus* Zone was not recognized. However, the *Discoaster druggii* Subzone was noted within the upper Rincon from 61 to 12 m below the base of the bentonite bed and included the earliest noted occurrence of *Helicopontosphaera ampliaperta* at the base of the subzone. The *Sphenolithus belemnos* Zone is present from 12 m below to 2 m above the bentonite bed. The base of the *Helicopontosphaera ampliaperta* Zone is placed 2 m above the top of the bentonite bed with the earliest occurrence of *Sphenolithus heteromorphus* noted 3 m above the bentonite bed. Thus, of the approximately 91 m of lower Monterey Shale that constitutes the type upper Saucesian, 89 m are considered to lie within the lower part of the *Helicopontosphaera ampliaperta* Zone of Bukry (1973a). Samples from the early Relizian strata within the lower Monterey at Los Sauces Creek also contain nannofossils indicative of the *Helicopontosphaera ampliaperta* Zone.

On the basis of previously published data from Warren and Newell (1976b) and new data outlined above, the writer's conclusion is that the Saucesian Stage includes all or parts of the following zones or subzones of Bukry (1973a), from oldest to youngest: *Discoaster deflandrei* Subzone of the *Triquetrorhabdulus carinatus* Zone, *Discoaster druggii* Subzone of the *Triquetrorhabdulus carinatus* Zone, *Sphenolithus belemnos* Zone, *Helicopontosphaera ampliaperta* Zone.

Previous reports of occurrences of *Sphenolithus belemnos* and *Triquetrorhabdulus carinatus* within strata of Relizian age (Lipps, 1967a, b, 1968) are erroneous in that the strata from which these species were recovered are actually from the type Saucesian or from the Saucesian of Graves Creek at Atascadero, San Luis Obispo County, California (Lipps and Kalisky, 1972).

Relizian Stage

This stage was originally defined on the basis of Foraminifera and considered to be of early Miocene (Burdigalian and Langhian) age by Kleinpell (1938).

The following workers have referred directly or indirectly to the Relizian Stage and its calcareous nannofossil associations: Bandy and Ingle (1970), Bramlette and Wilcoxon (1967), Bukry et al (1973), Lipps (1967a, b, 1968), and Lipps and Kalisky (1972).

Data or implications from authors cited above suggest that the Relizian Stage may incude all or parts of the following nannoplankton zones of Bukry (1973a): *Triquetrorhabdulus carinatus* Zone, *Sphenolithus belemnos* Zone, *Helicopontosphaera ampliaperta* Zone, *Sphenolithus heteromorphus* Zone.

The writer has not had available for study material from the type section of the Relizian stage in Reliz Canyon, Monterey County, California. However, in other sections examined, the weight of evidence suggests that the Relizian is equivalent to the upper part of the *Helicopontosphaera ampliaperta* Zone. For example, at the Naples Bluff section, Santa Barbara County, California, *Helicopontosphaera ampliaperta* occurs sporadically from just above the bentonite bed at the base of the continuum (also base of the late Saucesian) up to the base of the prominent 3-m chert bed which marks the boundary between the Relizian and Luisian Stages. *Sphenolithus heteromorphus* also occurs sporadically through the same section, but persists upward above the chert bed throughout strata assigned to the Luisian Stage on the east side of Dos Pueblos Canyon at Naples Bluff. These same relations have been observed in Relizian and Luisian strata at El Capitan Beach, Santa Barbara County, the Bonita Canyon–Newport Lagoon area, Orange County, and in numerous subsurface sections in southern California.

It is the conclusion of the writer that the Relizian Stage includes only the upper part of the *Helicopontosphaera ampliaperta* Zone of Bukry (1973a).

Luisian Stage

The Luisian was originally defined on the basis of Foraminifera and considered to be of middle Miocene age by Kleinpell (1938).

The following workers have referred directly or indirectly to the Luisian Stage and its calcareous nannoplankton associations: Bandy and Ingle (1970), Berggren and van Couvering (1974), Bukry et al (1973), Lipps (1967a, b, 1968), Lipps and Kalisky (1972), Martini (1971), Warren (1972), Wilcoxon (1969), and Wornardt (1973).

Data or implications from authors cited suggest that the Luisian Stage may include all or parts of the following nannoplankton zones of Bukry (1973a): *Helicopontosphaera ampliaperta* Zone, *Sphenolithus heteromorphus* Zone, *Discoaster exilis* Zone.

Samples from a continuum of Luisian strata at Quailwater Creek, Highland School District, San Luis Obispo County, California, a few miles east of the type Luisian, were examined by the writer; *Sphenolithus heteromorphus* and *Cyclococcolithina neogammation* are present, but neither *Helicopontosphaera ampliaperta* nor *Discoaster kugleri* were noted in these Luisian samples. Other continua of Luisian strata at Newport Lagoon, Orange County, and at El Capitan Beach and Naples Bluff, Santa Barbara County, California, also contain *Sphenolithus heteromophus* in the absence of *Helicopontosphaera ampliaperta* and *Discoaster kugleri*. Wilcoxon (1969, and *in* Wornardt, 1973) reported the occurrence of *Discoaster kugleri* in strata of Luisian age from Newport Lagoon and in association with *Sphenolithus heteromorphus*. Lipps (1968), Lipps and Kalisky (1972), and Warren (1972) have reported *Discoaster kugleri* as occurring only in the latest early Mohnian Stage at that locality. It is assumed that these conflicting data must be attributable to differences in species concepts among these workers. J. H. Newell (personal commun.) and the writer have, in addition, noted rare occurrences of *Discoaster kugleri s. str.* in samples that, on the basis of benthonic Foraminifera from various surface and subsurface localities in southern California, could only be assigned to the early Mohnian. Because of convincing evidence from several localities, it is the writer's opinion that the Luisian Stage includes only the *Sphenolithus heteromorphus* Zone of Bukry (1973a).

Mohnian Stage

The Mohnian was originally defined on the basis of Foraminifera and considered to be of late Miocene age by Kleinpell (1938).

The folowing workers have referred directly or indirectly to the Mohnian Stage and its calcareous nannofossil associations: Bandy and Ingle (1970), Berggren and van Couvering (1974), Bukry et al (1973), Lipps (1967a, b, 1968), Lipps and Kalisky (1972), Martini (1971), Martini and Bramlette (1963), Warren (1972), Wilcoxon (1969), and Wornardt (1973).

Data or implications from the authors cited suggest that the Mohnian Stage may include all or parts of the following nannoplankton zones of Bukry (1973a): *Discoaster exilis* Zone, *Catinaster coalitus* Zone, *Discoaster hamatus* Zone, *Discoaster neohamatus* Zone, *Discoaster quinqueramus* Zone.

No nannoplankton data are as yet available from the type section of the Mohnian Stage in Topanga Canyon, Los Angeles County, California. However, at Newport Lagoon, Orange County, California, the writer has reported the highest occurrence of *Cyclococcolithina neogammation* in the early Mohnian and the highest occurrence of *Discoaster kugleri* in the latest early Mohnian (Warren, 1972). In addition, the writer has noted *Discoaster hamatus* in a few samples form different localities in southern California all of which also contained well-developed late Mohnian foraminiferal faunules. Lipps (1968) and Lipps and Kalisky (1972) reported *Discoaster bollii* from late Mohnian strata in California. Consequently, it is interpreted here that the Mohnian Stage includes all or parts of the following nannoplankton zones of Bukry (1973a): *Discoaster exilis* Zone, *Catinaster coalitus* Zone, *Discoaster hamatus* Zone.

This does not imply that younger nannoplankton zones may not be eventually documented to be included within the limits of the Mohnian Stage, but at this time the writer recognizes no present evidence that either the *Discoaster neohamatus* or *Discoaster quinqueramus* Zones are equivalent to the Mohnian.

Since core material was made available from the Experimental Mohole, Guadelupe Island Site, several workers have indicated or implied the presence of Relizian or Luisian sedimentary strata within the section penetrated (Bandy, 1972a, b; Bandy and Ingle, 1970; Lipps and Kalisky, 1972; Parker, 1964). The writer has examined processed core material from the Experimental Mohole and on the basis of foraminiferal, nannoplankton, and silicoflagellate data, has concluded that the oldest microfossiliferous sediments found are of early Mohnian age. The Foraminifera noted in the deepest part of the hole are typical of an extremely deep-water facies of the early Mohnian which has been recognized in the offshore area of southern California. Nannoplankton present in the lowest part of the section penetrated indicate presence of the *Coccolithus miopelagicus* Subzone of the *Discoaster exilis* Zone of Bukry (1973a) which has been recognized within the early Mohnian of numerous surface and subsurface sections in southern California by the writer and his associates. Martini (1971) also reported that the lowermost microfossiliferous sediments in the Mohole are assigned to his *Discoaster exilis* Zone (NN 6) which is equivalent to the *Coccolithus miopelagicus* Subzone of Bukry (1973a). Silicoflagellates, although not abundant in the lower part of the Mohole, suggest assignment to the upper *Corbisema triacantha* Zone or *Distephanus longispinus* Zone of Bukry and Foster (1974) both of which zones are present, at least in part, within the early Mohnian strata at Newport Lagoon, Orange County, California; Luisian strata at Newport Lagoon are assigned to the *Distephanus octacanthus* and lower *Corbisema triacantha* Zone of Bukry and Foster (1974). Therefore, it is the writer's opinion that no sediments of Relizian nor Luisian age were found in the Experimental Mohole.

Delmontian Stage

The Delmontian was originally defined on the basis of Foraminifera and considered to be of late Miocene or Pliocene (Sarmatian) age by Kleinpell (1938). Pierce (1972) prepared data to support his contention that the type Delmontian Stage is actually equivalent in part to the late Mohnian Stage of Kleinpell (1938). The controversy in reference was discussed in the sections of this supplement devoted to reviews of the Mohnian and Delmontian Stages. Regardless of the merits of both sides of this question, there is more or less unanimous agreement that there exists a considerable body of sediments within the Los Angeles basin lying above definite Mohnian strata and below the basal Repettian Stage of Natland (1952) that has been traditionally assigned to the Delmontian Stage of

Kleinpell (1938) by commercial paleontologists. Wissler (1943) subdivided this section into all or parts of his divisions A, B, and C of the late Miocene. Wissler's divisions D and E are late and early Mohnian, respectively, and his division F is Luisian. Therefore, most commercial paleontologists recognize section in the Los Angeles basin that they assign to the Delmontian Stage and consisting of all the divisions A and B and all or part of division C of Wissler (1943).

At this time, there are no published data to clearly indicate the relation of the Delmontian Stage of Kleinpell (1938) to the nannoplankton zonation of Bukry (1973a). However, nannofossil data from the Mohnian and Repettian Stages discusssed herein suggest that the Delmontian Stage might include all or parts of the nannoplankton zones of Bukry (1973a) which are, from oldest to youngest: *Discoaster neohamatus* Zone, *Discoaster quinqueramus* Zone, *Ceratolithus tricorniculatus* Zone.

Repettian Stage

This stage was originally defined on the basis of Foraminifera and considered to be of early Pliocene age by Natland (1952).

Reference to the Repettian Stage and indirectly to its calcareous nannoplankton associations have been made by Martini (1971) and Martini and Bramlette (1963).

Data from these authors suggest that the Repettian Stage of Natland (1952) may be equated in part with the *Reticulofenestra pseudoumbilica* Zone of Bukry (1973a).

The stratotype of the Repettian Stage is in the Repetto Hills, Los Angeles County, California. No published data relating to nannofossil associations of the type section are known to the writer. However, subsurface data accumulated by the writer and J. H. Newell (personal commun.) from wells drilled in the Los Angeles basin suggest that the Repettian Stage there may be equated with all or parts of the nannoplankton zones of Bukry (1973a, 1975) as follows: *Ceratolithus tricorniculatus* Zone, *Reticulofenestra pseudoumbilica* Zone, *Discoaster tamalis* Subzone of the *Discoaster brouweri* Zone, *Discoaster surculus* Subzone of the *Discoaster brouweri* Zone, *Discoaster pentaradiatus* Subzone of the *Discoaster brouweri* Zone.

To be more specific, the highest occurrence of *Discoaster pentaradiatus* has been observed to coincide with the highest occurrence of diagnostic benthonic Foraminifera indicative of the top of the Repettian Stage in the subsurface of the Los Angeles basin.

Venturian Stage

The Venturian was originally defined on the basis of Foraminifera and considered to be of middle Pliocene age by Natland (1952).

No published nannopankton data are available on the type section of the Venturian Stage in Wheeler Canyon, Ventura County, nor are any available on other known sections of this stage. By implication, Bandy and Wilcoxon (1970) indicated that this stage must be older than the top of the *Cyclococcolithina macintyrei* Subzone of the *Discoaster brouweri* Zone of Bukry (1973a). Based on observations of this writer and of J. H. Newell (personal commun.), the Venturian Stage as manifested in the Los Angeles basin is here equated with the greater part, but not all, of the *Cyclococcolithina macintyrei* Subzone of the *Discoaster brouweri* Zone of Bukry (1973a).

Wheelerian Stage

This stage was originally defined on the basis of Foraminifera and considered to be of late Pliocene age by Natland (1952).

No published nannoplankton data are available on the type section of the Wheelerian Stage in Wheeler Canyon, Ventura County. Bandy and Wilcoxon (1970) have indicated that the lower part of the Wheelerian Stage at Balcom Canyon, Los Angeles County, California, may be equated with the uppermost part of the *Discoaster brouweri* Zone of Bukry (1973a). This correlation is in agreement with additional observations of the writer and J. H. Newell (personal commun.). The Wheelerian Stage in the opinion of the writer is very tentatively considered to include all or parts of the following nannoplankton zones or subzones of Bukry (1973a), from oldest to youngest: *Cyclococcolithina macintyrei* Subzone of the *Discoaster brouweri* Zone, *Crenalithus doronicoides* Zone, *Gephyrocapsa oceanica* Zone, *Emiliania huxleyi* Zone.

Hallian Stage

The Hallian was originally defined on the basis of Foraminifera and considered to be of Pleistocene age by Natland (1952).

No published nannoplankton data are available on the type section of the Hallian Stage in Hall Canyon, Ventura County. Based on observations by the writer and J. H. Newell (personal commun.), the Hallian Stage is very tentatively equated with the upper part of the *Emiliania huxleyi* Zone of Bukry (1973a).

Summary and Conclusions

In previous paragraphs, data from various sources have been cited to support the writer's interpretations as to the relations of the calcareous nannoplankton zones of Bukry (1973a, 1975) with the Cenozoic marine stages of California. Many of the opinions expressed herein are considerably different from some presented in previously published literature. Wherever possible, discussion of the nannoplankton zone-stage relations have been intentionally limited to type sections of the stages under consideration or to other nearby or well-known exposures where there has been little disagreement as to age or stage of the sections involved. These relations are illustrateed in two charts, one for the Paleogene (Fig. 3) and another for the Neogene (Fig. 4). Also illustrated on these charts is an integration of the nannoplankton zones of Bukry (1973a, 1975) and Martini (1971), and the planktonic foraminiferal zones of Banner and Blow (1965), Blow (1969), and Berggren (1971). From these charts, it is possible to equate the Cenozoic marine stages of California with those of the Atlantic and Gulf Coastal Provinces (see Murray, 1961), Europe (see Berggren, 1971), and New Zealand (see Jenkins, 1971b). These correlations are illustrated on a third chart (Fig. 5) which reflects the views the writer developed from personal observations, discussions with numerous colleagues, and previously published data of many workers.

It is hoped that this chapter will generate other data and opinions, either concurring or dissenting, that will ultimately result in a larger body of literature on nannoplankton biostratigraphy in California and other areas of the Pacific Coast.

Acknowledgments

J. H. Newell has contributed most significantly to this chapter by way of new data, confirmation of old data, and discussion of various nannofossil associations. Others who have been of great assistance mainly by way of meaningful discussions over the years are O. L. Bandy, David Bukry, Ann Tipton Donnelly, B. C. Jones, R. M. Kleinpell, M. L. Natland, S. F. Percival, and L. J. Simon.

Taxonomy

R. M. KLEINPELL and ANN TIPTON

Notes on Formal and Pentative Phyletic Revisions and Usages in Systematics

Much detailed biostratigraphic and biochronologic information regarding ocurrences of foraminiferal taxa in the West Coast middle Tertiary has accumulated since 1938, with the result that some of the phyletic relations are much better understood today than at that time. Inferred phylogenies of lineages such as the uvigerinids, the siphogenerinids, the bolivinids, some buliminids, the nonionids, and the valvulinerids are now under better discipline. It is the aim of this section to emend and improve past systematic usage insofar as possible, in order to better reflect the apparent biologic relations of the taxa involved.

A phyletic chart of uvigerinids and siphogenerine derivitives, and also a valvulinerid phyletic chart, both showing zonal ranges of the species in these various lineages, appear here (Fig. 6). Known extensions in the zonal ranges of bolivinid and nonionid species and of *Bulimina, Cassidulina* (a stock with cold-temperature optimum and euryopic as a whole), *Pullenia,* and some *Eponides* will be apparent in the congregations for the zones listed in the section devoted to a review of the stages and zones. The same applies to lagenid, heterohelicid, and anomalinid species other than valvulinerids, although most lagenid species are far too long-ranging in time for consideration in zonally diagnostic congregations except for a few cryptogenetic species that seem to be province invaders with provincial extinctions. Such species as *Eponides umbonatus, Gyroidina soldanii,* and certain *Anomalina, Cibicides,* and various uniserial lines appear to be long-ranging in time and essentially morphotypes of local ecologic significance, much as *Discorbis* and the elphidiids are indicative of shallow water rather than of time. Also, arenaceous species are on the whole too long ranging to be significant in zonal congregations, although that may not apply to all species of *Ammodiscus, Clavulina, Cyclammina, Gaudryina, Spiroplectammina, Textularia, Tritaxilina,* and *Trochammina.* The stratigraphic occurrence of these species in the California middle Tertiary seems to be about as previously reported (Kleinpell, 1938). Calcareous imperforate forms seem to be too spotty in occurrence to be of much significance in zone or even stage congregations although their presence usually points to local current-sheltered, rocky-bottom environmental conditions. *Pyrgo* seems to have a cold-temperature optimum.

Inevitably, other major taxonomic problems remain unsolved. For example, the genus *Bolivina,* admittedly polyphyletic, is retained as such in this study, encompassing *Brizalina* of other authors, because it is thought that to distinguish the two is merely to replace one polyphletic genus with two. Although both *Bolivina* and *Brizalina* appear to descend from one or another species of *Virgulina,* still other species, assigned to one or the other of these genera, appear possibly to be of heterohelicid origins through some species of *Plectofrondicularia,* thereby paralleling the derivation of *Textularia* from *Spiroplectammina* in a stock with agglutinated tests. Moreover, as between *Bolivina* and *Brizalina,* the problem of generic or subgeneric status remains unresolved. As another example, the actual phyletic position of certain species formerly assigned to *"Pulvinulinella"* or subsequently to *"Epistominella,"* still seems to remain undemonstrated and the validity of their new generic assignments a matter of some question, so for present purposes particular usages of diverse authors have been followed herein without further comment. *"Rotalia" garveyensis* Natland probably needs to be reassigned to a different, perhaps a new, genus, but present information is insufficient to resolve the question of generic position.

The following additions and revisions constitute an attempt at a taxonomic "progress report" wherever this appears to be possible. Much obviously remains to be done in this area, which will require further supplementing and revising as work continues. Beyond this, and except for occasional pertinent comment in the general script — especially with reference to revisions in age of particular occurrences — the systematics have been left in the form presented by Kleinpell (1938, p. 184-357). For the data pertinent to the types herewith figured, including the hypotypes, the reader is referred to the plate legends, and to the pages introductory to those plate legends.

Kingdom PROTISTA
Phylum PROTOZOA
Class SARCODINA BUTSCHLI 1882
Order FORAMINIFERA
Family NONIONIDAE
Genus *Nonion* Montford, 1808
Nonion kernensis Kleinpell

Nonion incisum (Cushman and Parker, 1931, p. 7, pl. 1, figs. 26a, b.
Nonion incisum (Cushman) var. *kernensis* Kleinpell, 1938, p. 232-233.

This species may be an ancestor of *N. multicameratum* Cushman and Kleinpell. Neither seems to belong to the *incisum* lineage.

Nonion multicameratum Cushman and Kleinpell

Nonion pizarrensis W. Berry var. *multicameratum* Cushman and Kleinpell, 1934, p. 4, pl. 1, figs. 10a,b.
Nonion multicameratum Cushman and Kleinpell, A.D. Warren, 1972, p. 35.

This widespread species (see Pl. IV) appears to be restricted to a neritic facies in both senses of the term "neritic." In California it also appears to be restricted stratigraphicallly to the two lower zones of the Mohnian Stage.

Nonion pizarrensis W. Berry

Nonion pizarrensis W. Berry, Cushman and Kleinpell, 1934, p. 4, pl. 1, Figs. 9a,b.

Of the synonyms listed by Kleinpell (1938, p. 234), only the above-cited specimen, from the upper Relizian at Henry Ranch, San Luis Obispo County, is clearly assignable to *N. pizarrensis*. Kleinpell's listed specimens from the Temblor Formation, from member 4 of the type Monterey Shale, and from Contra Costa County, are removed from *N. pizarrensis* and referred to *N. multicameratum* Cushman and Kleinpell.

Nonion weaveri Kleinpell and Tipton, n. sp.

Nonion costiferum var,. of Warren, 1972, fig. 2, p. 29.
Not "*Nonion costiferum* var. = *Nonion costifera* (Cushman) of Cushman and Laiming, 1931 . . . ," of Warren 1972, p. 35.

Test longer than broad, periphery keeled, as many as 20 chambers in the last-formed coil, usually umbilicate; sutures distinct and notably limbate throughout and slightly raised above the general surface and forming an acute angle with the periphery; in apertural view, the apertural face is heart-shaped or broadly triangular, the aperture a small semicircular opening at the base of the apertural face.

Length up to 1 mm; breadth up to 0.60 mm; thickness up to 0.40 mm, but usually somewhat more compressed.

Holotype—RMK 101, from the Newport sequence, (Fig. 1, sample 5).
Paratype—RMK 102, from locality M519C, upper Devilwater Silt Member of Monterey Shale (Table 6). Collectors: Carlos Key, H. H. Heikkila, and G. M. MacLeod.

This is a distinctive species in the lineage *N. ynezianum–N. incisum–N. costiferum–N. mediocostatum–N. montereyanum carmeloensis–N. montereyanum montereyanum–N. schencki,* occurring in the higher stratigraphic reaches of such related species as *N. costiferum,* especially in the upper Luisian and lower Mohnian Stages. Its most distincitve traits are its usually completely limbate sutures and, in well-preserved specimens, its thin keel. The sutures especiallly render it an aesthetically pleasing appearance. It may be distinguished from all the other species of this lineage by the continuation of its limbate sutures to the periphery of the test without their becoming notably incised, and its lack of costae, either of the *N. costiferum* or *N. mediocostatum* kind.

This species is named for Donald W. Weaver in honor of his extensive contributions to the Tertiary foraminiferology, stratigraphic paleontology, and geology of the California Coast Ranges over a period of many years. This is also the "*Nonion costiferum* Cushman var. *weaveri* n. var." of Table 6 which, following completion of the Table 6 check list, was compared with the holotype and proved to be conspecific.

Family BULIMINIDAE
Subfamily VIRGULININAE
Genus *Bolivina* d'Orbigny 1839
Bolivina brevior brevior Cushman

Bolivina brevior Cushman, 1925, p. 31, 32, pl. 5, figs. 8a, b; Kleinpell, 1938, p. 268, pl. VI, fig. 5.

Bolivina brevior dunlapi Kleinpell

Bolivina dunlapi Kleinpell, 1938, p. 271, pl. XV, fig. 2.

Except for its costae, this small form is very similar to *Bolivina brevior. Bolivina dunlapi* Kleinpell is here reinterpreted as the costate subspecies of *B. brevior,* an evolutionary relationship possibly parallel with those of *B. advena ornata* Cushman and *B. floridana striatella* Cushman (see below) to *B. advena advena* and *B. floridana floridana,* respectively. Kleinpell (1938, p. 271) compared *B. dunlapi* to *B. californica,* but it is markedly more flattened than the latter.

Bolivina churchi almgreni
Kleinpell and Tipton, n. sp., n. subsp.

Subspecies differing from *B. churchi churchi* (see below) in having two sharp costae near median line, extending approximately one-half the length of the test.

Holotype (Pl. IX, Fig. 14a, b), from H. Hoots' upper Modelo Formation, Murietta and Roblar Streets, north of Valley Vista Blvd., south of Ventura Blvd., Los Angles County, in diatomite. RMK Coll. No. 105.
Cotype (Pl. IX, Fig. 13), same sample as holotype. RMK Coll. No. 106.

This subspecies is named for A. A. Almgren, who collected the sample from which the type specimens were selected.

Bolivina churchi churchi
Kleinpell and Tipton, n. sp., n. subsp.

Bolivina cuneiformis Kleinpell, R. M. Kleinpell *in* Woodring, Stewart, and Richards, 1940, p. 127-128.

Test moderately compressed, broad for the genus, greatest breadth near apertural end; wall finely perforate; periphery acute with narrow limbate keel; chambers low, distinct, but only slightly inflated, except near median line, where they are drastically inflated to form double rows of nobs; sutures slightly depressed and limbate, almost sigmoid near the median line on each chamber's nob, then curving downward to periphery; aperture a small narrow loop rising from base of apertural face.

Holotype (Pl. IX, Fig. 11), from basal McLure Shale Member, east side of Sulphur Canyon, 1 mi. (1.6 km) south of Pipeline West, Kings County, California (see Church, 1972). RMK Coll. No. 107.
Cotype (Pl. IX, fig. 12), same sample as holotype. RMK Coll. No. 108.

This species and subspecies are named in honor of Clifford C. Church, who collected and discussed (Church, 1972) the assemblage from which the type specimens were selected.

Bolivina floridana floridana Cushman

Bolivina floridana Cushman, Cushman and Parker, 1931, p. 9. pl. 2, fig. 2; Kleinpell, 1938, p. 271, pl. XII, fig. 1.

Bolivina floridana striatella Cushman

Bolivina advena Cushman var. *striatella* Cushman, 1925, p. 30, pl. 5, figs. 3a, b; Kleinpell, 1938, p. 266, pl. XV, fig. 1.

When well-preserved specimens of so-callled *B. advena striatella* are examined, the sutures are clearly seen to be crenulated, as in *B. floridana,* rather than straight, as in *B. advena*. Except for the striae, *striatella* is very similar to *B. floridana,* and is here interpreted as a subspecies of the latter. A. D. Warren points out, furthermore, that there is a serious question as to whether the California species that has been referred to *B. floridana* is actually synonymous with the typical *B. floridana* so named from the Gulf Coast-Atlantic seaboard area.

Bolivina goudkoffi Rankin

Bolivina goudkoffi Rankin, MS *in* Cushman and Kleinpell, 1934, p. 22, pl. 4, fig. 4; Kleinpell, 1938, p. 272.
Bolivina granti Rankin, Pierce (in part), 1956, p. 1307.

This species differs from *B. goudkoffi* especially in having a somewhat rhomboid transverse section, so that the periphery tends to be rather sharp instead of broadly rounded. Of the two, it is the more closely related to *B. hughesi* Cushman. The figures of paratypes of the two species (pl. 4, fig. 3, of *B. granti,* and fig. 5, of *B. goudkoffi*) in Cushman and Kleinpell (1934) are misleading and questionable as to conspecificity with the two holotypes figured, though they reflect the possibility that *B. granti* may be so closely related to *B. hughesi* as to be a subspecies of the latter with principally sutural differences (see Cushman and Kleinpell, 1934, p. 22) as the most distinctive features. The sutures of *B. granti* are less depressed than those of *B. hughesi* (Cushman).

Bolivina hootsi Rankin

Bolivina hootsi Rankin, MS *in* Cushman and Kleinpell, 1934, p. 18, 21, pl. 4, figs. 1a, b; Kleinpell, 1938, p. 273.

Originally described as "related to *Bolivina seminuda* Cushman" (Cushman and Kleinpell, 1934, p. 21), this form is clearly in the *B. seminuda* lineage as are the related species *B. foraminata* and *B. rankini*.

Bolivina hughesi Cushman

Bolivina hughesi Cushman, 1926b, p. 43-44, pl. 6, figs. 4a, b; Kleinpell, 1938, p. 273.
Bolivina benedictensis Pierce, 1956, p. 1307, pl. 142, fig. 9.
Not *Bolivina hughesi* Cushman, of Pierce, 1956, p. 1307-1308, pl. 143, fig. 2.

Original Holotype (Cushman Coll. No. 5584); *not* hypotype (USC No. 3619).

That the original holotype of this species, as figured by Cushman in 1926 with its "whole test twisted strongly, periphery rounded . . . sutures distinct, later ones distinctly depressed" could be the identical specimen (see Cushman and Todd, 1937) referred to as the holotype by Pierce (1956, p. 1307) through the figure and description "by Ruth Todd in Cushman's *A Monography of the Foraminiferal Subfamily Virgulininae,* p. 117, pl. 14, fig. 7a, b" is quite inconceivable; though how the mischance of such a switch in specimens may have occurred is far from clear. A possible clue may be in the supplement to Cushman's original description, "There is a tendency to develop one or more lobes near the inner margin on the lower line of each chamber, a feature which becomes stronger in some specimens at different horizons in well samples, and making the species tend toward the conditions seen in *Bolivina decussata* H. B. Brady." Yet there is no such "tendency" to be found in his preceding description of, nor in his accompanying picture of, the type specimen, even though there is a faintly discernible trace of such a lobing on the *outer, not* the "inner margin," of the tiny second or third chamber in the very early stage of the holotype figured by Cushman in 1926.

All of the synonyms of his new species *B. benedictensis,* as listed by Pierce (1956, p. 1307), are instead conspecific with Cushman's holotype of *B. hughesi* as described and figured in 1926; so that it would seem that "*B. benedictensis*" becomes a *nomen nodum.* None of the synonymic forms listed by Pierce (1956, p. 1307-1308) under "*B. hughesi* Cushman" are conspecific with his USC Hypotype 3619 as figured (pl. 143, fig. 2), which appears to be a new and distinct species, as will be formally noted below under *B. wissleri* n. sp. The senior writer can only add that, to the best of his knowledge, all the cotypes of *B. hughesi* sent to West Coast students in the late 1920s were in every way conspecifical with the holotype as figured by Cushman (1926b), pl. 6, figs. 4a, b.

Bolivina mulleri Kleinpell and Tipton n. sp.

Test medium-sized for the genus, only slightly compressed, nearly conical in overall test form, broadest near apertural end; periphery subacute; chambers indistinct, not at all inflated; sutures slightly curved, flush, obscured by ornamentation; test finely perforate, covered with numerous high, sharp costae, extending nearly to top of test; aperture an elongate opening on inner face of final chamber.

Holotype (Pl. VI, Figs. 7a, b), from sample SBC 37, Naples Bluff section, Santa Barbara County, *Siphogenerina reedi* Zone, lower Luisian, Monterey Shale.

Paratype (Pl. VI, Figs. 6a, b), from sample SBC 37, Naples Bluff section, Santa Barbara County, *Siphogenerind reedi* Zone, lower Luisian, Monterey Shale.

This species is distinguished from other California Tertiary bolivinids by its high sharp costae, which extend nearly to the final chamber. In these respects, it differs from *B. conica* Cushman. The costae differ from those of *B. interjuncta* Cushman in that they rarely anastomose. It is finely perforate, thus unlike *B. costata,* and has more abundant costae than the latter.

Bolivina mulleri is named in honor of the late Siemon W. Muller, professor at Stanford University, stratigraphic paleontologist, friend and associate of the senior writer, and teacher of the junior writer.

Bolivina sinuata alisoensis Cushman and Adams

Bolivina sinuata Galloway and Wissler var. *alisoensis* Cushman and Adams, 1935, p. 19-20, pl. 3, figs. 5a, b; Kleinpell, 1938, p. 282.

Not *Bolivina sinuata* Galloway and Wissler var. *alisoensis* Cushman and Adams, of Pierce, 1956, p. 1307 (in part?).

Not *Bolivina sinuata alisoensis* Woodring, Bramlette, and Kleinpell, 1936, p. 141, of Pierce, 1956, p. 1307, *as synonymized by Pierce.*

This subspecies of *Bolivina sinuata* is much closer to the species *sensu stricto* of Galloway and Wissler than to *B. wissleri* n. sp., with which, under the heading of "*Bolivina hughesi* Cushman," it has been considered conspecific by Pierce (1956, p. 1307-1308). Though it seems to appear for the first time at horizons in the lower Mohnian Stage, its upward range in the column is not so clearly defined. Forms still very close to it, and perhaps conspecific with it, have been found in strata apparently as young as early Pliocene.

Bolivina wissleri Kleinpell and Tipton, n. sp.

Bolivina hughesi Pierce (not Cushman) 1956, p. 1307-1308, pl. 143, fig. 2.

Holotype RMK 103: From Newport (Fig. 1) sample TM 14.

Cotype RMK 104: From Newport (Fig. 1) sample NEW 68.

Paratype USC No. 3619.

"Test elongate, about 2 or 3 times as long as broad, slightly compressed," roundish "in cross section, sides tapering, somewhat twisted about the longitudinal axis" although not notably so as in the very twisted test of *B. hughesi* Cushman; "initial end rounded in the megalospheric forms and bluntly pointed in the microspheric forms; chambers biserial, slightly inflated, about 7 to 10 pairs, the last three pairs making up about half of the test, increasing in size uniformly; sutures" somewhat "compressed, crenulated and tend to develop a lobate character, making an angle of about 80 or 90 degrees with the periphery; wall finely perforated, four longitudinal rows of weak nodular projections on the later portion of the test; aperture elongate slit bordered by a narrow lip on the inner face of the ultimate chamber."

This species is most closely related to *B. floridana* Cushman of Cushman and Parker (1931, p. 9, pl. 2, fig. 2) and of Kleinpell (1938, pl. XII, fig. 1). Although both these West Coast species appear to be in the lineage of the Pleistocene end-form which is *B. sinuata* Galloway and Wissler, the crenualation of the sutures is much more faint and the ornamentation far less pronounced than in either *B. sinuata* or in *B. sinuata alisoensis* Cushman and Adams. Nevertheless, the crenulated appearance of the sutures is definite in both *B. floridana* and *B. wissleri,* rather than being pseudocrenulation such as sometimes develops in poorly preserved specimens of bolivines owing to differential weathering of chamber walls at their bases.

B. wissleri is named in honor of Stanley G. Wissler whose work on the Pliocene and upper Miocene foraminiferal sequences in southern California, especially in the subsurface of the Los Angeles basin, is still basic.

Subfamily UVIGERININAE

Genus *Galliherina* Kleinpell and Tipton n. gen.

Genotype *Galliherina uvigerinaformis* (Cushman and Kleinpell)

Test elongate, generally rounded in section; triserial; chambers inflated; wall calcareous, finely perforate; aperture narrow and elongate, somewhat curved and with a slight lip.

In addition to the genotype, this genus includes *G. delreyensis* (Cushman and Galliher). The aperture and uvigerine plan of growth of these forms suggest that the lineage may have arisen from uvigerinellid stock rather than from *Bulimina*. It seems sufficiently distinct to designate as a separate genus.

The genus is named in honor of the late E. W. Galliher for his contributions to the micropaleontology of the California Miocene.

Galliherina delreyensis (Cushman and Galliher)

Bulimina delreyensis Cushman and Galliher, 1934, p. 25, pl. 4, figs. 8a, b; Kleinpell, 1938, p. 253; Warren, 1972, p.29.

Not *Bulimina delreyensis* Cushman and Galliher, of Woodring and Bramlette, 1950, p. 22.

Not *Bulimina delreyensis* Cushman and Galliher, of Patsy Beckstead Smith, 1960, p. 482, table 3, p. 469.

Galliherina uvigerinaformis doanei Kleinpell and Tipton n. subsp.

Subspecies differing from the typical in having very subdued striae. Some specimens are almost entirely smooth.

Holotype: Pl. XI, Fig. 1, Miocene formation, Newport Lagoon, sample NEW 68, lower Mohnian Stage. RMK Coll. No. 111.

Paratype: Pl. XI, Fig. 2, Miocene formation, Newport Lagoon, sample NEW 51, lower Mohnian Stage. RMK Coll. No. 112.

This subspecies is named in honor of the late George Doane, a leading pioneer micropaleontologist in the Tertiary of the California Coast Ranges.

Gallliherina uvigerinaformis uvigerinaformis (Cushman and Kleinpell)

Bulimina uvigerinaformis Cushman and Kleinpell, 1934, p. 5-9, pl. 1, figs. 14a, b; Kleinpell, 1938, p. 259-260.

Galliherina uvigerinaformis warreni Kleinpell and Tipton n. subsp.

Bulimina delreyensis Woodring and Bramlette (not Cushman and Galliher), 1950, p. 22.
Bulimina delreyensis Patsy Beckstead Smith (not Cushman and Galliher), 1960, p. 482, table 3, p. 469.

Subspecies differs from the typical in having more numerous costae that are somewhat more continuous across the sutures.

Holotype: Pl. XI, Fig. 6 Monterey Shale, Newport Lagoon, sample NEW 63, lower Mohnian Stage. RMK Coll. No. 113.
Cotype: Pl. XI, Fig. 5, Monterey Shale, Newport Lagoon, sample NEW 63, lower Mohnian Stage. RMK Coll. No. 114.

This subspecies is named for A. D. Warren, from whose recent collections at Newport Lagoon the type specimens were selected (see Warren, 1972).

Genus *Uvigerina d' Orbigny 1828*
Uvigerina obesa obesa (Cushman)

Uvigerina (Uvigerinella) obesa Cushman, 1926c, p. 59, pl. 8, figs. 3a-c, 7.
Uvigerinella obesa Cushman, Cushman and Parker, 1931, p. 10, pl. 2, fig. 4; Kleinpell, 1938, p. 290, pl. IX, fig. 15.

The genus *Uvigerinella* Cushman, as typified by *U. californica* (genotype), its subspecies, and *U. nudocostata* Cushman, has an aperture distinct from that of *Uvigerina,* being loop-shaped with a raised collarlike rim, rather than being entirely raised and terminal, with a neck and lip, as in *Uvigerina*. The apertures of the species *obesa* and its relatives *obesa impolita* and *sparsicostata* (see below) are closer to those of *Uvigerina* than of typical *Uvigerinella*. In addition, *obesa, obesa impolita,* and *sparsicostata* are apparently not part of the *Uvigerinella californica* lineage, which appears cryptogenetically in lower Zemorrian strata. They are instead inferred to be descendants of the endemic West Coast uvigerinid stock of the Eocene and Oligocene, probably a costate form such as *Uvigerina gardnerae* Cushman or *Uvigerina kernensis* Barbat and von Estorff (see Tipton et al, 1973, p. 39, fig. 11).

For these reasons, the species and subspecies *obesa, obesa impolita,* and *sparsicostata* are removed from *Uvigerinella* and placed in *Uvigerina*.

Uvigerina obesa impolita (Cushman and Laiming)

Uvigerinella obesa Cushman var. *impolita* Cushman and Laiming, 1931, p. 111, pl. 12, figs. 11ab; Kleinpell, 1938, p. 291, pl. VII, fig. 8.

Uvigerina sparsicostata (Cushman and Laiming)

Uvigerinella sparsicostata Cushman and Laiming, 1931, p. 112, pl. 12, figs. 12a, b; Kleinpell, 1938, p. 292; Tipton, Kleinpell, and Weaver, 1973, p. 55, pl. 6, figs. 5a, b; 9a, b.

Uvigerina subperegrina Cushman and Kleinpell

Uvigerina subperegrina Cushman and Kleinpell, 1934, p. 12, pl. 2, figs. 9-11; Kleinpell, 1938, p. 298-299.
Uvigerinella aff. *U. obesa* Cushman, Kleinpell, 1938, p. 291, pl. XV, fig. 7.

The Luisian forms formerly referred to "*Uvigerinella* aff. *U. obesa*" are conspecific with *Uvigerina subperegrina*. Thus, the latter seems to have evolved from *U. obesa* during Luisian time, rather than in the Mohnian, as previously supposed.

Genus *Siphogenerina* Schlumberger 1883
Siphogenerina obispoensis Kleinpell and Tipton, n. sp.

Test medium-sized for the genus; very elongate, slender; nearly all of test uniserial after very short, indistinct, early triserial stage; chambers not at all inflated; sutures straight; wall ornamented by 11 or 12 sharp, high, straight costae that are not at all affected by sutures. Costae extend to top of test; aperture terminal with short neck and lip.

Holotype: Pl. XVI, Fig. 20, from EC 14, Edwards Canyon section of Highland homocline area, San Luis Obispo County, *Siphogenerina reedi* Zone, lower Luisian Stage. RMK Coll. No. 115.
Paratype: Pl. XVI, Fig. 18, from EC 13, Edwards Canyon section, as above. RMK Coll. No. 116.
Paratype: Pl. XVI, Fig. 19, from EC 14, as above. RMK Coll. No. 117.

This species is distinguished by its slender test and high straight costae unaffected by the sutures, which are also straight. It is smaller and more slender, with thinner costae, than *S. collomi* and has fewer costae than *S. reedi*. Its costae are higher and fewer than those of *S. branneri*, with no tendency to anastomose.

Its slender form and straight sutures suggest that *S. obispoensis* may be a descendant of the Saucesian *S. tenua*.

New Genera of West Coast Siphogenerinid Foraminifera

Tipton et al (1973, p. 37-39, fig. 11) pointed out that the siphogenerine Foraminifera of the West Coast Neogene are a polyphyletic group, whether they are assigned to *Siphogenerina* Schlumberger or to *Rectuvigerina* Matthews. They were said to comprise at least five separate lineages that arose during early Zemorrian time from native lineages of *Uvigerina*. Only one of these five lineages attaining the siphogenerine grade is interpreted to include the genotype of *Siphogenerina (S. raphanus–S. costata;* see Matthews, 1945; Bandy, 1952); that is, the lineage of *S. transversa, S. nodifera, S. kleinpelli,* etc. (see fig. 11, Tipton et al, 1973), apparently descendants of *Uvigerina gallowayi* Cushman and *Uvigerina gesteri* Barbat and von Estorff. Since this lineage apparently is most closely related to the genotype, it retains the name *Siphogenerina*.

The remaining four lineages were less sequential, and were all extinct by mid-Saucesian time. Nonetheless, they are phyletically and biochronologically significant and should be designated taxonomically to show their varying origins. They are here designated as new genera *Atwillina, Beckina, Estorffina,* and *Laimingina*.

Genus *Atwillina* Kleinpell and Tipton, n. gen.
Genotype *Atwillina pseudococoaensis* (Cushman and Kleinpell)

Test elongate, slender, circular in section; triserial throughout much of test length, with a short final uniserial stage; wall calcareous, finely perforate; suture distinct, chambers somewhat inflated; aperture in the adult terminal, circular, with a slight neck.

This genus descends from the Refugian *Uvigerina atwilli* Cushman and Simonson, and retains the uvigerine appearance of its ancestor except for the short late uniserial stage. As far as is known, it became extinct during late Zemorrian time.

Atwillina pseudococoaensis (Cushman and Kleinpell)

Siphogenerina pseudococoaensis Cushman and Kleinpell, 1934, p. 13, pl. 2, figs. 14a, b; Tipton, Kleinpell, and Weaver, 1973, p. 59, pl. 7, figs. 6a, b, 7a, b, 8a, b.

Genus *Beckina* Kleinpell and Tipton, n. gen.
Genotype *Beckina hornadayi* Tipton and Kleinpell n. sp.

Test elongate, circular in section; triserial throughout roughly one-half of test, then irregularly uniserial; wall calcareous, finely perforate; chambers somewhat inflated, sutures distinct where not obscured by ornamentation; aperture terminal, circular, with a short neck.

Beckina is named in honor of R. Stanley Beck, noted California micropaleontologist. The ancestors of this lineage may have been part of one of the endemic lineages of *Uvigerina,* or the lineage may be truly cryptogenetic, an invader. Its relation to the lineage of *Siphogenerina multicostata* Cushman and Jarvis (genotype of *Rectuvigerina* Matthews) is poorly understood. *Beckina* apparently became extinct in mid-Saucesian time.

Beckina hornadayi Kleinpell and Tipton, n. sp.

Siphogenerina multicostata Cushman and Jarvis, Kleinpell, 1938, p. 302, pl. V, fig. 7.
Not *Siphogenerina multicostata* Cushman and Jarvis, Cushman, 1929, p. 95, pl. 13, fig. 38.

Test elongate, roughly three times as long as broad, reaching greatest breadth just above midsection of test; circular in section; triserial through roughly one-half of test, then irregularly uniserial; chambers somewhat inflated, distinct; test ornamented by numerous (approximately 18) moderately high, broad costae, irregular but generally passing from base of test to top without interruption at sutures; aperture terminal, circular, with a short neck.

Holotype RMK 118. Pl. XIII, Fig. 3. Loc. RMK 1938, table XII, 23; Temblor Formation ("Vaqueros"), Sunset Valley District, California.
Cotype: *"Siphogenerina multicostata"* of Kleinpell, 1938, pl. V, fig. 7, no. 497203, U. S. National Museum, lower Saucesian, Temblor Formation ("Vaqueros"), Sunset Valley District, California. Loc. RMK 1938, table XII, 23, p. 60, 302.
Paratype RMK 119. Pl. XIII, Fig. 12. From Ellwood oil field, Doty 4 well at 3,162 ft of depth, subsurface Rincon Shale, lower member, 4 ft above top of Vaqueros Sandstone.

This species is distinct from *Siphogenerina multicostata* Cushman and Jarvis in having fewer costae and in having costae that, though irregular, generally pass from base to top of test uninterruptedly. The costae of *S. multicostata* are typically confined to individual chambers.

In addition to the Sunset Valley occurrences, *Beckina hornadayi* is found at Los Sauces Creek, Ventura County (Rincon Shale, upper Zemorrian), and in the upper Galloway Formation near Point Arena (lower Saucesian). Note that in Kleinpell (1938, p. 302) the reference to the Cuyama Gorge Vaqueros occurrence should have been to Table XII, *not* Table XIV as shown there.

This species is named in honor of Gordon Hornaday, industrial consultant and long-time curator of the micropaleontological collections at the University of California at Berkeley, whose consistently excellent contributions to West Coast foraminiferology have been manifold.

Beckina fulmeri Kleinpell and Tipton, n. sp.

Siphogenerina sp. Tipton, Kleinpell, and Weaver, 1973, p. 59, pl. 7, figs. 9a, b; 10a, b.

Test at maturity two to two-and-one-half times as long as thick in diameter; triserial through early half of mature test, usually with two uniserial end-chambers that place it in a siphogenerine evolutionary grade; wall finely perforate; chambers somewhat inflated, sutures distinct and somewhat broadly depressed, rendering the periphery of the test a slightly but definitely lobulate appearance; surface ornamented by about 14 to 17 or so very fine and low thin costae, some of them discontinuous in the very earliest whorl or two but continuous across the sutures from chamber-whorl to chamber-whorl and from chamber to chamber throughout most of the test, the costae nevertheless following the inflated chamber surfaces into and again out of the broadly depressed sutures; aperture terminal and central with a slight neck and lip.

Holotype: Siphogenerina sp. of Tipton, Kleinpell, and Weaver, 1973, U. C. Mus. Paleon. No. 47528 (*not* U. C. Mus. Paleon. "No. 37428" as mistakenly recorded in Tipton, Kleinpell, and Weaver, 1974, legend for plate 1 opposite p. 150), p. 59, pl. 7, figs. 10a, b. U. C. Loc D-4317, Texaco PUP 1 well, drilling depth 3,903 ft.

This is a widespread Oligocene species in the subsurface of the southern San Joaquin Valley, though, as with Oligocene foraminifers from this region generally, it is seldom found in surface sections except as indefinitely identifiable poorly preserved forms. It is apparently most closely related to *Beckina hornadayi* n. sp. It may be, although not clearly so, descended from *"Uvigerina* sp." of Tipton, Kleinpell, and Weaver 1973, p. 57, pl. 7, fig. 4a, b, from the subsurface Kreyenhagen Shale, a late Eocene and Oligocene(?) species which occurs sporadically in the subsurface of the San Joaquin Valley, California.

The species is named for Charles V. Fulmer in honor of his significant contributions to the faunal relations pertaining between the late Eocene, Oligocene, and early Miocene formations of California and the Pacific Northwest.

Genus *Estorffina* Kleinpell and Tipton, n. gen.
Genotype *Estorffina mayi* (Cushman and Parker)

Test elongate, slender, circular in section; triserial and in some specimens biserial chamber arrangement precedes final uniserial stage that makes up approximately one-half of test length; wall calcareous, finely perforate; chambers inflated, distinct; aperture circular, terminal, with short neck and slight lip.

This genus apparently descends from some early form of the *Uvigerina obesa* (Cushman) stock, perhaps via *U. obesa impolita* (Cushman and Laiming). *Estorffina* is named in honor of F. E. von Estorff for his most valuable and critical studies of foraminifers in this part of the California column.

Estorffina mayi (Cushman and Parker)

Siphogenerina mayi Cushman and Parker, 1931, p. 10-11, pl. 2, figs. 7a, b; Fairchild, Wesendunk and Weaver, 1969, p. 57, pl. 13, figs. 3a, b; Tipton, Kleinpell, and Weaver, 1973, p. 58, pl. 7, figs. 11a, b, 12a, b.

Since several types of ambiguities are involved in the original localities of reported occurrences of *"Siphogenerina" mayi* (see Kleinpell, 1938, p. 292, 302) and the species still is not known from the horizon of the "Valvulineria Silt," the writers have designated two new paratypes while retaining the original name of the species for the conspecific synonym from the lower Temblor "Siphogenerina Silt" locality wherein the holotype of the species derives.

Tipton et al (1973) discussed the similarities of *"S." mayi* and *S. fredsmithi* Garret, and placed the latter questionably in synonymy with *mayi*. Lamb (1964, p. 468-470) synonymized the two asserting that "the distinction is very slight, and we are probably dealing with interbreeding populations of the same species interval."

In view of the change in genus involved, the writers decline to synonymize *Siphogenerina fredsmithi* with *Estorffina mayi* (the latter has priority) until the phyletic history of the Gulf Coast form can be investigated. The two species may be products of parallel evolution in two separate lineages.

Paratype RMK 120. Pl. XIII, Figs. a, b. Loc. RMK 1938, table XI, 17; middle Santos (upper "B Shale" of Mahoney) Shale Member, type Temblor Formation of Carneros Creek (see also Curran, 1943). Upper Zemorrian.

Paratype UC 47477: *"Siphogenerina mayi"* of Tipton, Kleinpell, and Weaver 1973, p. 58, pl. 7, figs. 12a, b. U.C. loc. D-4385 (T. H. Purman HUB Cymric 1 well, subsurface upper "lower Santos shale," Temblor Formation, depth 5,338 ft). Upper Zemorrian.

Genus *Laimingina* Kleinpell and Tipton, n. gen.
Genotype *Laimingina smithi* (Kleinpell)

Test elongate, circular in section; triserial through much of test length, with short final uniserial stage; wall calcareous, finely perforate; chambers somewhat inflated; sutures distinct to obscured in highly ornamented forms; aperture central, terminal, with short neck.

This genus represents the development of a uniserial plan of growth from the lineage of *Uvigerina sparsicostata* (Cushman and Laiming). It is named in honor of Boris Laiming, co-author of the ancestral species.

Laimingina smithi (Kleinpell)

Siphogenerina Smith, W. M. Smith, 1930, p. 7.
Siphogenerina smithi Kleinpell, 1938, p. 304, pl. VI, figs. 1, 2.
Holotype No. 6109 Stanford Univ. Micropaleo. Type Coll. Loc. Doty 4 well, depth 3,162 ft.
Cotype RMK 121, plate XIV, figs. 7a, b. Loc. lower member Rincon Shale, Doty 4 well, Ellwood oil field, depth 3,162 ft, 4 ft above top of Vaqueros sandstone.
Paratype No. 6110 Stanford Univ. Micropaleo. Type Coll. Loc. RMK 1938, table VI, CM-64; pl. VI fig. 2.

Location Tables of Cited Foraminifera Samples in California

Table 1. Check List of Foraminifera from Temblor-Monterey Sequence, Indian Creek, Highland Homocline, San Luis Obispo County, California

STAGE	RELIZIAN							LUISIAN																									
ZONE	S. branneri							S. reedi									S. nuciformis										S. collomi					
SPECIES / SAMPLE NUMBER	CS†	HM-1	HM-2	HM-3	HM-3A	HM-4	HM-4A	HM-5	HM-6	HM-7	HM-8	HM-9	HM-10	HM-11	HM-12	HM-13	HM-14	HM-15	HM-16	HM-17	HM-18	HM-19	HM-20	HM-21	HM-22	HM-23	HM-24	HM-25	HM-26	HM-27	HM-28	HM-29
Baggina californica Cushman	–	–	–	–	–	–	–	–	–	–	–	–	–	–	–	–	–	–	–	–	–	–	–	–	–	–	F	C	C	C	–	–
Baggina robusta robusta Kleinpell	–	–	–	–	–	–	–	–	R	–	–	F	F	F	F	A	–	–	–	–	–	–	–	–	–	–	–	–	–	–	–	–
Bolivina advena advena Cushman	C	F	C	F	R	R	–	R	R	R	A	–	–	–	–	–	R	R	R	R	R	R	R	R	–	F	R	R	R	R	F	–
Bolivina advena ornata Cushman	R	R	F	C	R	–	–	–	R	–	R	–	–	–	–	–	F	F	F	F	F	R	R	F	–	F	–	–	–	–	–	–
Bolivina adelaidana Cushman and Kleinpell	R	–	–	–	–	–	–	–	–	–	–	–	–	–	–	–	–	–	–	–	–	–	–	–	–	–	–	–	–	–	–	–
Bolivina brevior brevior Cushman	R	–	F	–	–	–	–	R	C	R	R	–	–	–	–	R	C	C	C	C	C	R	R	C	–	–	F	F	F	F	–	–
Bolivina brevior dunlapi Kleinpell	–	–	–	–	–	–	–	–	–	–	–	–	–	–	–	–	R	R	R	R	R	R	R	R	–	–	–	–	–	–	–	–
Bolivina californica Cushman	R	–	–	F	–	–	–	–	R	R	–	–	–	–	–	F	R	R	R	R	R	R	R	R	–	–	–	–	–	–	–	–
Bolivina conica Cushman	R	C	F	C	F	R	–	C	C	F	C	–	–	–	–	F	C	C	C	C	C	R	R	C	–	R	–	–	–	–	–	–
Bolivina floridana Cushman	F	F	C	C	R	R	–	C	C	–	C	–	–	–	–	–	–	–	–	–	–	–	–	–	–	–	–	R	R	R	–	–
Bolivina floridana striatella Cushman	–	–	–	–	–	–	–	–	F	–	R	–	–	–	–	–	R	R	R	R	R	R	R	R	R	R	–	R	R	R	–	–
Bolivina imbricata imbricata Cushman	F	–	R	–	–	–	–	–	–	–	F	–	–	–	–	–	A	A	A	A	A	R	R	A	–	C	–	–	–	–	–	–
Bolivina marginata marginata Cushman	C	–	–	–	–	–	–	–	C	–	C	–	–	–	–	–	F	F	F	F	F	R	R	F	–	R	–	–	–	–	–	–
Bolivina perrini Kleinpell	–	–	R	–	–	–	–	–	–	–	–	–	–	–	–	–	–	–	–	–	–	–	–	–	–	–	–	–	–	–	–	–
Bolivina salinasensis Kleinpell	–	–	–	–	–	–	–	R	R	–	R	–	–	–	–	–	–	–	–	–	–	–	–	–	–	–	–	–	–	–	–	–
Bolivina tumida tumida Cushman	C	–	–	–	–	–	–	C	A	C	C	–	–	–	–	–	R	R	R	R	R	R	R	R	–	R	–	–	–	–	–	–
Bulimina ovata d'Orbigny	–	–	–	–	–	–	–	R	–	–	–	–	–	–	–	–	–	–	–	–	–	–	–	–	–	–	–	–	–	–	–	F
Bulimina ovula d'Orbigny	–	–	–	–	–	–	–	–	–	–	–	–	–	–	–	R	–	–	–	–	–	–	–	–	–	–	–	–	–	–	–	–
Bulimina pseudoaffinis Kleinpell	–	–	–	–	–	–	R	–	–	–	–	–	–	–	–	–	–	–	–	–	–	–	–	–	–	–	–	–	–	–	–	–
Bulimina pseudotorta Cushman	–	–	–	–	–	–	–	–	–	–	–	A	A	A	A	C	–	–	–	–	–	–	–	–	–	–	–	R	–	–	–	–
Buliminella californica Cushman	R	–	–	–	–	–	–	–	–	–	–	–	–	–	–	–	–	–	–	–	–	–	–	–	–	–	–	–	–	–	–	–
Buliminella curta Cushman	R	–	–	–	–	–	–	–	–	–	–	–	–	–	–	–	–	–	–	–	–	–	–	–	–	–	–	–	–	–	–	–
Buliminella henryana Cushman and Kleinpell	R	–	–	–	–	–	–	–	–	–	–	–	–	–	–	–	–	–	–	–	–	–	–	–	–	–	–	–	–	–	–	–
Buliminella subfusiformis Cushman	C	–	–	–	–	–	–	–	–	–	–	–	–	–	–	–	R	R	R	R	R	R	R	R	–	–	–	–	–	–	C	–

Cancris baggi Cushman and Kleinpell	R	–	–	–	–	–	–	–	–	–	–	–	–	–	–	–	–	–	–	–	–	–	–	–	–	–	–	–	–	–	–	–
Cassidulina crassa d'Orbigny	–	–	–	–	–	–	–	–	R	–	–	–	–	–	–	–	–	–	–	–	–	–	–	–	–	–	–	–	–	–	–	–
Cassidulina laevigata carinata Cushman	–	–	–	–	–	–	–	R	–	–	–	–	–	–	–	–	–	–	–	–	–	–	–	–	–	–	–	–	–	–	–	–
Cassidulina margareta Karrer	R	–	–	–	–	–	–	–	–	–	–	–	–	–	–	–	–	–	–	–	–	–	–	–	–	–	–	–	–	–	–	–
Cassidulina panzana Kleinpell	–	–	–	–	–	–	–	–	R	–	–	–	–	–	–	–	–	–	–	–	–	–	–	–	–	–	–	R	R	R	–	–
Cassidulina pulchella d'Orbigny	R	–	–	–	–	–	–	–	–	–	–	–	–	–	–	–	–	–	–	–	–	–	–	–	–	–	–	–	–	–	–	–
Dentalina obliqua (Linné)	–	–	–	–	–	–	–	–	F	–	–	–	–	–	–	–	–	–	–	–	–	–	–	–	–	–	–	–	–	–	–	–
Dentalina roemeri Neugeboren	–	–	–	–	–	–	–	–	R	–	–	–	–	–	–	–	–	–	–	–	–	–	–	–	–	–	–	–	–	–	R	–
Globigerina spp.	F	C	C	F	C	F	R	C	F	F	A	–	–	–	–	–	R	R	R	R	R	R	R	R	–	–	–	–	–	–	F	–
Hemicristellaria beali (Cushman)	–	–	–	–	–	–	–	–	–	–	–	R	R	R	R	R	–	–	–	–	–	–	–	–	–	–	F	–	–	–	–	–
Lagena (?) cf. L. acuticosta Reuss	R	–	–	–	–	–	–	–	–	–	–	–	–	–	–	–	–	–	–	–	–	–	–	–	–	–	–	–	–	–	–	–
Nodogenerina advena advena (Cushman and Laiming)	R	–	–	–	–	–	–	–	–	–	–	–	–	–	–	R	–	–	–	–	–	–	–	–	–	–	–	–	–	–	–	–
Nonion costiferum (Cushman)	C	–	–	–	–	–	–	–	–	–	–	–	–	–	–	F	F	F	F	F	F	R	R	F	–	F	F	C	C	C	F	C
Nonion incisum (Cushman)	–	–	–	R	–	–	–	–	–	–	–	–	–	–	–	–	–	–	–	–	–	–	–	–	–	–	–	R	R	R	–	?
Nonion medio-costatum (Cushman)	F	–	–	–	–	–	–	–	–	–	–	–	–	–	–	–	–	–	–	–	–	–	–	–	–	–	–	R	R	R	R	?
Plectofrondicularia miocenica miocenica Cushman	F	–	–	–	–	–	–	–	–	–	–	–	–	–	–	–	–	–	–	–	–	–	–	–	–	–	–	–	–	–	–	–
Pulvinulinella subperuviana Cushman	–	–	R	–	–	–	–	–	–	–	–	–	–	–	–	–	–	–	–	–	–	–	–	–	–	–	–	R	R	R	–	–
Robulus hughesi Kleinpell	–	–	–	–	–	–	–	–	R	–	–	–	–	–	–	–	–	–	–	–	–	–	–	–	–	–	–	–	–	–	–	–
Robulus miocenicus (Chapman)	–	–	–	–	–	–	–	–	R	–	–	–	–	–	–	–	–	–	–	–	–	–	–	–	–	–	–	–	–	–	–	–
Robulus reedi Kleinpell	R	–	–	–	–	–	–	–	–	–	–	–	–	–	–	–	–	–	–	–	–	–	–	–	–	–	–	–	–	–	–	–
Robulus smileyi Kleinpell	–	–	–	–	–	–	–	R	R	–	–	–	–	–	–	–	F	F	F	F	F	R	R	F	–	F	R	–	–	–	–	–
Siphogenerina branneri (Bagg)	–	–	–	–	–	–	–	–	–	–	–	–	–	–	–	C	–	–	–	–	–	–	–	–	–	–	–	–	–	–	–	–
Siphogenerina collomi Cushman	–	–	–	–	–	–	–	–	–	–	–	–	–	–	–	–	–	–	–	–	–	–	–	–	–	–	F	–	–	–	–	–
Siphogenerina kleinpelli Cushman	–	–	–	–	–	–	–	–	R	–	–	–	–	–	–	R	–	–	–	–	–	–	–	–	–	–	F	–	–	–	–	–
Siphogenerina nuciformis Kleinpell	–	–	–	–	–	–	–	–	–	–	–	–	–	–	–	–	–	–	–	–	–	–	–	–	–	F	A	–	–	–	–	–
Siphogenerina reedi Cushman	–	–	–	–	–	–	–	–	F	–	–	–	–	–	–	F	–	–	–	–	–	–	–	–	–	–	F	–	–	–	–	–
Uvigerina obesa obesa (Cushman)	R	C	–	–	–	F	–	R	R	–	–	–	–	–	–	–	C	C	C	C	C	R	R	C	–	C	–	–	–	–	–	–
Uvigerina subperegrina Kleinpell	–	–	–	–	–	–	–	–	F	–	–	–	–	–	–	–	C	C	C	C	C	R	R	C	–	A	–	–	–	–	–	–
Uvigerinella californica californica Cushman	F	–	–	–	–	–	–	–	F	–	–	–	–	–	–	–	R	R	R	R	R	R	R	R	–	F	–	A	A	A	C	–
Uvigerinella californica ornata Cushman	–	–	–	–	–	–	–	R	R	–	–	–	–	–	–	–	F	F	F	F	F	R	R	F	–	R	–	–	R	–	–	–
Uvigerinella nudocostata Cushman	–	–	–	–	–	–	–	–	–	–	–	–	–	–	–	R	–	–	–	–	–	–	–	–	–	–	–	F	F	F	–	–
Valvulineria californica appressa Cushman	R	R	–	A	–	C	–	C	F	A	A	A	A	A	A	C	F	F	F	F	F	R	R	F	R	R	R	C	C	C	–	–
Valvulineria californica californica Cushman	–	–	–	–	–	–	–	F	R	F	R	R	R	R	R	–	A	A	A	A	A	R	R	A	C	C	A	F	F	F	–	–

Table 1. Continued

STAGE	RELIZIAN							LUISIAN																								
ZONE	S. branneri							S. reedi									S. nuciformis										S. collomi					
SPECIES / SAMPLE NUMBER	CS†	HM-1	HM-2	HM-3	HM-3A	HM-4	HM-4A	HM-5	HM-6	HM-7	HM-8	HM-9	HM-10	HM-11	HM-12	HM-13	HM-14	HM-15	HM-16	HM-17	HM-18	HM-19	HM-20	HM-21	HM-22	HM-23	HM-24	HM-25	HM-26	HM-27	HM-28	HM-29
Valvulineria californica obesa Cushman	F	—	—	C	—	R	—	C	C	A	A	C	C	C	C	F	F	F	F	F	F	R	R	F	F	F	C	F	F	F	R	—
Valvulineria depressa Cushman	A	C	C	A	F	C	F	C	A	C	A	C	C	C	C	F	—	—	—	—	—	—	—	—	—	—	—	—	—	—	—	—
Valvulineria miocenica Cushman	—	—	—	—	—	—	—	R	R	—	—	R	R	R	R	—	C	C	C	C	C	R	R	C	C	C	C	A	A	A	A	—
Valvulineria ornata Cushman	—	—	R	R	—	—	—	F	R	F	R	R	R	R	R	—	F	F	F	F	F	R	R	F	F	F	F	C	C	C	C	—
Valvulineria williami Kleinpell	R	—	—	—	—	—	—	—	—	—	—	—	—	—	—	—	—	—	—	—	—	—	—	—	—	—	—	—	—	—	—	—
Virgulina californiensis Cushman	F	—	—	F	—	—	—	C	—	F	—	—	—	—	—	—	R	R	R	R	R	R	R	R	—	—	—	—	—	—	—	—

† Clay shale east side of Indian Creek, stratigraphically below Stake No. 11.

All HM samples are from Highland homocline Temblor-Monterey formational sequence, collected from bed of Indian Creek or extensions of creek-bed strata onto western canyon wall, unless noted as coming from east side of Indian Creek where outcrops are stratigraphically less continuous. Samples collected by W. D. Kleinpell, M. N. Bramlette, K. E. Lohman, and R. M. Kleinpell.

Clay Shale —About 3,230 ft above granite-Temblor Formation contact, from dark argillaceous shale with buff, hard, conchoidal, calcareous concretions, east side of Indian Creek.

HM-1 —3,375 ft above granite-Temblor Formation contact.
HM-2 —3,388 ft above granite-Temblor Formation contact.
HM-3 —3,423 ft above granite-Temblor Formation contact.
HM-3A —Approximately same horizon as HM-3 but on east side of Indian Creek.
HM-4 —About 3,450 ft above granite-Temblor Formation contact; stratigraphic horizon indefinite, inaccurate other than within 10-ft thick interval of strata.
HM-4A —Approximately same horizon as HM-4 but on east side of Indian Creek.
HM-5 —3,450 ft above granite-Temblor Formation contact.
HM-6 —About 3,590 ft above granite-Temblor Formation contact.
HM-7 —3,601 ft above granite-Temblor Formation contact.
HM-8 —3,612 ft above granite-Temblor Formation contact.
HM-9 —3,629 ft above granite-Temblor Formation contact.
HM-10 —3,639 ft above granite-Temblor Formation contact.
HM-11 —3,659 ft above granite-Temblor Formation contact.
HM-12 —3,669 ft above granite-Temblor Formation contact.
HM-13 —3,698 ft above granite-Temblor Formation contact.
HM-14 —3,728 ft above granite-Temblor Formation contact.
HM-15 —3,854 ft above granite-Temblor Formation contact.
HM-16 —3,862 ft above granite-Temblor Formation contact.
HM-17 —4,078 ft above granite-Temblor Formation contact.
HM-18 —4,088 ft above granite-Temblor Formation contact.
HM-19 —4,098 ft above granite-Temblor Formation contact.
HM-20 —4,250 ft above granite-Temblor Formation contact.
HM-21 —4,313 ft above granite-Temblor Formation contact.
HM-22 —4,323 ft above granite-Temblor Formation contact.
HM-23 —4,355 ft above granite-Temblor Formation contact.
HM-24 —4,365 ft above granite-Temblor Formation contact.
HM-25 —4,417 ft above granite-Temblor Formation contact.
HM-26 —4,421 ft above granite-Temblor Formation contact.
HM-27 —4,441 ft above granite-Temblor Formation contact.
HM-28 —4,467 ft above granite-Temblor Formation contact.
HM-29 —4,535 ft above granite-Temblor Formation contact.

Table 2. Check List of Foraminifera from Temblor-Monterey Sequence, Quailwater Creek, Highland Homocline, San Luis Obispo County, California

STAGE	R**	LUISIAN													?
ZONE	S.b.*	S. reedi				S. nuciformis			S. collomi						?
SPECIES / SAMPLE NUMBER	HMC-30	HMC-31	HMC-32	HMC-33	HMC-34	HMC-35	HMC-36	HMC-37	HMC-38	HMC-39	HMC-40	HMC-41	HMC-43	HMC-44	? HMC-47
Baggina californica Cushman	—	—	—	—	—	—	—	—	—	C	—	—	—	—	—
Baggina robusta robusta Kleinpell	R	C	—	—	—	—	—	—	—	—	—	—	—	—	—
Bolivina advena advena Cushman	A	—	—	—	C	—	—	—	R	—	F	F	—	—	—
Bolivina advena ornata Cushman	—	—	—	R	F	—	—	—	—	—	R	—	—	—	—
Bolivina brevior brevior Cushman	F	R	—	F	R	—	—	—	—	R	C	—	—	—	—
Bolivina brevior dunlapi Kleinpell	—	R	R	—	—	—	—	—	—	—	—	—	—	—	—
Bolivina californica Cushman	C	—	—	—	—	—	—	—	—	—	—	—	—	—	—
Bolivina conica Cushman	—	—	—	R	C	—	R	C	F	F	R	—	—	—	—
Bolivina floridana Cushman	A	—	—	—	—	—	—	R	R	—	R	—	—	—	—
Bolivina floridana striatella (Cushman)	C	—	—	—	—	—	—	C	R	R	—	—	—	—	—
Bolivina imbricata imbricata Cushman	—	C	—	C	C	C	—	—	—	—	—	—	—	—	—
Bolivina marginata marginata Cushman	R	—	—	—	—	—	—	—	—	—	—	—	—	—	—
Bolivina perrini Kleinpell	F	—	—	—	—	—	—	—	—	—	—	—	—	—	—
Bolivina salinasensis Kleinpell	—	—	—	R	R	—	—	—	—	—	—	—	—	—	—
Bolivina tumida tumida Cushman	F	—	—	—	—	—	—	—	—	—	—	R	—	—	—
Bolivina cf. vaughani Natland	—	—	—	—	—	—	—	—	—	—	F	—	—	—	—
Bulimina montereyana delmonteensis Kleinpell	—	—	—	—	—	—	—	—	R	—	—	—	—	—	—
Bulimina ovula d'Orbigny	—	—	—	—	—	—	—	—	—	R	—	—	—	—	—
Bulimina pseudotorta Cushman	—	—	—	R	—	—	—	—	—	—	—	—	—	—	—
Buliminella brevior Cushman	—	—	—	R	—	—	—	—	—	—	—	—	—	R	—
Buliminella curta Cushman	—	—	—	—	F	—	—	—	R	—	A	—	—	A	—
Buliminella elegantissima (d'Orbigny)	—	—	—	—	—	—	—	—	—	—	—	—	—	—	A
Buliminella henryana Cushman and Kleinpell	—	—	—	—	—	—	—	—	—	—	R	—	—	F	—
Buliminella subfusiformis Cushman	R	—	R	R	C	—	—	—	—	—	C	—	—	—	—
Cancris brongniartii (d'Orbigny)	—	—	—	—	F	F	—	—	—	—	—	—	—	—	—
Cassidulina crassa d'Orbigny	—	—	—	—	—	—	—	—	—	—	R	—	—	—	—

Cassidulina laevigata carinata Cushman	—	—	—	F	—	—	—	—	—	—	—	—	—	—	—
Cassidulina margareta Karrer	R	—	—	—	—	—	—	—	—	—	—	—	—	—	—
Cassidulina panzana Kleinpell	—	—	—	—	—	—	—	—	R	—	—	—	—	—	—
Dentalina obliqua (Linné)	—	—	—	—	R	F	—	—	—	—	—	—	—	—	—
Dentalina roemeri Neugeboren	R	—	—	—	—	—	—	—	—	—	—	—	—	—	—
Elphidium (?) sp.	—	—	—	—	—	—	—	—	—	—	—	—	—	—	A
Globigerina spp.	F	—	—	—	R	—	—	—	—	—	—	—	—	—	—
Hemicristellaria beali (Cushman)	—	F	—	C	—	—	—	—	—	—	—	—	—	—	—
Marginulina subbullata Hantken?	R	R	—	—	—	—	—	—	—	—	—	—	—	—	—
Nodogenerina advena advena (Cushman and Laiming)	R	—	—	—	—	—	—	—	—	—	—	—	—	—	—
Nonion costiferum (Cushman)	—	F	C	A	A	F	—	C	C	C	A	C	R	C	—
Nonion incisum (Cushman)	—	—	—	—	—	—	—	—	—	—	—	—	R	F	—
Nonion medio-costatum (Cushman)	—	—	—	—	—	—	—	—	—	—	—	R	R	A	—
Nonionella miocenica Cushman	R	—	—	—	—	—	—	—	—	—	—	—	—	—	—
Pulvinulinella subperuviana Cushman	F	—	—	C	—	—	—	—	—	—	—	—	—	—	—
Robulus miocenicus (Chapman)	—	—	—	—	R	C	—	—	—	—	—	—	—	—	—
Robulus smileyi Kleinpell	R	F	—	C	A	A	—	—	F	R	R	—	—	—	—
Siphogenerina branneri (Bagg)	—	F	—	—	—	—	—	—	—	—	—	—	—	—	—
Siphogenerina collomi Cushman	—	—	—	—	—	—	—	—	C	—	—	—	—	—	—
Siphogenerina kleinpelli Cushman	—	R	—	—	—	—	—	—	R	—	—	—	—	—	—
Siphogenerina nuciformis Kleinpell	—	—	—	—	—	C	—	—	R	—	—	—	—	—	—
Siphogenerina reedi Cushman	—	—	—	—	—	—	—	—	R	—	—	—	—	—	—
Uvigerina obesa obesa (Cushman)	A	—	—	—	A	F	—	F	F	—	—	—	—	—	—
Uvigerina subperegrina Kleinpell	—	—	—	—	C	C	—	C	R	—	—	—	—	—	—
Uvigerinella californica californica Cushman	F	—	—	—	—	—	—	—	—	—	—	A	—	—	—
Uvigerinella californica ornata Cushman	R	—	—	—	R	—	—	—	—	—	—	R	—	—	—
Uvigerinella nudocostata Cushman	—	—	—	—	—	—	—	R	R	—	—	—	—	—	—
Valvulineria californica appressa Cushman	C	A	C	C	F	R	—	—	—	R	—	—	—	—	—
Valvulineria californica californica Cushman	—	C	—	C	A	A	C	A	A	C	—	—	—	—	—
Valvulineria californica obesa Cushman	C	C	F	F	F	F	F	F	F	R	—	—	—	R	—
Valvulineria depressa Cushman	C	F	—	F	F	—	—	—	—	—	—	—	—	—	—
Valvulineria miocenica Cushman	—	F	R	C	C	A	C	A	A	A	A	R	—	—	—
Valvulineria ornata Cushman	R	R	R	F	F	F	R	F	C	F	F	A	—	—	—
Virgulina californiensis Cushman	F	R	—	R	R	—	—	—	—	—	—	—	—	—	—

* S. branneri
** Relizian

Table 2. Continued

All HMC samples are from exposures along Quailwater Creek, from Temblor-Monterey formational sequence, where deformation has tilted them differentially into structure known as Highland homocline. Samples collected by W. D. Kleinpell, M. N. Bramlette, K. E. Lohman, and R. M. Kleinpell. Stratigraphic horizon used as datum for all but lowermost three samples is a conspicuous and gently northward-dipping sage-colored limestone ledge in shale of same sage-green color, stratigraphically above lowermost chert bed (flinty, white, 1 ft thick) exposed in bed of Quailwater Creek. At 384 ft stratigraphically above this limestone-ledge datum, a conspicuously white-weathering bed of greenish volcanic ash crops out in creek bed, 53 ft stratigraphically above lowest diatomite in this column and 43 ft stratigraphically above most prominent and lowest sandstone bed (massive, white, fine-grained, 3 ft thick) in column. Stratigraphic horizons of HMC foraminiferal samples may be recalculated (from footage figures given here) downward from either of these two higher horizons which (though a few sandy stringers occur a few feet lower and some diatomite beds still occur somewhat higher), here approximate most mappable contact between Monterey Shale and Santa Margarita Formation.

HMC-30— About 50 ft stratigraphically below HMC-31; in mudstone, east side of creek.

HMC-31— From shaly mudstone at base of a rather high bluff, upper (overlying) part of which exposes stream to fill (including angular shale pebbles). Stratigraphic horizon below HMC-32 indefinite as to footage, but several hundred feet below HMC-34.

HMC-32— From 6 to 7 ft stratigraphically below lowest chert bed (flinty, white, 1 ft thick) exposed in bed of Quailwater Creek, stratigraphically between samples HMC-31 and HMC-33 but footage indefinite because of poor continuity of outcrops. Creek makes fairly sharp turn toward west; exposures are in large bluff on east side of creek at bend. Buff carbonate ledges as well as chert bed present, dip steeply to north; dips become much flatter upstream but downstream, in paper-thin to platy siliceous shale, dips attain to vertical and in places more (i.e., overturned) with an observable swing in strike of these beds.

HMC-33— About 60 ft below HMC-34. Shaly marls with siliceous stringers; dips nearly vertical.

HMC-34— From light chrome-greenish porous "soapy" clay; immediately below conspicuous outcropping limestone bed or "ledge" used as datum horizon for samples higher in section, and 384 ft below conspicuous ash bed at or near Monterey-Santa Margarita contact downstream to north (thus, in turn, about 330 to 331 ft below lowest diatomite, 320 to 321 ft below base of lowest prominent massive 3 ft thick white fine-grained sandstone bed in section).

HMC-35— Sage-colored shaly clay, 5 ft above HMC-34.

HMC-36— Thin parting in extremely hard siliceous shale and white chert, 45 ft above HMC-34.

HMC-37— 88 ft above HMC-34.

HMC-38— 91 ft above HMC-34; brown calcareous clay shale partings in lighter colored platy to conchoidally fracturing siliceous shale and chert.

HMC-39— About 107 ft above HMC-34; brown clay shale in siliceous shale.

HMC-40— 132 ft above HMC-34.

HMC-41— 149 ft above HMC-34; foraminiferal molds only in overlying white laminated cherts.

HMC-43— 212 ft above HMC-34; soft sage-colored parting in cherts (below one calcareous bed).

HMC-44— 219 ft above HMC-34 and about 5 ft below lowest conspicuous bed of phosphatic oolites (sporbo) which recurs in more sandy laminae in sequence of flinty white cherts, paper-thin siliceous shales, a few soft white more calcareous(?) siliceous shales and perhaps diatomite stringers about 107 ft thick. Sequence is overlain by diatomite beginning from about 326 to 331 ft above HMC-34 and is best exposed on east side of creek.

HMC-47— 326 ft above HMC-34, in low bluffs of white diatomite on west side of creek, interbedded with thin stringers of fine-grained white sandstone. At 336 ft above HMC-34, 3 ft thick massive bed of fine-grained white sandstone appears, and at 384 ft above HMC-34 white-weathering greenish bed of volcanic ash appears and continues 50 ft downstream (about 30 ft thick) to area of poor outcrops, then white sandstone.

Table 3. Check List of Foraminifera from Temblor-Monterey Sequence, Edwards Canyon, Highland Homocline, San Luis Obispo County, California, Showing Sequence Through Type Section of Luisian Stage

STAGE	RELIZIAN			LUISIAN																	
ZONE	S. branneri			Siphogenerina reedi											S. nuciformis			S. collomi			
SUBZONE				lower								upper									
SPECIES / SAMPLE NUMBER	EC-1	EC-2	EC-3	EC-4	EC-5	EC-6	EC-7	EC-8	EC-10	EC-11	EC-12	EC-13	EC-14	EC-15	EC-16	EC-17	EC-18	EC-19	EC-20	EC-21	EC-22
Baggina californica Cushman	—	—	—	—	—	—	—	—	—	—	—	—	—	—	—	—	—	—	C	F	F
Baggina robusta robusta Kleinpell	—	—	—	F	—	R	R	R	—	F	—	—	—	—	—	—	—	—	—	R	—
Bolivina advena advena Cushman	—	—	A	—	—	R	—	—	R	F	F	C	F	F	F	R	—	A	A	A	A
Bolivina advena ornata Cushman	—	—	—	—	—	R	—	—	—	R	—	F	—	—	—	R	—	—	—	—	—
Bolivina brevior brevior Cushman	—	—	R	—	—	F	R	—	—	F	C	A	—	—	R	—	—	—	R	F	R
Bolivina brevior dunlapi Kleinpell	—	—	—	—	—	—	—	—	—	—	—	R	—	—	—	—	—	R	R	—	R
Bolivina californica Cushman	—	—	—	—	—	—	—	—	—	R	R	F	—	—	—	—	—	R	—	—	—
Bolivina conica Cushman	—	—	—	—	—	F	—	—	—	—	—	F	R	R	—	—	—	R	R	—	—
Bolivina floridana Cushman	—	—	R	—	—	R	—	—	—	—	—	R	—	—	—	—	—	R	R	F	A
Bolivina floridana striatella (Cushman)	—	—	F	—	—	R	—	—	—	F	R	R	—	—	—	—	—	F	C	A	R
Bolivina imbricata imbricata Cushman	—	R	F	—	—	A	C	F	F	C	F	C	—	C	C	C	—	—	—	—	—
Bolivina marginata marginata Cushman	—	F	C	R	—	F	R	R	—	R	R	F	—	F	R	—	—	—	—	—	—
Bolivina perrini Kleinpell	—	R	—	—	—	—	—	—	—	—	—	—	—	—	—	—	—	—	—	—	—
Bolivina salinasensis Kleinpell	—	—	—	—	—	—	—	—	—	—	—	F	—	—	—	—	—	—	—	—	—
Bolivina tumida tumida Cushman	—	—	—	—	—	F	—	—	—	—	—	—	—	—	—	—	—	—	—	R	—
Bolivina cf. vaughani Natland	—	—	—	—	—	—	—	—	—	—	—	—	—	R	—	—	—	—	R	—	—
Bulimina montereyana delmonteensis Kleinpell	—	—	—	—	—	—	—	—	—	—	—	F	F	—	—	R	—	—	—	—	—
Bulimina ovula d'Orbigny	—	—	—	R	—	—	—	—	—	—	—	—	—	—	—	—	—	—	—	—	—
Bulimina pseudotorta Cushman	—	—	—	A	F	—	—	—	—	—	—	—	—	—	—	R	—	—	—	—	—
Buliminella brevior Cushman	—	—	—	—	—	—	—	—	—	—	—	—	—	—	—	R	—	—	—	—	—
Buliminella californica Cushman	—	—	—	—	—	—	—	—	—	R	—	—	R	—	—	—	—	—	—	—	—
Buliminella curta Cushman	—	—	—	—	—	—	R	—	—	R	—	R	F	—	—	—	—	R	—	F	F
Buliminella henryana Cushman and Kleinpell	—	—	—	—	—	—	—	—	—	R	—	—	R	—	—	—	—	—	—	—	—
Buliminella subfusiformis Cushman	—	—	—	—	—	F	—	—	—	F	C	A	A	—	—	—	—	A	R	R	—
Cassidulina crassa d'Orbigny	—	—	—	—	—	—	—	—	—	—	—	—	—	—	—	—	—	—	—	R	—
Cassidulina laevigata carinata Cushman	—	—	—	—	—	—	—	—	—	—	—	—	—	—	—	—	—	—	F	—	—

Species																					
Cassidulina limbata Cushman and Hughes	—	—	—	—	—	—	—	—	—	—	—	—	—	—	—	—	—	—	R?	—	—
Cassidulina panzana Kleinpell	—	—	—	—	—	—	—	—	—	—	—	—	—	—	—	—	—	—	R	—	—
Cassidulina pulchella d'Orbigny	—	—	—	—	—	—	—	—	—	—	—	—	—	—	—	—	—	—	—	R	—
Dentalina obliqua (Linné)	—	—	—	—	—	R	R	F	—	C	—	C	—	—	—	—	R	R	C	—	—
Dentalina roemeri Neugeboren	—	—	—	—	—	—	—	F	—	F	—	—	—	—	—	—	—	—	—	—	—
Elphidium granti Kleinpell	R	—	—	—	—	—	—	—	—	—	—	—	—	—	—	—	—	—	—	—	—
Frondicularia foliacea Schwager	—	—	—	—	—	—	—	—	—	—	R	—	—	—	—	—	—	—	—	—	—
Globigerina spp.	—	—	—	—	—	—	—	—	—	—	F	R	—	—	—	F	R	—	—	—	—
Hemicristellaria beali (Cushman)	—	—	—	F	—	F	C	C	F	C	C	C	—	A	C	C	—	F	F	R	—
Marginulina subbullata Hantken?	—	—	—	—	—	—	R	R	—	R	R	—	—	—	—	—	—	—	—	—	—
Nodogenerina advena advena (Cushman and Laiming)	—	—	—	—	—	—	—	R	—	F	—	—	—	—	—	—	—	—	—	—	—
Nonion costiferum (Cushman)	—	F	—	F	—	F	—	C	R	C	R	C	C	R	R	R	—	—	—	F	F
Nonion incisum (Cushman)	—	R	—	—	—	—	—	—	—	—	—	—	R	—	—	—	—	—	—	—	—
Nonion medio-costatum (Cushman)	—	—	—	—	—	—	R	R	—	—	—	—	F	—	—	—	—	F	—	—	R
Plectofrondicularia californica Cushman and Stewart	—	—	—	—	—	—	—	—	R	—	—	—	—	—	—	—	—	—	—	—	—
Pulvinulinella subperuviana Cushman	—	—	—	—	—	—	—	—	—	R	F	—	—	—	—	—	—	R	F	—	R
Robulus cf. americanus (Cushman)	—	—	—	—	—	—	—	—	—	F	—	—	—	—	—	—	—	—	—	—	—
Robulus hughesi Kleinpell	—	—	—	—	—	—	—	F	—	—	—	—	—	—	—	—	—	—	—	—	—
Robulus miocenicus (Chapman)	—	—	—	—	—	—	—	—	R	R	R	C	—	—	—	—	—	—	—	—	—
Robulus smileyi Kleinpell	—	F	—	R	—	—	F	C	C	C	C	C	—	—	R	—	—	C	F	F	C
Siphogenerina branneri (Bagg)	—	—	—	—	—	—	C	C	—	A	—	—	R	F	—	—	—	—	—	—	—
Siphogenerina collomi Cushman	—	—	—	—	—	—	—	—	—	—	—	—	—	—	—	—	—	R	C	A	—
Siphogenerina kleinpelli Cushman	—	—	—	—	—	—	F	F	—	F	—	R	R	R	R	R	F	R	C	F	—
Siphogenerina nuciformis Kleinpell	—	—	—	—	—	—	—	—	—	—	—	—	—	—	C	C	A	F	A	F	—
Siphogenerina obispoensis n. sp.	—	—	—	—	—	—	—	—	—	—	—	C	F	R	—	—	—	—	—	—	—
Siphogenerina reedi Cushman	—	—	—	—	—	—	F	R	—	F	—	F	R	C	C	C	F	R	R	R	—
Uvigerina obesa obesa (Cushman)	—	—	R	—	—	—	—	—	—	F	—	—	—	—	—	F	—	R	—	—	—
Uvigerinella californica californica Cushman	—	—	—	—	—	—	—	—	—	—	—	—	—	—	—	—	—	—	—	R	R
Valvulineria californica appressa Cushman	—	R	C	F	C	A	C	C	C	C	F	F	F	F	F	R	—	R	F	R	R
Valvulineria californica californica Cushman	—	—	—	F	F	F	C	R	F	F	R	A	C	A	A	A	A	A	A	A	A
Valvulineria californica obesa Cushman	—	F	F	A	C	C	F	F	C	C	A	C	F	R	C	C	C	C	C	C	C
Valvulineria depressa Cushman	—	C	A	F	F	C	C	C	C	C	C	F	F	R	—	—	—	—	—	—	—
Valvulineria miocenica Cushman	—	—	—	—	—	—	—	—	—	—	—	—	—	A	C	R	R	—	F	C	A
Valvulineria ornata Cushman	—	—	R	R	R	R	F	F	—	—	R	—	F	C	C	C	R	R	F	C	C
Valvulineria vilardeboana (d'Orbigny)	—	—	—	R?	—	—	R	—	—	F	—	—	—	—	—	—	R	—	—	—	—
Virgulina californiensis Cushman	—	—	R	—	—	F	R	—	—	—	R	A	A	—	—	R	—	A	—	R	R

Table 3. Continued

All EC samples are from Highland homocline Temblor-Monterey formational sequence, collected from "Edwards Canyon" (named after M. G. Edwards who showed this gulley, with its excellent sequence of continuous exposures in Monterey Shale, to the collectors), a prominent northwest-draining gulley in extreme southeast corner of Sec. 21, T28S, R14E, San Luis Obispo County. All but lowermost three samples constitute a sampling of Monterey Shale sequence selected as type section of Luisian Stage. Base of Luisian Stage here is horizon of prominent lithologic change from clay shale below to organic shale above (approximately equivalent to the Temblor-Monterey contact) at head of gulley in extreme northeast corner of Sec. 28, T28S, R14E (see Kleinpell, 1938, p. 122, Fig. 11). Datum horizon used in stratigraphic allocation of EC samples is top of a conspicuous white arkosic sandstone where it is overlain by pebbly brown sandstone of granitic detritus bearing *Turritella ocoyana*, a large oyster, and other pelecypods, conspicuously exposed a short distance southward and downslope from brow of dip-slope hill at head of Edwards Canyon. Collectors: M. N. Bramlette, K. E. Lohman, R. M. Kleinpell.

EC-1 — In course "granitic" sandstone, Temblor Formation, 202 ft above top of white arkosic sandstone used here as stratigraphic datum horizon.
EC-2 — About 25 ft above EC-1, in siltstone immediately beneath ledge of coarse sandstone.
EC-3 — 325 ft above "top of white arkosic sandstone" datum and about 25 ft above highest sandstone beds here; in shaly claystone which contains only one notable thin and discontinuous sandstone lentil in more or less silty clay shale and mudstone poorly cropping out northward downhill into Edwards Canyon.
EC-4 — 590 ft above datum horizon, about 10 ft above prominent lithologic change from clay (Temblor) to organic (Monterey) shale in extreme northeast corner or Sec. 28, T28S, R14E; in Monterey Shale.
EC-5 — 615 ft above datum horizon.
EC-6 — 659 ft above datum horizon, in "punky" organic shale.
EC-7 — 669 ft above datum horizon.
EC-8 — 681 ft above datum horizon, immediately below some cream-colored stringers of very fine sandstone. At 691 ft above datum horizon, sample EC-9 from highest sandstone stringers bears large foraminifers tentatively identified in field as *Gypsina* sp.
EC-10— 706 ft above datum horizon.
EC-11— 739 ft above datum horizon, in "punky" shale becoming notably diatomaceous.
EC-12— 839 ft above datum horizon, in massive "punky" diatomaceous shale.
EC-13— 846 ft above datum horizon, in "punky" shale with some tiny pelecypods.
EC-14— 936 ft above datum horizon, in "punky" shale.
EC-15— 1,009 ft above datum horizon, above bed of white cherty limestone with few sporbo and sand grains.
EC-16— 1,033 ft above datum horizon.
EC-17— 1,040 ft above datum horizon, above bed composed almost entirely of Foraminifera.
EC-18— 1,142 ft above datum horizon, in "punky" diatomaceous shale.
EC-19— 1,152 ft above datum horizon, in bedded shale becoming mortarlike in texture.
EC-20— 1,188 ft above datum horizon, in "punky organic" shale.
EC-21— 1,256 ft above datum horizon.
EC-22— 1,314 ft above datum horizon, in whitish-buff "organic" shale; below prominent hard ledge of chert at 1,370 ft above datum horizon, above which are scattered horizons of vitreous chert concretions with much diatomite, above which in turn are beds of buff phosphatic-pellet (sporbo)-bearing fine-grained sandstone. At 1,485 ft above datum horizon is base of bed of volcanic ash 25 ft or more thick, succeeded upward by soft tuffaceous shaly and fine-grained thin sandy beds, and finally coarse gray sandstone, base of which is at 1,564 ft above datum horizon and approximates base of mappable Santa Margarita Formation which here concordantly and conformably overlies Monterey Shale.

Table 4. Check List of Foraminifera from Monterey Formation in Coastal Bluffs near Naples, West of Mouth of Dos Pueblos Creek, Santa Barbara County, California

SPECIES / SAMPLE NUMBER	SBC 1	SBC 2	SBC 3	SBC 4	SBC 5	SBC 6	SBC 7	SBC 8	SBC 9	SBC 10	SBC 11	SBC 12	SBC 14	SBC 15	SBC 16	SBC 17	SBC 18	SBC 19	SBC 22	SBC 23	SBC 24	SBC 25	SBC 26	SBC 27	SBC 28	SBC 29	SBC 30	SBC 31	SBC 32	SBC 33a	SBC 33b	SBC 34	SBC 35
STAGE	SAUCESIAN																								RELIZIAN								
	MIDDLE		UPPER																						LOWER								Up
ZONE	Plectofrondicularia miocenica		Uvigerina obesa																						Siphogenerina hughesi								S. branneri
Anomalina glabrata Cushman and Laiming	—	—	—	—	—	—	—	—	—	—	—	F	R	—	—	—	—	—	—	—	—	—	—	—	—	—	—	—	—	—	—	—	—
Anomalina sp	—	—	—	—	—	—	—	—	—	—	—	—	—	—	—	—	R	—	F	R	—	R	—	—	—	—	—	—	—	—	—	—	—
Baggina cancriformis Kleinpell	—	—	—	—	—	—	—	—	—	—	—	—	—	—	—	—	—	—	—	—	—	—	—	—	—	—	C	—	—	—	—	—	—
Baggina robusta Kleinpell	—	—	—	—	—	—	—	—	F	R	—	—	C	—	—	—	—	—	—	—	—	—	—	—	—	—	—	—	—	—	—	—	—
Baggina aff. B. robusta Kleinpell	—	—	F	F	R	R	—	R	R	—	—	—	—	—	—	—	—	—	—	—	—	—	—	—	—	R	—	—	—	—	—	—	—
Bolivina advena Cushman	—	—	F	F	R	R	R	C	C	F	F	—	—	—	—	F?	—	F	C	—	—	—	—	R	—	—	F	—	—	—	—	C	—
Bolivina californica Cushman	—	—	—	—	—	—	—	—	—	—	—	R	—	—	—	—	—	R?	—	R	F	F	F	C	F	—	R	F	F	R	F	—	F
Bolivina floridana floridana Cushman	—	—	—	—	—	R	R	—	R	R	—	C	F	—	F	—	A	—	—	F	C	C	C	C	C	C	A	C	C	F	C	C	C
Bolivina floridana striatella Cushman	—	—	—	—	—	—	—	—	—	—	—	—	—	—	—	—	—	—	—	—	—	—	—	—	—	—	—	—	—	R	—	—	—
Bolivina imbricata Cushman	—	—	—	—	—	—	—	—	—	—	—	—	—	—	—	—	—	—	—	—	—	—	—	—	—	—	—	—	—	R	—	—	—
Bolivina marginata Cushman	—	—	—	—	—	—	—	—	—	—	—	—	—	—	—	—	—	—	—	—	—	—	—	—	F	R	—	F	—	F	C	—	R
Bulimina alligata Cushman and Laiming	—	—	F	R	—	F	—	—	—	—	—	—	—	—	—	—	—	R	—	—	—	—	—	—	—	—	—	—	—	—	—	—	—
Bulimina ovata d'Orbigny	—	—	—	—	—	—	—	—	—	—	—	—	—	—	—	—	—	—	—	—	—	—	—	—	—	—	R	R	—	—	—	—	—
Bulimina ovula d'Orbigny	—	—	—	—	—	—	—	—	—	—	—	—	—	—	—	—	—	—	—	—	—	—	—	—	—	—	—	—	—	—	—	—	R
Bulimina pseudoaffinis Kleinpell	—	—	—	—	—	—	—	—	—	—	—	—	—	—	—	—	—	—	—	—	—	—	—	—	—	—	—	—	—	—	—	—	R
Buliminella curta Cushman	—	—	A	F	R	R	F	C	F	F	R	F	F	—	—	—	A	—	—	F	F	F	F	F	—	—	—	—	—	—	—	—	—
Buliminella subfusiformis Cushman	—	—	F	—	—	—	—	—	—	—	—	—	—	—	—	—	—	—	—	—	—	—	—	—	C	F	C	C	F	C	F	—	F
Cassidulina crassa d'Orbigny	—	—	—	—	—	—	—	—	—	—	—	—	—	—	—	—	—	—	—	—	—	—	—	—	—	—	R?	—	—	—	—	—	—
Cassidulina laevigata carinata Cushman	—	—	—	—	—	—	—	R	—	—	—	—	—	—	—	—	—	—	—	—	—	—	—	—	—	—	—	—	—	—	—	—	—
Cassidulina williami Kleinpell	—	—	F	—	—	—	—	C	F	F	—	R	F	—	—	—	—	R	—	—	—	F	F	F	—	—	R	—	—	R	R	—	—

Cibicides americanus americanus (Cushman)	—	—	—	—	—	R	—	—	—	—	—	—	—	—	—	—	—	R	—	—	—	—	—	—	—	—	—	—	—	—	—	—	—
Cibicides americanus crassiseptus Cushman and Laiming	—	—	—	—	—	—	—	—	—	—	—	—	—	—	—	—	—	R	—	—	—	—	—	—	—	—	—	—	—	—	—	—	—
Cibicides floridanus (Cushman)	—	—	—	R	—	R	—	—	—	—	—	—	—	—	—	—	R	R	—	—	R?	—	—	—	—	—	—	—	—	—	—	—	—
Dentalina pauperata d'Orbigny	—	—	—	R	—	—	—	—	—	—	—	—	—	—	—	—	—	—	—	—	—	—	—	—	—	—	—	—	—	—	—	—	—
Epistominella minuta (Cushman and Laiming)	—	—	C	—	—	—	—	—	—	—	—	—	—	—	—	—	—	—	—	—	—	—	—	—	—	—	—	—	—	—	—	—	—
Eponides tenera H. B. Brady	—	—	—	—	—	—	—	—	—	—	—	—	—	—	—	—	—	—	—	—	—	—	—	—	R	R	—	—	—	—	—	—	R
Frondicularia foliacea Schwager	—	—	R	—	—	—	—	—	—	—	—	—	—	—	—	—	—	—	—	—	—	—	—	—	—	—	—	—	—	—	—	—	—
Gaudryina sp	—	—	—	R	—	—	—	—	—	—	—	—	—	—	—	—	—	—	—	—	—	—	—	—	—	—	—	—	—	—	—	—	—
Hemicristellaria beali (Cushman)	—	—	—	—	—	—	—	—	—	—	—	—	—	—	—	—	—	—	—	—	—	—	—	—	—	—	R?	—	—	—	—	—	—
Lagena sulcata apiculata Cushman	—	—	—	—	—	—	—	—	—	—	—	—	—	—	—	—	—	—	—	—	—	—	—	—	—	—	—	—	—	R	—	—	—
Nodogenerina advena Cushman and Laiming	—	—	R	R	—	R?	—	—	—	—	—	F	R	—	—	—	F	R	F	C	R	F	F	F	—	R	—	—	—	—	—	—	—
Nodosaria tympaniplectriformis Schwager	—	—	—	—	—	—	—	—	—	—	—	—	—	—	—	—	R	—	R	—	—	—	—	—	—	—	—	—	—	—	—	—	—
Nonion costiferum (Cushman)	F	—	—	R	—	—	—	—	—	—	—	—	—	—	—	—	—	R	—	—	—	—	R	R	R	F	F	F	R	—	F	F	R
Nonion incisum (Cushman)	—	—	—	—	—	—	—	—	R	—	—	—	—	—	—	—	—	—	—	—	—	—	—	—	R	—	—	—	—	—	—	—	—
Nonionella miocenica Cushman	—	—	—	—	—	—	—	—	—	—	—	—	—	—	—	—	—	—	—	—	—	—	—	R	R	—	—	—	—	—	—	—	—
Planulina baggi Kleinpell	—	—	—	—	—	—	—	—	—	—	—	R	—	—	—	—	—	—	—	—	—	—	—	F	—	—	—	—	—	—	—	—	—
Plectofrondicularia miocenica Cushman	—	—	F	F	F	R	—	R	—	—	—	R	—	—	—	—	R	R	F	R	—	R	R	R	R	R	F	F	—	—	R	—	F
Plectofrondicularia miocenica directa Cushman and Laiming	—	—	R	—	—	—	—	—	—	—	—	—	—	—	—	—	R	—	—	—	—	—	—	—	—	—	—	R	—	—	—	—	—
Robulus mayi Cushman and Parker	—	—	—	—	—	—	—	R	—	—	—	—	—	—	—	—	—	—	F	—	—	—	—	R	—	—	—	—	—	—	—	—	—
Robulus simplex (d'Orbigny)	—	—	A	C	C	C	—	—	—	—	—	F	—	—	—	—	—	C	F	—	—	F	R	R	C	C	—	F	—	—	—	—	R
Siphogenerina branneri (Bagg)	—	—	—	—	—	—	—	—	—	—	—	—	—	—	—	—	—	—	R?	—	—	R	—	—	—	—	—	—	—	—	—	—	—
Siphogenerina cymricensis Tipton, Kleinpell and Weaver (?)	—	—	—	—	—	—	—	—	—	—	—	C	R	—	—	—	—	—	—	—	—	—	—	—	—	—	—	—	—	—	—	—	—
Siphogenerina kleinpelli Cushman	—	—	R	—	—	—	—	—	—	—	—	R	F	—	—	—	—	C	F	—	—	—	—	—	—	—	—	—	—	—	—	—	—
Siphogenerina transversa Cushman	—	—	C	C	R	A	—	R	—	—	—	—	R?	—	—	F	—	R	F	—	—	—	—	R?	—	—	—	—	—	—	—	—	—
Siphogenerina aff. S. transversa Cushman	—	—	R	—	—	—	—	—	—	—	—	—	—	—	—	—	—	—	—	—	—	—	—	—	—	—	—	—	—	—	—	—	—
Suggrunda kleinpelli Bramlette	—	—	R	—	—	—	—	R	—	—	—	R	—	—	—	—	—	—	—	R	R	F	F	F	R	—	R	—	—	—	—	—	R
Uvigerina obesa obesa (Cushman)	—	R?	R	A	A	—	—	—	—	—	—	—	—	—	R	R	R	F	C	F	—	F	—	F	C	F	—	—	—	—	—	—	C
Uvigerina obesa impolita (Cushman and Laiming)	R	—	R	—	R	C	—	R	—	—	—	—	—	—	—	—	—	—	—	R	—	C	C	F	—	—	—	—	—	—	—	—	—
Uvigerinella californica Cushman	—	—	R	—	—	—	—	F	C	R	R	C	—	—	—	—	—	—	—	—	—	F	F	R	F	C	—	C	A	C	F	F	
Uvigerinella californica gracilis Cushman and Kleinpell	—	—	—	—	—	—	—	—	—	—	—	—	—	—	—	—	—	—	—	—	—	—	—	—	—	R	R?	—	—	—	—	—	R
Uvigerinella californica ornata Cushman	—	—	—	—	—	—	—	—	—	—	—	—	—	—	—	—	—	—	—	—	—	—	—	—	R	R	—	R	—	—	R	—	—

STRATIGRAPHIC DISTANCE IN FEET ABOVE THE BENTONITE BED	-393	-325	27	34	38	79	94	99	111	150	196	241	268	298	322	322	338	350	395	413	423	424	427	432	439	460	474	534	544	Matrix 554	Clast 554	564	574
Valvulineria californica obesa Cushman	—	—	—	—	—	—	—	—	—	—	—	—	—	—	—	—	—	—	—	—	—	—	—	—	—	F	—	—	—	—	—	—	
Valvulineria depressa Cushman	—	—	—	—	—	—	—	A	A	A	C	A	F	R?	F	F	C	F	F	C	C	A	A	A	C	C	A	C	C	—	C	C	C
Valvulineria ornata Cushman	—	—	—	—	—	—	—	—	—	—	—	—	—	—	—	—	—	—	—	—	—	—	—	—	—	—	—	—	—	—	—	—	F
Valvulineria williami Kleinpell	—	—	—	—	—	—	—	F	—	R	—	—	—	—	—	—	—	—	—	—	—	—	—	—	—	—	—	—	—	—	—	—	—
Virgulina californiensis Cushman	—	—	—	—	—	—	—	—	—	—	—	—	—	—	—	—	—	—	—	—	—	—	—	—	R	—	—	—	—	—	—	—	R

Table 5. Check List of Foraminifera from Monterey Shale in Coastal Bluffs near Naples, East of Mouth of Dos Pueblos Creek, Sanda Barbara County, California.

STAGE	LUISIAN													MOHNIAN							DELMONTIAN								
	LOWER					MIDDLE			UPPER					LOWER							UPPER					LO.		UPPER	
ZONE	Siphogenerina reedi					Siphogenerina nuciformis			Siphogenerina collomi					Bolivina modeloensis	Bolivina barbarana						Bolivina wissleri		Bol. foraminata / Bolivina goudkoffi					Bolivina obliqua	
SPECIES / SAMPLE NUMBER	SBC 36	SBC 37	SBC 38	SBC 39	SBC 40	SBC 41	SBC 42	SBC 43	SBC 44	SBC 45	SBC 46	SBC 47	SBC 48	SBC 49	SBC 50	SBC 51	SBC 52	SBC 54	SBC 55	SBC 56	SBC 57	SBC 58	SBC 59	SBC 60	SBC 61	SBC 62	SBC 63	SBC 64	SBC 66
Anomalina salinasensis Kleinpell	R	C	R	F	F	C	F	C	C	F	F	—	—	—	—	—	—	—	—	—	—	—	—	—	—	—	—	—	—
Baggina californica californica Cushman	—	—	R	R	—	—	R	R	C	R	C	C	—	F	—	—	—	—	—	—	—	—	—	—	—	—	—	—	—
Baggina robusta Kleinpell	R	—	—	—	—	—	—	—	—	—	—	—	—	—	—	R?	—	—	—	—	—	—	—	—	—	—	—	—	—
Baggina aff. B. robusta Kleinpell	—	—	—	—	—	—	—	—	—	—	—	—	R	—	—	—	—	—	—	—	—	—	—	—	—	—	—	—	—
Bolivina advena Cushman	—	—	—	—	—	—	—	—	—	R	R	—	R	—	—	—	—	—	—	—	—	—	—	—	—	—	—	—	—
Bolivina cf. B. barbarana Cushman and Kleinpell	—	—	—	—	—	—	—	—	—	—	—	—	—	—	—	F	C	F	—	—	—	—	—	—	—	—	—	—	—
Bolivina bramlettei Kleinpell	—	—	—	—	—	—	—	—	—	—	—	—	—	—	—	—	—	—	—	R	F	—	—	—	—	—	—	—	—
Bolivina brevior Cushman	—	—	—	—	—	—	—	—	—	—	—	—	—	—	—	R	—	R	—	—	—	—	—	—	—	—	—	—	—
Bolivina californica Cushman	R	—	—	R	R	—	F	—	R	—	R	—	—	F	F	R	—	—	—	—	—	—	—	—	—	—	—	—	—
Bolivina conica Cushman	—	—	F	F	C	R	—	—	—	—	—	—	—	—	—	—	—	—	—	—	—	—	—	—	—	—	—	—	—
Bolivina cuneiformis Kleinpell	—	—	—	—	—	—	—	—	—	—	—	—	—	F	—	—	—	—	—	—	—	—	—	—	—	—	—	—	—
Bolivina aff. B. cuneiformis Kleinpell	—	—	—	—	—	—	—	—	—	—	—	—	—	R	—	—	—	—	—	—	—	—	—	—	—	—	—	—	—
Bolivina decurtata Cushman	—	—	—	—	—	—	—	—	—	—	—	—	—	—	C	—	—	—	—	—	—	—	—	—	—	—	—	—	—
Bolivina floridana Cushman	—	—	—	—	—	—	—	—	C	—	F	F	R?	—	—	—	—	—	—	—	—	—	—	—	—	—	—	—	—
Bolivina floridana striatella Cushman	—	—	—	—	—	—	—	—	—	—	—	—	—	—	—	—	—	—	—	—	—	—	—	—	R	—	—	—	—
Bolivina girardensis Rankin	—	—	—	—	—	—	—	—	—	—	—	—	—	—	—	—	R	F	—	F	R	—	R	F	R	—	R	—	—

Bolivina hootsi Rankin	—	—	—	—	—	—	—	—	—	—	—	—	—	—	—	—	—	—	—	—	—	—	—	—	R	—	—	—
Bolivina imbricata Cushman	C	C	F	—	—	R	—	—	—	—	—	—	—	—	—	—	—	—	—	—	—	—	—	—	—	—	—	—
Bolivina malagaensis Kleinpell	—	—	—	—	—	—	—	—	—	—	—	—	—	—	—	—	—	—	—	—	—	—	—	—	—	—	—	R
Bolivina marginata gracillima Cushman	—	—	—	—	—	—	—	—	—	—	—	—	—	F	—	—	—	—	—	—	—	—	—	—	—	—	—	—	—
Bolivina modeloensis Cushman and Kleinpell	—	—	—	—	—	—	—	—	—	—	—	—	—	F	—	—	—	—	—	—	—	—	—	—	—	—	—	—	—
Bolivina mulleri n. sp.	—	R	—	—	—	R	—	—	—	—	—	—	—	—	—	—	—	—	—	—	—	—	—	—	—	—	—	—	—
Bolivina obliqua Barbat and Johnson	—	—	—	—	—	—	—	—	—	—	—	—	—	—	—	—	—	—	—	—	—	—	—	—	—	—	—	—	C
Bolivina parva (?) Cushman and Galliher	—	—	—	—	—	—	—	—	—	—	—	R?	—	R	—	—	—	—	—	—	—	—	—	—	—	—	—	—	—
Bolivina pseudospissa Kleinpell	—	—	—	—	—	—	—	—	—	—	—	—	—	—	—	F	—	R	F	C	C	—	F	C	C	C	—	—	—
Bolivina salinasensis Kleinpell	—	—	—	—	—	—	—	F	—	—	—	—	R	—	—	—	—	—	—	—	—	—	—	—	—	—	—	—	—
Bolivina foraminata Stewart and Stewart	—	—	—	—	—	—	—	—	—	—	—	—	—	—	—	—	—	—	—	—	R	—	R	R	R	—	—	—	—
Bolivina seminuda seminuda Cushman	—	—	—	—	—	—	—	—	—	—	—	—	—	—	—	—	—	—	—	—	—	R	R	—	—	—	—	—	—
Bolivina sinuata sinuata Galloway and Wissler	—	—	—	—	—	—	—	—	—	—	—	—	—	—	R?	—	R	—	—	—	—	—	—	R	—	—	—	—	—
Bolivina subadvena spissa Cushman	—	—	—	—	—	—	—	—	—	—	—	—	—	—	—	—	—	—	—	R	F	—	—	—	—	—	—	—	—
Bolivina wissleri n. sp.	—	—	—	—	—	—	—	—	—	—	—	—	—	—	—	R	—	R	—	F	F	—	—	C	—	C	—	—	—
Bolivina woodringi Kleinpell	—	—	—	—	—	—	—	—	—	—	—	—	—	—	—	R	—	—	—	R	—	—	—	—	—	R	—	—	—
Bolivina sp.	—	—	R	—	—	—	—	C	—	—	—	—	—	—	—	—	—	—	—	—	—	—	—	—	—	—	—	—	—
Bulimina montereyana Kleinpell	—	—	—	—	—	—	—	—	—	—	—	—	—	—	—	—	—	—	—	R?	—	F	—	—	—	—	—	R?	—
Bulimina ovata d'Orbigny	—	—	—	—	—	—	—	—	—	—	—	—	—	—	—	—	—	—	—	—	—	—	—	—	—	—	—	—	R
Bulimina ovula d'Orbigny	—	—	—	—	—	R	—	—	—	—	—	—	—	—	—	—	—	—	—	R	—	—	R	F	—	R	—	—	—
Buliminella brevior Cushman	—	—	—	—	—	—	—	—	—	—	—	—	—	R	F	—	C	R	—	R	—	—	—	—	—	—	—	—	—
Buliminella californica Cushman	—	—	—	—	—	—	—	—	—	—	—	R	—	—	—	—	—	—	—	—	—	—	—	—	—	—	—	—	—
Buliminella curta Cushman	R	—	R	—	F	R	R	—	—	—	R	—	F	—	—	—	—	—	—	—	—	—	—	—	—	—	—	—	—
Buliminella semihispida Kleinpell	—	—	—	—	—	—	—	—	—	—	—	—	—	—	—	—	—	—	—	R	—	—	—	—	R	—	F	—	—
Buliminella subfusiformis Cushman	—	—	—	—	—	—	—	—	R	—	—	R	—	—	—	—	—	—	—	R	F	—	—	R	—	—	—	—	—
Cancris brongniarti (d'Orbigny)	—	—	—	—	—	—	—	—	—	—	—	—	F	—	—	—	—	—	—	—	—	—	—	—	—	—	—	—	—
Cassidulina cf. C. barbarana Cushman and Kleinpell	—	—	—	—	—	—	—	—	—	—	—	—	—	—	F	F	C	—	F	—	—	—	R	C	—	R	—	—	—
Cassidulina crassa d'Orbigny	—	—	—	—	—	R	F	C	C	C	F	—	—	F	—	—	—	—	—	—	—	—	—	—	—	—	—	—	—
Cassidulina delicata Cushman	—	—	—	—	—	—	—	—	—	—	—	—	—	—	—	—	—	—	—	—	—	—	—	—	—	—	—	—	R
Cassidulina monicana Cushman and Kleinpell	—	—	—	—	—	—	—	—	—	—	—	—	—	R	—	—	—	—	—	—	—	—	—	—	—	—	—	—	—
Cassidulina panzana Kleinpell	—	F	R	—	F	—	—	—	—	—	—	—	—	—	—	—	—	—	—	—	—	—	—	—	—	—	—	—	—
Cassidulina williami Kleinpell	—	—	—	—	—	—	—	—	—	R	—	—	F	—	—	—	—	—	—	—	R?	—	—	—	—	—	—	—	—
Cassidulina sp.	—	—	—	—	—	—	—	—	—	—	—	—	—	—	—	—	—	—	—	F	—	—	—	R	—	—	—	—	—

Chilostomella cf. C. ovoidea Reuss	— — — — —	— — —	— — — — —	—	—	— — — — —	— — — — — — —	—	F
Cibicides illingi (Nuttall)	— — — — —	— — —	— — — — —	R	—	— — — — —	— — — — — — —	—	—
Dentalina barnesi Rankin	— — — — —	— — —	— — — R —	R	—	— — — — —	— — — — — — —	—	—
Dentalina consobrina d'Orbigny	— — R R F	— R R	R — R? — —	—	—	— — — — —	— — — — — — —	—	—
Dentalina pauperata d'Orbigny	— — — — —	— — —	R R — R —	—	—	— — — — —	— — — — — — —	—	—
Discorbinella valmonteensis Kleinpell	— — — — —	— — —	— — — — —	—	—	F F — R F	— F — — R — —	—	—
Elphidium granti Kleinpell	— — — R —	— — —	— — — — —	—	—	— — — — —	— — — — — — —	—	—
Epistominella capitanensis (Cushman and Kleinpell)	— — — — —	— — —	— — — — R	F	—	— R — — —	— — — — — — —	—	—
Epistominella pacifica (Cushman)	— R — — —	— R —	— — — — —	—	—	— — — — —	— — C C F — —	—	—
Epistominella subperuviana (Cushman)	— — F — —	R — —	— — — — —	—	F	— — F F —	F — R — — — —	—	—
Epistominella spp.	R — — — —	— — —	— — — — —	—	—	— — — — —	— — — — — — —	—	F
Eponides keenani Cushman and Kleinpell	— — — — —	— — —	— — — R C	—	—	— — — — —	— — — — — — —	—	—
Eponides rosaformis Cushman and Kleinpell	R — F R R	— C C	F R R F —	F	F	— — — — —	— — — — — — —	—	—
Frondicularia foliacea Schwager	— — — — —	— — —	— — — R —	—	—	— — — — —	— — — — — — —	—	—
Gallinerina uvigerinaformis uvigerinaformis (Cushman and Kleinpell)	— — — — —	— — —	— — — — —	—	A	— — — — —	— — — — — — —	—	—
Guttulina sp.	R — — — —	— — —	— — — — —	—	—	— — — — —	— — — — — — —	—	—
Gyroidina multicamerata (Kleinpell)	— — — — —	— — R	— — — — —	—	F	— — — — C	F — — — — R —	—	—
Gyroidina soldanii rotundimargo Stewart and Stewart	— R — — —	F — —	F — — — —	—	—	— — — — —	— — — — — — —	—	—
Nodogenerina advena Cushman and Laiming	— — — — —	— — —	R — — C F	R	—	— — — — —	— — — — — — —	—	—
Nodogenerina irregularis Kleinpell	— — — — —	— — —	— — — — —	—	—	R R — — R	— — R — — — —	—	—
Nodosaria parexilis Cushman and Stewart	— — — — —	— — —	R — — — R	—	—	— — — — —	— — — — — — —	—	—
Nodosaria tympaniplectriformis Schwager	— — — — —	— — —	— — — — —	—	—	— R — R —	F — R F — — —	—	—
Nonionella miocenica Cushman	— — R — —	— — —	— — — — —	—	—	— — — — —	— — — — — — —	—	F
Planularia cushmani Kleinpell	— — — — —	— — —	— — — — —	—	—	— — — — R	— — F — R — —	—	—
Planularia dubia Kleinpell	— — — — —	— — —	R — R — —	—	—	— — — — —	— — — — — — —	—	—
Planulina cf. P. ariminensis d'Orbigny	— — — — —	— — —	R — R R —	—	—	— — — — —	— — — — — — —	—	—
Planulina ornata (d'Orbigny)	— — — — —	— — —	— — — R R	—	—	R — — — —	— — — — — — —	—	—
Pullenia miocenica Kleinpell	F C C F F	C F C	F — C — R?	—	—	— — — — —	— — — — — — —	—	—
Buliminella semih									

Pullenia miocenica globula Kleinpell	F	–	–	–	–	–	–	–	–	F	F	–	–	–	–	–	–	–	–	–	–	–	–	–	–	–	–	–	–	–
Pullenia moorei Kleinpell	–	–	–	–	–	–	–	–	–	–	–	–	–	–	R?	–	–	–	–	–	–	–	–	–	–	–	–	–	–	–
Pullenia multilobata Chapman	–	–	–	–	–	–	–	–	–	–	R	–	–	–	–	–	–	–	–	–	–	–	–	–	–	–	–	–	–	–
Robulus miocenicus (Chapman)	–	R	–	–	–	–	–	–	–	–	–	–	–	–	–	–	–	–	–	–	–	–	–	–	–	–	–	–	–	–
Robulus nikobarensis cushmani Galloway and Wissler	–	–	–	–	–	–	–	–	–	–	–	–	–	–	–	R	–	–	–	–	–	–	R	–	–	–	–	–	–
Robulus simplex d'Orbigny	R	–	R	–	–	–	–	R	–	–	–	–	–	–	–	–	–	–	–	–	–	–	–	R	–	–	–	–	–
Robulus smileyi Kleinpell	–	–	–	–	–	–	–	–	–	R	–	–	–	–	–	–	–	–	–	–	–	–	–	–	–	–	–	–	–
Rotalia garveyensis Natland	–	–	–	–	–	–	–	–	–	–	–	–	–	–	–	–	–	–	–	–	–	–	R	R	R	–	–	–	–
Siphogenerina branneri (Bagg)	F	R	C	–	F	–	–	–	–	–	–	–	–	–	–	–	–	–	–	–	–	–	–	–	–	–	–	–	–
Siphogenerina collomi Cushman	–	–	–	–	–	–	–	R	F	–	–	–	–	–	–	–	–	–	–	–	–	–	–	–	–	–	–	–	–
Siphogenerina kleinpelli Cushman	R	R	–	–	–	–	–	–	–	–	–	–	–	–	–	–	–	–	–	–	–	–	–	–	–	–	–	–	–
Siphogenerina nuciformis Kleinpell	–	–	–	–	–	R	F	F	C	–	–	–	–	–	–	–	–	–	–	–	–	–	–	–	–	–	–	–	–
Siphogenerina cf. S. nuciformis Kleinpell	–	–	–	–	–	R	–	–	–	–	–	–	–	–	–	–	–	–	–	–	–	–	–	–	–	–	–	–	–
Siphogenerina reedi Cushman	F	F	C	R?	–	–	F	R	R	–	–	–	–	–	–	–	–	–	–	–	–	–	–	–	–	–	–	–	–
Sphaeroidina variabilis Reuss	–	–	–	–	–	–	–	–	–	–	–	F	R	–	–	–	–	–	–	–	–	–	–	–	F	–	–	–	–	–
Suggrunda kleinpelli Bramlette	–	–	–	–	–	–	–	–	–	–	–	–	–	–	R	–	–	–	–	–	–	–	–	–	–	–	–	–	–	–
Uvigerina hootsi Rankin	–	–	–	–	–	–	–	–	–	–	–	–	–	C	–	R	R	A	–	–	R	–	–	–	–	–	–	–	–
Uvigerina subperegrina Cushman and Kleinpell	–	–	–	–	–	–	–	–	–	–	–	–	–	A	R	–	F	–	R	A	A	F	F	C	F	R?	A?	R	–
Uvigerinella californica Cushman	–	R	R	–	F	–	–	–	–	–	–	F	C	–	–	–	–	–	–	–	–	–	–	–	–	–	–	–	–
Uvigerinella californica gracilis Cushman and Kleinpell	–	–	–	–	–	–	–	–	–	–	–	R	–	–	–	–	–	–	–	–	–	–	–	–	–	–	–	–	–
Uvigerinella californica ornata Cushman	C	–	R	–	–	–	–	–	–	–	–	–	–	–	–	–	–	–	–	–	–	–	–	–	–	–	–	–	–
Uvigerinella californica parva Kleinpell	–	–	–	–	–	–	–	–	–	–	R	–	–	–	–	–	–	–	–	–	–	–	–	–	–	–	–	–	–
Uvigerinella nudocostata Cushman	–	–	–	–	–	–	–	–	–	–	R?	R	R	–	–	–	–	–	–	–	–	–	–	–	–	–	–	–	–
Valvulineria araucana (d'Orbigny)	–	–	–	–	–	–	–	–	–	–	–	–	–	–	–	R	–	–	–	–	R	–	R	R	–	–	–	–	–
Valvulineria californica appressa Cushman	C	C	–	–	–	–	–	F	–	–	–	–	–	–	–	–	–	–	–	–	–	–	–	–	–	–	–	–	–
Valvulineria californica californica Cushman	–	R	–	–	–	–	R	F	F	F	C	–	–	–	–	–	–	–	–	–	–	–	–	–	–	–	–	–	–
Valvulineria californica obesa Cushman	C	F	C	F	C	R	C	F	F	F	F	–	–	–	–	–	–	–	–	–	–	–	–	–	–	–	–	–	–
Valvulineria miocenica Cushman	–	–	–	–	–	–	–	–	–	–	–	R	C	–	–	–	–	–	–	–	–	–	–	–	–	–	–	–	–
Valvulineria ornata Cushman	–	–	–	–	–	F	–	–	–	–	–	–	–	–	–	–	–	–	–	–	–	–	–	–	–	–	–	–	–

STRATIGRAPHIC DISTANCE IN FEET ABOVE BENTONITE BED	714	728	747	749	752	774	806	846	871	896	912	930	934	962	1,065
Valvulineria cf. V. williami Kleinpell	–	–	–	–	–	–	–	–	R	–	–	–	–	–	–
Virgulina californiensis Cushman	R	–	–	–	–	–	–	–	–	–	–	–	–	–	–
Virgulina californiensis grandis Cushman and Kleinpell	–	–	–	–	–	–	–	–	–	–	–	–	–	–	–

STRATIGRAPHIC DISTANCE IN FEET ABOVE BENTONITE BED	1,097	1,112	1,147	1,178	1,198	1,225	1,250	1,265	1,280	1,342	1,347	1,376	1,401	1,449
Valvulineria cf. V. williami Kleinpell	–	–	–	–	–	–	–	–	–	–	–	–	–	–
Virgulina californiensis Cushman	–	–	R	–	–	F	–	R	–	–	–	–	–	–
Virgulina californiensis grandis Cushman and Kleinpell	R	–	–	–	R	–	–	–	–	–	R	–	–	C

Table 6. Check List of Miocene Foraminifera from Shale Point Area, (after Kay, 1955).

STAGE	RELIZIAN	LUISIAN					MOHNIAN	RELIZIAN		LUSIAN	LUSIAN									MOHNIAN				
ZONE	Siphogenerina branneri	Siphogenerina reedi			Siphogenerina nuciformis		lower	Siphogenerina branneri		S. reedi	Siphogenerina reedi			S. nuciformis	Siphogenerina collomi					lower				
FORMATION	Undiff. Gould Sh. and Devilwater Silt	Undiff. Gould Sh. and Lower Devilwater Silt			UPPER DEVIL-WATER SILT		MCDONALD	Undiff. Gould Sh. and Devilwater Silt			Undiff. Gould Sh. and Lower Devilwater Silt			UPPER DEVILWATER SILT						TWISSELMAN SAND		McDONALD SHALE		
SPECIES / SAMPLE NUMBER	M511A	M511B	M511C	M511D	M511E	M511F	M506	M510	M498	M498A	M518B	M523	M519A	M519B	M519C	M519D	M529A	M532F	M532E	M536	M513	M533	M535	M512
Anomalina salinasensis Kleinpell	—	—	A	F	—	—	—	—	—	—	—	F	C	—	—	—	R	F	—	—	—	—	—	—
Baggina californica Cushman	—	—	C	R	—	R	—	—	—	—	R	—	—	A	A	R	—	—	—	—	—	—	—	—
Baggina robusta robusta Kleinpell	—	—	F	—	—	—	—	F	R	R	F	—	A	—	—	—	F	—	—	—	—	—	—	—
Bolivina advena advena Cushman	—	—	A	—	—	—	—	F	—	—	—	F	R	F	F	—	—	—	—	—	—	—	—	—
Bolivina advena ornata Cushman	—	—	—	—	—	—	—	F	F?	—	R	—	F	—	—	—	—	—	—	—	—	—	—	—
Bolivina brevior Cushman	—	—	—	—	—	—	—	—	—	—	—	—	R	—	—	—	—	—	—	—	—	—	—	—
Bolivina conica Cushman	—	—	—	—	—	—	—	—	—	—	—	C	—	—	—	—	R	R?	—	—	—	—	—	—

Bolivina floridana striatella Cushman	—	R A A	R A	—	F —	R	— C F	A	A A A R —	C* —	— — —
Bolivina hughesi Cushman	—	— — —	— —	—	— —	—	— — —	—	— — — — —	— —	— A —
Bolivina imbricata imbricata Cushman	R?	A — —	R —	—	A A	—	— C A	R	— — R R —	R* —	— — —
Bolivina marginata marginata Cushman	C?	— — —	— —	—	— —	—	— — R?	—	— — — — —	— —	— — —
Bolivina salinasensis Kleinpell	—	— — —	— —	—	F —	—	— — —	—	— — — F —	— —	— — —
Bolivina sinuata alisoensis Cushman and Adams	—	— — —	— —	—	— —	—	— — —	—	— — — — —	— —	— F —
Bolivina tumida tumida Cushman	R	F — A	— —	—	F R	—	— C —	—	— R C F —	— —	R — —
Bulimina montereyana Kleinpell	—	— — —	— —	—	— —	—	— — —	—	— — — — —	— —	R C F
Bulimina montereyana delmonteensis Kleinpell	—	— — —	— —	—	— —	—	— — —	—	— — — — —	— —	— R —
Bulimina ovata d'Orbigny	—	— — —	— —	—	— —	—	— — —	—	— R — R? —	— —	— — —
Bulimina pseudotorta Cushman	—	F C —	R —	—	A —	—	A R C	—	— — A R —	— —	— — —
Buliminella brevior Cushman	—	— — —	— —	—	— —	—	— — R	—	— — — — —	— —	— — —
Buliminella californica Cushman	—	— — R?	— —	—	R? —	—	— — —	R?	— F — F —	— —	— — —
Buliminella curta Cushman	—	F — R	— F	—	— F	—	— — —	R	R — — — —	— —	— — —
Buliminella subfusiformis Cushman	—	F R R	R C	—	R R	—	— R —	—	— F R — —	— R	— A —
Cassidulina crassa d'Orbigny	—	— A —	— —	—	— —	—	— F? R	—	— — — — —	— —	— — —
Cassidulina sp. cf. C. limbata Cushman and Hughes	—	— — —	— —	—	— —	—	— F —	—	— — — R —	— —	— — —
Cassidulina sp. cf. C. modeloensis Rankin	—	— — —	— —	R	— —	—	— — —	—	— — — — —	R? —	R — —
Cassidulina panzana Kleinpell	—	— R —	F —	—	— —	R?	— — R	—	— R F — —	— —	— — —
Cibicides illingi (Nuttall)	—	— — —	— —	—	— —	—	— — —	—	— — — — —	— —	— R —
Dentalina obliqua (Linne)	—	— F —	— —	—	R —	R	R — F	F	F R R — —	— —	— — —
Dentalina roemeri Neugeboren	—	— — —	— —	—	R —	—	— — —	—	— — — — —	— —	— — —
Dentalina sp. aff. D. roemeri Neugeboren	—	— — —	— —	—	— —	R	F — —	—	— — R — —	— —	— — —
Epistominella sp. aff. E. capitanensis (Cushman and Kleinpell)	—	— F? —	— —	—	— —	—	— — —	—	— — R? — —	— —	— — —
Epistominella gyroidinaformis (Cushman and Goudkoff)	—	— — —	— —	R	— —	—	— — —	—	— — — — —	— R?	C C —
Epistominella pacifica (Cushman)	—	— — —	— —	—	— —	—	— — —	—	— — — — —	— —	R R —
Epistominella subperuviana (Cushman)	—	R — —	— R	—	F R	—	— — —	F	— — — — —	— —	— — —
Eponides rosaformis Cushman and Kleinpell	—	— C R	— —	—	— —	—	— — —	—	— — R R —	— —	— — —
Frondicularia foliacea Schwager	—	— — —	— —	—	R? —	—	— — —	A	— R — — —	— —	— — —

Species											
Globobulimina galliheri (Kleinpell)	—	— — —	— —	—	— —	—	— — —	—	— — — — —	— —	R — —
Gyroidina multicamerata (Kleinpell)	—	— — R	— —	—	— —	—	— — R	—	— R — — —	— —	— — —
Hemicristellaria beali Cushman	R?	— R —	— —	—	A R	R	F — C	R	F F — — —	— —	— — —
Nodogenerina advena Cushman and Laiming	—	— — —	— —	—	— —	R	R — R	A	F — — — —	— —	— — —
Nonion costiferum (Cushman)	—	R C —	— C	R	A A	R	— R F	F	A F R R —	— —	— — —
Nonion costiferum (Cushman) var. weaveri n. var.**	—	— —	— —	—	— —	—	— — —	—	F — — — —	— —	— — —
Nonion multicameratum Cushman and Kleinpell	—	— —	— —	—	— —	—	— — —	—	— — — — —	F —	— — A
Nonionella miocenica Cushman	—	— — —	— —	—	— R	—	— — —	—	— R R F —	— R	F — —
Planulina sp. cf. P. ariminensis d'Orbigny	—	— — —	— —	—	— —	—	— — —	—	R — — — —	— —	— — —
Pullenia miocenica globula Kleinpell	—	— — —	— —	—	— —	—	— R —	—	— — — — —	— —	— — —
Pullenia miocenica miocenica Kleinpell	—	— A R	R —	—	R —	—	R — F	—	— — F R —	— —	— — —
Robulus miocenicus (Chapman)	—	— R —	— —	—	R A	—	— — F	A	— R F — —	— —	— — —
Robulus smileyi Kleinpell	—	— C —	— —	—	R F	R	— — F	C	A F A — —	R? —	— — —
Siphogenerina branneri (Bagg)	—	F C —	— —	—	C —	—	F? R C	—	— — — — —	— —	— — —
Siphogenerina collomi Cushman	—	— — —	— —	—	— —	—	— — —	—	F — — — R	— —	— — —
Siphogenerina kleinpelli Cushman	—	— — —	— —	—	— —	—	F — C	—	F — R — —	— —	— — —
Siphogenerina nuciformis Kleinpell	—	— — —	— —	—	— —	F?	— — —	C	A F R F R	— —	— — —
Siphogenerina reedi Cushman	—	C R —	— —	—	— —	F	R? R F	R	R R F R —	— —	— — —
Suggrunda sp. cf. S. kleinpelli Bramlette	—	— — —	— —	—	— —	—	— — —	—	— R — — —	— —	— — —
Uvigerina carmeloensis Cushman and Kleinpell	—	— — —	— —	R?	— —	—	— — —	—	— — — — —	— F?	A A R?
Uvigerina hootsi Rankin	—	— — —	— —	—	— —	—	— — —	—	— — — — —	— —	F — —
Uvigerina joaquinensis Kleinpell	—	— — —	— —	—	— —	—	— — —	—	— — — — —	— —	— R? —
Uvigerina segundoensis Cushman and Galliher	—	— — —	— —	—	— —	—	— — —	—	— — — — —	— R?	— — R?
Uvigerina subperegrina Kleinpell	—	— — —	— —	—	— —	—	— — —	—	— — R — —	— —	— R? —
Uvigerinella californica californica Cushman	—	— C? —	— R	—	R? —	—	— — C	R	— R — — —	— —	— — —
Uvigerinella californica ornata Cushman	—	— R? R?	C C	—	R —	—	F F —	F	A A — — —	— —	— — —
Uvigerinella nudocostata Cushman	—	— — —	— —	—	— —	—	— — —	C	— — — — —	— —	— — —
Valvulineria californica appressa Cushman	—	C R —	— —	—	C —	—	— — —	C	— — F — —	— —	— — —

Valvulineria californica californica Cushman	—	R?	—	R	C	A	—	—	—	F	—	C	F	C	A	A	C	C	R	—	—	—	—	—
Valvulineria californica obesa Cushman	—	A	C	A	F	F	—	F	—	F	F	F	A	C	C	F	F	—	—	—	—	—	—	—
Valvulineria depressa Cushman	C	—	—	—	—	—	—	—	—	—	—	—	—	—	—	—	—	—	—	—	—	—	—	—
Valvulineria miocenica Cushman	—	—	—	—	—	—	—	—	—	—	—	—	—	F?	F	C	—	—	—	R*	—	—	—	—
Valvulineria ornata Cushman	—	—	—	—	R	—	—	—	—	—	—	—	—	A	A	—	F	—	—	—	—	—	—	—
Virgulina californiensis Cushman	A	C	—	C	—	—	—	R	—	—	—	C	—	—	—	—	—	A	—	—	—	R	—	—
Virgulina californiensis grandis Cushman and Kleinpell	—	—	A	—	—	—	—	—	—	—	—	—	—	—	—	—	—	—	—	—	—	—	R	—

These checklisted localities are referred to and their faunal assemblages are listed in p. 52-127 of Carlos E. Key (1955), "Biostratigraphy of the Bitterwater - Packwood Creek Area, Kern County, California." The sample locations are shown on the accompanying geologic map (Plate 1, Carlos E. Key, op. cit.). The localities are also listed in the Register of Localities of Leland Stanford Junior University. The geology of the area is treated in Heikkila and MacLeod (1951).

*Reworked.
**This is conspecific with Nonion weaveri of Taxonomy section of this paper.

M-511A—Packwood quadrangle; undifferentiated Gould Shale and Devilwater Silt, 400 ft above top of Carneros Sandstone.
M-511B—Packwood quadrangle; undifferentiated Gould Shale and lower Devilwater Silt, 900 ft above top of Carneros Sandstone.
M-511C—Packwood quadrangle; undifferentiated Gould Shale and lower Devilwater Silt, 1,000 ft above top of Carneros Sandstone.
M-511D—Packwood quadrangle; undifferentiated Gould Shale and lower Devilwater Silt, 1,100 ft above top of Carneros Sandstone.
M-511E—Packwood quadrangle; upper Devilwater Silt, 1,225 ft above top of Carneros Sandstone.
M-511F—Packwood quadrangle; upper Devilwater Silt, 1,525 ft above top of Carneros Sandstone.
M-506—Packwood quadrangle; McDonald Shale, approx. 30 ft above McDonald-Cretaceous contact.
M-510—Packwood quadrangle; undifferentiated Gould Shale and Devilwater Silt, about 20 ft above top of upper Temblor Sandstone.
M-498—Packwood quadrangle; undifferentiated Gould Shale and Devilwater Silt, about 25 ft above top of upper Temblor Sandstone.
M-498A—Packwood quadrangle; undifferentiated Gould Shale and Devilwater Silt, about 125 ft stratigraphically above M-498.
M-518B—Packwood quadrangle; undifferentiated Gould Shale and lower Devilwater Silt, about 511 ft above top of upper Temblor Sandstone.
M-523—Shale Point quadrangle; undifferentiated Gould Shale and lower Devilwater Silt, about 637 ft above top of upper Temblor Sandstone.
M-519A—Packwood quadrangle; undifferentiated Gould Shale and lower Devilwater Silt, about 568 ft above top of upper Temblor Sandstone.
M-519B—Packwood quadrangle; upper Devilwater Silt, about 775 ft above top of upper Temblor Sandstone.
M-519C—Packwood quadrangle; upper Devilwater Silt, about 25 ft above M-519B.
M-519D—Packwood quadrangle; upper Devilwater Silt, about 75 ft above M-519C.
M-529A—Shale Point quadrangle; upper Devilwater Silt, exact stratigraphic position unknown.
M-532F—Shale Point quadrangle; upper Devilwater Silt, exact stratigraphic position unknown.
M-532E—Shale Point quadrangle; upper Devilwater Silt, about 84 ft below base of McDonald Shale.
M-536—Packwood quadrangle; Twisselman Sandstone, overlying Cretaceous on southwest flank of Raven Pass anticline, about 80 ft below base of McDonald Shale.
M-513—Packwood quadrangle; Twisselman Sandstone, shale parting about 500 ft below top of unit.
M-533—Shale Point quadrangle; McDonald Shale, 1 to 2 ft above contact of McDonald Shale with Point of Rocks Sandstone.
M-535—Shale Point quadrangle; McDonald Shale, about 223 ft above contact of McDonald Shale with Point of Rocks Sandstone.
M-512—Taken in middle portion of McDonald Shale; top of hill on axis of anticline, in Sec. 34 (at western side), T26S, R18E, M.D.B.&M.

Figures to Accompany Kleinpell et al

FORMATION	MONTEREY SHALE																																
EPOCH	M. MIOCENE					LATE MIOCENE																											
STAGE	LUISIAN					MOHNIAN																											
	UPPER					LOWER															UPPER												
ZONE	Siphogenerina collomi					Brizalina modeloensis			Bulimina uvigerinaformis												Bolivina hughesi												
SUBZONE									Concavella gyroidinaformis					Brizalina woodringi																			
SAMPLE NUMBER	NEW 5	NEW 17	NEW 21	Tm 9	NL 8	Tm 14	Tm 16	NEW 27	Tm 17	Tm 18	NEW 30A	NEW 40	NEW 42	NEW 43	NEW 45	NEW 48	NEW 51	NEW 63	NEW 68	NEW 69 NEW 57	NEW 58	NEW 74	NEW 61	NEW 86	NEW 90	NEW 94	WNPB 7 NEW 95	WNPB 12	WNPB 13	WNPB 18	WNPB 20	WNPB 30	WNPB 36
FOOTAGE ABOVE BASE OF EXPOSED MONTEREY SHALE	227	368	402	442	464	473	492	514	538	605	655	697	717	722	737	761	789	795	827	830	842	862	866	924	958	998	1006	1034	1047	1108	1133	1248	1300
Brizalina advena ornata	F	A	A																														
Brizalina imbricata	A	R																															
Nonion costiferum var.[1]	R	R	F						F[1]																								
Pullenia miocenica	R	F	A	A	A																												
Valvulineria californica californica	A	C	C	C	C																												
Valvulineria miocenica	A	F	C	R	F																												
Brizalina californica var.[9]		R	F	F	R		R		F		R	F	R																				
Siphogenerina reedi		VR	R		F																												
Baggina californica			F	C		R			F		F	F	F																				
Dentalina obliqua			R	F	R																												
Anomalina salinasensis				A	R																												
Brizalina advena striatella [8]				R	A																												
Saracenaria beali				R	F																												
Siphogenerina collomi					F																												
Bolivina hughesi var.[8]						C								F	C	C	R	R		R	R												
Brizalina bramlettei						R					R								C	C	F	C	F	F				R	F			F	F
Brizalina modeloensis						C	C	C																									
Nonion costiferum [1]						F[1]	R[1]	C[1]			R																						
Uvigerina hootsi						C	C	C	R	F	A	C					F	R	F	F		R											
Virgulina (Virgulinella) miocenica						R	R	C																									
Cassidulina monicana							R	C	R		R																						
Bulimina uvigerinaformis[2]								C[2]	F[3]		F[3]	R[2]					C[4]	C[3]	C[5]	C[4]													
Concavella gyroidinaformis								C	R	F	C	A	F																				
Megastomella capitanensis								F																									
Nonion multicameratum								C[7]																									
Suggrunda californica								C																									
Hanzawaia illingi												R		F	C	C																	
Uvigerina angelina												R	R					C		F													
Pullenia moorei													F																				
Brizalina californica [9]															F	C		F	C	F													
Brizalina girardensis																		VR	R	VR	F	R	F		F			F	F				
Brizalina woodringi																		A	C	C	C	F	A	F	C	F		C	F	R	C	C	C
Discorbinella valmonteensis var.[10]																		R	C	F													
Nonion goudkoffi																		R	F	F	F	F	R			R		R	C				
Bolivina hughesi[8]																			F	F		R	C										
Brizalina benedictensis [8]																			R	R	C		A					F			VR		
Brizalina granti [8]																			F	F	A							A			A	C	A
Brizalina vaughani																			C	C	F	F		F		F		F	F			R	R
Cassidulina modeloensis																			C	A	R		C		F			F					
Hopkinsina magnifica																			F	F	F		C										
Brizalina decurtata																					R			F		F			F			C	
Discorbinella valmonteensis [10]																					C	C	C	C		C			R	VR			
Bolivina sinuata [8]																					R	R	F		C	F		C	R			F	
Cassidulinella renulinaformis																						F											
Buliminella semihispida																									F			F	R				
Uvigerina hannai																									C			C			F		A
Rotalia garveyensis																												R					
Bulimina delreyensis[6]																													C[6]				
Nonionella sp.																															F	F	C
Brizalina rankini																																	F

Frequency Symbols:

- A – ABUNDANT
- C – COMMON
- F – FEW
- R – RARE
- VR – VERY RARE

An alternative view of the significance of this column, involving different interpretations of the phyletic relationships of some of the species recorded, and some differing interpretations of some of the Stage and Zone boundaries when criteria for an Oppelian zonation are applied to the same sequence for purposes of age and correlation.

	Siphogenerina collomi Zone	Bolivina modeloensis Zone	Bolivina barbarana Zone	Bolivina wissleri Zone	Bolivina goudkoffi Zone	Bolivina foraminata Zone	Bolivina obliqua Zone	
LUISIAN		MOHNIAN				DELMONTIAN		

Footnotes in explanation of the numbers added to the species names and occurrences in the above checklist.

1. A variant of Nonion costiferum (Cushman). Both N. mediocostatum and N. costiferum are very closely related species; variation in these instances are toward N. montereyanum; in the material examined, no nonionids were found in either sample Tm-14 or Tm-16. The "variety" seems as compressed as N. mediocostatum and shows signs of a keel. It seems to be sufficiently distinctive to be described as a new species in the accompanying TAXONOMY section.
2. Galliherina uvigerinaformis uvigerinaformis; new genus (new generic name).
3. Galliherina uvigerinaformis warreni; new subspecies.
4. Galliherina uvigerinaformis doanei; new subspecies.
5. All three subspecies of Galliherina uvigerinaformis were found in this sample: G. u. doanei (C); G. u. warreni (F); G. u. uvigerinaformis (R).
6. Galliherina delreyensis; new generic assignment.
7. In addition to specimens of Nonion multicameratum, common in this sample, a few N. montereyanum carmeloensis were also found in this sample.
8. See TAXONOMY section of this Supplement for additional differences in taxonomic usages. Note: many of the generic assignments in the above checklist, not referred to in the TAXONOMY section, are also not followed in this Supplement (for example: "Bolivina" vs. "Brizalina"; "Saracenaria"; "Concavella"; "Megastomella"; and "Hanzawaia").

9, and 10. Both varieties left "lumped" with the respective species sensu stricto in the accompanying text.

dgc 1975

FIG. 1 — Two alternative interpretations of age-significance of foraminiferal sequence at Newport, California, representing views of A. D. Warren (top; see Warren, 1972) and R. M. Kleinpell (bottom).

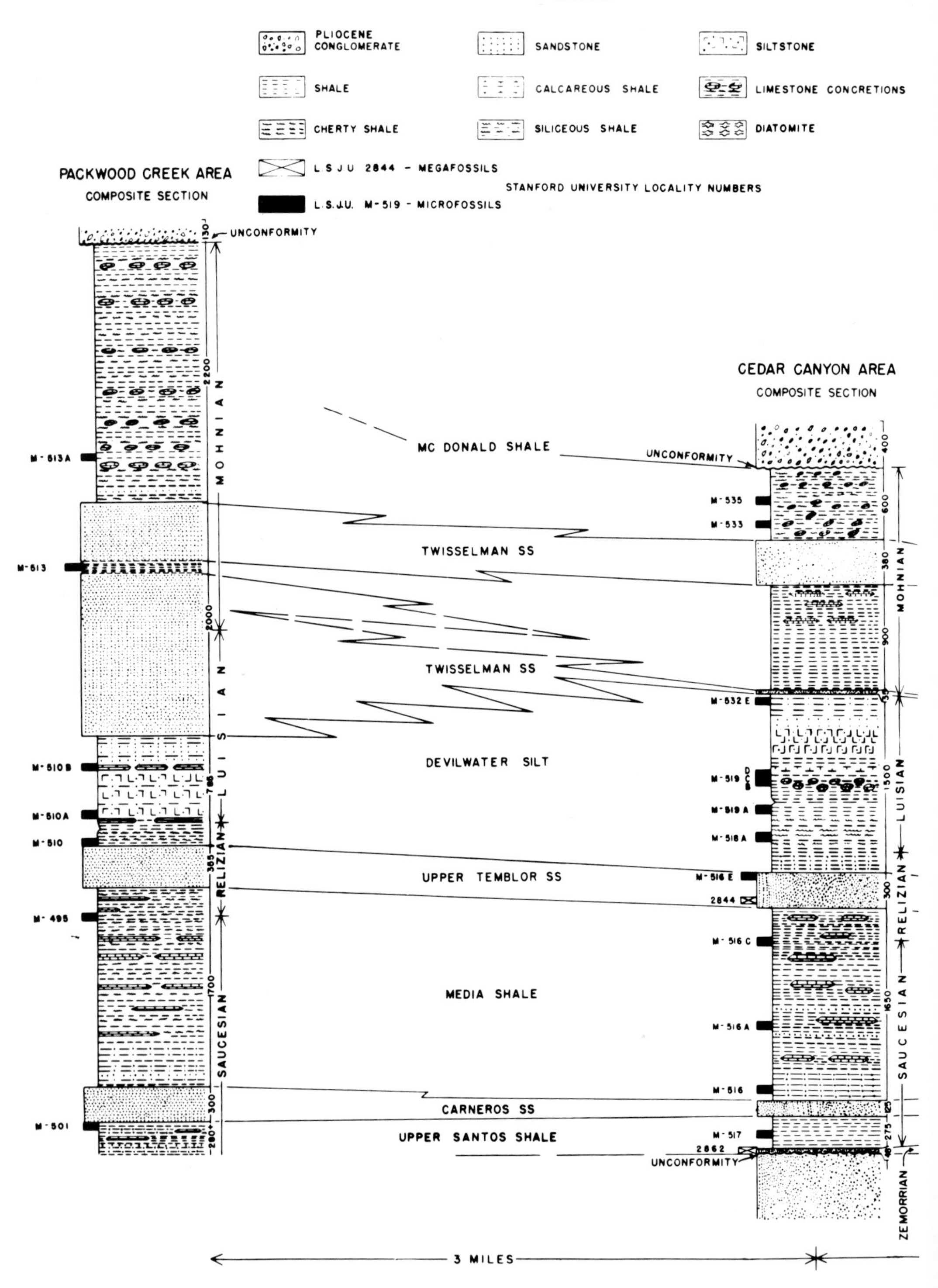
LEGEND
PLIOCENE CONGLOMERATE
SANDSTONE
SILTSTONE
SHALE
CALCAREOUS SHALE
LIMESTONE CONCRETIONS
CHERTY SHALE
SILICEOUS SHALE
DIATOMITE
L.S.J.U. 2844 - MEGAFOSSILS
STANFORD UNIVERSITY LOCALITY NUMBERS
L.S.J.U. M-519 - MICROFOSSILS
PACKWOOD CREEK AREA
COMPOSITE SECTION
CEDAR CANYON AREA
COMPOSITE SECTION
UNCONFORMITY
MOHNIAN
LUISIAN
RELIZIAN
SAUCESIAN
ZEMORRIAN
MC DONALD SHALE
TWISSELMAN SS
DEVILWATER SILT
UPPER TEMBLOR SS
MEDIA SHALE
CARNEROS SS
UPPER SANTOS SHALE
M-513A
M-513
M-510B
M-510A
M-510
M-495
M-501
M-535
M-533
M-532 E
M-519
M-519 A
M-518 A
M-516 E
2844
M-516 C
M-516 A
M-516
M-517
2862
3 MILES

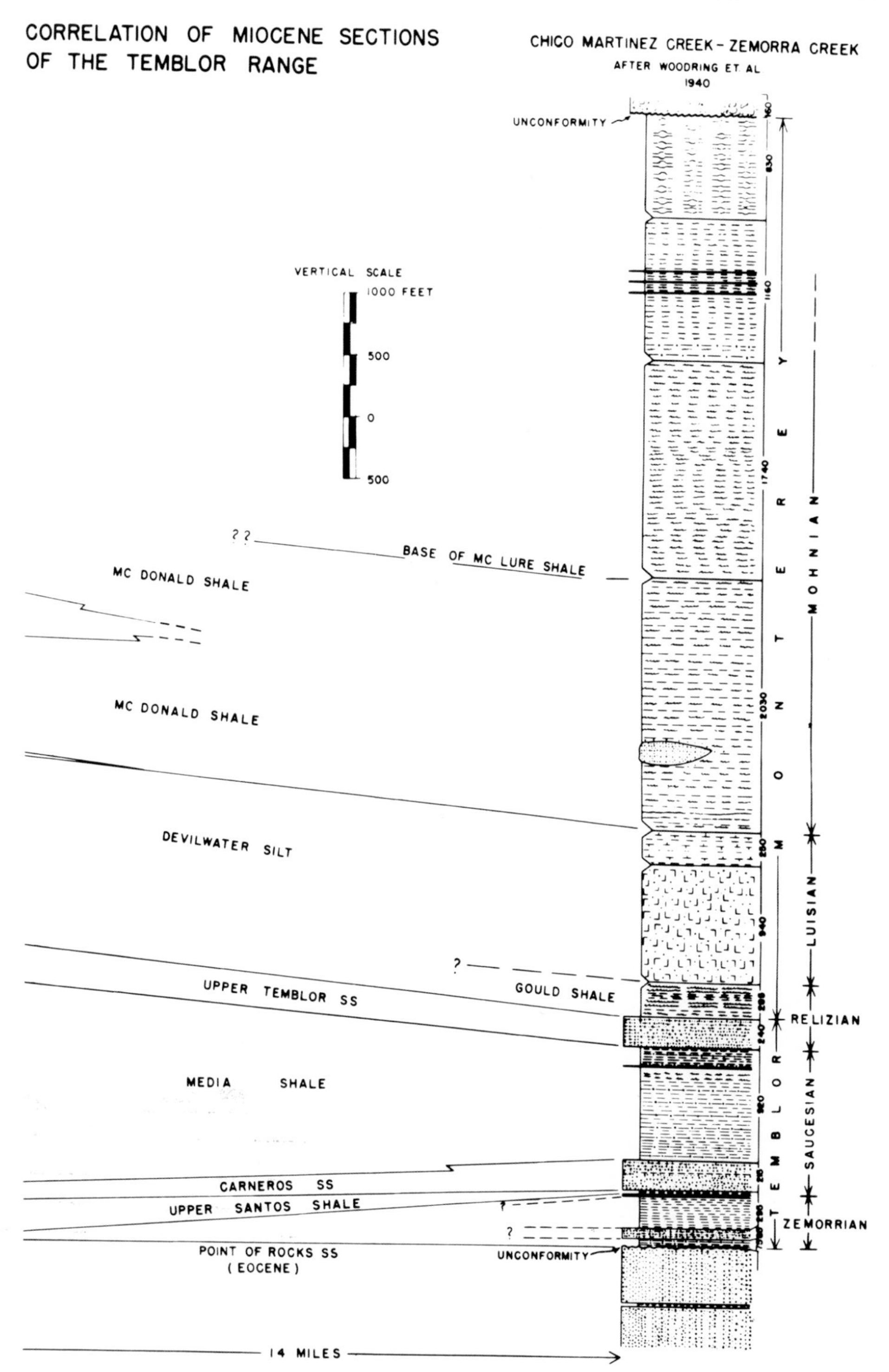

FIG. 2 — Correlation of Miocene sections of Temblor Range, California, from Key (1955).

EST. MM YRS. B.P. AFTER BUKRY (1975)	PALEOGENE LOW LATITUDE NANNOPLANKTON ZONATION OF BUKRY (1973, 1975)			STANDARD PALEOGENE NANNOPLANKTON ZONATION OF MARTINI (1971)
	SERIES OR SUBSERIES	ZONES AND SUBZONES		ZONES
	LOWER MIOCENE (part)	Triquetrorhabdulus carinatus (part)	Discoaster deflandrei	NN 1 Triquetrorhabdulus carinatus
23.0	OLIGOCENE		Cyclicargolithus abisectus	
		Sphenolithus ciperoensis	Dictyococcites bisectus	NP 25 Sphenolithus ciperoensis
			Cyclicargolithus floridanus	NP 24 Sphenolithus distentus
		Sphenolithus distentus		NP 23 Sphenolithus predistentus
		Sphenolithus predistentus		
		Helicopontosphaera reticulata	Reticulofenestra hillae	NP 22 Helicopontosphaera reticulata
			Coccolithus formosus	NP 21 Ericsonia ? subdistichus
			Coccolithus subdistichus	
38.0	UPPER EOCENE	Discoaster barbadiensis	Isthmolithus recurvus	NP 20 Sphenolithus pseudoradians
				NP 19 Isthmolithus recurvus
			Chiasmolithus oamaruensis	NP 18 Chiasmolithus oamaruensis
42.0	MIDDLE EOCENE	Reticulofenestra umbilica	Discoaster saipanensis	NP 17 Discoaster saipanensis
			Discoaster bifax	NP 16 Discoaster tani nodifer
		Nannotetrina quadrata	Coctolithus staurion	NP 15 Chiphragmalithus alatus
			Chiasmolithus gigas	
			Discoaster strictus	
		Discoaster sublodoensis	Rhabdosphaera inflata	NP 14 Discoaster sublodoensis
49.0	LOWER EOCENE		Discoasteroides kuepperi	
		Discoaster lodoensis		NP 13 Discoaster lodoensis
		Tribrachiatus orthostylus		NP 12 Marthasterites tribrachiatus
		Discoaster diastypus	Discoaster binodosus	NP 11 Discoaster binodosus
			Tribrachiatus contortus	NP 10 Marthasterites contortus
53.0	PALEOCENE	Discoaster multiradiatus	Campylosphaera eodela	
			Chiasmolithus bidens	NP 9 Discoaster multiradiatus
		Discoaster nobilis		NP 8 Heliolithus riedeli
		Discoaster mohleri		NP 7 Discoaster gemmeus
		Heliolithus kleinpelli		NP 6 Heliolithus kleinpelli
		Fasciculithus tympaniformis		NP 5 Fasciculithus tympaniformis
60.0		Cruciplacolithus tenuis		NP 4 Ellipsolithus macellus
				NP 3 Chiasmolithus danicus
				NP 2 Cruciplacolithus tenuis
				NP 1 Markalius inversus
63.0				

IFIC COAST PALEOGENE STAGES, LETTER ONES AND ORIGINAL PROVINCIAL AGE SSIGNMENTS AFTER KLEINPELL (1938), ENCK AND KLEINPELL (1936), LAIMING 941), MALLORY (1959) and *GOUDKOFF (1945)		
STAGES	LETTER ZONES	SERIES AND SUBSERIES
Lower aucesian		LOWER MIOCENE (part)
Zemorrian		LOWER MIOCENE (part)
Refugian	R	UPPER EOCENE or LOWER OLIGOCENE
Narizian	A1	UPPER EOCENE
	A2	UPPER EOCENE
	A3	UPPER EOCENE
Jlatisian	B1A to B4	MIDDLE EOCENE
Penutian	C	LOWER EOCENE
Bulitian	D	PALEOCENE
Ynezian	E	PALEOCENE
?	?	PALEOCENE
Cheneyan	*A1	PALEOCENE
	*A2	PALEOCENE

PALEOGENE PLANKTONIC FORAMINIFERAL ZONES, EUROPEAN STAGES AND ESTIMATED ABSOLUTE AGES AFTER BANNER AND BLOW (1965), BLOW (1969) AND BERGGREN (1972)			
MM YRS. B.P.	EPOCH / SERIES	EUROPEAN STAGES	PLANKTONIC FORAMINIFERAL ZONES
	EARLY MIOCENE (part)	Aquitanian (part)	N.4 G. quadrilobatus primordius / Globorotalia kugleri
22.5	LATE OLIGOCENE	Chattian	P.22/N.3 Globigerina angulisuturalis
			?
			P.21/N.2 G. angulisuturalis / Globorotalia opima opima
			?
			P.20/N.1 Globigerina ampliapertura
32.0	EARLY OLIGOCENE		?
		Rupelian	P.19 Globigerina sellii / Pseudohasterina barbadoensis
		Lattorfian	P.18 Globigerina tapuriensis
37.5	LATE EOCENE	Bartonian / Priabonian	P.17 Globigerina gortanii / Globorotalia centralis
			?
			P.16 Cribrohantkenina inflata
			?
			P.15 Globigerapsis mexicana
43.0	MIDDLE EOCENE	Lutetian	P.14 Truncorotaloides rohri - Globigerinita howei
			?
			P.13 Orbulinoides beckmanni
			?
			P.12 Globorotalia lehneri
			?
			P.11 Globigerapsis kugleri
			?
			P.10 Hantkenina aragonensis
49.0	EARLY EOCENE	Ypresian	P.9 Acarinina densa
			P.8 Globorotalia aragonensis
			P.7 Globorotalia formosa
			P.6 b Globorotalia subbotinae - Pseudohastigerina wilcoxensis
53.5	PALEOCENE	Thanetian	P.6 a Globorotalia velascoensis / G. subbotinae
			P.5 Globorotalia velascoensis
			P.4 Globorotalia pseudomenardii
			P.3 Globorotalia pusilla pusilla - G. angulata
60.0		Danian	?
			P.2 Globorotalia uncinata - Globigerina spiralis
			?
			P.1 d G. compressa - G. inconstans - G. trinidadensis
			?
			P.1 c Globorotalia pseudobulloides
			?
			P.1 b Globigerina triloculinoides
			?
			P.1 a Globigerina eobulloides
65.0			

FIG. 3 — Tentative relations of Paleogene calcareous plankton zonations with marine stages of Pacific Coast.

EST. MM YRS. B.P. AFTER BUKRY (1975)	NEOGENE LOW LATITUDE NANNOPLANKTON ZONATION OF BUKRY (1973,1975)			STANDARD NEOGENE NANNOPLANKTON ZONATION OF MARTINI (1971)
	SERIES OR SUBSERIES	ZONES AND SUBZONES		ZONES
	HOLOCENE	Emiliania huxleyi		NN 21 Emiliania huxleyi
0.2	PLEISTOCENE	Gephyrocapsa oceanica	Ceratolithus cristatus	NN 20 Gephyrocapsa oceanica
			Emiliania ovata	NN 19 Pseudoemiliania lacunosa
		Crenalithus doronicoides	Gephyrocapsa caribbeanica	
			Emiliania annula	
1.8	UPPER PLIOCENE	Discoaster brouweri	Cyclococcolithina macintyrei	NN 18 Discoaster brouweri
			Discoaster pentaradiatus	NN 17 Discoaster pentaradiatus
			Discoaster surculus	NN 16 Discoaster surculus
			Discoaster tamalis	
3.0	LOWER PLIOCENE	Reticulofenestra pseudoumbilica	Discoaster asymmetricus	NN 15 Reticulofenestra pseudoumbilica
			Sphenolithus neoabies	
		Ceratolithus tricorniculatus	Ceratolithus rugosus	NN 14 Discoaster asymmetricus NN 13 Ceratolithus rugosus
5.0	UPPER MIOCENE		Ceratolithus acutus	NN 12 Ceratolithus tricorniculatus
			Triquetrorhabdulus rugosus	
		Discoaster quinqueramus	Ceratolithus primus	NN 11 Discoaster quinqueramus
			Discoaster berggrenii	
		Discoaster neohamatus	Discoaster neorectus	NN 10 Discoaster calcaris
			Discoaster bellus	
11.0	MIDDLE MIOCENE	Discoaster hamatus	Catinaster calyculus	NN 9 Discoaster hamatus
			Helicopontosphaera kamptneri	
		Catinaster coalitus		NN 8 Catinaster coalitus
		Discoaster exilis	Discoaster kugleri	NN 7 Discoaster kugleri
			Coccolithus miopelagicus	NN 6 Discoaster exilis
		Sphenolithus heteromorphus		NN 5 Sphenolithus heteromorphus
15.0	LOWER MIOCENE	Helicopontosphaera ampliaperta		NN 4 Helicopontosphaera ampliaperta
		Sphenolithus belemnos		NN 3 Sphenolithus belemnos
		Triquetrorhabdulus carinatus	Discoaster druggii	NN 2 Discoaster druggii
			Discoaster deflandrei	NN 1 Triquetrorhabdulus carinatus
23.0	UPPER OLIGOCENE (part)		Cyclicargolithus abisectus	

PACIFIC COAST NEOGENE STAGES, LETTER DIVISIONS AND ORIGINAL PROVINCIAL AGE ASSIGNMENTS AFTER NATLAND (1952), KLEINPELL (1938) AND WISSLER (1941)			
STAGES		WISSLER (1941)	SERIES AND SUBSERIES
Hallian		Pico	PLEISTOCENE
Wheelerian		Pico	UPPER PLIOCENE
Venturian			MIDDLE PLIOCENE
Repettian		Upper	LOWER PLIOCENE
		Mid.	
		Lower	
Delmontian		Div. A	UPPER MIOCENE
		Div. B	
		Div. C	
nian	Upper	Div. D	
	Lower	Div. E	
Luisian		Div. F	MIDDLE MIOCENE
Relizian			
cesian	Upper		LOWER MIOCENE (part)
	Middle		
	Lower		
Upper Zemorrian			

NEOGENE PLANKTONIC FORAMINIFERAL ZONES, EUROPEAN STAGES AND ESTIMATED ABSOLUTE AGES AFTER BANNER AND BLOW (1965), BLOW (1969), BERGGREN (1972) AND BERGGREN AND VAN COUVERING (1974)			
MM YRS. B.P.	EPOCH / SERIES	EUROPEAN STAGES	PLANKTONIC FORAMINIFERAL ZONES
	PLEISTOCENE	Tyrrhenian	N.23 Globigerina calida/ Sphaeroidinella dehiscens excavata
0.2		Milazzian	
0.6		Sicilian	
0.8		Emilian	N.22 Globorotalia truncatulinoides
1.0		Calabrian	
1.8	PLIOCENE	Piacenzian	N.21 Globorotalia tosaensis
3.5		Zanclian	N.20 Globorotalia multicamerata - Pulleniatina obliquiloculata
			N.19 Sphaeroidinella dehiscens - Globoquadrina altispira
			? N.18 G. tumida - S. subdehiscens paenedehiscens ?
5.0	LATE MIOCENE	Messinian	N.17 Globorotalia tumida plesiotumida
6.5		Tortonian	?
			N.16 Globorotalia acostaensis - G. merotumida
			?
			N.15 Globorotalia continuosa
11.0	MIDDLE MIOCENE	Serravalian	N.14 Globigerina nepenthes / Globorotalia siakensis
			N.13 Sphaeroidinellopsis subdehiscens - G. druryi
			N.12 Globorotalia fohsi
			N.11 Globorotalia praefohsi
13.0		Langhian	N.10 Globorotalia peripheroacuta
			N.9 Orbulina suturalis - Globorotalia peripheroronda
			N.8 Globigerinoides sicanus - Globigerinatella insueta
14.5	EARLY MIOCENE	Burdigalian	N.7 G. insueta - Globigerinoides quadrilobatus trilobus
			N.6 G. insueta - Globigerinita dissimilis
19.0		Aquitanian	N.5 G. dehiscens praedehiscens - G. dehiscens dehiscens
			?
			N.4 G. quadrilobatus primordius / Globorotalia kugleri
22.5	LATE OLIGOCENE (part)	Chattian (part)	P.22/N.3 Globigerina angulisuturalis

FIG. 4 — Tentative relations of Neogene calcareous plankton zonations with marine stages of Pacific Coast.

FIG. 5 — Inferred relations among Cenozoic marine stages of California, U.S. Atlantic and Gulf Coastal Provinces, Europe, and New Zealand.

INFERRED RELATIONSHIPS OF THE CENOZOIC MARINE STAGES OF CALIFORNIA TO THOSE OF THE U. S. ATLANTIC AND GULF COASTAL PROVINCE, EUROPE AND NEW ZEALAND BASED ON DISTRIBUTION OF NANNOPLANKTON AND PLANKTONIC FORAMINIFERA

SERIES / SUBSERIES	ESTIMATED "ABSOLUTE" AGES IN MILLIONS OF YEARS B. P.	CALIFORNIA	ATLANTIC AND GULF COASTAL PROVINCE	EUROPE	NEW ZEALAND
PLEISTOCENE	1.8	Hallian Wheelerian	Late Wisconsinan Bradyan Early Wisconsinan Sangamon Illinoisan Yarmouthian Kansan Aftonian	Tyrrhenian Milazzian Sicilian Emilian Calabrian	Castlecliffian Okehuan Nukumaruan Huatawan
PLIOCENE	5.0	Venturian Repettian	Nebraskan Foleyan	Piacenzian Zanclian	Waitotaran Opoitian
UPPER MIOCENE	11.0	Delmontian	Foleyan Clovellyan	Messinian Tortonian	Kapitian Tongaporutuan
MIDDLE MIOCENE	15.0	Mohnian Luisian	Clovellyan Duck Lakian	Serravalian Langhian	Waiauan Lillburnian
LOWER MIOCENE	23.0	Relizian Saucesian	Duck Lakian Napoleonvillian Anahuacian	Burdigalian Aquitanian	Clifdenian Altonian Awamoan Hutchinsonian Otaian
OLIGOCENE	38.0	Zemorrian	Chickasawhayan Vicksburgian	Chattian Rupelian Lattorfian	Waitakian Duntroonian Whaingaroan
UPPER EOCENE	42.0	Refugian	Jacksonian	Bartonian Priabonian	Runangan Kaiatan
MIDDLE EOCENE	49.0	Narizian	Claibornian	Lutetian	Bortonian Porangan
LOWER EOCENE	53.0	Ulatisian Penutian Bulitian	Sabinian	Ypresian	Heretaungan Mangaorapan Waipawan
PALEOCENE	63.0	Ynezian Cheneyan	Sabinian Midwayan	Thanetian Danian	Teurian

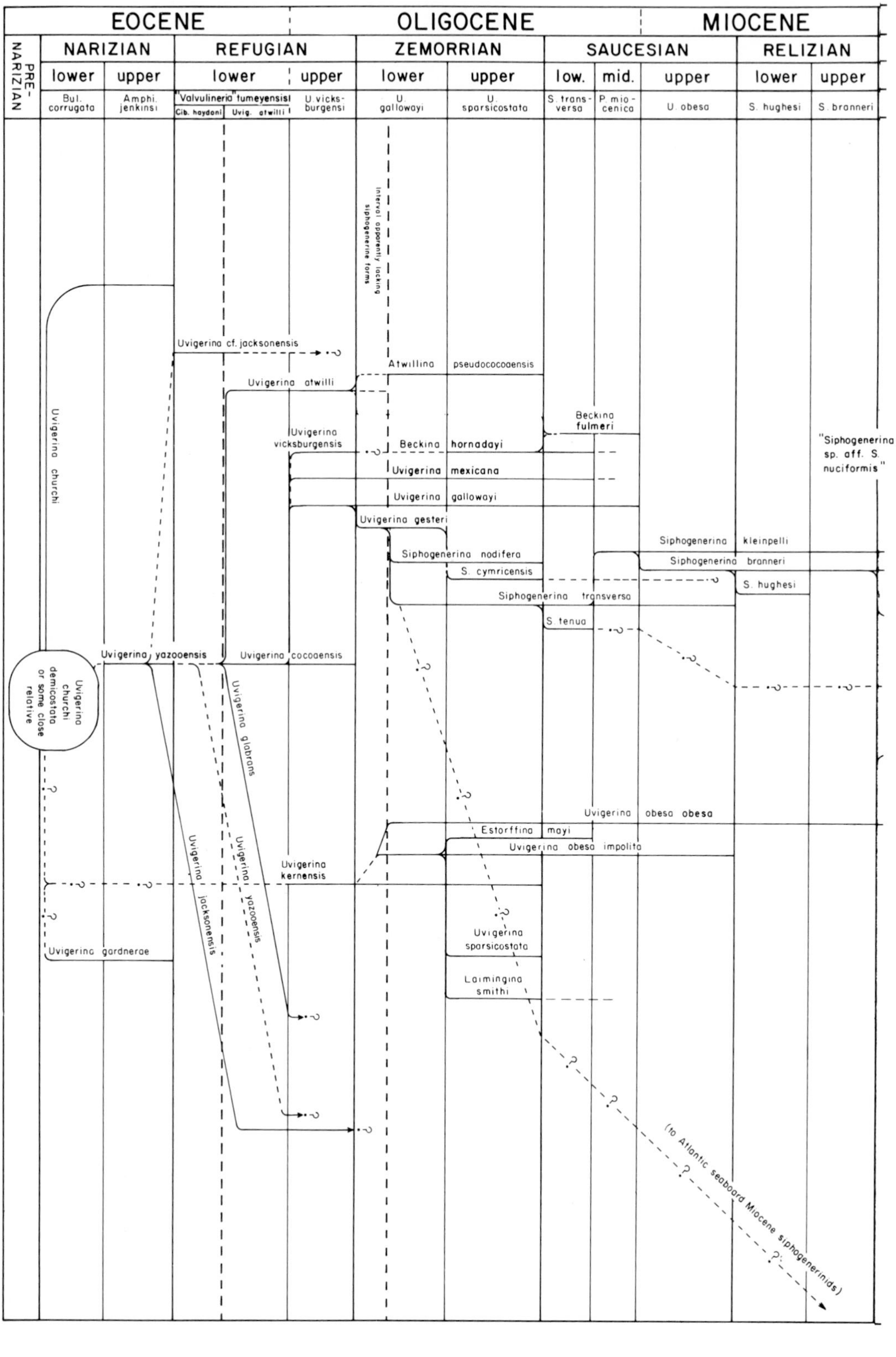

EOCENE					OLIGOCENE					MIOCENE	
PRE-NARIZIAN	NARIZIAN		REFUGIAN		ZEMORRIAN		SAUCESIAN			RELIZIAN	
	lower	upper	lower	upper	lower	upper	low.	mid.	upper	lower	upper
	Bul. corrugata	Amphi. jenkinsi	"Valvulineria" tumeyensis (Cib. haydoni / Uvig. atwilli)	U. vicksburgensi	U. gallowayi	U. sparsicostata	S. transversa	P. miocenica	U. obesa	S. hughesi	S. branneri

Interval apparently lacking siphogenerine forms

Uvigerina cf. jacksonensis

Atwillina pseudococoaensis

Uvigerina atwilli

Uvigerina churchi

Beckina fulmeri

Uvigerina vicksburgensis

Beckina hornadayi

"Siphogenerina sp. aff. S. nuciformis"

Uvigerina mexicana

Uvigerina gallowayi

Uvigerina gesteri

Siphogenerina kleinpelli

Siphogenerina nodifera

Siphogenerina branneri

S. cymricensis

S. hughesi

Siphogenerina transversa

S. tenua

Uvigerina yazooensis

Uvigerina cocoaensis

Uvigerina churchi demicostata or some close relative

Uvigerina glabrans

Uvigerina obesa obesa

Estorffina mayi

Uvigerina obesa impolita

Uvigerina jacksonensis

Uvigerina yazooensis

Uvigerina kernensis

Uvigerina sparsicostata

Uvigerina gardnerae

Laimingina smithi

(to Atlantic seaboard Miocene siphogenerinids)

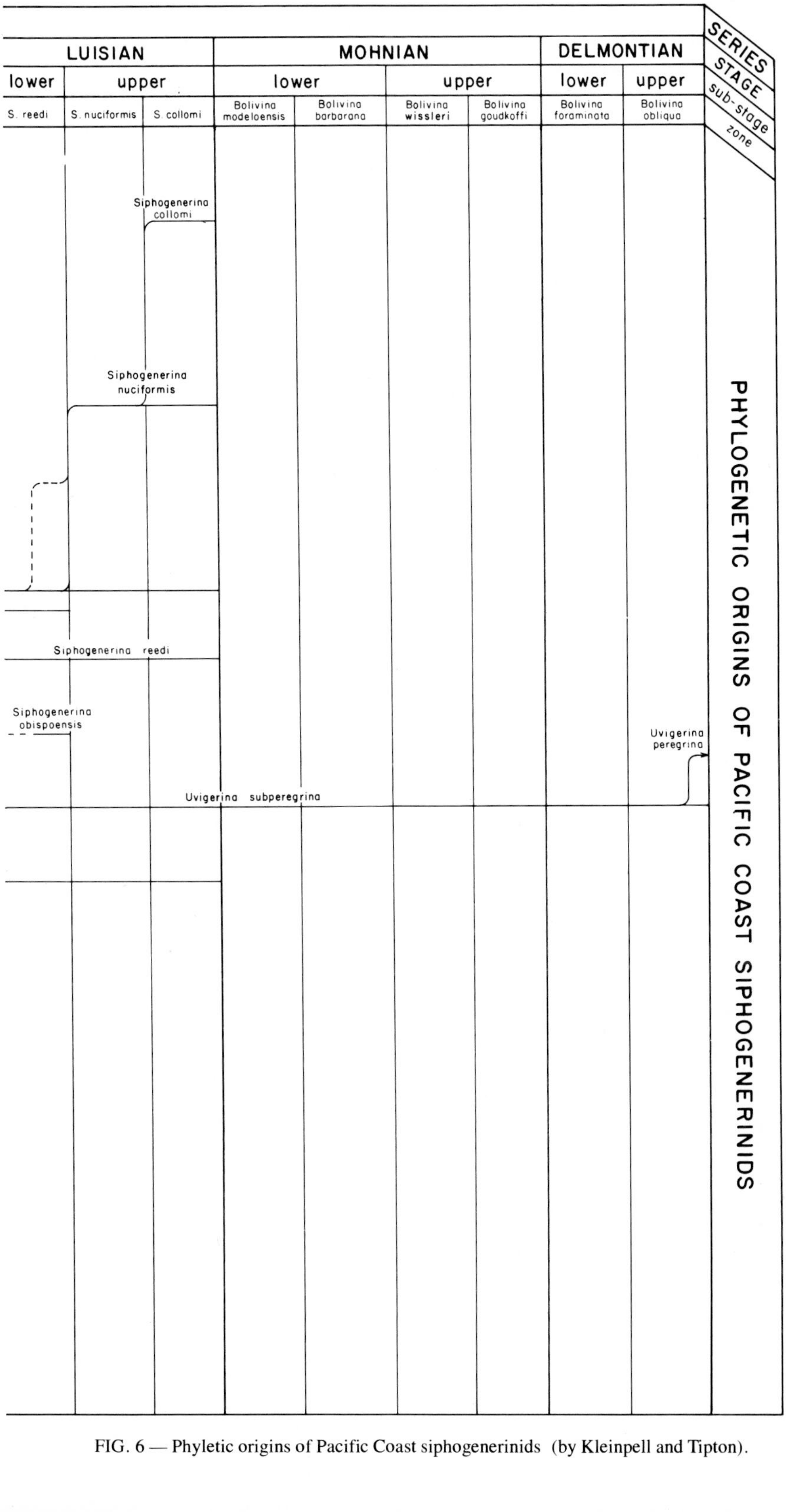

SERIES									
STAGE	LUISIAN			MOHNIAN				DELMONTIAN	
sub-stage	lower	upper		lower		upper		lower	upper
zone	S. reedi	S. nuciformis	S. collomi	Bolivina modeloensis	Bolivina barbarana	Bolivina wissleri	Bolivina goudkoffi	Bolivina foraminata	Bolivina obliqua

Siphogenerina collomi

Siphogenerina nuciformis

Siphogenerina reedi

Siphogenerina obispoensis

Uvigerina peregrina

Uvigerina subperegrina

PHYLOGENETIC ORIGINS OF PACIFIC COAST SIPHOGENERINIDS

FIG. 6 — Phyletic origins of Pacific Coast siphogenerinids (by Kleinpell and Tipton).

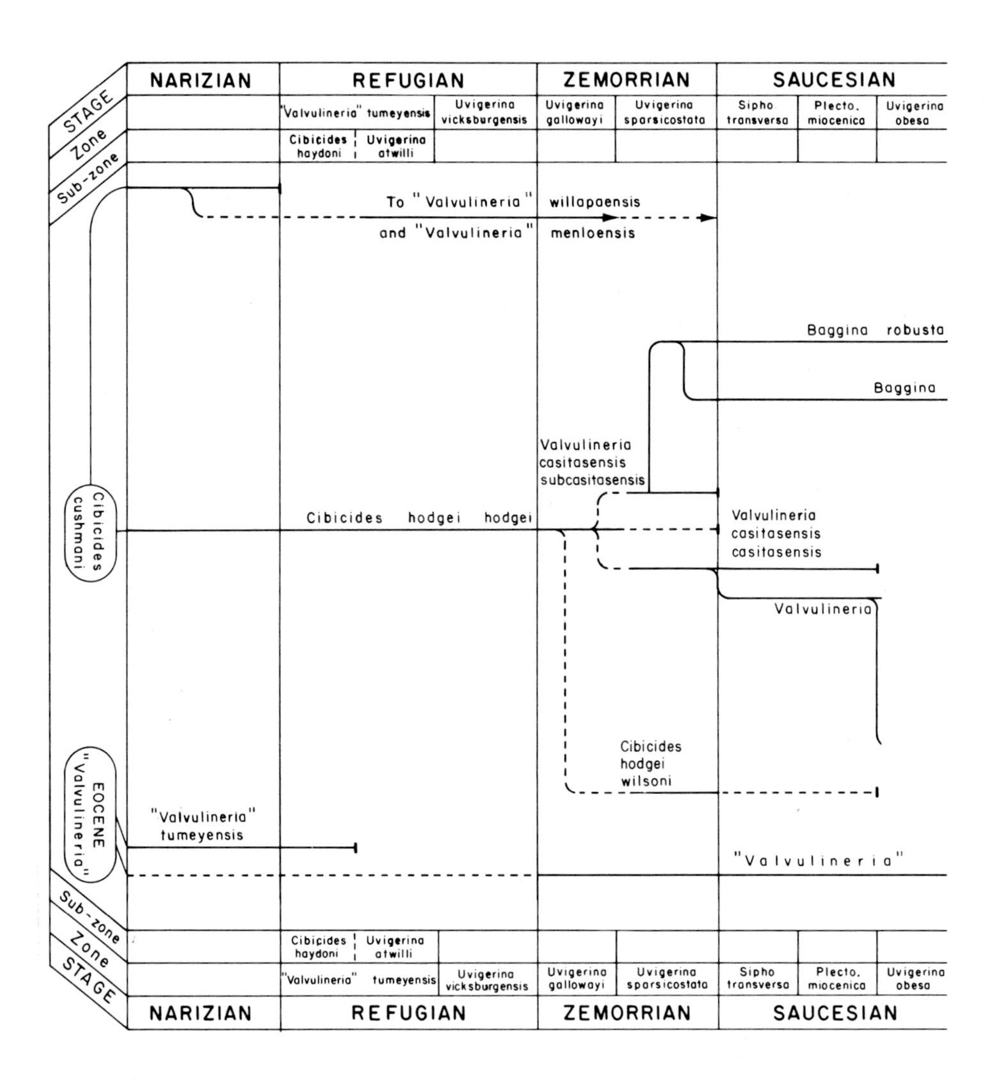
STAGE
Zone
Sub-zone
NARIZIAN
REFUGIAN
ZEMORRIAN
SAUCESIAN
"Valvulineria" tumeyensis
Uvigerina vicksburgensis
Uvigerina gallowayi
Uvigerina sparsicostata
Sipho transversa
Plecto. miocenica
Uvigerina obesa
Cibicides haydoni
Uvigerina atwilli
To "Valvulineria" willapaensis
and "Valvulineria" menloensis
Baggina robusta
Baggina
Valvulineria casitasensis subcasitasensis
Cibicides cushmani
Cibicides hodgei hodgei
Valvulineria casitasensis casitasensis
Valvulineria
Cibicides hodgei wilsoni
EOCENE "Valvulineria"
"Valvulineria" tumeyensis
"Valvulineria"

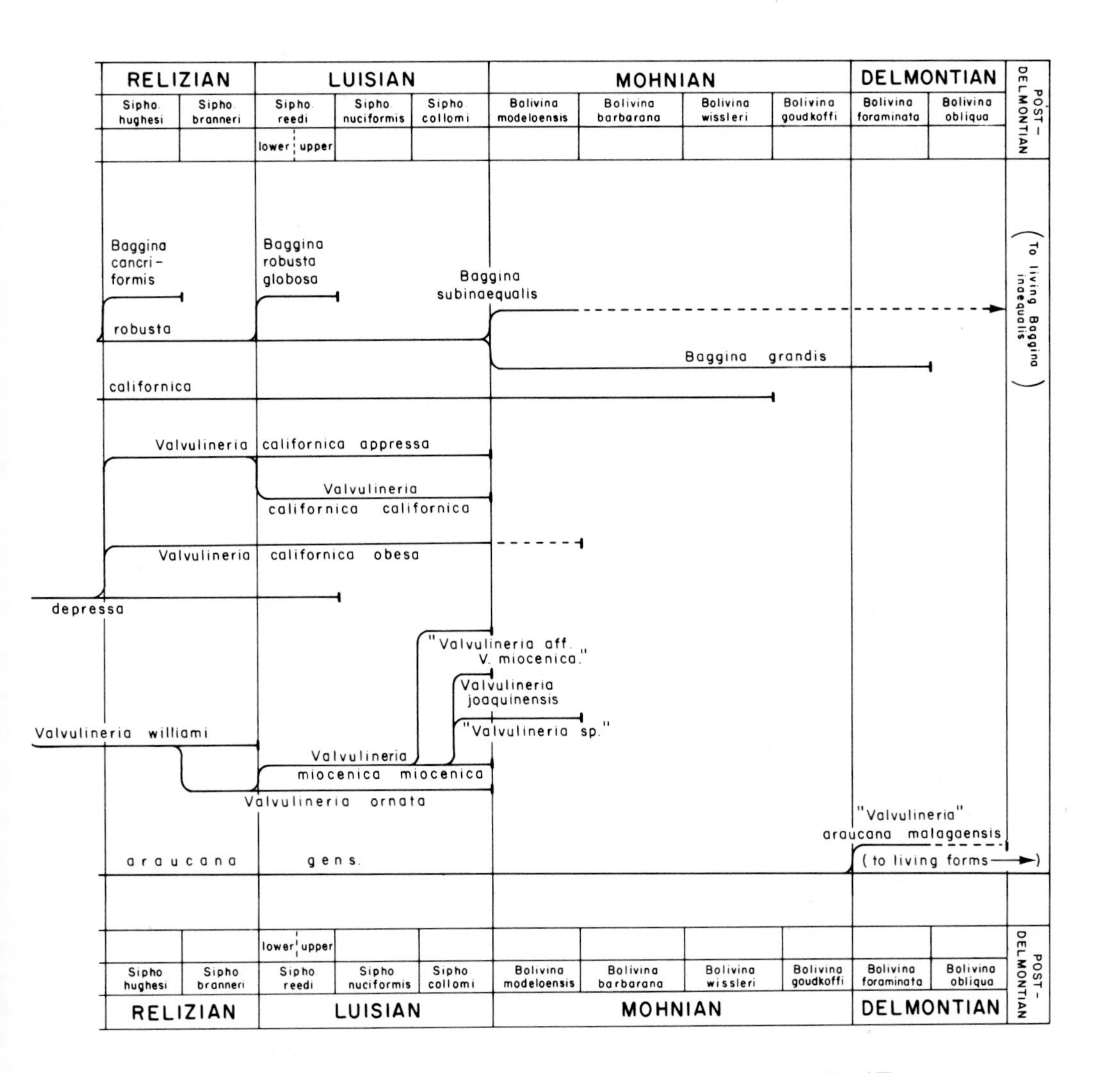

FIG. 7 — Phyletic origins of Pacific Coast valvulinerids (by Kleinpell and Tipton).

PLATES I - XXVI

In the following illustrations of California middle Tertiary foraminifers the reader will discern, in a phylogenetic sense, a notably uneven coverage. As mentioned at the outset, this has been inherent in the nature of the aim of the paper, since it constitutes a supplement. Only a few new species have needed figuring. However, where the phyletic relations of some of the long-known species have come under a more clearly defined discipline, additional figures have been deemed one of the most significant features in attempting to update previous work. Thus, with a few exceptions apparently of cryptogenetic origin, the plates are arranged in phyletic order, eclectically so, yet in general keeping with the basic classifications of Schubert and his successors Cushman, Galloway, and others. Illustrations are by Mary Taylor, except Plate XVII, Plate XIII, figures 6, 7, 9, 10, 11, 13, 14, and 15, and Plate XXIII, figures 1, 2, 3, and 4, which are by Margaret Moore Hanna.

As noted previously, relationships among the many bolivinid and nonionid species are often indicated, although for many, these are still far from clear. Though in no sense a true bioseries, a lineage from *Bolivina floridana* through others comparable in morphology to *B. sinuata sinuata* seems rather well defined and would appear to represent the true genus *Bolivina*. Whether this is in the form of a genus or a subgenus is not clear in view of the apparent polyphyletic origins of others referred to *Brizalina:* some of the latter are of buliminid origins through *Virgulina,* others apparently are of heterohelicid origins although this is more difficult to demonstrate. The lineage of *Bolivina seminuda,* apparently of northern origins, seems well defined for a half-dozen species. Degrees of preservation commonly affect the superficial appearance of the end-species of this lineage as has been depicted in the two pairs of specimens on Plate VIII (Figs. 5, 6, 9, 10). *Bolivina bramlettei* is clearly close to *B. modeloensis* (see Pl. VII) but to what degree is not clear (see also Cushman and Kleinpell, 1934, Pl. 2, Figs. 4a,b), but this merely brings up the age-old paleontologic problem of "splitting" and "lumping," which is again indicated in at least two or three other instances. Many forms of *Bolivina woodringi* show a few costae on the earliest chambers, others do not; the noncostate variants are figured here. *"Bolivina californica* var." of Warren (Pl. IX, Figs. 15, 16) seems indeed distinct from *B. californica californica,* but in view of the rarity of its records it is not clear whether this difference is subspecific, varietal, or a matter of individual variation. However, *"Nonion costiferum* var." of Warren shows consistent distinction from one end of the state to the other and consequently has been considered and described as a new species. Again, in the Zemorrian strata of the coastal areas, *Cibicides hodgei hodgei, C. hodgei wilsoni, Valvulineria casitasensis casitasensis,* and *V. casitasensis subcasitasensis* generally can be distinguished readily from each other in typical form, yet occasional individuals defy classification in other than open nomenclature. In effect these are "missing links" that are no longer "missing," and thereby serve to link the ancestry of the California Miocene genus *Valvulineria* with a radicle other than has been supposed.

Another item worth noting with reference to the plates is the distortion that naturally meets the micropaleontologist's eye, owing to the diversity of magnification employed in any one plate. Thus the plate does not present to him in its actual form the assemblage of specimens he views under his microscope, and he must exercise marked care and imagination in making his comparisons. This is simply a practical matter, owing to the diversity in size of the adults of many even closely related species. A comparison of *Bolivina parva* with other related bolivinids on Plate IX will bring this problem into bold relief. Our artists have depicted two plates of *Siphogenerina* (Pl. XVI and XVII) in one of which the diversity of magnification varies considerably and in the other is at a minimum. Both kinds of presentations—comparative size of specimens pictured, and number of species that may be figured on any single plate—have their respective practical significance. This involves a differential of 20 to 25% in efficiency of presentation.

Finally, the stratigraphic occurrences of the pictured individuals are indicated, either directly or indirectly, on the plate legends. The stratigraphic ranges as now known for the siphogenerines and valvulinerids are graphically depicted in Figures 6 and 7. Other species that appear to be of chronologic significance have, to the best of our ability, been noted as such in the lists of zonal congregations in the text.

For the older or Oligocene part of the column, formerly the "lower Miocene" of local usage, the number of species figured is minimal because so many of them have already been figured extensively in publications such as those of Rau (1948,1951), Sullivan (1962), Fairchild et al (1969), R. K. Smith (1971), Tipton et al (1973), and Fulmer (1975). For geologic distribution of the strata chronologically characterized by the foraminifer species depicted in the following plates, the reader is referred to the annual Pacific Section AAPG-SEPM-SEG Guidebooks prepared for the field trips that since about 1952 have replaced the old "bug-of-the-month-club" meetings (see Kleinpell 1971, 1972) of the SEPM in California. A host of additional biostratigraphic data will be found in the Pacific Section AAPG geologic cross sections 1-12 (1951-59), and the east-west trans-San Andreas fault cross sections AA through HH compiled by E. W. Christensen, T. W. Dibblee, Jr., E. A. Gribi, Jr., John L. Kilkenny, R. M. Kleinpell, O. F. Kotick, Gladys P. Louke, Harold L. Reade, Jr., et al, for the Pacific Section AAPG Committee for Cross Sections chaired by Max B. Payne, co-chairman Richard J. Walters, and Otto Hackel (1967).

Type numbers given in the following plate legends (as to some extent in the Taxonomy section also) refer to the deposition of the figured specimens in the private collection of the senior writer. References to "RMK 1938" are simply abbreviated references to the book being updated in the present supplement. References to the localities ("Loc.") from which the specimens figured on the following plates have been collected are to the locality descriptions recorded in Figure 1 or in Tables 1 through 6 of the present supplement when shown in Arabic numerals, but to the tables in the 1938 work when shown in Roman numerals. The latter are *not* to be confused with the Roman numerals of the plates themselves in the present supplement which, other than in the plate legends, are shown only without other bibliographic context at various places in the script, or in the section on Taxonomy in cases of the type specimens of new species. Locality numbers prefixed with the letters LSJU refer to collecting localities listed under these numbers in the Leland Stanford Junior University Register of Localities, as are also the locality numbers prefixed with the letter M, shown here in Table 6 and which are also Leland Standford Junior University locality numbers (see Key, 1955, p. 52, and Geological Map Pl. 1). Other localities and type numbers listed, such as UC (for University of California, Berkeley), those of the U. S. National Museum, and the California Academy of Sciences (CAS) will be self-explanatory.

PLATE I

STRATIGRAPHICALLY SIGNIFICANT ROTALIA, PULLENIA, AND LAGENIDAE FROM CALIFORNIA MIOCENE

Figure	
1a-c	*Rotalia garveyensis* Natland, ×141. Loc. Fig. 1: WNPB 12. Hypotype RMK 122.
2a-c	*Rotalia garveyensis* Natland, ×94. Loc. Table 5: SBC 61. Hypotype RMK 123.
3a,b	*Pullenia miocenica* Kleinpell, ×113. Loc. LSJU 793 (Snedden, 1932, p. 44: *"Pullenia* cf. *bulloides";* RMK 1938, p. 339). Holotype RMK 124.
4a,b	*Robulus miocensicus* (Chapman), ×49. Loc. Key (1955, pl. 1) M521E, lower Devilwater Silt Member, Sec. 26, R18E, T27S. Hypotype RMK 124.
5	*Hemicristellaria beali* (Cushman), ×47. Loc. LSJU 791 (Snedden, 1932, p. 44: *"Cristelleria beali";* RMK 1938, p. 206). Hypotype RMK 125.
6	*Hemicristellaria beali* (Cushman), ×29. Loc. Table 6: M519C. Hypotype RMK 126.
7	*Frondicularia foliacea* Schwager, ×56. Loc. Table 6: M519C Hypotype RMK 127.
8	*Robulus smileyi* Kleinpell, ×56. Loc. LSJU 792 (Snedden, 1932, p. 44: "*Robulus* n. sp. Kleinpell (MS)"; RMK 1938, p. 203). Hypotype RMK 128.
9a,b	*Robulus smileyi* Kleinpell, ×51. Loc. Table 6: M519D. Hypotype RMK 129.
10	*Robulus mohnensis* Kleinpell, ×35. Loc. Table VII: 4. Cotype RMK 130.

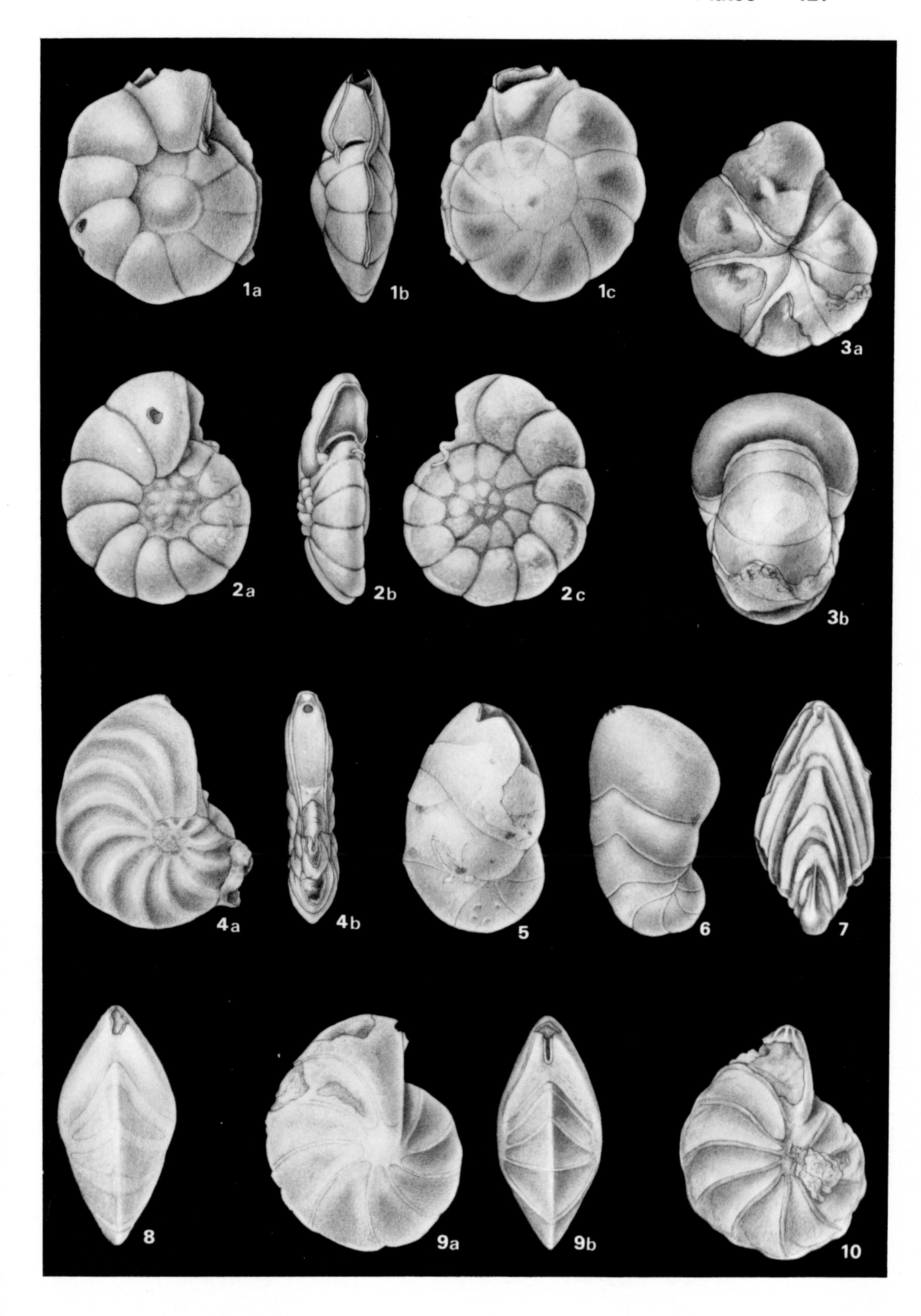
1a
1b
1c
3a
2a
2b
2c
3b
4a
4b
5
6
7
8
9a
9b
10

PLATE II

NONIONIDAE FROM CALIFORNIA MIOCENE

Figure	
1	*Nonion montereyanum montereyanum* Cushman and Galliher, ×71. Loc. Table V: 18 (CV-14). Hypotype RMK 131.
2	*Nonion schencki* Kleinpell, ×62. Loc. Table V: 30 (LSJU 335). Hypotype RMK 132.
3a,b	*Nonion weaveri* n. sp., ×94. Loc. Fig. 1: 5. Holotype RMK 101.
4	*Nonion montereyanum carmeloensis* Cushman and Galliher, ×51. Loc. Table V: 21 (CV-15). Hypotype RMK 133.
5	*Nonion weaveri* n. sp., ×66. Loc. Table XIII: LHF-8. Hypotype RMK 13.
6a,b	*Nonion weaveri* n. sp., ×80. Loc. Table 6: M519C, Paratype RMK 102.
7	*Nonion costiferum* (Cushman), ×62. Loc. Table 4: SBC 35. Hypotype RMK 135.
8	*Nonion weaveri* n. sp., ×75. Loc. Table XIII: LHF-8 Paratype RMK 136.
9a,b	*Nonion costiferum* (Cushman), ×62. Loc. Fig. 1: TM-14. Hypotype RMK 137.
10	*Nonion incisum (Cushman),* ×86. Loc. Table 4: SBC-28. Hypotype RMK 138.
11	*Nonion incisum (Cushman),* ×80. Loc. Table 1: HM-28. Topotype RMK 139.
12a,b	*Nonion costiferum (Cuxhman),* ×80. Loc. LSJU 763 (Snedden, 1932, p. 43, 44, 45: *"Nonion costifera";* RMK 1938, p. 230). Hypotype RMK 140.

1
2
3a
3b
4
5
6a
6b
7
8
9a
9b
10
11
12a
12b

PLATE III

NONIONIDAE FROM MOHNIAN AND DELMONTIAN STAGES, CALIFORNIA MIOCENE

Figure

1a,b *Nonion schencki* Kleinpell ×53. Loc. Table V: 23 (CV-16). Paratype RMK 141.

2a,b *Nonion schencki* Kleinpell, 2a, ×53; 2b, ×56. Loc. Table II: 00. 2a, Paratype RMK 142. 2b, Paratype RMK 143.

3a,b *Nonion montereyanum carmeloensis* Cushman and Galliher, ×86. Loc. Table IX: SOCO Piedmont 1 well, depth 2,215-2,218 ft. Hypotype RMK 144.

4a,b *Nonion montereyanum montereyanum* Cushman and Galliher, ×86. Loc. Table IX: SOCO Piedmont 1 well, depth 2,215-2,218 ft. Hypotype RMK 145.

5 *Nonion schencki* Kleinpell, ×59. Loc. Table V: 32 (LSJU 662). Paratype RMK 146.

6 *Nonion montereyanum montereyanum* Cushman and Galliher, ×56. Loc. Table V: 21 (CV-15). Hypotype RMK 147.

7a,b *Nonion montereyanum montereyanum* Cushman and Galliher, ×59. Loc. Table V: 23 (CV-16). Hypotype RMK 148.

8a,b *Nonion montereyanum montereyanum* Cushman and Galliher, ×71. Loc. Table V: 20 (LSJU 330). Hypotype RMK 149.

9a,b *Nonion montereyanum carmeloensis* Cushman and Galliher, ×75. Loc. Table V: 17 (CV-13). Hypotype RMK 150.

10a,b *Nonion montereyanum carmeloensis* Cushman and Galliher, ×71. Loc. Fig. 1: NEW 27. Hypotype RMK 151.

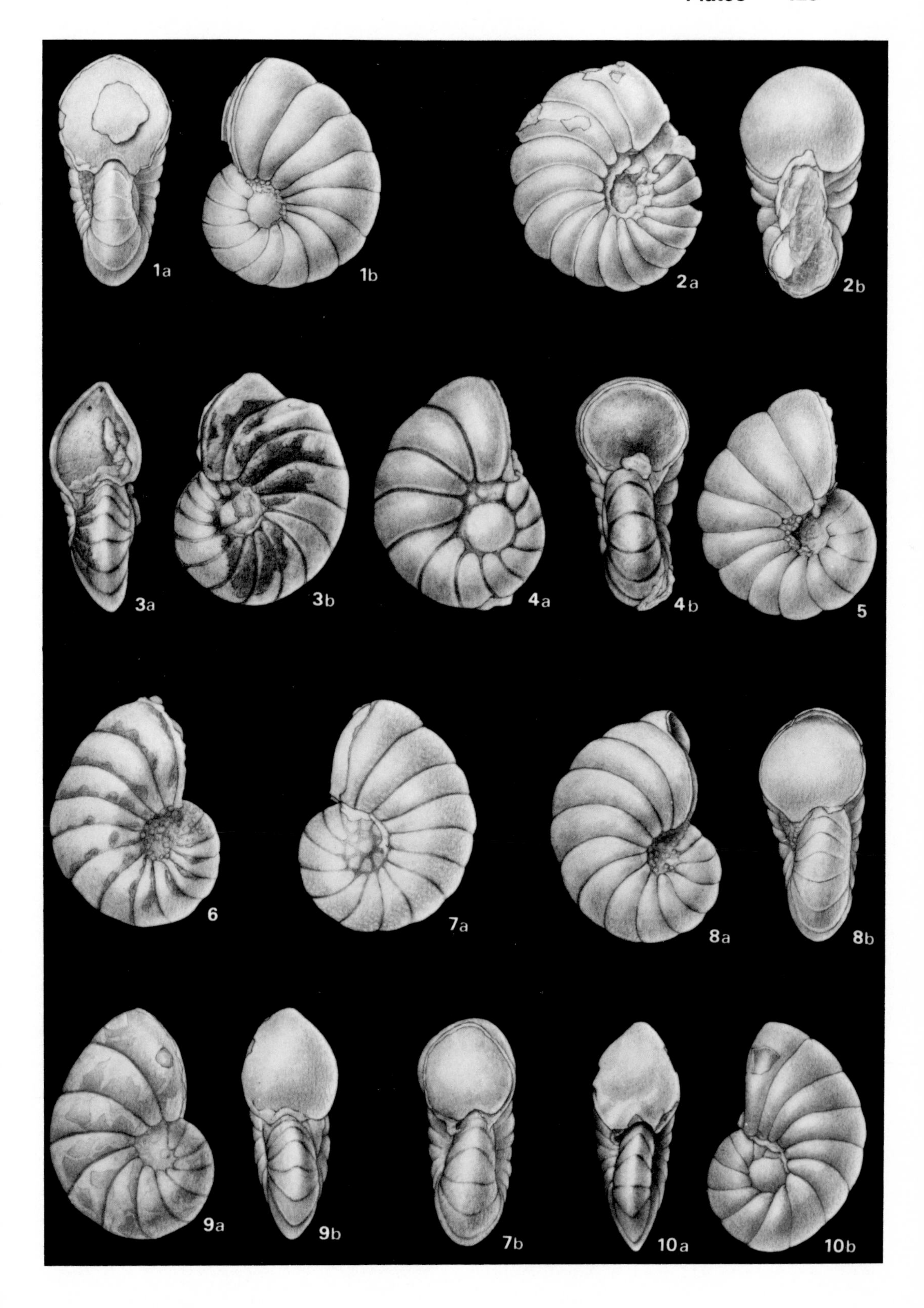
1a
1b
2a
2b
3a
3b
4a
4b
5
6
7a
8a
8b
9a
9b
7b
10a
10b

PLATE IV

VARIANTS OF NONION MULTICAMERATUM CUSHMAN AND KLEINPELL, MOHNIAN OF CALIFORNIA

Figure	
1a,b	Loc. Table XIII: P-5. Rodeo shale. Hypotype RMK 152. ×45.
2a,b	Loc. Fig. 1: NEW 27. Monterey Shale, Newport. Hypotype RMK 153. ×62.
3	Loc. Table V: 16 (LSJU 337). Type Monterey member 4. Hypotype RMK 154. ×56.
4	Loc. Table V: 14 (Coyote Gulch). Type Monterey member 4. Hypotype RMK 155. ×44.
5a,b	Loc. Table 6: M535. McDonald Shale. Hypotype RMK 156. ×66.
6	Loc. Table V: 16 (LSJU 337). Type Monterey member. 4 Paratype RMK 157. ×56.
7	Loc. Table V: 14 (Coyote Gulch). Type Monterey member 4. Hypotype RMK 158. ×51.
8a,b	Loc. Table V: 15 (LSJU 337). Type Monterey member 4. Paratype RMK 159. ×49.
9a,b	Loc. Table XIII: P-4. Rodeo shale. Hypotype RMK 160. ×53.
10a,b	Loc. Table XIII: P-4. Rodeo shale. Hypotype RMK 161. ×51.

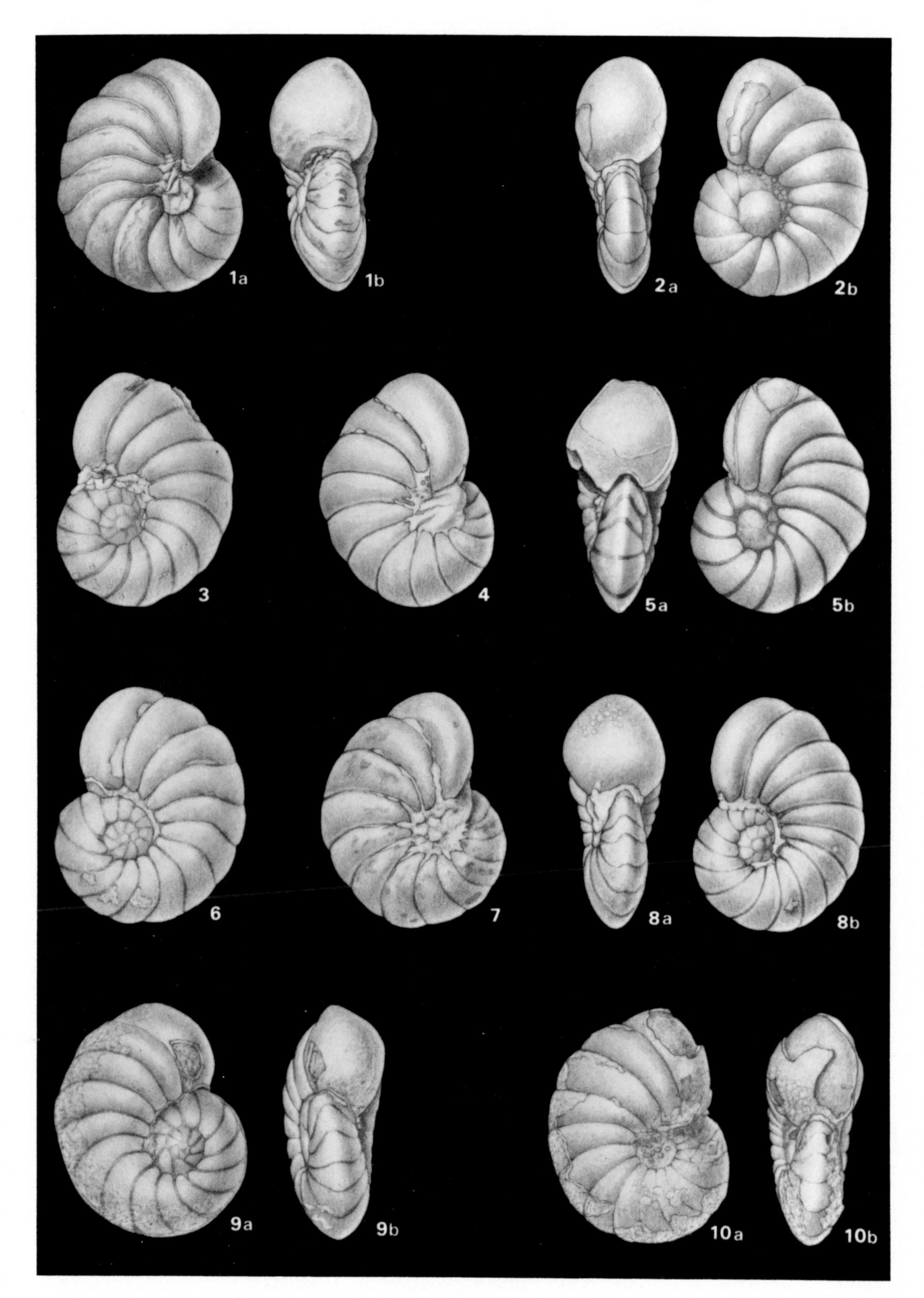
1a
1b
2a
2b
3
4
5a
5b
6
7
8a
8b
9a
9b
10a
10b

PLATE V

MISCELLANEOUS SPECIES FROM CALIFORNIA MIOCENE AND LOWER PLIOCENE

Figure	
1a-c	*Epistominella pacifica* (Cushman), ×86. Loc. Table 6: M535. Hypotype RMK 162.
2	*Plectofrondicularia californica* Cushman and Stewart, ×25. Loc. Repettian of Lomita Gulley, Palos Verdes Hills; collector R. M. Kleinpell. Hypotype RMK 163.
3a-c	*"Valvulineria" araucana* (d'Orbigny), ×66. Loc. Table 5: SBC 59. Hypotype RMK 164.
4	*Plectofrondicularia californica* Cushman and Stewart, ×32. Loc. Repettian of Rincon Point, Ventura–Santa Barbara County line; collector RMK. Hypotype RMK 165.
5a-c	*Nonionella* sp. cf. *N. davanaensis* Pierce, ×94. Loc. Fig. 1: WNPB 20. Hypotype RMK 166.
6	*Plectofrondicularia miocenica* Cushman, ×47. Loc. LSJU 768. (Snedden, 1932, p. 43; RMK 1938, p. 240). Hypotype RMK 167.
7a,b	*Nonion goudkoffi* Kleinpell, ×113. Loc. Fig. 1: WNPB 13. Hypotype RMK 168.
8	*Nonion montereyanum cameloensis* Cushman and Galliher, ×86. Loc. Table XIII: Selby 2 (undifferentiated Monterey Shale). Hypotype RMK 169.
9a,b	*Nonion kernensis* Kleinpell, ×62. Loc. Spring Canyon, Piru quadrangle. Hypotype RMK 170.

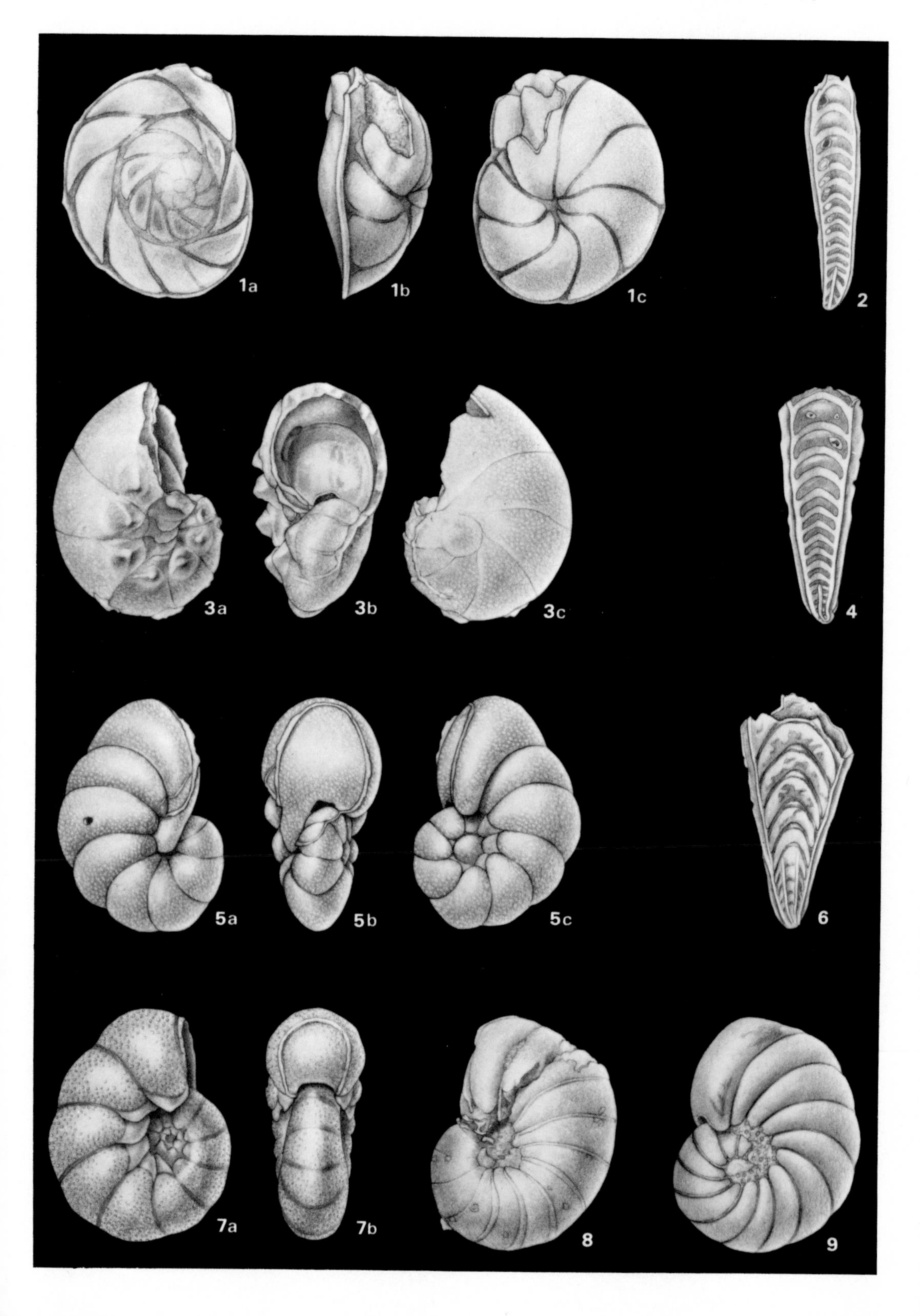
1a
1b
1c
2
3a
3b
3c
4
5a
5b
5c
6
7a
7b
8
9

PLATE VI

BOLIVINIDS FROM CALIFORNIA MIOCENE

Figure	
1	*Bolivina girardensis* Rankin, ×75. Loc. Table 5: SBC 60. Hypotype RMK 171.
2	*Bolivina marginata gracillima* Cushman, ×71 Loc. Table 5: SBC 49. Hypotype RMK 172.
3	*Bolivina imbricata* Cushman, ×62. Loc. Table 5 : SBC 37. Hypotype RMK 173.
4	*Bolivina imbricata* Cushman, ×47. Loc. Table 6: M510. Hypotype RMK 174.
5a-c	*Bolivina girardensis* Rankin, ×94. Loc. Table 5: SBC 54. Hypotype RMK 175.
6a,b	*Bolivina mulleri* n. sp., ×62. Loc. Table 5: SBC 37. Cotype RMK 110.
7a,b	*Bolivina mulleri* n. sp., ×62. Loc. Table 5: SBC 37. Holotype RMK 109.
8	*Bolivina* aff. *cuneiformis* Kleinpell, ×51. Loc. Table 5: SBC 43. Hypotype RMK 176.
9	*Bolivina cuneiformis* Kleinpell, ×102. Loc. Table 5: SBC 49. Hypotype RMK 177.
10	*Bolivina cuneiformis* Kleinpell, ×94. Loc. Table 5: SBC 49. Hypotype RMK 178.

1
2
3
4
5b
6b
7b
5a
5c
6a
7a
8
9
10

PLATE VII

BOLIVINIDS FROM CALIFORNIA MIOCENE

Figure	
1	*Bolivina advena* Cushman, ×59. Loc. Table 4: SBC 34. Hypotype RMK 179.
2	*Bolivina subadvena spissa* Cushman, ×56. Loc. Table 5: SBC 56. Hypotype RMK 180.
3	*Bolivina subadvena spissa* Cushman, ×38. Loc. corner Murietta and Roblar Sts., upper Modelo diatomite; collector A. A. Almgren. Hypotype RMK 181.
4	*Bolivina pseudospissa* Kleinpell, ×75. Loc. Table 5: SBC 54. Hypotype RMK 182.
5	*Bolivina woodringi* Kleinpell, ×38. Loc. Fig. 1: NEW 63. Hypotype RMK 183.
6	*Bolivina woodringi* Kleinpell, ×41. Loc. Table 5: SBC 56. Hypotype RMK 184.
7	*Bolivina* sp. cf. *B. woodringi* Kleinpell, ×80. Loc. corner Murietta and Roblar Sts., upper Modelo diatomite; collector A. A. Almgren. Hypotype RMK 185.
8	*Bolivina woodringi* Kleinpell, ×75. Loc. Table 5: SBC 51. Hypotype RMK 186.
9	*Bolivina modeloensis* Cushman and Kleinpell, ×47. Loc. Table XIII: SPD 4, upper Tice Shale. Hypotype RMK 187.
10	*Bolivina* cf. *B. hughesi* Kleinpell, ×141. Loc. basal McLure shale "one mile south of the pipeline road of the east side of Sulphur Spring Canyon" (Church, 1972, p. 76). Hypotype RMK 188.
11	*Bolivina brevior* Cushman, ×113. Loc. Table 5: SBC 54. Hypotype RMK 189.
12	*Bolivina modeloensis* Cushman and Kleinpell, ×35. Loc. Fig. 1: NEW 27. Hypotype RMK 190.
13	*Bolivina modeloensis* Cushman and Kleinpell, ×51. Loc. VII: 7 (Dry Canyon Road). Cotype RMK 191.
14a,b	*Bolivina bramlettei* Kleinpell, ×86. Loc. Table 5: SBC 57. Hypotype RMK 192.
15a,b	*Bolivina bramlettei* Kleinpell, ×71. Loc. Fig. 1: NEW 58. Hypotype RMK 193.
16	*Bolivina modeloensis* Cushman and Kleinpell, ×44. Loc. basal McLure shale "one mile south of the pipeline road on the east side of Sulphur Spring Canyon" (Church, 1972, p. 76). Hypotype RMK 194.

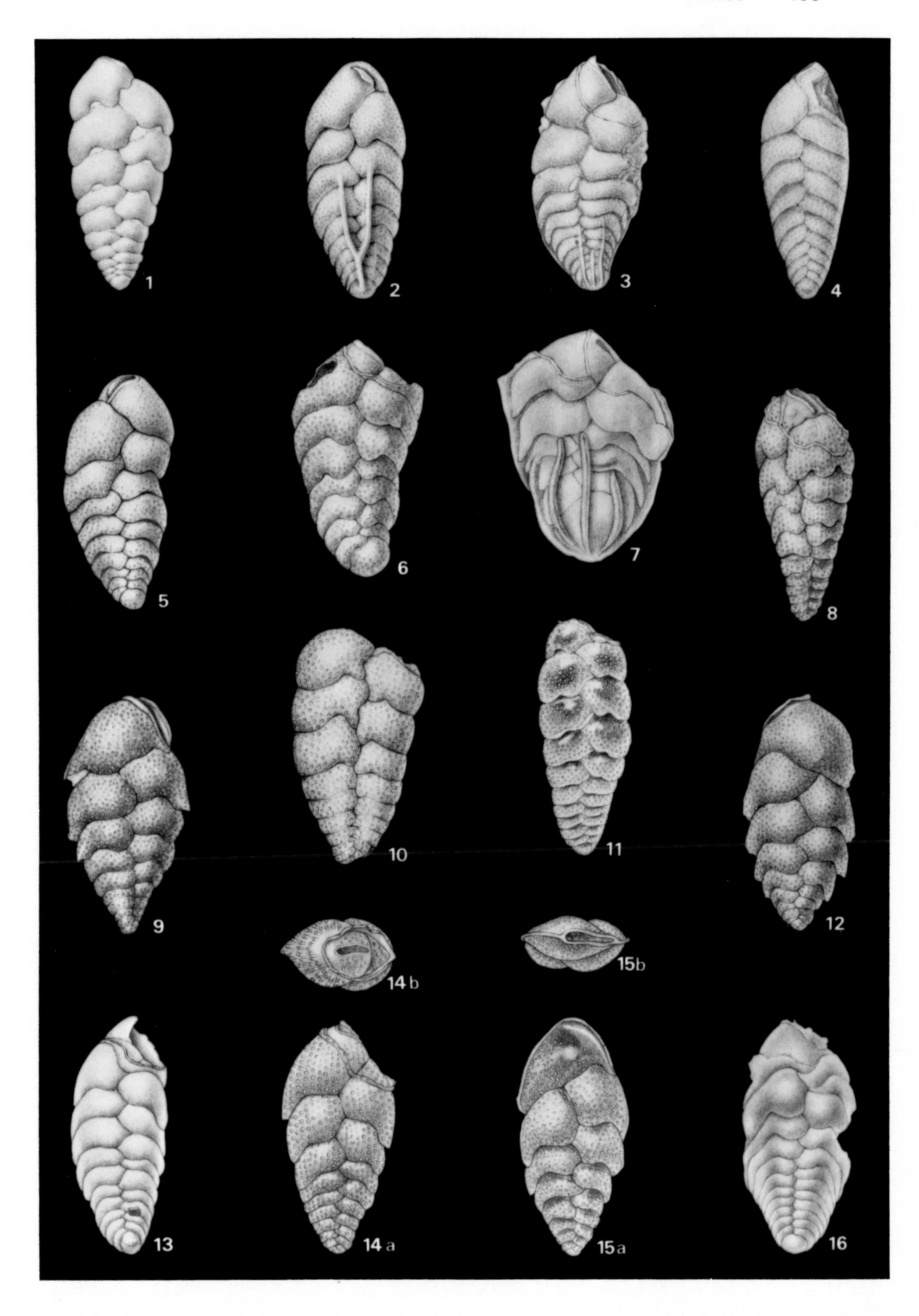
1
2
3
4
5
6
7
8
9
10
11
12
14 b
15b
13
14 a
15a
16

PLATE VIII

BOLIVINIDS FROM CALIFORNIA MIOCENE

Figure	
1	*Bolivina hootsi* Rankin, ×86. Loc. Table 5: SBC 62. Hypotype RMK 195.
2	*Bolivina rankini* Kleinpell, ×161. Loc. Fig. 1: WNPB. Hypotype RMK 196.
3	*Bolivina rankini* Kleinpell, ×80. Loc. Rankin sample 172 (Hoots, 1931, p. 113) Hypotype RMK 197.
4	*Bolivina rankini* Kleinpell, ×113. Loc. Table VII: 1. Cotype RMK 198.
5	*Bolivina seminuda seminuda* Cushman, ×71. Loc. Table V: 32 (LSJU 662). Hypotype RMK 199.
6	*Bolivina seminuda seminuda* Cushman, ×71. Loc. Table V: 31 (CV-26). Hypotype RMK 200.
7	*Bolivina foraminata* R. E. and K. C. Stewart, ×62. Loc. Table 5: SBC 59. Hypotype RMK 201.
8	*Bolivina foraminata* R. E. and K. C. Stewart, ×49. Loc. upper Monterey or Sisquoc diatomite at west side of Goleta Point. Hypotype RMK 202.
9	*Bolivina seminuda seminuda* Cushman, ×113. Loc. Table 5: SBC 59. Hypotype RMK 203.
10	*Bolivina seminuda seminuda* Cushman, ×86. Loc. Table 5: SBC 59. Hypotype RMK 204.
11	*Bolivina sinuata alisoensis* Cushman and Adams, ×44. Loc. Table 6: M535. Hypotype RMK 205.
12	*Bolivina sinuata sinuata* Galloway and Wissler, ×41. Loc. Table 5: SBC 60. Hypotype RMK 206.
13	*Bolivina floridana floridana* Cushman, ×66. Loc. Table 4: SBC 28. Hypotype RMK 207.
14a,b	*Bolivina wissleri* n. sp., ×56. Loc. Fig. 1: TM 14. Holotype RMK 103.
15	*Bolivina wissleri* n. sp., ×44. Loc. Fig. 1: NEW 68. Cotype RMK 104.
16	*Bolivina wissleri* n. sp.(?), ×29. Loc. Table 5: SBC 51. Hypotype RMK 208.

1
2
3
4
5
6
7
8
9
10
11
12
14b
13
14 a
15
16

PLATE IX

BOLIVINIDS FROM CALIFORNIA MIOCENE

Figure	
1	*Bolivina malagaenis* Kleinpell, ×80. Loc. Table 5: SBC 66. Hypotype RMK 209.
2	*Bolivina obliqua* Barbat and Johnson, ×71. Loc. Table IX: SOCO Piedmont 1, depth 1,995-2,000 ft. Hypotype RMK 210.
3	*Bolivina obliqua* Barbat and Johnson,×59. Loc. Table 5: SBC 66. Hypotype RMK 211.
4	*Bolivina obliqua* Barbat and Johnson, ×80. Loc. Table VII: 1 Hypotype RMK 212.
5	*Bolivina obliqua* Barbat and Johnson, ×71. Loc. Table VII: 1. Hypotype RMK 213.
6	*Bolivina parva* Cushman and Galliher, ×86. Loc. Table V: 23. Hypotype RMK 214.
7	*Bolivina hughesi* Cushman, ×38. Loc. Fig. 1: NEW 61. Hypotype RMK 215.
8	*Bolivina* cf. *hughesi* Cushman, ×56. Loc. Table V: 13 (CV 10). Hypotype RMK 216.
9	*Bolivina granti* Rankin, ×44. Loc. Fig. 1: NEW 58. Hypotype RMK 217.
10	*Bolivina goudkoffi* Rankin, ×71. Loc. Table IX: SOCO Piedmont 1, depth 2,215-2,218 ft. Hypotype RMK 218.
11	*Bolivin churchi churchi* n. sp., n. subsp., ×62. Loc. basal McLure shale "one mile south of the pipeline road of the east side of Sulphur Spring Canyon" (Church, 1972, p. 76). Holotype RMK 107.
12	*Bolivina churchi churchi* n. sp., n. subsp., ×53. Loc. basal McLure shale "one mile south of the pipeline road of the east side of Sulphur Spring Canyon" (Church, 1972, p. 76). Cotype RMK 108.
13	*Bolivina churchi almgreni* n. sp., n. subsp., ×62. Loc. corner Murietta and Roblar Sts., upper Modelo diatomite; collector A. A. Almgren. Cotype RMK 106.
14a,b	*Bolivina churchi almgreni* n. sp., n. subsp., ×45. Loc. corner Murietta and Roblar Sts., upper Modelo diatomite; collector A. A. Almgren. Holotype RMK 105.
15a,b	*Bolivina californica* Cushman var., ×102. Loc. Table 5: SBC 39. Hypotype RMK 219.
16a,b	*Bolivina californica* Cushman var., ×113. Loc. Fig. 1: NEW 40. Hypotype RMK 220.
17a,b	*Bolivina californica* Cushman, ×86. Loc. Table 5: SBC 50. Hypotype RMK 221.
18	*Bolivina californica* Cushman, ×94. Loc. Table 4: SBC 12. Hypotype RMK 222.

1
2
3
4
5
6
7
8
9
10
11
12
13
14 b
14 a
15 b
16 b
17 b
15 a
16 a
17 a
18

PLATE X

BULIMINIDAE FROM CALIFORNIA MIOCENE

Figure

1 *Virgulina californiensis grandis* Cushman and Kleinpell, ×51. Loc. Table 5: SBC 56. Hypotype RMK 223.

2 *Virgulina californiensis ticeensis* Cushman and Kleinpell, ×51. Loc. Table XIII: SPD 4. Upper Tice Shale. Cotype RMK 224.

3 *"Elipsoglandulina" fragilis* Bramlette, ×25. Loc. see USGS Prof. Paper 222, p. 60, pl. 22. "Metatype" (topotype) RMK 225. Collector M. N. Bramlette.

4 *Hopkinsina magnifica* Bramlette, ×23. Loc. see USGS Prof. Paper 222, p. 59-60, pl. 22. "Metatype" (topotype) RMK 226. Collector M. N. Bramlette.

5 *Virgulina delmonteensis* Cushman and Galliher, ×66. Loc. Table V: 30 (LSJU 335). Hypotype RMK 227.

6 *Bulimina montereyana montereyana* Kleinpell, ×32. Loc. Table 6: M535. Hypotype RMK 228.

7 *Bulimina montereyana montereyana* Kleinpell, ×31. Loc. Table II: 00. Hypotype RMK 229.

8 *Bulimina brevior* Cushman, ×75. Loc. Table 5: SBC 52. Hypotype RMK 230.

9 *Bulimina montereyana delmonteensis* Kleinpell, ×71. Loc. Table 6: M535. Hypotype RMK 231.

10 *Bulimina pseudoaffinis* Kleinpell, ×71. Loc. LSJU 738 (Snedden, 1932, p. 44: "*Bulimina* sp. A"; RMK 1938, p. 257). Hypotype RMK 232.

11 *Buliminella semihispida* Kleinpell, ×80. Loc. Table 5: SBC 61. Hypotype RMK 233.

12 *Virgulinella miocenica* Cushman and Ponton, ×41. Loc. Table XIII: P-5. Rodeo Shale. Hypotype RMK 234.

13 *Bulimina alligata* Cushman and Laiming, ×71. Loc. Table 4: SBC 3. Hypotype RMK 235.

14 *Bulimina pseudotorta* Cushman, ×49. Loc. LSJU 786 (Snedden, 1932, p. 46: "*Bulimina* cf. *ovula* . . .(in part)"; RMK 1938, p. 258). Preservation is poor but all of Snedden's forms tentatively assigned by him as shown appear instead to be Cushman's common upper Relizian and Luisian species. Hypotype RMK 236.

1
2
3
4
5
6
7
8
9
10
11
12
13
14

PLATE XI

UVIGERINE BULIMINIDAE FROM CALIFORNIA MIOCENE

Figure	
1	*Galliherina uvigerinaformis doanei* n. subsp., ×66. Loc. Fig. 1: NEW 68. Holotype RMK 111.
2	*Galliherina uvigerinaformis doanei* n. subsp., ×66. Loc. Fig. 1: NEW 51. Paratype RMK 112.
3	*Galliherina delreyensis* (Cushman and Galliher), ×51. Loc. Fig. 1: WNPB 13. Hypotype RMK 137.
4	*Galliherina uvigerinaformis doanei* n. subsp., ×29. Loc. contact between middle Modelo shale and upper Modelo sandstone, Fresno Canyon, western end of Sulphur Mountain, Ventura County. Hypotype RMK 238.
5	*Galliherina uvigerinaformis warreni* n. subsp., ×56. Loc. Fig. 1: NEW 63. Holotype RMK 113.
6	*Galliherina uvigerinaformis warreni* n. subsp., ×49. Loc. Fig. 1: NEW 63. Holotype RMK 114.
7	*Galliherina delreyensis* (Cushman and Galliher), ×51. Loc. Table V: 30 (LSJU 335). This Delmontian species often has been confused in the literature with the subspecies *G. u. warreni,* to which it is closely related, although the subspecies *warreni* to date is known only from stratigraphically lower horizons, (i.e., horizons within the Mohnian Stage). Hypotype RMK 239.
8	*Galliherina uvigerinaformis uvigerinaformis* (Cushman and Kleinpell), ×42. Loc. Table 5: SBC 50. Cotype RMK 240.
9	*Galliherina uvigerinaformis uvigerinaformis* (Cushman and Kleinpell), ×38. Loc. Fig. 1: NEW 27. Hypotype RMK 241.
10	*Galliherina uvigerinaformis uvigerinaformis* (Cushman and Kleinpell), ×42. Loc. Fig. 1: NEW 27. This specimen of the species *sensu stricto* shows signs of developing toward the subspecies *warreni,* but the costae are too low and too broad to be those of *warreni* even though atypically a few too numerous. Hypotype RMK 242.
11	*Uvigerinella californica gracilis* (Cushman and Kleinpell), ×71. Loc. Table 4: SBC 29. Hypotype RMK 243.
12	*Uvigerinella californica californica* Cushman, ×62. Loc. LSJU 736 (Snedden, 1932, p. 44: "*Uvigerina* sp. A"; RMK 1938, p. 288). Hypotype RMK 244.
13	*Uvigerinella californica californica* Cushman, ×80. Loc. Table 4: SBC 3. Hypotype RMK 245.
14	*Uvigerinella californica ornata* Cushman, ×80. Loc. Table 4: SBC 28. Hypotype RMK 246.
15	*Uvigerinella nudocostata* Cushman, ×71. Loc. Table 5: SBC 47. Hypotype RMK 247.

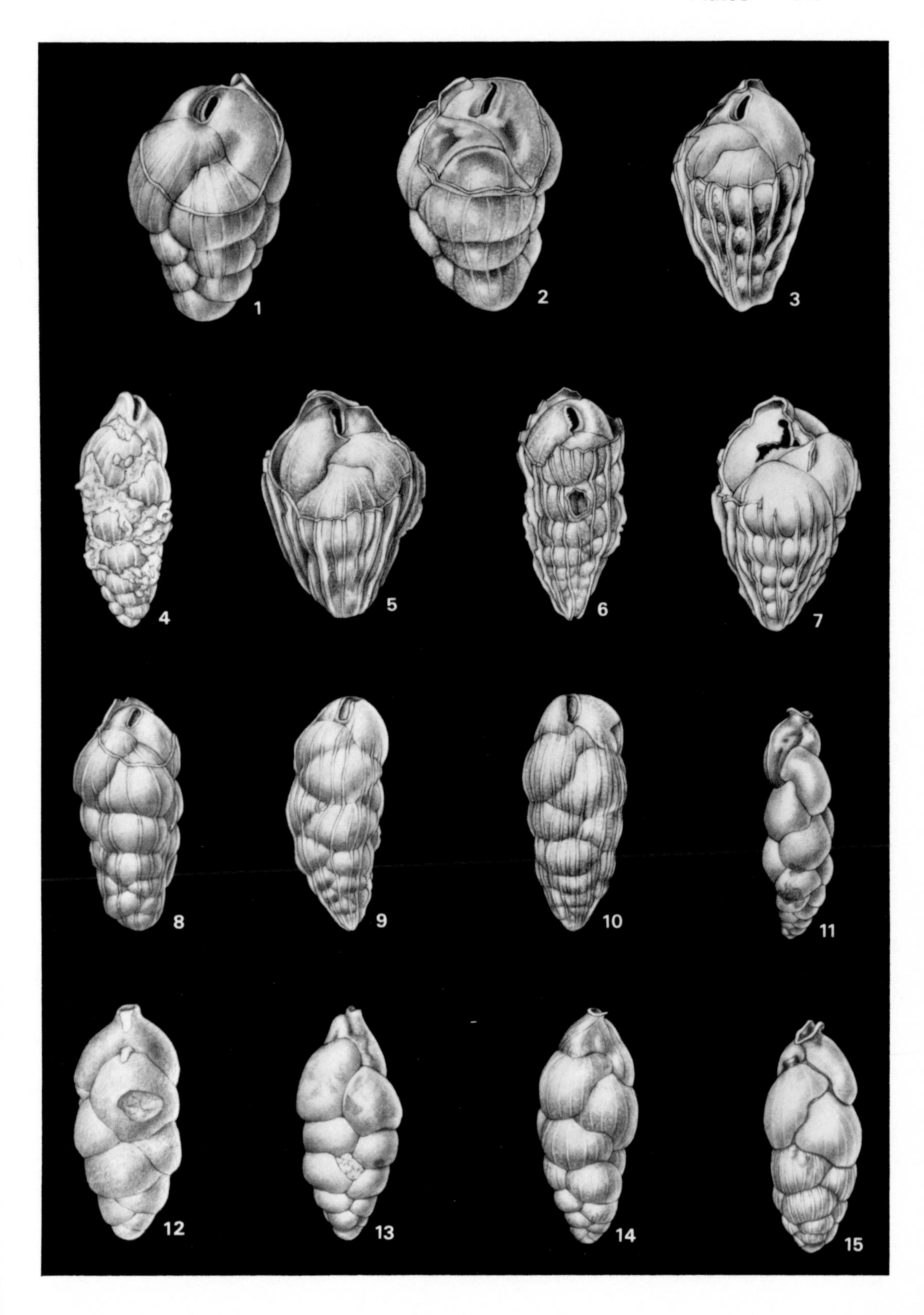
1
2
3
4
5
6
7
8
9
10
11
12
13
14
15

PLATE XII

SPECIES OF UVIGERINA FROM CALIFORNIA MIOCENE

Figure	
1	*Uvigerina modeloensis* Cushman and Kleinpell, ×38. Loc. Table VII: 4. Cotype RMK 248.
2	*Uvigerina hootsi* Rankin, ×47. Loc. Table 5: SBC 49. Hypotype RMK 249.
3	*Uvigerina hannai* Kleinpell, ×35. Loc. Table V: 23 (CV-16). Hypotype RMK 250.
4	*Uvigerina hannai* Kleinpell, ×53. Loc. Fig. 1: NEW 90. Hypotype RMK 251.
5	*Uvigerina carmeloensis* Cushman and Kleinpell, ×38. Loc. Table 6: M535. Hypotype RMK 252.
6	*Uvigerina carmeloensis* Cushman and Kleinpell, ×35. Loc. Table V: 14 (Coyote Gulch). Hypotype RMK 253.
7	*Uvigerina segundoensis* Cushman and Galliher, ×47. Loc. Table V: 22 (LSJU 336). Hypotype RMK 254.
8	*Uvigerina segundoensis* Cushman and Galliher, ×41. Loc. Table V: 21 (CV-15). Hypotype RMK 255.
9	*Uvigerina joaquinensis* Kleinpell, ×44. Loc. Table VI: CM-35. Some individuals within the population of this species, such as this specimen, lack the more bristly ornamentation of the more typical form and thereby approach *U. carmeloensis* with which the species is closely related. Hypotype RMK 256.
10	*Uvigerina subperegrina* Cushman and Kleinpell, ×49. Loc. Table 5: SBC 59. Hypotype RMK 257.
11	*Uvigerina subperegrina* Cushman and Kleinpell, ×59. Loc. Table 5: SBC 59. Hypotype RMK 258.
12	*Uvigerina obesa obesa* (Cushman), ×62. Loc. Table 4: SBC 4. Hypotype RMK 259.
13	*Uvigerina obesa impolita* (Cushman and Laiming), ×51. Loc. lower Modelo Shale, Los Sauces Creek (Snedden, unpub. master's thesis, Stanford Univ., 1931: "*Uvigerina peregrina*"). Hypotype RMK 260.
14	*Uvigerina obesa impolita* (Cushman and Laiming), ×51. Loc. lower Modelo Shale, Los Sauces Creek (Snedden, unpub. master's thesis, Stanford Univ., 1931: "*Uvigerina peregrina*"). Hypotype RMK 261.
15	*Uvigerina obesa impolita* (Cushman and Laiming), ×71. Loc. Table 4: SBC 8. Hypotype RMK 262.
16	*Uvigerina obesa obesa* Cushman, ×66. Loc. Table 4: SBC 27. Hypotype RMK 263.

1
2
3
4
5
6
7
8
9
10
11
12
13
14
15
16

PLATE XIII

BULIMINID SPECIES OF UVIGERINE AND SIPHOGENERINE GRADES
FROM CALIFORNIA OLIGOCENE AND MIOCENE

Figure

1 *Uvigerina hannai* Kleinpell, ×42. Loc. Table V: 21 (CV-15). Hypotype RMK 264.

2 *Uvigerina hootsi* Rankin, ×47. Loc. Table VII: 5. Hypotype RMK 265.

3 *Beckina hornadayi* n. sp., ×51. Loc. Table XII: 23. Holotype RMK 118.

4 *Uvigerina subperegrina* Cushman and Kleinpell, ×51. Loc. basal McLure shale "one mile south of the pipeline road on the east side of Sulphur Spring Canyon" (Church, 1972, p. 76). Hypotype RMK 266.

5 *Uvigerina* aff. *U. subperegrina* Cushman and Kleinpell, ×44. Loc. corner Murietta and Roblar Sts., upper Modelo diatomite; collector A. A. Almgren. The species here represented has costae atypically heavy and seemingly developing in the direction of *U. peregrina,* the Pliocene descendant of *U. subperegrina*. In conformation, these costae are very different from the coarse and heavy costae of *U. obesa impolita*. Hypotype RMK 267.

6 *Siphogenerina nodifera* Cushman and Kleinpell, ×34. Loc. Table XI: GH 13 (Chico Martinez Creek). Hypotype RMK 268.

7a,b *Siphogenerina nodifera* Cushman and Kleinpell, ×34. Loc. Table XI: JFM 7 (Carneros Creek; lower Zemorrian). Cotype RMK 269.

8 *Uvigerina gesteri* Barbat and von Estorff, ×49. Loc. Table XII: 13 (CAS loc. 27,278; lower Zemorrian). This species also occurs in the lower type Zemorrian and at the same horizons on Chico Martinez Creek and on Carneros Creek (RMK 1938, p. 44, 55). Cotype RMK 270.

9 *Siphogenerina nodifera* Cushman and Kleinpell, ×34. Loc. Table XI: JFM 13 (Carneros Creek; upper Zemorrian). Hypotype RMK 271.

10 *Siphogenerina nodifera* Cushman and Kleinpell, ×34. Loc. Table XI: JFM 16 (Carneros Creek, type Temblor Formation, upper Zemorrian). Hypotype RMK 272.

11a,b *Estorffina mayi* (Cushman and Parker), ×69. Loc. Table XI: JFM 17 (Carneros Creek, type Temblor Formation, upper Zemorrian). Paratype RMK 120.

12 *Beckina hornadayi* n. sp., ×35. Loc. Ellwood oil field Doty 4, depth 3,162 ft, 4 ft above top of Vaqueros Sandstone (upper Zemorrian). Paratype RMK 119.

13 *Siphogenerina nodifera* Cushman and Kleinpell, ×34. Loc. Table XI: JFM 16 (Carneros Creek, type Temblor Formation, upper Zemorrian). Hypotype RMK 273.

14a,b *Siphogenerina transversa* Cushman, ×34. Loc. Table XI: JFM 33; not recorded from this sample in 1938. Lower Saucesian. Hypotype RMK 274.

15a,b *Siphogenerina tenua* Cushman and Kleinpell, ×34. Loc. Table XI: JFM 33 (upper Santos Shale, "C shale" of Mahoney, middle type Temblor Formation, lower Saucesian). Cotype RMK 275.

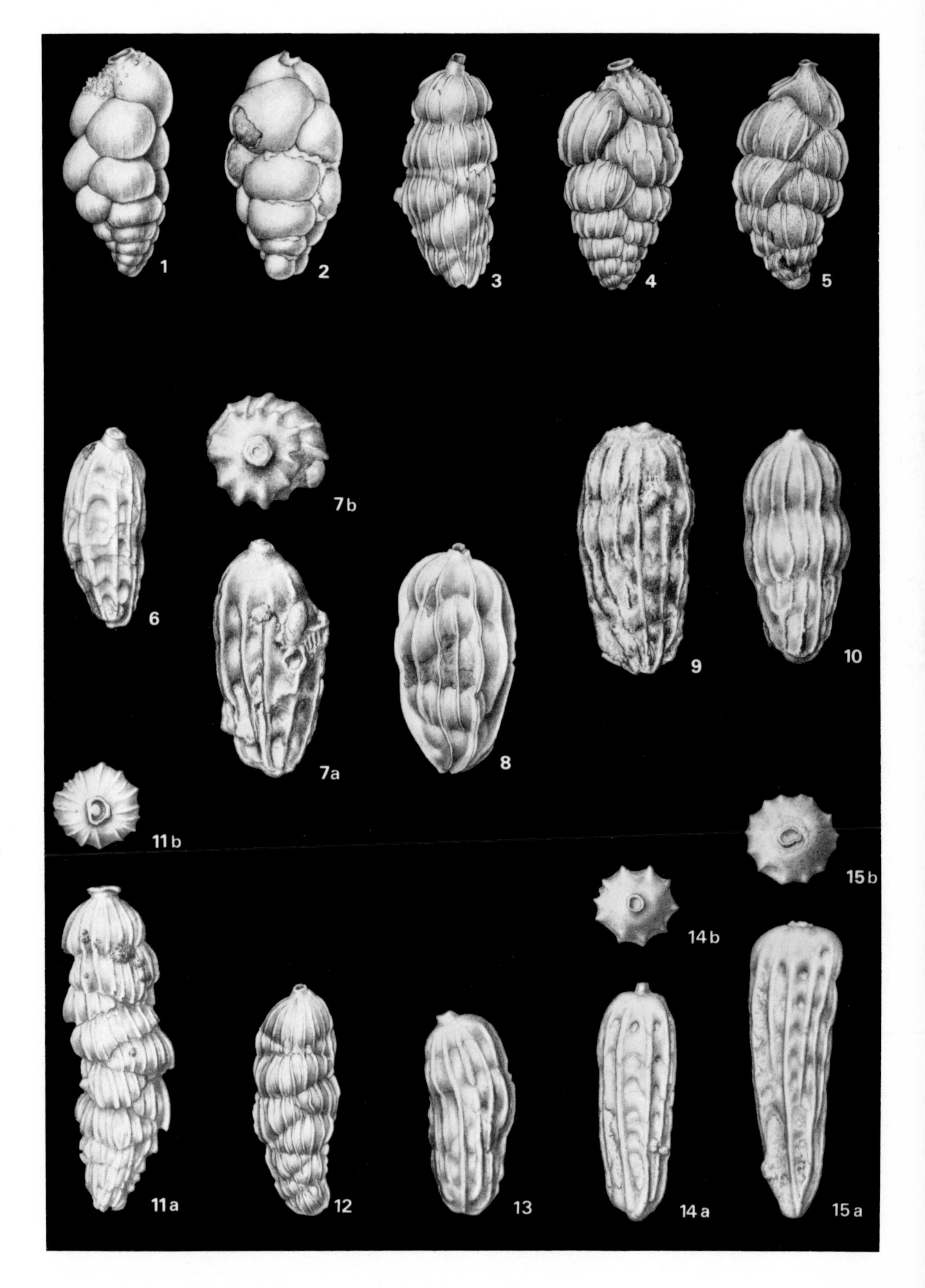
1
2
3
4
5
7b
6
7a
8
9
10
11b
15b
14b
11a
12
13
14a
15a

PLATE XIV

BULIMINID SPECIES OF UVIGERINE AND SIPHOGENERINE GRADES
FROM CALIFORNIA OLIGOCENE AND MIOCENE

Figure	
1a,b	*Uvigerina cocoaensis* Cushman, ×51. Loc. UC D-4335. Hypotype UC 47505.
2a,b	*Uvigerina gallowayi* Cushman, ×49. Loc. UC D-4260. Hypotype UC 47507.
3	*Uvigerina gesteri* Barbat and von Estorff, ×44. Loc. Table XII: 13 (CAS loc. 27,278; lower Zemorrian). Cotype RMK 276.
4a,b	*Siphogenerina nodifera* Cushman and Kleinpell, ×33. Loc. UC D-4386. Hypotype UC 47480.
5a,b	*Siphogenerina transversa* Cushman, ×33. Loc. UC D-4449. Hypotype UC 47490.
6a,b	*Uvigerina sparsicostata* (Cushman and Kleinpell), ×53. Loc. UC D-4342. Hypotype UC 47519.
7	*Laimingina smithi* (Kleinpell), ×33. Loc. Ellwood oil field Doty 4 well, depth 3,162 ft, 4 ft above top of Vaqueros Sandstone (upper Zemorrian). Cotype RMK 121.
8a,b	*Siphogenerina kleinpelli* Cushman, ×32. Loc. UC D-4429. Hypotype UC 47475.
9a,b	*Siphogenerina cymricensis* Tipton, Kleinpell, and Weaver, ×27. Loc. UC D-4365. Paratype UC 47472.
10a,b	*Siphogenerina forticostata* Tipton, Kleinpell, and Weaver, ×27. Loc. UC D-4383. Holotype UC 47481.
11a,b	*Siphogenerina tenua* Cushman and Kleinpell, ×25. Loc. UC D-4391. Hypotype UC 47489.
12a,b	*Uvigerina obesa impolita* (Cushman and Laiming), ×66. Loc. UC D-4368. Hypotype UC 47518.
13a,b	*Estorffina mayi* (Cushman and Parker), ×35. Loc. UC D-4385. Hypotype UC 47477.
14a,b	*Uvigerina atwilli* Cushman and Simonson, ×44. Loc. UC D-4327. Hypotype UC 47502.
15a,b	*Atwillina pseudococoaensis* (Cushman and Kleinpell), ×31. Loc. UC D-4341. Hypotype UC 47486.
16a,b	*Beckina fulmeri* n. sp., ×45. Loc. UC D-4317. Holotype UC 47528. (*not* UC 37528 as recorded in Tipton et al, 1974, Pl. 1, opposite p. 150).

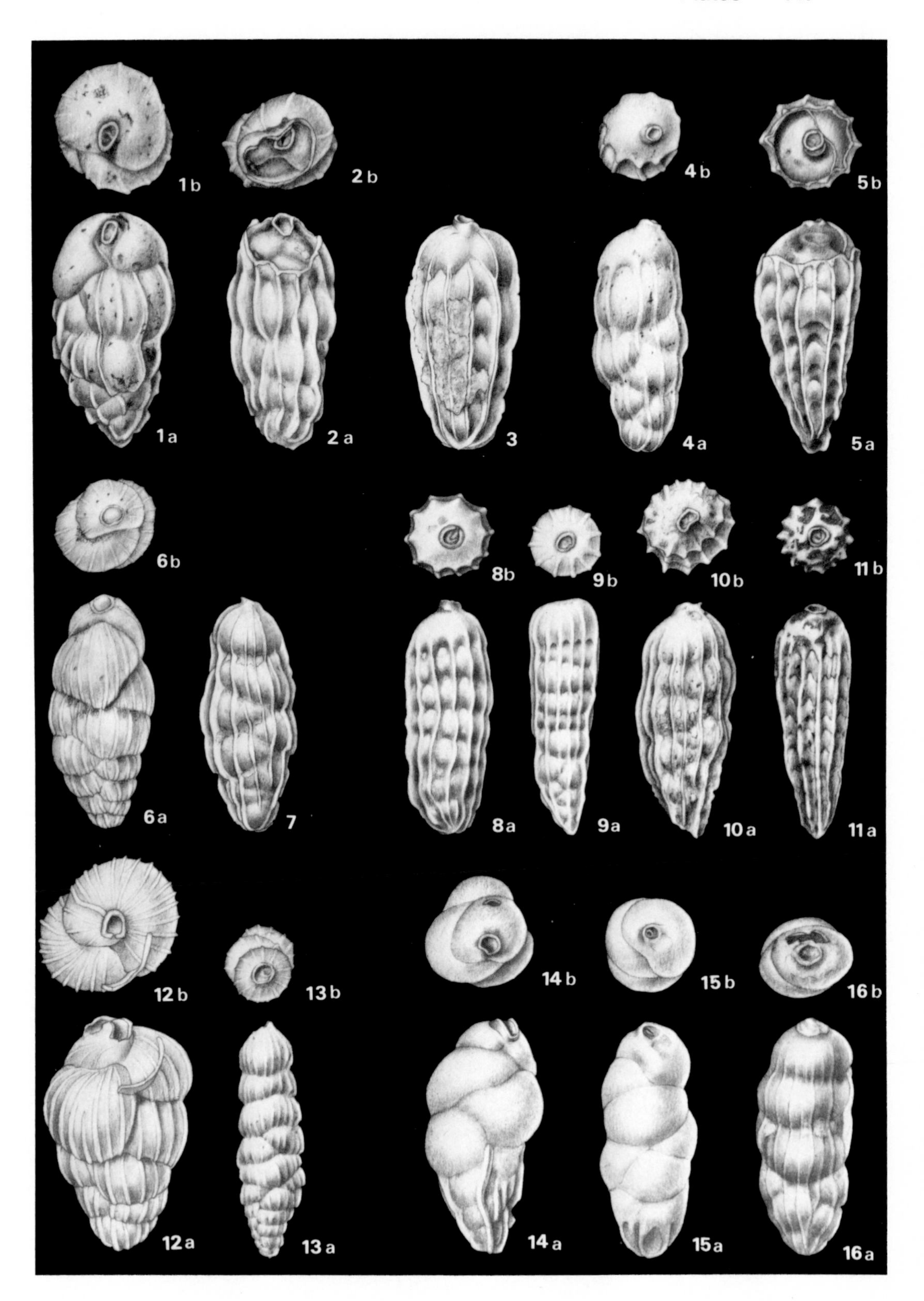
1b
2b
4b
5b
1a
2a
3
4a
5a
6b
8b
9b
10b
11b
6a
7
8a
9a
10a
11a
12b
13b
14b
15b
16b
12a
13a
14a
15a
16a

PLATE XV

SPECIES OF SIPHOGENERINA FROM TYPE SAUCESIAN AND FROM OTHER EARLY TO MIDDLE MIOCENE LOCALITIES IN SOUTHERN CALIFORNIA

Figure	
1	*Siphogenerina hughesi* Cushman, ×29. Loc. LSJU 779 (Snedden, 1932, p. 44, from "300 ft. above the 'bentonite bed' " to "450 ft. above the 'bentonite bed,' " the "bentonite bed" constituting the contact between the Rincon and the Monterey formations on Los Sauces Creek, Ventura County, California. Hypotype RMK 276.
2	*Siphogenerina hughesi* Cushman(?), ×29. Loc. LSJU 778 (Snedden, 1932, p. 44; also RMK 1938, p. 301). From 300 to 450 ft above Rincon-Monterey contact. Hypotype RMK 277.
3	*Siphogenerina hughesi* Cushman, ×36. Loc. FJP 206, Monterey Shale between mouths of Dos Pueblos and Las Varas Creeks. Collector Jay Phillips. Hypotype RMK 278.
4	*Siphogenerina hughesi* Cushman, ×41. Loc. San Rafael Mountains locality of Keenan (1932, p. 69; see RMK 1938, p. 301). Hypotype RMK 279.
5	*Siphogenerina branneri* (Bagg), ×26. Loc. LSJU 782 (Snedden, 1932, p. 44: at 150 ft above the "bentonite bed" and commonly to 550 ft above the "bentonite bed"). Specimens of this species, either well preserved or rolled and worn nearly smooth yet showing the roots of costae typical of *S. branneri* (but many of them inadvertently recorded in 1932 as *S. hughesi*) "continue to the uppermost zone studied" (i.e., to "Modelo" or Monterey beds of early Luisian age). See RMK 1938, p. 299-300, 301: these two species "(in part)." Hypotype RMK 280.
6	*Siphogenerina branneri* (Bagg), ×23. Loc. Table 6: M510. Hypotype RMK 281.
7	*Siphogenerina branneri* (Bagg), ×23. Loc. Table 6: M510. Hypotype RMK 282.
8	*Siphogenerina hughesi* Cushman(?), ×29. Loc. LSJU 776 (Snedden, 1932, p. 44, from 300 to 450 ft above the Rincon-Monterey contact; see RMK 1938, p. 300: this species "(in part)." Hypotype RMK 283.
9	*Siphogenerina transversa* Cushman, ×38. Loc. LSJU 766: "Modelo Shale of Los Sauces Creek (upper Saucesian)" of Snedden (1932), that is, the lower 300 ft of the Monterey formation on Los Sauces Creek, Ventura County. See RMK 1938, p. 306. Hypotype RMK 284.
10	*Siphogenerina cymricensis* Tipton, Kleinpell, and Weaver(?), ×44. Loc. Table 4: SBC 12. Perhaps two or three distinct species have been included within this taxon. See, for example, Tipton et al, 1973, p. 57, pl. 8, figs. 1a,b and 2a,b, for a possible third atypical variant. Hypotype RMK 285.
11	*Siphogenerian kleinpelli* Cushman, ×31. Loc. Table 4: SBC 3. Hypotype RMK 286.
12	*Siphogenerina kleinpelli* Cushman, ×33. Loc. Table 4: SBC 4. Hypotype RMK 287.
13	*Siphogenerina transversa* Cushman, ×33. Loc. Table 4: SBC 37. Hypotype RMK 288.
14	*Siphogenerina transversa* Cushman, ×56. Loc. LSJU 753 (Snedden, 1932, p. 43: from the higher *Uviginerella obesa* subzone, from 2,025 ft to 2,260 ft above the Vaqueros; this is Snedden's "top of Vaqueros," or the uppermost 235 ft of the Rincon Shale of the Los Sauces Creek, Ventura County; and p. 45: "the Rincon Shale of the Los Sauces Creek section"). See RMK 1938, p. 305. Hypotype RMK 289.
15	*Siphogenerina transversa* Cushman, ×47. Loc. LSJU 765. A rolled and worn specimen from the Rincon Shale of Los Sauces Creek. In rolled and worn individuals such as this, specimens of *S. transversa* tend to lose the high sharp edges characteristic of this species, although other aspects of the broad and relatively few costae, typical of the species, remain the same. Hypotype RMK 290.
16	*Siphogenerina transversa* Cushman, ×32. Loc. LSJU 758 (Snedden, 1932, p. 45; RMK 1938, p. 305). Rincon Shale of Los Sauces Creek. Hypotype RMK 291.
17	*Siphogenerina transversa* Cushman, ×47. Loc. LSJU 764 (Snedden, 1932, p. 43; see RMK 1938, p. 306). Lower 300 ft of Monterey Shale on Los Sauces Creek, Ventura County. Hypotype RMK 292.

1
2
3
4
5
6
7
8
9
10
11
12
13
14
15
16
17

PLATE XVI

SPECIES OF SIPHOGENERINA FROM LUISIAN STAGE,
MIDDLE MIOCENE OF CALIFORNIA

Figure	
1	*Siphogenerina collomi* Cushman, ×47. Loc. Fig. 1: NL 8. Hypotype RMK 293.
2	*Siphogenerina collomi* Cushman, ×29. Loc. Table 5: SBC 43. Hypotype RMK 294.
3	*Siphogenerina collomi* Cushman, ×23. Loc. Table 5: SBC 44. Hypotype RMK 295.
4	*Siphogenerina nuciformis* Kleinpell(?), ×25. Loc. Table V: 4 (LSJU 341). Costae on this specimen appear, in this view, atypically few for *S. nuciformis*. Hypotype RMK 296.
5	*Siphogenerina collomi* Cushman, ×19. Loc. Table V: 7 (LSJU 333). Hypotype RMK 297.
6	*Siphogenerina nuciformis* Kleinpell, ×23. Loc. Table V: 5 (HGK sample). Hypotype RMK 298.
7	*Siphogenerina nuciformis* Kleinpell, ×26. Loc. Table 5: SBC 43. Hypotype RMK 299.
8	*Siphogenerina nuciformis* Kleinpell, ×27. Loc. Table 5: SBC 44. Hypotype RMK 300.
9	*Siphogenerina nuciformis* Kleinpell, ×22. Loc. Table V: 7 (LSJU 333). Hypotype RMK 301.
10	*Siphogenerina nuciformis* Kleinpell, ×20. Loc. Table 6: M519C. Hypotype RMK 302.
11	*Siphogenerina reedi* Cushman, ×25. Loc. Table 5: SBC 37. Hypotype RMK 303.
12	*Siphogenerina reedi* Cushman, ×20. Loc. Table V: 4 (LSJU 341). Hypotype RMK 304.
13	*Siphogenerina reedi* Cushman, ×32. Loc. Table 6: M519D. Hypotype RMK 305.
14	*Siphogenerina reedi* Cushman, ×36. Loc. Table XIII: DC 1. Redwood Canyon Road, Diamond Canyon. Specimen typical of leached foraminifers of Claremont Shale, but with roots of the costae typical of those in *S. reedi*. Hypotype RMK 306.
15	*Siphogenerina* cf. *S. nuciformis* Kleinpell, ×31. Loc. Table 5: SBC 41. Specimen typical of those so designated by Weaver et al (1969, Pl. 29, Fig. 10) from Channel Islands where, as elsewhere, they appear at horizons in the *S. reedi* Zone (see also Fig. 6: "*Siphogenerina* sp. aff. *S. nuciformis*"). Hypotype RMK 307.
16	*Siphogenerina branneri* (Bagg), ×38. Loc Table 5: SBC 36. Hypotype RMK 308.
17	*Siphogenerina kleinpelli* Cushman, ×34. Loc. Table 4: SBC 19. Hypotype RMK 309.
18	*Siphogenerina obispoensis* n. sp., ×34. Loc. Table 3: EC 13. Type Luisian. Paratype RMK 116.
19	*Siphogenerina obispoensis* n. sp., ×31. Loc. Table 3: EC 14. Type Luisian. Paratype RMK 117.
20	*Siphogenerina obispoensis* n. sp., ×27. Loc. Table 3: EC 14. Type Luisian. Holotype RMK 115.

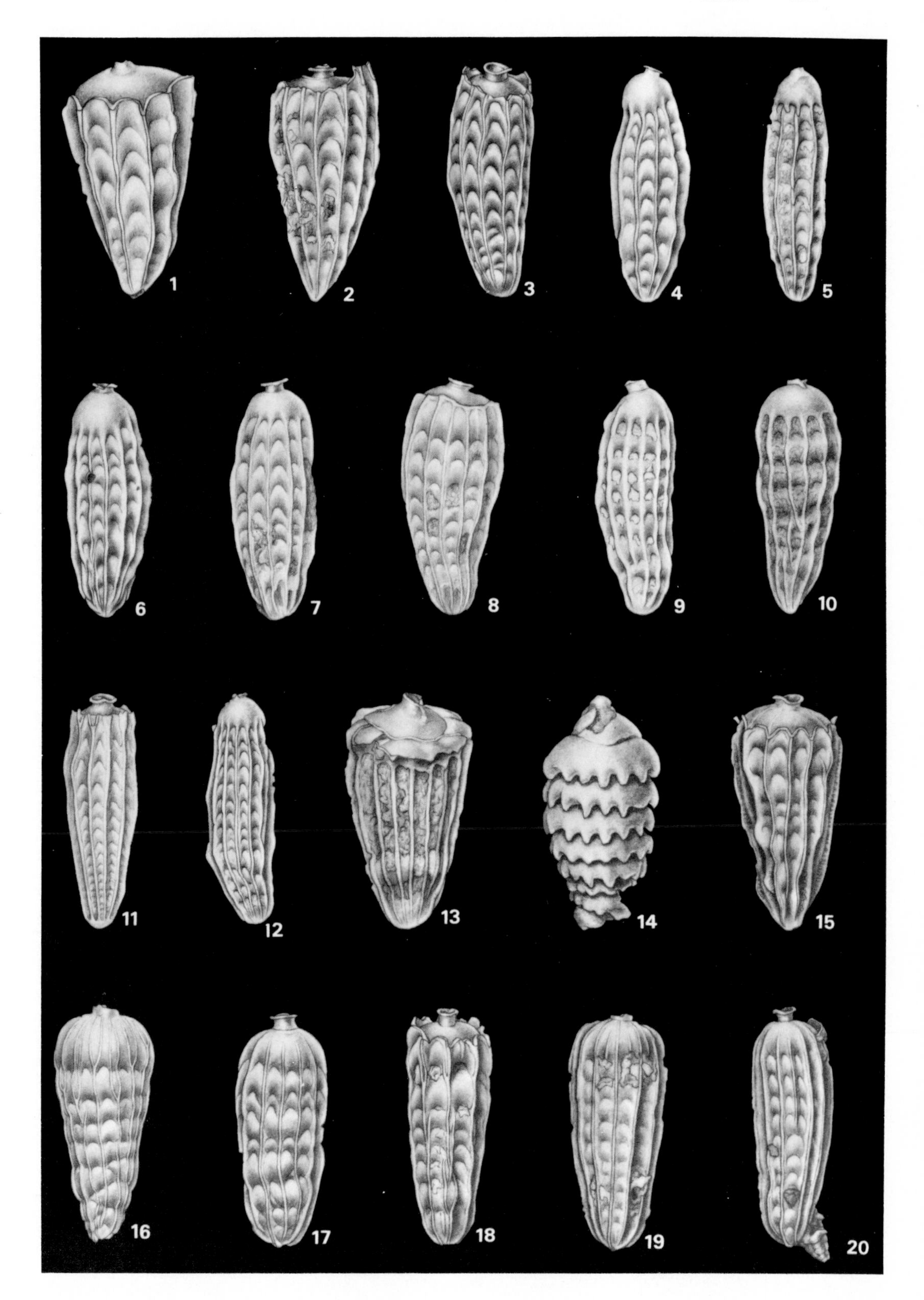
1
2
3
4
5
6
7
8
9
10
11
12
13
14
15
16
17
18
19
20

PLATE XVII

SPECIES OF SIPHOGENERINA FROM MIDDLE MIOCENE OF CALIFORNIA

Figure	
1	*Siphogenerina nuciformis* Kleinpell, ×34. Loc. Table VI: CM 40 ("Flood zone" Chico Martinez Creek). Hypotype RMK 310.
2	*Siphogenerina nuciformis* Kleinpell, ×34. Loc. Table 3: EC 16. Type Luisian. Hypotype RMK 311.
3a,b	*Siphogenerina collomi* Cushman, ×34. Loc. Table 3: EC 21. Type Luisian. Hypotype RMK 312.
4	*Siphogenerina collomi* Cushman, ×23. Loc. Table 3: EC 21. Type Luisian. Hypotype RMK 313.
5	*Siphogenerina collomi* Cushman, ×34. Loc. Table 3: EC 21. Type Luisian. Hypotype RMK 314.
6a,b	*Siphogenerina reedi* Cushman, ×34. Loc. Table 3: EC 16. Type Luisian. Hypotype RMK 315.
7a,b	*Siphogenerina reedi* Cushman, ×34. Loc. Table II: 4. Collectors: Schenck and Farish. Hypotype RMK 316.
8	*Siphogenerina nuciformis* Kleinpell, ×34. Loc. Table VI: CM 40 ("Flood zone," Chico Martinez Creek). Hypotype RMK 317.
9a,b	*Siphogenerina nuciformis* Kleinpell, ×34. Loc. Table 3: EC 18. Type Luisian. Hypotype RMK 318.
10	*Siphogenerina nuciformis* Kleinpell, ×23. Loc. Table 3: EC 18. Type Luisian. Hypotype RMK 319.
11	*Siphogenerina branneri* (Bagg), ×34. Loc. Table 3: EC 7. Type Luisian. Hypotype RMK 320.
12a,b	*Siphogenerina branneri* (Bagg), ×23. Loc. Bagg's Henry Ranch locality; collected by J. C. Branner (upper Relizian). Hypotype RMK 321.
13a,b	*Siphogenerina kleinpelli* Cushman, ×23. Loc. Table I: RC-8, shale partings in uppermost type Vaqueros Formation. Hypotype RMK 322.
14	*Siphogenerina obispoensis* n. sp., ×34. Loc. Table 3: EC 13. Type Luisian. Paratype RMK 323.
15	*Siphogenerina branneri* (Bagg), ×23. Loc. Bagg's Henry Ranch locality; collected by J. C. Branner (upper Relizian). Hypotype RMK 324.
16a,b	*Siphogenerina branneri* (Bagg), ×34. Loc. Table VI: CM 45 (Gould Shale Member, Chico Martinez Creek). Hypotype RMK 325.

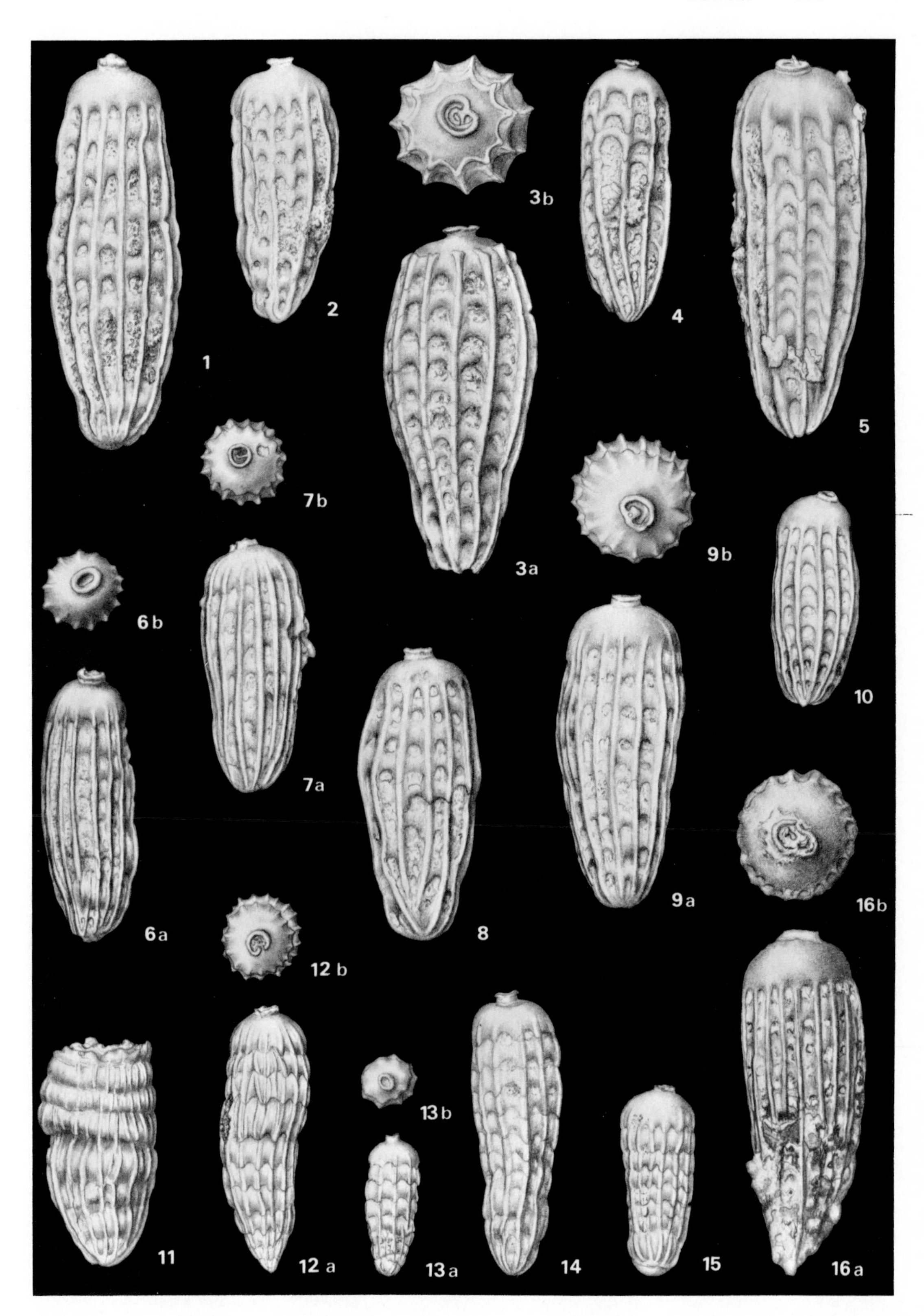
1
2
3b
3a
4
5
6b
6a
7b
7a
8
9b
9a
10
11
12 b
12 a
13b
13a
14
15
16b
16a

PLATE XVIII

SPECIES OF CASSIDULINA FROM CALIFORNIA MIOCENE

Figure

1a-c *Cassidulina modeloensis* Rankin, ×71. Loc. Fig. 1: NEW 57. Hypotype RMK 326.

2a-c *Cassidulina modeloensis* Rankin(?), ×56. Loc. basal McLure shale, "one mile south of the pipeline road on the east side of Sulphur Spring Canyon" (Church, 1972, p. 76). Preservation of these specimens render it difficult to determine whether *C. cushmani* and *C. laevigata carinata* may be present at this locality. Hypotype RMK 327.

3a-c *Cassidulina monicana* Cushman and Kleinpell, ×80. Loc. Table VII: 7 (Dry Canyon). Cotype RMK 328.

4a-c *Cassidulina panzana* Kleinpell, ×56. Loc. Table 6: M531E. Hypotype RMK 329.

5a-c *Cassidulina delicata* Cushman, ×86. Loc. Table 5: SBC 66. Hypotype RMK 330.

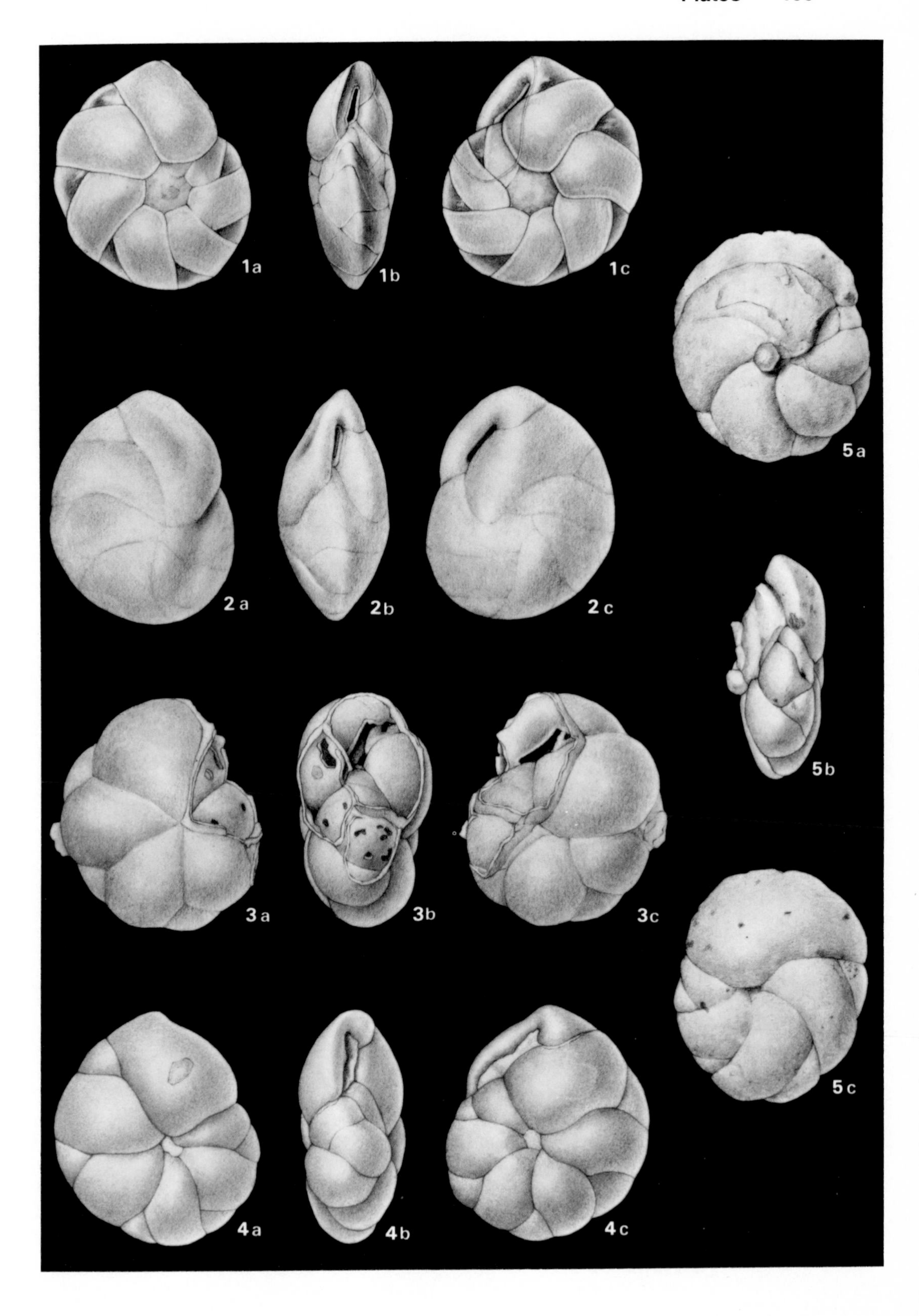
1a
1b
1c
2a
2b
2c
3a
3b
3c
4a
4b
4c
5a
5b
5c

PLATE XIX

EPISTOMINELLIDS AND A DISCORBINELLA FROM CALIFORNIA MIOCENE

Figure

1a-c *Epistominella purisima* (Bramlette), ×62. Loc. see USGS Prof. Paper 222, p. 60-61, pl. 23. "Metatype" (Topotype) RMK 331 (Collector M. N. Bramlette).

2a-c *Epistominella capitanesis* (Cushman and Kleinpell), ×94. Loc. VII: 7 (Dry Canyon). Hypotype RMK 332.

3a-c *Epistominella gyroidinaformis* (Cushman and Goudkoff), ×102. Loc Fig. 1: NEW 40. Hypotype RMK 333.

4a-c *Epistominella gyroidina formis* (Cushman and Goudkoff), ×113. Loc. Table XIII: SPD 4 (Tice Shale). Hypotype RMK 334.

5a-c *Disorbinella Valmoteensis* Kleinpell, ×31. Loc. Fig. 1: NEW 58. Hypotype RMK 335.

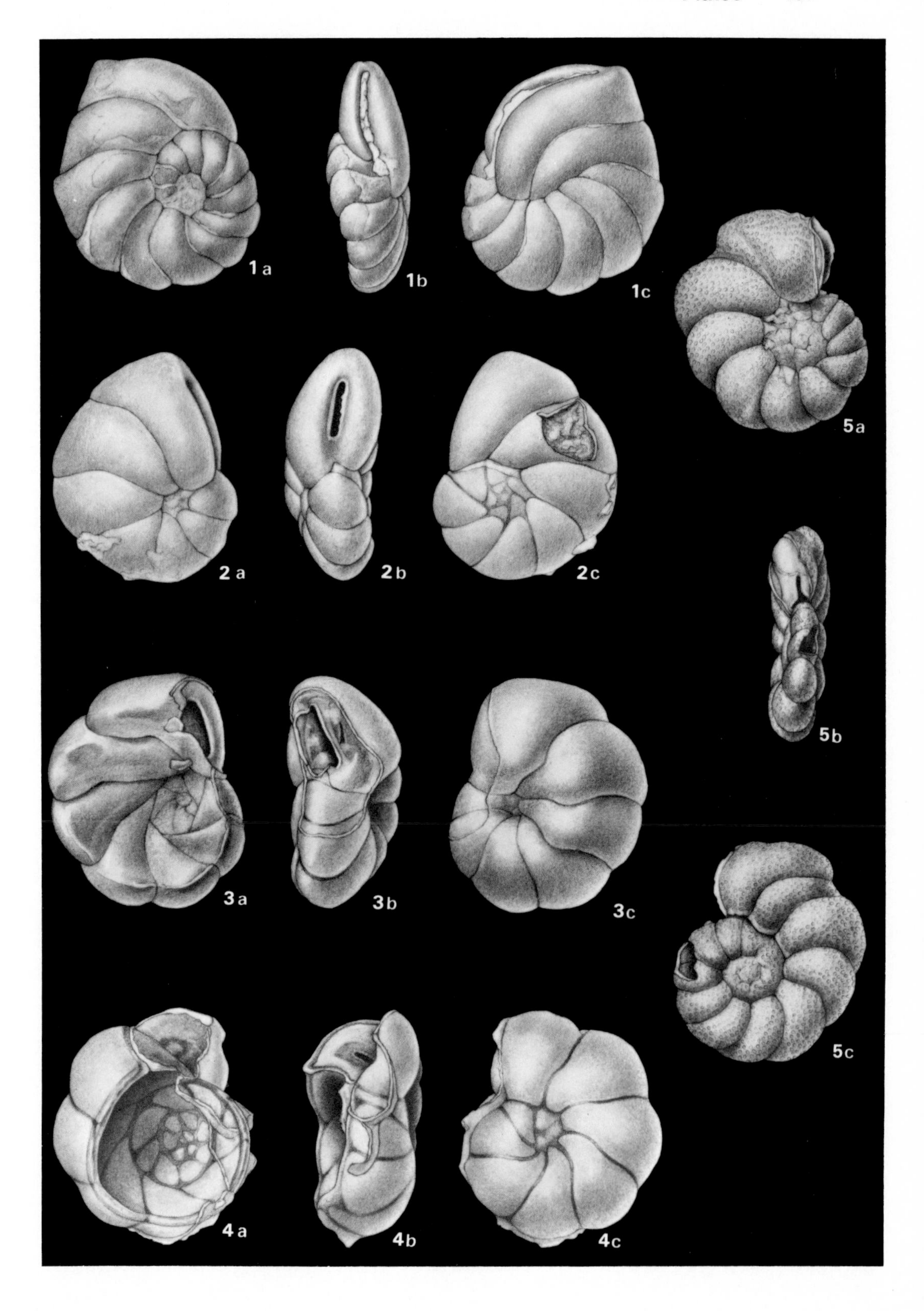
1a
1b
1c
2a
2b
2c
3a
3b
3c
4a
4b
4c
5a
5b
5c

PLATE XX

THREE ANOMALINID SPECIES FROM CALIFORNIA MIOCENE

Figure	
1a-c	*Cibicides illingi* (Nuttall), ×51. Loc. Fig. 1: NEW 45. Hypotype RMK 336.
2a-c	*Cibicides illingi* (Nuttall), ×45. Loc Table 5: SBC 49. Hypotype RMK 337.
3a-c	*Cibicides illingi* (Nuttall), ×62. Loca Table 6: M535. Hypotype RMK 338.
4a-c	*Anomalina salinasensis* Kleinpell, ×53. Loc. Table 6: M521E. Hypotype RMK 339.
5a,b	*Planulina* sp. cf. *P. ariminensis* d'Orbigny, ×64. Loc. Table 6: M519C. Hypotype RMK 340.

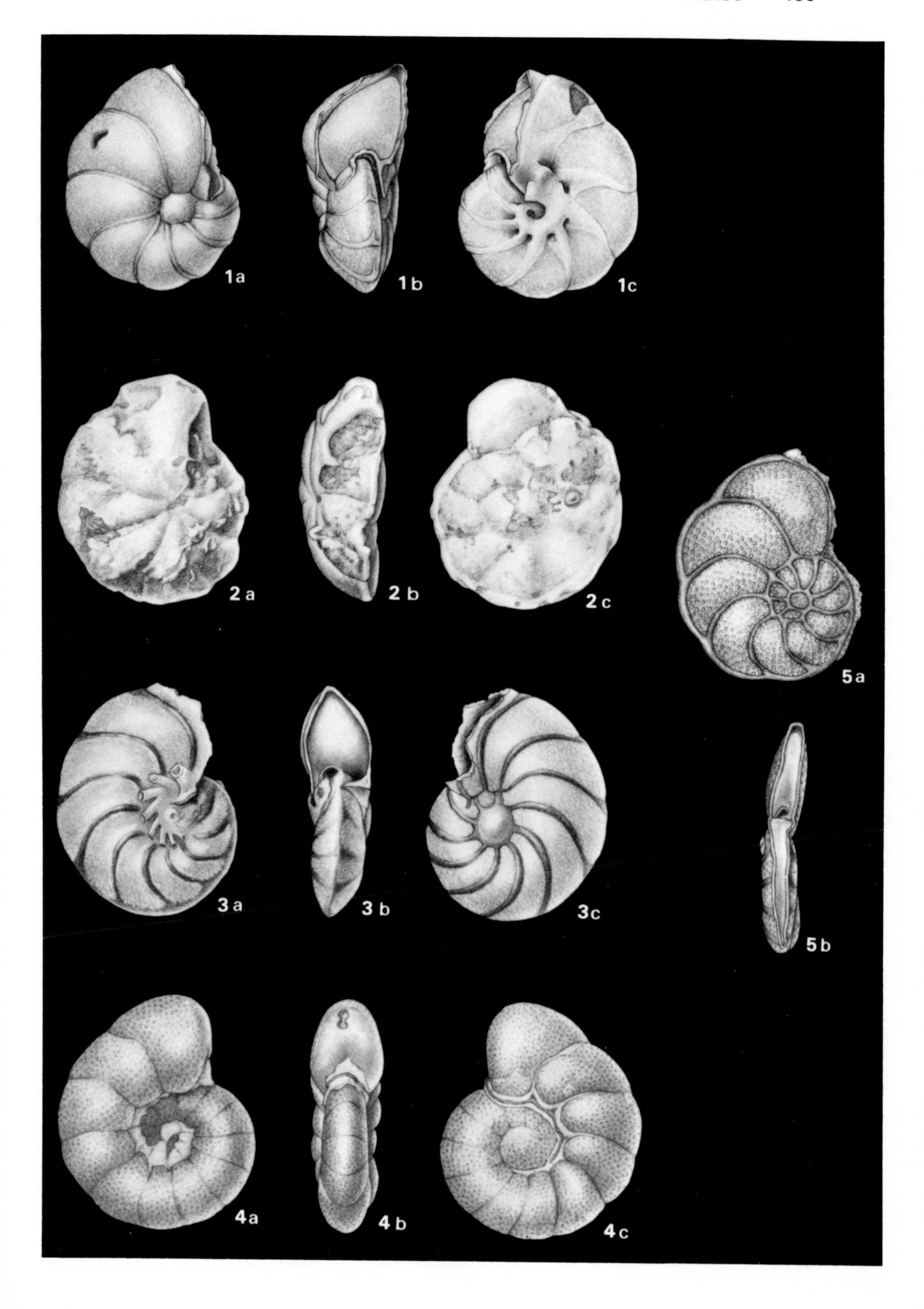
1a
1b
1c
2a
2b
2c
5a
3a
3b
3c
5b
4a
4b
4c

PLATE XXI

MISCELLANEOUS ANOMALINID SPECIES FROM CALIFORNIA MIOCENE

Figure	
1a-c	*Baggina* aff. *B. robusta* Kleinpell, ×59. Loc Table 4: SBC 9.
2a-c	*Cancris brogniarti* (d'Orbigny), ×62. Loc. Table 5: SBC 48. Hypotype RMK 342.
3a-c	*Cibicides americanus crassiseptus* Cushman and Laiming, ×113. Loc. Table 4: SBC 22. Hypotype RMK 343.
4a-c	*Cibicides americanus americanus* (Cushman), ×71. Loc. Table 4: SBC 22. Hypotype RMK 344.
5a-c	*Cibicides floridanus* (Cushman), ×44. Loc. Table 4: SBC 6. Hypotype RMK 345.

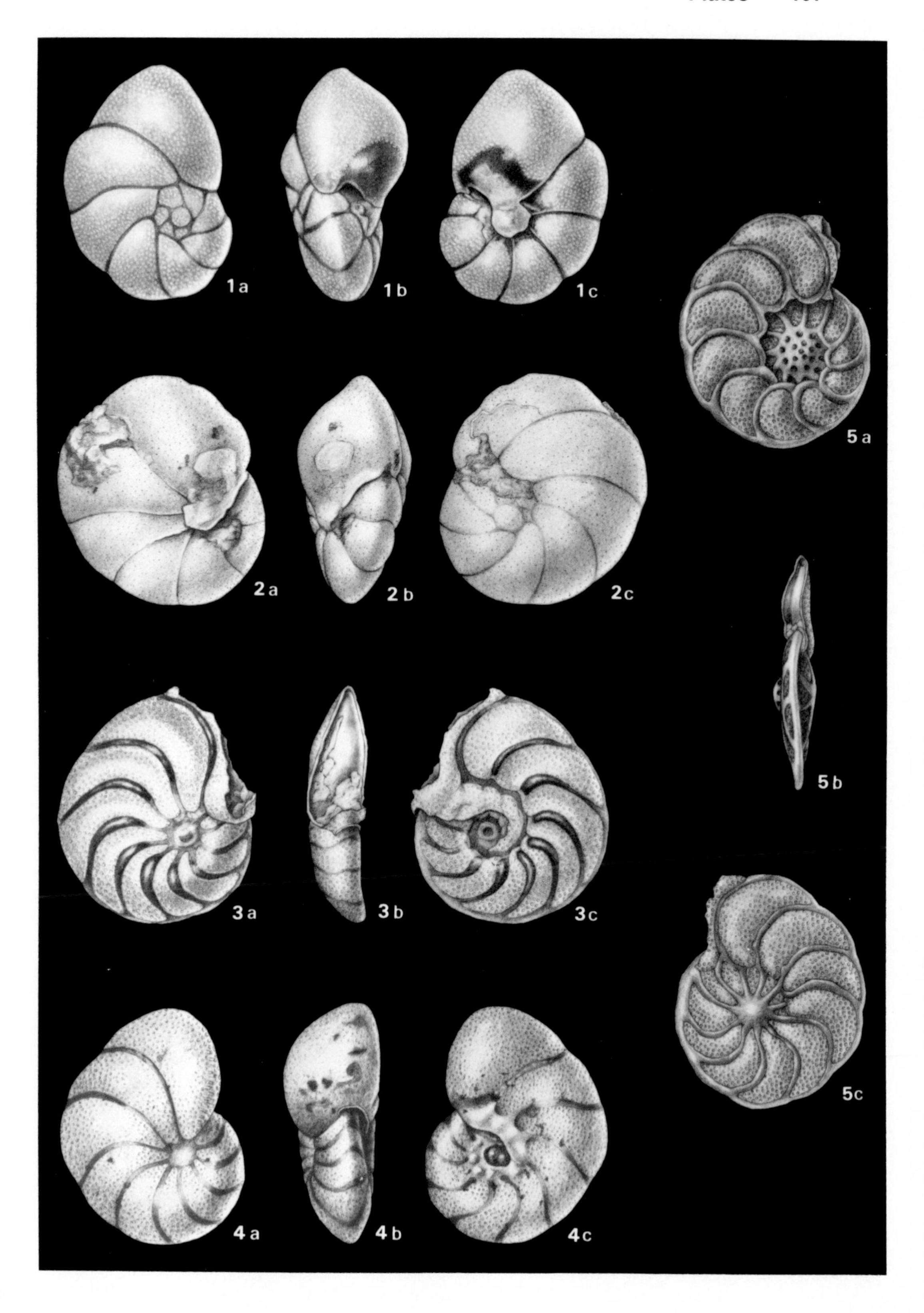
1a
1b
1c
2a
2b
2c
3a
3b
3c
4a
4b
4c
5a
5b
5c

PLATE XXII

SPECIES OF BAGGINA FROM MIOCENE OF CALIFORNIA

Figure

1a-c *Baggina californica* Kleinpell, ×56. Loc. Table 6: M519D. Hypotype RMK 346.

2a-c *Baggina robusta* Kleinpell, ×62. Loc. Table 4: SBC 9. Hypotype RMK 347.

3a-c *Baggina* aff. *B. robusta* Kleinpell, ×80. Loc. Table 4: SBC 3. Hypotype RMK 348.

4a-c *Baggina robusta* Kleinpell, ×51. Loc. LSJU 734 (Snedden, 1932, p. 44, 46 "*Valvulineria* species A"; RMK 1938, p. 325). Monterey ("Modelo") Shale of Los Sauces Creek, Ventura County. Hypotype RMK 349.

5a-c *Baggina subinequalis* Kleinpell, ×75. Loc. basal McLure shale, "one mile south of the pipeline road on the east side of Sulphur Spring Canyon" (Church, 1972, p. 76). Collector, C. C. Church. Hypotype RMK 350.

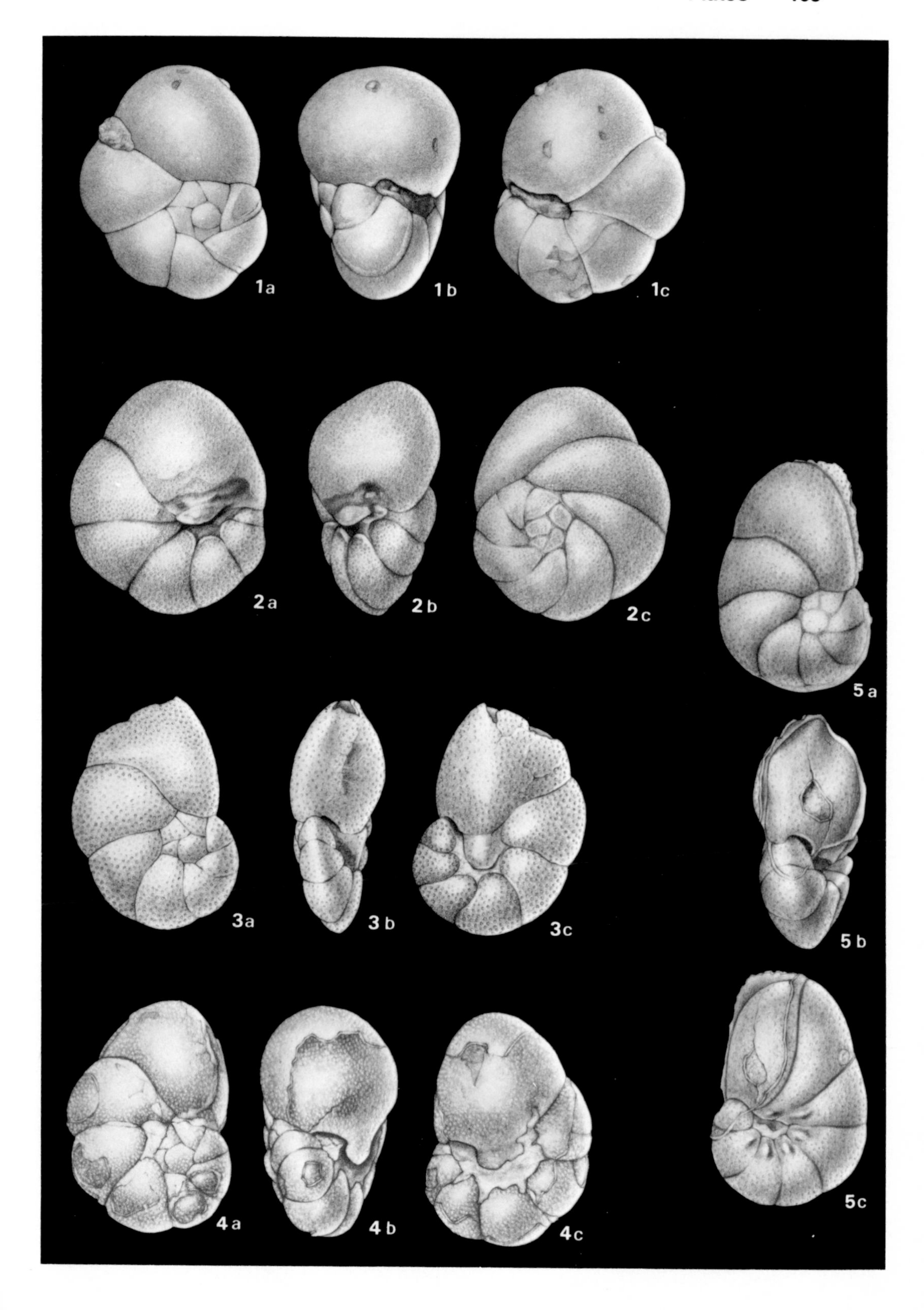
1a
1b
1c
2a
2b
2c
5a
3a
3b
3c
5b
4a
4b
4c
5c

PLATE XXIII

SPECIES OF CALIFORNIA MIDDLE TERTIARY VALVULINERID LINEAGE

Figure	
1	*Valvulineria californica appressa* Cushman, ×34. Loc. HM 8 (lower Luisian). Topotype RMK 351.
2	*Valvulineria californica obesa* Cushman, ×34. Loc. Table VI: CM 47. Basal Gould shale 2 ft above top of Temblor button bed, Zemorra Creek (upper Relizian). Hypotype RMK 352.
3	*Valvulineria californica obesa* Cushman, ×34. Loc. Table 3: EC 4 (type Luisian). Hypotype RMK 353.
4	*Valvulina californica californica* Cushman, ×34. Loc. Table 3: EC 20 (type Luisian). Hypotype RMK 354.
5a-c	*Valvulineria depressa* Cushman, ×75. Loc. Table 4: SBC 9 (upper Saucesian). Hypotype RMK 355.
6a-c	*Valvulineria williami* Kleinpell, ×125. Loc. Table 4: SBC 10 (upper Saucesian). Hypotype RMK 356.
7a-c	*Cibicides hodgei hodgei* Cushman and Schenck, ×53. Loc. LSJU 1436 (Zemorrian; see RMK 1938, p. 33, 111, 354, pl. II, figs. 10, 11, 13). Collectors Hollis D. Hedberg, R. M. Kleinpell, and H. G. Schenck. Hypotype RMK 357.
8a-c	*Valvulineria casitasensis casitasensis* Cushman and Laiming, ×62. Loc. Cushman and Laiming (1931): locality number 11 (Fig. 5) in the lower member of the type Rincon Shale of Los Sauces Creek (upper Zemorrian). Cotype RMK 358 (collected and identified by Boris Laiming).

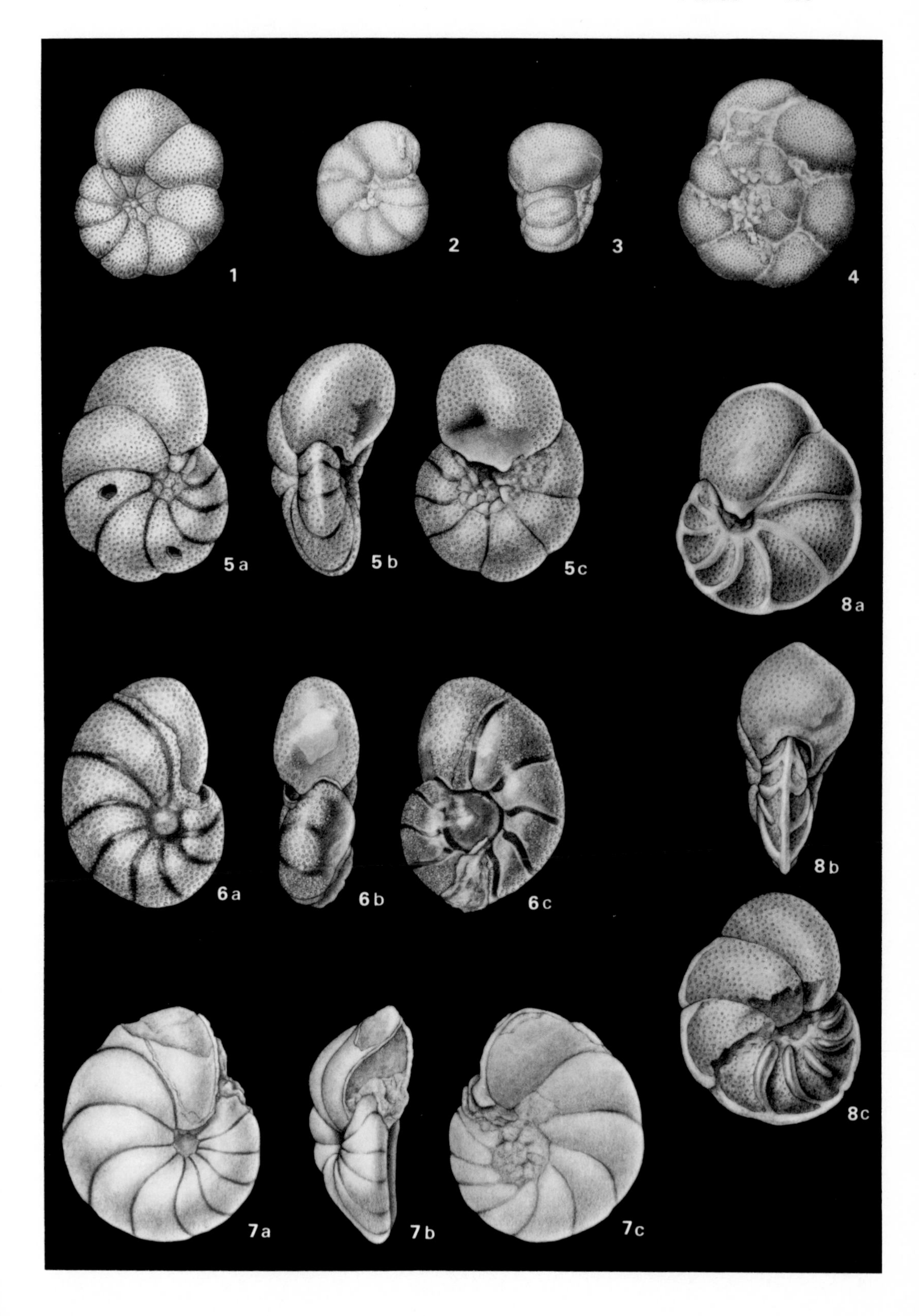
1
2
3
4
5a
5b
5c
8a
8b
6a
6b
6c
8c
7a
7b
7c

PLATE XXIV

MIOCENE SPECIES OF VALVULINERIA FROM NAPES BLUFFS SEQUENCE, SANTA BARBARA COUNTY, CALIFORNIA

Figure

1a-c	*Valvulineria miocenica miocenica* Cushman, ×71. Loc. Table 5: SBC 47. Hypotype RMK 359.
2a-c	*Valvulineria ornata* Cushman, ×62. Loc. Table 5: SBC 41. Hypotype RMK 360.
3a-c	*Valvulineria californica obesa* Cushman, ×62. Loc. Table 5: SBC 36. Hypotype RMK 361.
4a-c	*Valvulineria depressa* Cushman, ×80. Loc. Table 4: SBC 8. Hypotype RMK 362.
5a-c	*Valvulineria californica californica* Cushman, ×51. Loc. Table 5: SBC 43. Hypotype RMK 363.

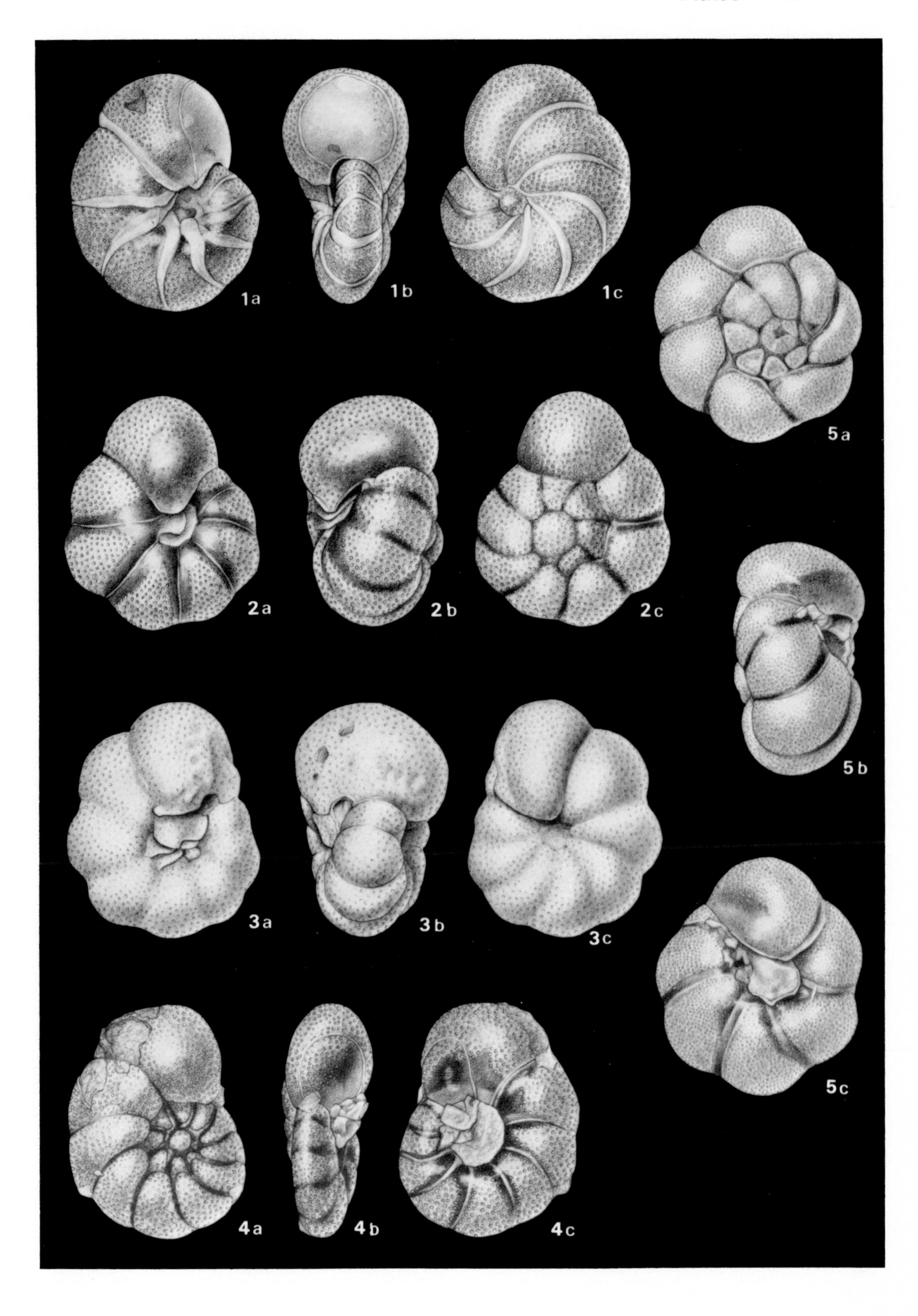
1a
1b
1c
2a
2b
2c
3a
3b
3c
4a
4b
4c
5a
5b
5c

PLATE XXV

MIOCENE SPECIES OF VALVULINERIA FROM TRANSVERSE AND CENTRAL COAST RANGES, CALIFORNIA

Figure

1a-c *Valvulineria californica californica* Cushman, ×39. Loc. Table 6: M519D. Hypotype RMK 364.

2a-c *Valvulineria californica appressa* Cushman, ×45. Loc. LSJU 785 (Snedden, 1932, p. 44, as recorded, and also "*Valvulineria* sp.," both from 650 ft above the bentonite bed), Monterey ("Modelo") Shale of Los Sauces Creek, Ventura County (upper Relizian). See also RMK 1938, p. 309. Hypotype RMK 365.

3a-c *Valvulineria californica obesa* Cushman, ×71. Loc. Table 6: M519D. Hypotype RMK 366.

4a-c *Valvulineria californica obesa* Cushman, ×45. Loc. LSJU 730. Snedden master's thesis sample 67. Monterey ("Modelo") Shale of Los Sauces Creek from 300 to 650 ft above the top of the type Rincon Shale (unrecorded by Snedden, 1932). Hypotype RMK 367.

5a-c *Valvulineria californica californica* Cushman, ×45. Loc.: shale interbedded wtih basaltic lava flows of the middle or upper Topanga Formation on Santa Barbara Island (Luisian). Collected by George L. Richards, Jr. and Luis Kemnitzer (see RMK 1938, p. 164). Hypotype RMK 368.

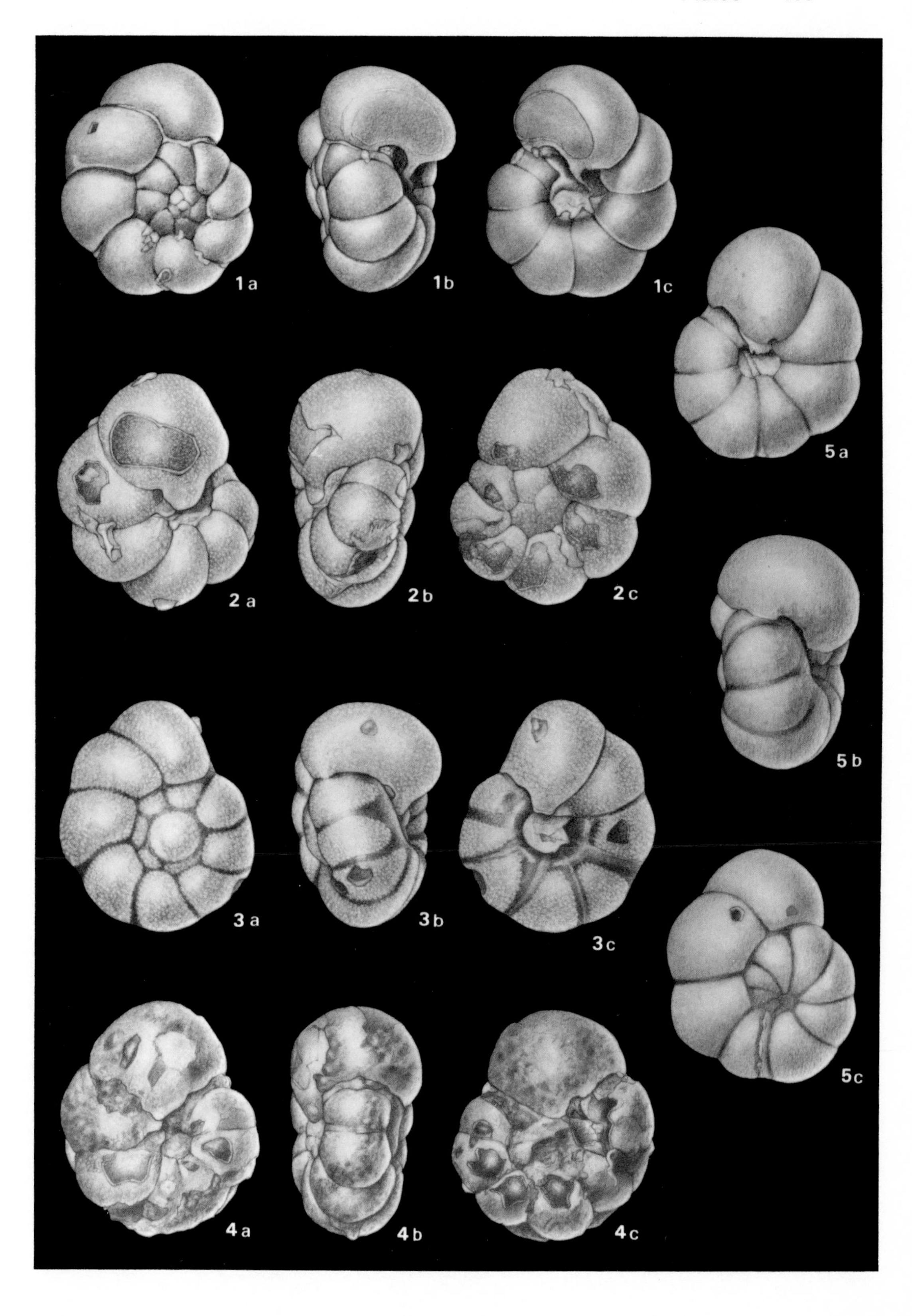
1a
1b
1c
2a
2b
2c
3a
3b
3c
4a
4b
4c
5a
5b
5c

PLATE XXVI

MIOCENE SPECIES OF VALVULINERIA FROM COAST RANGES, CALIFORNIA

Figure

1a-c *Valvulineria miocenica miocenica* Cushman, ×56. Loc. Table V: 9 (LSJU 339). Lower member type Monterey Shale. Hypotype RMK 369.

2a-c *Valvulineria californica californica* Cushman, ×41. Loc. Table V: 9 (LSJU 339). Lower member type Monterey Shale. Hypotype RMK 370.

3a-c *Valvulineria californica californica* Cushman, ×36. Loc. Table XIII: LHF 9. Claremont Shale. Hypotype RMK 371.

4a-c *Valvulineria californica appressa* Cushman, ×51. Loc. Table 5: SBC 36. Monterey Shale of Naples Bluffs section. Hypotype RMK 372.

5a-c *Valvulineria californica californica* Cushman, ×38. Loc. Table V: 9 (LSJU 339). Lower member type Monterey Shale. Hypotype RMK 373.

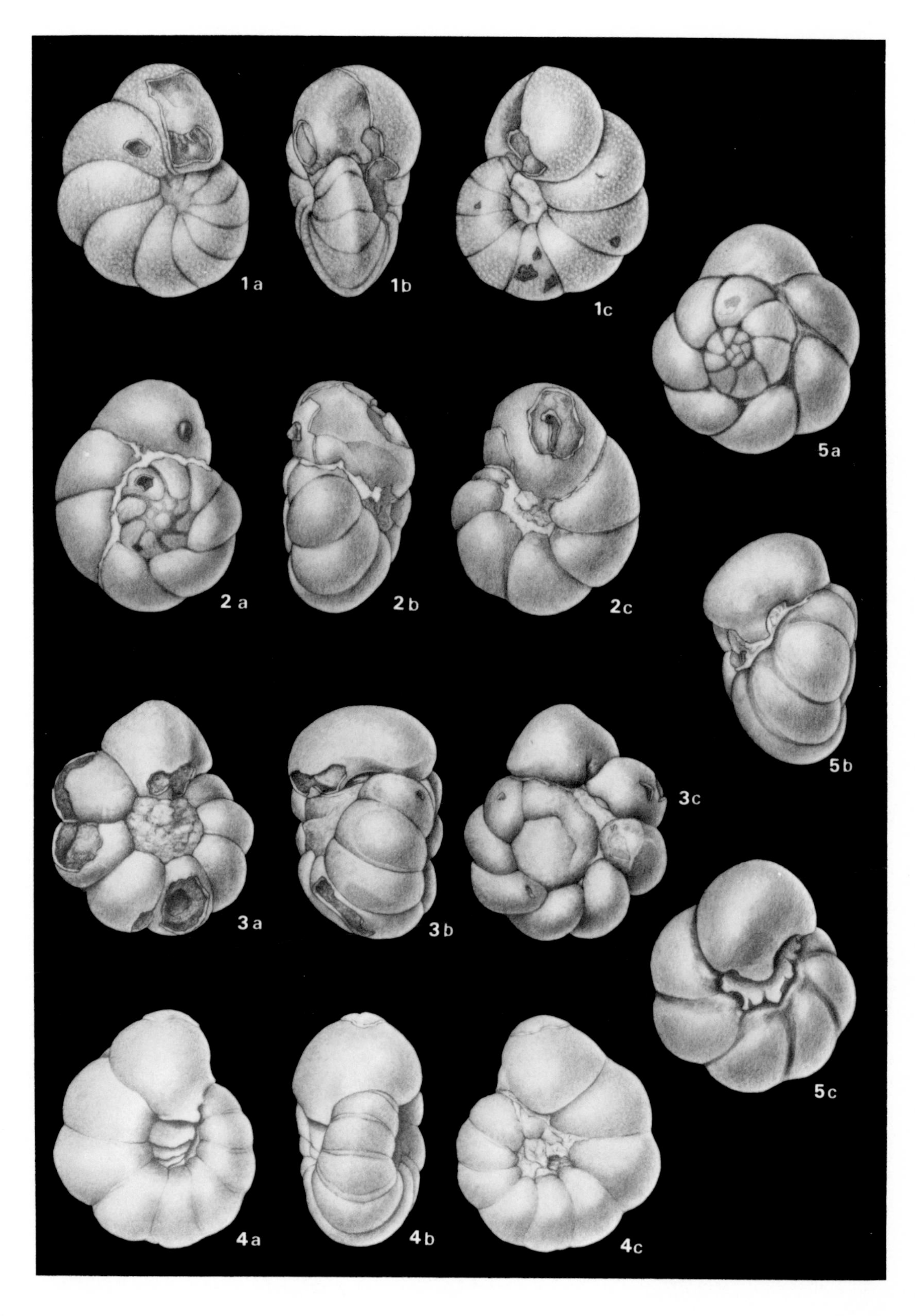
1a
1b
1c
5a
2a
2b
2c
5b
3c
3a
3b
5c
4a
4b
4c

Selected References

Addicott, W. O., 1967, Age of the Skooner Gulch Formation, Mendocino County, California: U.S. Geol. Survey Bull. 1254-C, 11 p.

——— 1970, Miocene gastropods and biostratigraphy of the Kern River area, California: U.S. Geol. Survey Prof. Paper 642, p. 1-174.

——— 1972, Provincial middle and late Tertiary molluscan stages, Temblor Range, California, *in* E. H. Stinemeyer, ed., The Pacific Coast Miocene biostratigraphic symposium: Bakersfield, Calif., Pacific Sec. SEPM Symposium Proc., p. 1-26.

Anderson, F. M., 1905, A stratigraphic study in the Mount Diablo Range of California: California Acad. Sci. Proc., 3d ser., Geology, v. 2, p. 155-248.

——— 1908, A further stratigraphic study in the Mount Diablo Range of California: California Acad. Sci. Proc., 4th ser., v. 3, p. 38-39.

——— 1911, The Neocene deposits of Kern River, California, and the Temblor basin: California Acad. Sci. Proc., 4th ser., v. 3, p. 73-148.

——— and B. Martin, 1914, Neocene record in the Temblor basin, California, and Neocene deposits of the San Juan district, San Luis Obispo County: California Acad. Sci. Proc., 4th ser., v. 4, p. 40-44.

Arnold, R., 1906, The Tertiary and Quaternary pectens of California: U.S. Geol. Survey Prof. Paper 47, p. 264.

——— and R. Anderson, 1907, Geology and oil resources of the Santa Maria oil district, Santa Barbara County, California: U.S. Geol. Survey Bull. 322, 161 p.

——— ——— 1910, Geology and oil resources of the Coalinga district, California: U.S. Geol. Survey Bull. 398, 354 p.

——— and H. R. Johnson, 1910, Preliminary report on the McKittrick-Sunset oil region, Kern and San Luis Obispo Counties: U.S. Geol. Survey Bull. 406, p. 1-225.

Bagg, R. M., 1905, Miocene Foraminifera from the Monterey Shale of California: U.S. Geol. Survey Bull. 268, 78 p.

Bailey, T. L., 1935, Lateral change of fauna in the lower Pleistocene: Geol. Soc. America Bull., v. 46, p. 489-502.

Bandy, O.L., 1952, The genotype of *Siphogenerina:* Cushman Found. Foram. Research Contr., v. 3, p. 17-18.

——— 1960, The geological significance of coiling ratios in the foraminifer *Globigerina pachyderma* (Ehrenberg): Jour. Paleontology, v. 34, no. 4, p. 671-681.

——— 1967, Foraminiferal definition of the boundaries of the Pleistocene in southern California, U.S.A., *in* Progress in oceanography: Oxford, England, Pergamon, v. 4, p. 27- 49.

——— ed., 1970, Radiometric dating and paleontologic zonation: Geol. Soc. America Spec. Paper 124, 247 p.

——— 1971, Recognition of upper Miocene Neogene Zone 18, experimental Mohole, Guadalupe site: Nature, v. 233, p. 476-478.

——— 1972a, Late Paleogene-Neogene planktonic biostratigraphy and its geologic implications, California, *in* E. H. Stinemeyer, ed., The Pacific Coast Miocene biostratigraphic symposium: Bakersfield, Calif., Pacific Sec. SEPM Symposium Proc., p. 37-51.

——— 1972b, Neogene planktonic foraminiferal zones, California, and some geologic implications: Palaeogeography, Palaeoclimatology, Palaeoecology, v. 12, p. 131-150.

——— 1972c, Upper Paleogene planktonic correlations *in* D. W. Weaver, ed., Central Santa Ynez Mountains, Santa Barbara County, California: Pacific Sec. AAPG-SEPM Guidebook, p. 55-58.

——— and R. Arnal, 1957, Some new Tertiary and Recent Foraminifera from California and the eastern Pacific: Cushman Found. Foram. Research Contr., v. 8, pt. 2, p. 54-58.

——— ——— 1960, Concepts of foraminiferal paleoecology: AAPG Bull., v. 44, p. 1921-1932.

——— ——— 1964, Middle Tertiary foraminiferal paleoecology, San Joaquin Valley, California (abs.): AAPG Bull., v. 48, p. 516.

——— ——— 1968, Middle Tertiary planktonic foraminiferal facies, San Joaquin basin, California (abs.): AAPG Bull. v. 52, p. 518-519.

——— ——— 1969, Middle Tertiary basin development, San Joaquin Valley, California: Geol. Soc. America Bull., v. 80, p. 783-870.

——— and J. C. Ingle, Jr., 1970, Neogene planktonic events and radiometric scale, California: Geol. Soc. America Spec. Paper 124, p. 131-172.

——— and R. Kolpack, 1963, Foraminiferal and sedimentological trends in the Tertiary section of Tecolote Tunnel, California: Micropaleontology, v. 9, p. 117-170.

——— and J. A. Wilcoxon, 1970, The Pliocene-Pleistocene boundary, Italy and California: Geol. Soc. America Bull., v. 81, p. 2939-2948.

——— J. C. Ingle, and R. C. Wright, 1971, *Globorotaloides suteri* Bolli subspecies *relizensis* n. subsp.: Jour. Foram. Research, v. 1, p. 15-16.

——— R. W. Morin, and R. W. Wright, 1969, Definition of the *Catapsydrax stainforthi* zone in the Saucesian Stage, California: Nature, v. 222, p. 468-469.

Banner, F. T., and W. H. Blow, 1965, Progress in the planktonic foraminiferal biostratigraphy of the Neogene: Nature, v. 208, p. 1164-1166.

Barbat, W. F., 1943, *in* California Div. Mines Bull. 118, p. 483, 588.

——— and F. E. von Estorff, 1933, Lower Miocene Foraminifera from the southern San Joaquin Valley, California: Jour. Paleontology, v. 7, p. 164-174.

——— and F. L. Johnson, 1934, Stratigraphy and Foraminifera of the Reef Ridge Shale, upper Miocene, California: Jour. Paleontology, v. 8, p. 3-17.

——— and A. L. Weymouth, 1931, Stratigraphy of *Borophagus littoralis* locality, California: California Univ. Pubs., Dept. Geol. Sci. Bull., v. 23, p. 25-36.

Barron, J. A., 1975, Marine diatom biostratigraphy of the upper Miocene-lower Pliocene strata of southern California: Jour. Paleontology, v. 49, p. 619-632.

Beck, S. R., compiler, 1952, Correlation chart of Oligocene, Miocene, Pliocene, and Pleistocene in San Joaquin Valley and Cuyama Valley areas, *in* Field trip routes, oil fields, Geology: AAPG-SEPM-SEG Joint Ann. Mtg. Guidebook, p.104.

Berggren, W. A., 1969a, Paleogene biostratigraphy and planktonic Foraminifera of northern Europe: 1st Internat. Conf. Planktonic Microfossils Proc., Leiden, Netherlands, E. J. Brill, v. 1, p. 121-160.

——— 1969b, Cenozoic chronostratigraphy, planktonic foraminiferal zonation and the radiometric time scale: Nature, v. 224, p. 1072-1075.

——— 1971, Tertiary boundaries and correlations, *in* B. M. Funnell and W. R. Riedel, eds. The micropaleontology of oceans: Cambridge, England, Cambridge Univ. Press, p. 693-809.

——— 1972, A Cenozoic time-scale—Some implications for regionsl geology and paleobiogeography: Lethaia, v. 5, no. 2, p. 195-215.

——— and J. A. van Couvering, 1974, The late Neogene: Amsterdam, Netherlands, Elsevier Pub. Co., p. 3-215.

Berry, D. R., 1971 Paleoecology and biostratigraphy of the foraminiferal sequence at Grimes Canyon, Ventura County, California: Master's thesis, Univ. California, Berkeley.

Berry, W. B. N., 1964, The Middle Ordovician of the Oslo region, Norway: Norsk. Geol. Tidsskr., v. 44, p. 61-170.

——— 1966, Zones and zones—with exemplification from the Ordovician: AAPG Bull., v. 50, p. 1487-1500.

——— 1968, Growth of a prehistoric time scale: San Francisco, W. H. Freeman, 158 p.

——— 1974, Erben's "inventory in stratigraphy"— a model from the California Tertiary foraminifer succession: Newsletter Stratigraphy v. 3, no. 2, p. 65-72.

Blackwelder, Eliot, 1909, The valuation of unconformities: Jour. Geology, v. 17, p. 289- 300.

Blacut, G., and R. M. Kleinpell, 1969, A stratigraphic sequence of benthonic smaller Foraminifera from the La Boca Formation, Panama Canal Zone: Cushman Found. Foram. Research Contr., v. 20, pt. 1, p. 1-22.

Blake, W. P., 1855, Remarks in conclusion [of Conrad's report on the fossil shells collected in California]: U.S. Pacific R.R. Expl., (U.S. 33d Cong., 1st Sess., H Ex. Doc. 129); appendix to the preliminary geological report of W. P. Blake, p. 20-21; 1856: Am. Jour. Sci., ser. 2, v. 21, p. 270-272; 1857, Neues Jahrb., p. 241-244.

Blow, W. H., 1959, Age, correlation, and biostratigraphy of the upper Tocuyo (San Lorenzo) and Pozon formations eastern Falcon, Venezuela: Bulls. Am. Paleontology, v. 39, no. 178, 235 p.

——— 1969, Late middle Eocene to Recent planktonic foraminiferal biostratigraphy: 1st Internat. Conf. Planktonic Microfossils Proc., Leiden, Netherlands, E. J. Brill, v. 1, p. 199-422.

Bolli, H. M., 1957a, The genera *Globigerina* and *Globorotalia* in the Paleocene-lower Eocene Lizard Springs Formation of Trinidad B.W.I.: U.S. Natl. Mus. Bull. 215, p. 61-81.

——— 1957b, Planktonic Foraminifers from the Oligocene-Miocene Cipero and Lengua Formations of Trinidad, B.W.I.: U.S. Natl. Mus. Bull. 215, p. 97-123.

——— 1957c, Planktonic Foraminifera from the Eocene Navet and San Fernando Formations of Trinidad, B.W.I.: U.S. Natl. Mus. Bull. 215, p. 155-171.

——— 1966, Zonation of Cretaceous to Pliocene marine sediments based on planktonic Foraminifera: Bol. Inf., v. 9, p. 3-32.

Bowen, O. E., 1965, Stratigraphy, structure, and oil possibilities in Monterey and Salinas quadrangles, California (abs.): AAPG Bull., v. 49, p. 1081.

——— 1969, Geologic map of the Monterey quadrangle: California Dept. Natural Resources Div. Mines Open-File Map.

Brabb, E. E., D. Bukry, and R. L. Pierce, 1971, Eocene (Refugian) nannoplankton in the Church Creek Formation near Monterey, central California: U.S. Geol. Survey Prof. Paper 750-C, p. C44-C47.

Bramlette, M.N., 1946, Monterey Formation of California and origin of its siliceous rocks: U.S. Geol. Survey Prof. Paper 212, 55 p.

——— and W. R. Riedel, 1954, Stratigraphic value of discoasters and some other microfossils related to Recent

coccolithophores: Jour. Paleontology, v. 28, p. 385-403.
——— and F. R. Sullivan, 1961, Coccolithophorids and related nannoplankton of the early Tertiary in Californai: Micropaleontology, v. 7, p. 129-174.
——— and J. A. Wilcoxon, 1967, Middle Tertiary calcareous nannoplankton of the Cipero section, Trinidad, W.I.: Tulane Studies Geology, v. 5, no. 3, p. 93-132.
Bukry, D., 1971, Cenozoic calcareous nannofossils from the Pacific Ocean: San Diego Soc. Nat. History Trans., v. 16, no. 14, p. 303-327.
——— 1973a, Low-latitude coccolith biostratigraphic zonation, *in* N. T. Edgar et al, Initial reports of the Deep Sea Drilling Project, vol. 15: Washington, D.C., U.S. Govt. Printing Office, p. 685-703.
——— 1973b, Coccolith and silicoflagellate stratigraphy, Deep Sea Drilling Project Leg 18, eastern North Pacific, *in* L. D. Kulm et al, 1973, Initial reports of the Deep Sea Drilling Project, vol. 18: Washington, D.C., U.S. Govt. Printing Office, p. 817-832.
——— 1975, Coccolith and silicoflagellate stratigraphy northwestern Pacific Ocean, Deep Sea Drilling Project, vol. 32: Washington, D.C., U.S. Govt. Printing Office, p. 677-701.
——— and J. H. Foster, 1974, Silicoflagellate zonation of Upper Cretaceous to lower Miocene deep-sea sediments: U.S. Geol. Survey Jour. Research, v. 2, no. 3, p. 303-310.
——— and M. P. Kennedy, 1969, Cretaceous and Eocene coccoliths at San Diego, California: California Div. Mines and Geology Spec. Rept. 100, p. 33-43.
——— E. E. Brabb, and J. G. Vedder, 1973, Correlation of Tertiary nannoplankton assemblages from the Coast and Peninsular Ranges of California: 2d Latin America Geol. Cong. Preprint, p. 1-35.
Canfield, C. R., 1939, Subsurface stratigraphy of Santa Maria Valley oil field and adjacent parts of Santa Maria Valley, California: AAPG Bull., v. 23, p. 45-81.
Carson, C. M., 1965, The Rincon Formation in western Santa Ynez Mountains, Santa Barbara County, California: Coast Geol. Soc., Pacific Sec. SEPM Guidebook, p. 38-41.
Chapman, F., 1896, On some Pliocene Ostracoda from near Berkeley, California: California Dept. Geology Bull., v. 2, p. 93-100.
——— 1900, Foraminifera from the Tertiary of California: California Acad. Sci. Proc., v. 3, geology 1, p. 241-
Church, C. C., 1972, The *Uvigerina subperegrina* Zone of the basal McClure Shale of Reef Ridge, Kings County, California, *in* West side central San Joaquin Valley: Pacific Sec. AAPG-SEPM-SEG Guidebook, p. 75-76.
Cifelli, R. D., 1951, Eocene Foraminifera from the Point of Rocks area, California: Master's thesis, Univ. California, Berkeley, p. 1-94.
——— 1969, Radiation of Cenozoic planktonic Foraminifera: Syst. Zoology, v. 18, no.2, p. 154-168.
Clark, B. L., 1915, Fauna of the San Pablo Group of middle California: California Univ. Pubs., Dept. Geol. Sci. Bull., v. 8, no. 22, p. 385-572.
Clark, J. C., 1966a, Tertiary stratigraphy of the Felton-Santa Cruz Mountains, California: PhD thesis, Stanford Univ., 184 p.
——— 1966b, Tertiary stratigraphy of the Felton-Santa Cruz area, Santa Cruz Mountains, California (abs.): Dissert. Abs., v. 27, no. 4, p. 1184-b.
——— and J. D. Rietman, 1973, Oligocene stratigraphy, tectonics, and paleogeography southwest of the San Andreas fault, Santa Cruz Mountains and Gabilan Range, California Coast Ranges: U.S. Geol. Survey Prof. Paper 783, 18 p.
——— et al, 1974, Preliminary geologic map of the Monterey and Seaside 7.5 minute quadrangles, Monterey County, California: U.S. Geol. Survey Misc. Field Studies Map MF-577, 2 sheets.
Clark, L. M., and A. Clark, 1934, The Vaqueros in the Temblor Range: Unpub. paper read before Pacific Sec. AAPG, Nov. 9, 1934.
Cole, W. S., 1957, Variation in American Oligocene species of *Lepidocyclina:* Bulls. Am. Paleontology, v. 38, no. 166, p. 31-51.
——— 1961, An analysis of certain taxonomic problems in the larger Foraminifera: Bulls. Am. Paleontology, v. 43, no. 197, p. 373-407.
——— 1967, A review of American species of Miogypsinids (larger Foraminifera): Cushman Found. Foram. Research Contr., v. 18, p. 99-117.
Cook, T. D., 1950, Eocene Foraminifera from the Devils Den area, California: Master's thesis, Univ. California, Berkeley, p. 1-116.
Cummings, J. C., R. M. Touring, and E. E. Brabb, 1962, Geology of the northern Santa Cruz Mountains: California Div. Mines Bull 181, p. 179-220
Curran, J. F., 1943, Eocene stratigraphy of Chico Martinez Creek area, Kern County, California: AAPG Bull., v. 27, p. 1361-1386.
Cushman, J. A., 1925a, Three new species of *Siphogenerina* from the Miocene of California: Cushman Lab. Foram. Research Contr., v. 1, pt. 1, p. 2-3.
——— 1925b, Apertural characters in Cristellaria with descriptions of a new species: Cushman Lab. Foram. Research Contr., v. 1, pt. 1, p. 24-25.
——— 1925c, Some Texturlariidae from the Miocene of California: Cushman Lab. Foram. Research Contr., v. 1,

pt. 2, p. 29-34.
——— 1925d, *Siphogenerina hughesi,* a new species from California: Cushman Lab. Foram. Research Contr., v. 1, pt. 2, p. 36.
——— 1926a, Miocene species of *Nonionina* from California: Cushman Lab. Foram. Research Contr., v. 1, pt. 4, p. 88-92.
——— 1926b, Some Pliocene *Bolivinas* from California: Cushman Lab. Foram Research Contr., v. 2, pt. 2, p. 40-47.
——— 1926c, Foraminifera of the typical Monterey of California: Cushman Lab. Foram. Research Contr., v. 2, pt. 3, p. 53-66.
——— 1926d, Foraminifera of the genera *Siphogenerina* and *Pavonina:* U.S. Natl. Mus. Proc., v. 67, art. 25, p. 1-24.
——— 1929, A late Tertiary fauna of Venezuela and other related regions: Cushman Lab. Foram. Research Contr., v. 5, pt. 4, p. 77-101.
——— and B. C. Adams, 1935, New late Tertiary *Bolivinas* from California: Cushman Lab. Foram. Research Contr., v. 11, pt. 1, p. 16-20.
——— and E. W. Galliher, 1934, Additional new Foraminifera from the Miocene of California: Cushman Lab. Foram. Research Contr., v. 10, pt. 1, p. 24-26.
——— and R. M. Kleinpell, 1934, New and unrecorded Foraminifera from the Miocene of California: Cushman Lab. Foram. Research Contr., v. 10, pt. 1, p. 17-23.
——— and B. Laiming, 1931, Miocene Foraminifera from Los Sauces Creek, Ventura County, California: Jour. Paleontology, v. 5. p. 79-120.
——— and F. L. Parker, 1931, Miocene Foraminifera from the Temblor of the east side of the San Joaquin Valley, California: Cushman Lab. Foram. Research Contr., v. 7, p. 1-16.
——— and H. G. Schenck, 1928, Two foraminiferal faunules from the Oregon Tertiary: California Univ. Pubs., Dept. Geol. Sci. Bull., v. 17, p. 305-324.
——— and R. R. Simonson, 1944, Foraminifera from the Tumey Formation, Fresno County, California: Jour. Paleontology, v. 18, p. 186-203.
——— and Ruth Todd, 1937, A monograph of the foraminiferal subfamily Virgulininae: Cushman Lab. Foram. Research Spec. Pub. 9, p. 1-228.
Davies, A. M., 1934, Tertiary faunas, a text-book for oilfield paleontologists and students of geology: London, England, Thomas Murby & Co., v. 2, 252 p.
Dibblee, T. W., Jr., 1950, Geology of southwestern Santa Barbara County, California: California Div. Mines Bull. 150, 95 p.
——— 1969, Regional geologic map of southern Coast Ranges near San Andreas fault from Cholame to Maricopa, Santa Barbara, San Luis Obispo, Kern, and Kings Counties, California: U.S. Geol. Survey Open-File Map.
Diener, C., 1925, Grundzuge der Biostratigraphie: Franz Deuticke, Leipzig and Vienna.
Domning, D. P., 1972, Sirenians in the West Coast Miocene stratigraphy, *in* E. H. Stinemeyer, ed., The Pacific Coast Miocene biostratigraphic symposium: Bakersfield, Calif., Pacific Sec. SEPM Symposium Proc., p. 146-149.
Driver, H. L., 1928, Foraminiferal section along Adams Canyon, Ventura County, California: AAPG Bull., v. 12, p. 753-756.
——— 1948, Genesis and evolution of Los Angeles basin, California: AAPG Bull., v. 32, p. 109-125.
Dymond, J. R., 1966, Potassium-argon geochronology of deep-sea sediments: Science, v. 152, no. 3726, p. 1239-1241.
Eames, F. E., et al, 1962, Fundamentals of mid-Tertiary stratigraphical correlation: Cambridge, England, Cambridge Univ. Press, 163 p.
——— et al, 1968, Some larger Foraminifera from the Tertiary of Central America: Paleontology, v. 11, p. 283-305.
Edwards, A. R., 1971, A calcareous nannoplankton zonation of the New Zealand Paleogene: 2d Planktonic Conf. Proc., 1970, v. 1, p. 381-419.
Edwards, L. N., 1972, Notes on the Vaqueros and Rincon formations, *in* D. W. Weaver, ed., Central Santa Ynez Mountains, Santa Barbara County, California: Pacific Sec. AAPG Fall Guidebook.
Eldridge C. H., and R. Arnold, 1907, The Santa Clara valley, Puente Hills, and Los Angeles oil districts, southern California: U.S. Geol. Survey Bull. 309, 266 p.
Emery, K. O., et al, 1952, Submarine geology off San Diego California: Jour. Geol., v. 60, p. 511-548.
English, W. A., 1916, Geology and oil prospects of Cuyama Valley, California: U.S. Geol. Survey Bull. 621, p. 191-215.
——— 1918, Geology and oil prospects of the Salinas Valley-Parkfield area, California: U.S. Geol. Survey Bull., 691-H., p. 219-250.
——— 1921, Geology and petroleum resources of northwestern Kern County, California: U.S. Geol. Survey Bull. 721, 48 p.
——— 1926, Geology and oil resources of Puente Hills region, southern California: U.S. Geol. Survey Bull. 768,

110 p.
Evans, D. L., 1928, A new Tertiary horizon in Ventura County, California: Master's thesis, Stanford Univ.
Fairbanks, H. W., 1904, San Luis, California: U.S. Geol. Survey Geol. Atlas Folio 104.
Fairchild, W. W., P. R. Wesendunk, and D. W. Weaver, 1969, Eocene and Oligocene Foraminifera from the Santa Cruz Mountains, California: California Univ. Pubs. Geol. Sci., v. 81, 144 p.
Fenton, C. L., and A. F. Fenton, 1928, Ecologic interpretation of some biostratigraphic terms: Am. Midland Naturalist, v. 11, no. 11, p. 20-22.
Fergusson, G. C., 1943, Correlation of oil field formations on east side San Joaquin Valley: California Div. Mines Bull. 118, p. 239-246.
Foss, C. D., and R. Blaisdell, 1969, Guidebook, geology and oil fields, west side, southern San Joaquin Valley: Pacific Sec. AAPG-SEPM-SEG, p.33-43.
Fox, L. S., 1929, Structural features of the east side of the San Joaquin Valley, California: AAPG Bull., v. 13, p. 101-108.
Frizzell, D. L., and R. Blackwelder, 1933, Preliminary analysis of the type Lincoln fauna (Oligocene) of Washington: Micropaleontology Bull., v. 4, no. 2, p. 53-63.
Fulmer, C. V., 1954, Stratigraphy and paleontology of the type Blakeley Formation of Washington (abs.): Geol. Soc. America Bull., v. 65, p. 1340-1341.
——— 1975, Stratigraphy and paleontology of the type Blakeley and Blakeley Harbor Formations, *in* D. W. Weaver et al, eds., Future energy horizons of the Pacific Coast: Pacific Sec. AAPG-SEPM-SEG, p. 210-271.
Galliher, E. W., 1931, Stratigraphic position of the Monterey formation: Micropaleontology Bull., v. 2, no. 4, p. 71-74.
——— 1932, Geology and physical properties of building stone from Carmel Valley, California: California Div. Mines, Rept. of State Mineralogist v. 28, no. 1, p. 14-41.
Galloway, J. J., and S. G. Wissler, 1927, Pleistocene Foraminifera from the Lomita quarry, Palos Verdes Hills, California: Jour. Paleontology, v. 1, p. 35-87.
Garrison, L. E., 1959, Miocene Foraminifera from the Temblor Formation north of Coalinga, California: Jour. Paleontology, v. 33, p. 662-669.
Gartner, S., 1973, Absolute chronology of the late Neogene calcareous nannofossil succession in the equatorial Pacific: Geol. Soc. America Bull., v. 84, p. 2021-2034.
Gibson, J. M., 1974, Distribution of planktonic Foraminifera and calcareous nannoplankton, Late Cretaceous to early middle Eocene, Santa Ynez Mountains, California (abs.): Geol. Soc. America Abs. with Program, v. 6, p. 181-182.
——— 1976, Distribution of planktonic foram. and calcareous nannoplankton, Late Cretaceous and early Paleogene, Santa Ynez Mountains, California: Jour. Foram. Research, v. 6.
——— and P. L. Steineck, 1972, Age and correlation of the Ulatisian and Narizian Stages, California (reply): Geol. Soc. America Bull., v. 83, p. 2225-2232.
Glaessner, M. F., 1945, Principles of micropaleontology: Melbourne Australia, Melbourne Univ. Press, 296 p.
Goudkoff, P. O., 1945, Stratigraphic relations of Upper Cretaceous in Great Valley, California: AAPG Bull, v. 29, p. 956-1007.
Graham, J. J., and C. W. Drooger, 1952, An occurrence of *Miogypsina* in California: Contrib. Cushman Found. Foram. Research, v. 3, p. 21-22.
Haller, C. R., 1971, Neogene foraminiferal faunas of the Humboldt Bay region: PhD thesis, Univ. California, Berkeley.
——— 1980, Pliocene biostratigraphy of California: this volume.
Hamlin, H., 1904, Water resources of Salinas Valley: U.S. Geol. Survey Water Supply Paper 89, 91 p.
Hay, W. W., 1967, Zonation of the middle-upper Eocene interval, *in* W. W. Hay et al, Calcareous nannoplankton zonation of the Cenozoic of the Gulf Coast and Caribbean-Antillean area, and transoceanic correlation: Gulf Coast Assoc. Geol. Socs. Trans., v. 17, p. 428-480.
——— and H. P. Mohler, 1967, Calcareous nannoplankton from early Tertiary rocks at Pont Labau, France, and Paleocene–early Eocene correlations: Jour. Paleontology, v. 41, p. 1505-1541.
——— ——— and Mary Wade, 1966, Calcareous nannofossils from Nalchik (northwest Caucasus): Eclogae Geol. Helvetiae, v. 59, no. 1, p. 379-400.
Heikkila, H. H., and G. M. MacLeod, 1931, Geology of Bitterwater Creek area, Kern County, California: California Div. Mines Spec. Rept. 6, 21 p.
Hickernell, R. L., 1972, Mr. Richard (Dick) L. Pierce, *in* E. H. Stinemeyer, ed., The Pacific Coast Miocene biostratigraphic symposium: Bakersfield, Calif., Pacific Sec. SEPM Symposium Proc., unnumbered intro. page.
Hill, M. L., S. A. Carlson, and T. W. Dibblee, Jr., 1958, Stratigraphy of Cuyama Valley – Caliente Range area, California: AAPG Bull., v. 42, p. 2973-3000.
Holman, W. H., 1958, Correlation of producing zones of Ventura basin oil fields (California), *in* J. W. Higgins, ed., A guide to the geology and oil fields of the Los Angeles and Ventura regions; Pacific Sec. AAPG Guidebook, p. 191-199.

Hoots, H. W., 1931, Geology of the eastern part of the Santa Monica Mountains, Los Angeles County, California: U.S. Geol. Survey Prof. Paper 165-C, 134 p.
Hornaday, G. R., 1972, Oligocene smaller Foraminifera associated with an occurrence of *Miogypsina* in California: Jour. Foram. Research, v. 2, no. 1, p. 35-46.
——— and F. J. Phillips, 1972, Paleogene correlations, Santa Barbara area, California; an alternative view: Pacific Sec. AAPG-SEPM Guidebook, June 1972 Field Trip, p. 59-71.
Hornibrook, N. de B., and A. R. Edwards, 1971, Integrated planktonic foraminiferal and calcareous nannoplankton datum levels in the New Zealand Cenozoic: 2d Internat. Planktonic Conf. Proc., 1970, v. 1, p. 649-657.
Hubbert, M. K., 1963, Are we retrogressing in science?: Geol. Soc. America Bull., v. 74, p. 365-378.
Hudson, F. S., and E. K. Craig, 1929, Geologic age of the Modelo Formation: AAPG Bull., v. 13, p. 509-518.
Huxley, J. C., 1958, Evolutionary processes and taxonomy with special reference to grades: Uppsala Univ. Arsskrift, no. 6, p. 21-39.
Ingle, J. C., Jr., 1963a, Paleoecologic, sedimentary and structural history of the late Tertiary Capistrano embayment, California (abs.): AAPG Bull., v. 47, p. 361.
——— 1963b, Miocene-Pliocene paleoecology of San Fernando basin, California (abs.): AAPG Bull., v. 47, p. 1771-1772.
——— 1967, Foraminiferal biofacies variation and the Miocene-Pliocene boundary in southern California: Bulls. Am. Paleontology v. 52, no. 236, p. 210-394.
——— 1972, Biostratigraphy and paleoecology of early Miocene through early Pleistocene benthonic and planktonic Foraminifera, San Joaquin hills– Newport Bay, Orange County, California, *in* E. H. Stinemeyer ed., The Pacific Coast Miocene biostratigraphic symposium: Bakersfield, Calif., Pacific Sec. SEPM Symposium Proc., p 255-283.
——— 1973a, Biostratigraphy and paleoecology of early Pleistocene benthonic and planktonic Foraminifera, San Joaquin hills-Newport Bay-Dana Point area, Orange County, California, *in* Miocene sedimentary environments and biofacies, southeastern Los Angeles basin: AAPG-SEPM-SEG, Guidebook, S.E.P.M. Trip 1, p. 18-38.
——— 1973b, Summary comments on Neogene biostratigraphy, physical stratigraphy and paleooceanography in the marginal northeastern Pacific Ocean, *in* L. D. Kulm et al, Initial reports of the Deep Sea Drilling Project, vol. 18, Washington, D.C., U.S. Govt. Printing Office, p. 949-960.
Jeletzky, J. A., 1973, Age and depositional environments of Tertiary rocks of Nootka Island, British Columbia (92°E); mollusks versus foraminifers: Canadian Jour. Earth Sci., v. 10, p. 331-365.
Jenkins, D. C., 1966, Planktonic foraminiferal zones and new taxa from the Danian to lower Miocene of New Zealand: New Zealand Jour. Geology and Geophysics, v. 8, p. 1088-1126.
——— 1971a, The reliability of some Cenozoic planktonic foraminiferal "datum-planes" used in biostratigraphic correlation: Jour. Foram. Research, v. 1, no. 2, p. 82-86.
——— 1971b, New Zealand Cenozoic planktonc Foraminifera: New Zealand Geol. Survey Paleontology Bull., 42, p. 1-278.
Keenan, M. F., 1932, The Eocene Sierra Blanca Limestone at the type locality in Santa Barbara County, California: San Diego Soc. Nat. History Trans., v. 7, no. 8, p. 53-84.
Kennett, J. P., 1965, Recognition and correlation of the Kapitean Stage (upper Miocene, New Zealand): New Zealand Jour. Geology and Geophysics, v. 10, p. 1051-1063.
——— 1968, Paleooceanographic aspects of the foraminiferal zonation in the upper Miocene–lower Pliocene of New Zealand: Comm. on Mediterranean Neogene Stratigraphy Proc., 4th Sess., Bologna, 1967, p. 143-156.
Key, C. E., 1955, Biostratigraphy of the Bitterwater-Packwood Creek area, Kern County, California: Master's thesis, Stanford University.
Kinney, D. M., et al, 1954, Geology of the Los Angeles basin, *in* Geology of Southern California: California Div. Mines Bull. 170, chap. 2.
Kleinpell, R. M., 1930, Zonal distribution of the Miocene Foraminifera in Reliz Canyon, California: Micropaleontology Bull., v. 2, no. 2, p. 27-32.
——— 1933a, Miocene Foraminifera from Reliz Canyon, Monterey County, California: Geol. Soc. America Bull., v. 44, p. 165.
——— 1933b, Miocene Foraminifera from Contra Costa County, California (abs.): Geol. Soc. America Proc. 1933, p. 390.
——— 1934, Difficulty of using cartographic terminology in historical geology: AAPG Bull., v.18, p. 374-379.
——— 1938, Miocene stratigraphy of California: AAPG Pub., p. 1-450.
——— 1948, Miocene–Pliocene boundary in California as a typical example of series-epoch boundary problems in correlation: Geol. Soc. America Bull., v. 59, p. 1387-1388.
——— 1964, Zoogeographic facies phenomena in Cenozoic correlation by fossils: Geol. Soc. America Spec. Paper 82, p. 109-110.
——— 1971, California's early "oil bug" profession: Jour. of the West, v. 10, no. 1, p. 72-101.
——— 1972, Some of the historical context in which a micropaleontological stage classification of the Pacific Coast middle Tertiary has developed, *in* E. H. Stinemeyer, ed., The Pacific Coast Miocene biostratigraphic

symposium: Bakersfield, Calif., Pacific Sec. SEPM Symposium Proc., p. 89-110.
——— 1977, Oligocene: New York, McGraw—Hill Yearbook Sci. and Technology, p. 355-358.
——— and D. W. Weaver, 1963, Oligocene biostratigraphy of the Santa Barbara embayment, California, part I: California Univ. Pubs. Geol. Sci., v. 43, p. 1-77.
Krueger, M. L., 1936, The Sycamore Canyon Formation, California (abs.): AAPG Bull., v. 20. p. 1520.
Lamb, J. L., 1964, The stratigraphic occurrences and relationships of some mid-Tertiary Uvigerinas and Siphogenerinas.; Micropaleontology, v. 10, p. 457-476.
Lamb, T. N., 1972, Geology of the Coronado Islands, northwestern Baja California, Mexico, (abs.): G.S.A. program for 1972, Cordilleran section, Honolulu, Hawaii.
Lawson, A. C., 1914, Description of the San Francisco, Concord, San Mateo, and Hayward quadrangles: U.S. Geol. Survey Geol. Atlas 193, San Francisco Folio, 24 p.
Leupold, W., and I. M. van der Vlerk, 1931, The Tertiary; *in* Feestbundel: K. Martin, Leidsche Geol. Mededeelingen, pt. 5, p. 611-648.
Lipps, J. H., 1964, Miocene planktonic Foraminifera from Newport Bay California: Tulane Studies Geology and Paleontology, v. 2, no. 4, p. 109-133.
——— 1965, Oligocene in California?: Nature, v. 208, p. 885-886.
——— 1967a, Planktonic Foraminifera, intercontinental correlation and age of California mid-Cenozoic microfaunal stages: Jour. Paleontology, v. 41, p. 994-1005.
——— 1967b, Miocene calcareous plankton, Reliz Canyon, California, *in* Gabilan Range and adjacent San Andreas fault: Pacific Sec. AAPG-SEPM Guidebook, October 1967, p. 54-60.
——— 1968, Mid-Cenozoic calcareous nannoplankton from western North America: Nature, v. 218, no. 5137, p. 1151-1152.
——— and M. Kalisky, 1972, California Oligo-Miocene calcareous nannoplankton biostratigraphy and paleoecology, *in* The Pacific Coast Miocene biostratigraphic symposium: Bakersfield, Calif., Pacific Sec. SEPM Symposium Proc., p. 239-254.
Loeblich, A. R., Jr., 1958, Danian Stage of Paleocene in California: AAPG Bull., v. 42, p. 2260-2261.
Loel, W., and W. H. Corey, 1932, The Vaqueros Formation, lower Miocene of California: California Univ. Pubs., Dept. Geol. Sci. Bull., v. 22, no. 3, p. 31-410.
Lohman, K E., 1974, Lower middle Miocene marine diatoms from Trinidad: Verhandl. Naturf. Ges. Basle, v. 84, no. 1, p. 326-360.
Louderback, G. D., 1913, The Monterey Series in California: California Univ. Pubs., Dept. Geol. Sci. Bull., v. 7, p. 177-241.
Lutz, G. C., 1951, The Sobrante Sandstone: California Univ. Pubs. Geol. Sci., v. 28, p. 367-406.
Lyell, C., 1833, Principles of geology, v. 3: J. Murray, London, 398 p. 109 p. apend.
Mallory, V. S., 1959, Lower Tertiary biostratigraphy of the California Coast Ranges: AAPG Pub., p. 1-416.
——— 1970, Lower Tertiary Foraminifera from the Media Agua Creek drainage area, Kern County, California: Seattle, Washington, Thomas Burke Memorial, Washington State Museum, p. 1-182.
Martin, L., 1964, Upper Cretaceous and lower Tertiary Foraminifera from Fresno County, California: Jahrb. Geol. Bundesanstalt, Sonderband 9, p. 1-128.
Martin, L. T., 1943, Eocene Foraminifera from the type Lodo Formation, Fresno County, California: Stanford Univ. Pubs. Geol. Sci., v. 3, no. 3, p. 93-125.
Martini, E., 1971, Standard Tertiary and Quarternary calcareous nannoplankton zonation: 2d Internat: Planktonic Conf. Proc., Rome, 1970, v. 2, p. 739-786.
——— and M. N. Bramlette, 1963, Calcareous nannoplankton from the experimental Mohole drilling: Jour. Paleontology, v. 37, p. 845-856.
Mathews, R. D., 1945, *Rectuvigerina,* a new genus of Foraminifera from a restudy of *Siphogenerina:* Jour. Paleontology, v. 19, no. 6, p. 588-606.
McKeel, D. R., and J. H. Lipps, 1972, Calcareous plankton from the Tertiary of Oregon: Palaeogeography, Paleoclimatology Palaeoecology, v. 12, p. 75-93.
——— ——— 1975, Eocene and Oligocene planktonic Foraminifera from the central and southern Oregon Coast Range: Jour. Foram. Research, v. 5, no. 4, p. 249-269.
Merriam, C. W., 1941, Fossil Turritellas from the Pacific Coast region of North America: California Univ. Pubs. Geol. Sci., v. 26, p. 1-114.
Merriam, J. C., 1904, A note on the fauna of the lower Miocene in California: California Univ. Pubs., Dept. Geol. Sci. Bull., v. 3, p. 377-381.
Mills, J.S., *in* Webster 1934, p. 1967.
Mohler, H. P., and W. W. Hay, 1967, Zonation of the Paleocene-lower Eocene interval, *in* W. W. Hay et al, Calcareous nannoplankton zonation of the Cenozoic of the Gulf Coast and Caribbean-Antillean area, and transoceanic correlation: Gulf Coast Assoc. Geol. Socs. Trans., v. 17, p. 428-480.
Munier-Chalmas, and de Lapparent, 1893, Note sur le Nomenclature des Terrains Sedimentaires: Bull. Soc. Geol. France, ser. 3, v. 21, p. 438-488.
Murray, G. E., 1961, Geology of the Atlantic and Gulf Coastal Province of North America: New York, Harper & Bros., p. 1-659.

Natland, M. L., 1933, The temperature and depth-distribution of some recent and fossil Foraminifera in the southern California region: Scripps Inst. Oceanog. Bull., Tech. Ser., v. 3, no. 10, p. 225-230.
——— 1952, Pleistocene and Pliocene stratigraphy of southern California: PhD thesis, Univ. California, Los Angeles, p. 1-165.
——— 1953, Correlation of Pleistocene and Pliocene Stages in southern California: Pacific Petroleum Geologist Newsletter, Feb., p. 2.
——— 1957, Paleoecology of West Coast Tertiary sediments: Geol. Soc. America Mem. 67, v. 2, p. 543-572.
Oakeshott, Gordon, 1964, The San Andreas fault revisited: Geol. Soc. America Cordilleran Sec. Mtg.
Oppel, A., 1856-58, Die Juraformation Englands, Frankreich und des Sudwestlichen Deutschlands: Stuttgart, Wurttemb. Naturwiss. Veren Jahreh, 857 p.
Parker, F. L., 1964, Foraminifera from the experimental Mohole drilling near Guadalupe Island, Mexico: Jour. Paleontology, v. 38, p. 617-636.
Patet, A., 1972, A subsurface study of the foraminiferal fauna of the Vaqueros, the Rincon, and the lower Monterey formations from the Elwood oil field in Santa Barbara County, California, *in* E. H. Stinemeyer, ed., The Pacific Coast Miocene biostratigraphic symposium: Bakersfield Calif., Pacific Sec. SEPM Symposium Proc., p. 150-157.
Phillips, F. J., 1972a, Miocene biofacies in the Saltos Shale and Whiterock Bluff Shale of Caliente Mountain: Pacific Petroleum Geologist Newsletter, v. 26, no. 4-5, p. 7.
——— 1972b, Age and correlation of the Eocene Ulatisian and Narizian Stages, California; discussion: Geol. Soc. America Bull., v. 83, p. 2217-2224.
——— A. Tipton, and R. Watkins, 1974, Outcrop studies of the Eo-Oligocene Tumey Formation, Monocline Ridge, Fresno County, California, *in* G. R. Hornaday, ed., The Paleogene of the Panoche Creek–Cantua Creek Area: Pacific Sec. SEPM, Fall 1974 Guidebook, p. 99-131.
Pierce, R. L., 1956, Upper Miocene Foraminifera and fish from the Los Angeles area, California: Jour. Paleontology, v. 30, p. 1288-1314.
——— 1970, Preliminary revaluation of Late Miocene biostratigraphy of California (abs.): AAPG Bull., v. 54, p. 559.
——— 1972, Revaluation of the late Miocene biostratigraphy of California; summary of evidence, *in* E. H. Stinemeyer, ed., The Pacific Coast Miocene biostratigraphic symposium: Bakersfield, Calif., Pacific Sec. SEPM Symposium Proc., p. 334-340.
Popenoe, W. P., and R. M. Kleinpell, 1978, Age and stratigraphic significance for Lyellian correlation of the fauna of the Viga Formation, Luzon, Philippines: California Acad. Sci. Occasional Paper 129, p. 1-73.
Postuma, J. A., 1971, Manual of planktonic Foraminifera: New York, Elsevier, 420 p.
Rau, W. W., 1948, Foraminifera from the Miocene Astoria Formation in southwestern Washington: Jour. Paleontology, v. 22, p. 774-782.
——— 1951, Tertiary Foraminifera from the Willapa River Valley of southwestern Washington: Jour. Paleontology, v. 25, p. 417-453.
——— 1958, Stratigraphy and foraminiferal zonation in some new Tertiary rocks of southwest Washington: U.S. Geol. Survey Oil and Gas Inv. Chart, OC-57.
——— 1964, Foraminifera from the northern Olympic Peninsula, Washington: U.S. Geol. Survey Prof. Paper 374-G, p. G1-G33.
——— 1966, Stratigraphy and Foraminifera of the Satsop River area, southern Olympic Peninsula, Washington: Washington Div. Mines and Geology Bull., no. 53, p. 3-66.
——— 1967, Geology of the Wynootchee Valley quadrangle, Grays Harbor County, Washington: Washington Div. Mines and Geology Bull., no. 56, p. 1-55.
——— 1975, Foraminifera and biostratigraphy of the Alsea Formation of western Oregon, *in* D. W. Weaver et al, eds., Future energy horizons of the Pacific Coast: Pacific Sec. AAPG-SEPM-SEG, p. 409-416.
Redwine, L. E., et al, 1952, Cenozoic correlation section paralleling north and south margins western Ventura basin from Point Conception to Ventura and Channel Islands, California: Pacific Sec. AAPG, Correlation Chart.
Reed, R. D., 1925, The post-Monterey disturbance in the Salinas Valley, California: Jour. Geology, v. 33, p. 588-607.
——— 1926, Miocene paleogeography in the central Coast Ranges: AAPG Bull., v. 10, p. 130-137.
——— 1933, Geology of California: AAPG Pub., 355 p.
——— 1935a, Miocene orogenies in the California Coast Ranges (abs): Geol. Soc. America Cordilleran Sec., Papers with Abs., p. 6-7.
——— 1935b, Miocene breccias of the Santa Barbara district (abs): Geol. Soc. America Cordilleran Sec., Papers with Abs., p. 36-37.
——— and J. S. Hollister, 1936, Structural evolution of southern California: AAPG, 157 p.; also AAPG Bull., v. 20, p. 1533-1721; reprinted, 1951.
Richards, G. L., 1935a, *Astrodapsis* faunal zones of California upper Miocene and lower Pliocene formations (abs): Paleont. Soc., Pacific Coast Br., Papers with Abs., April, 1935, p. 45.
——— 1935b, Revision of some California species of Astrodapsis: San Diego Soc. Nat. History Trans., v. 8, no.9, p. 63.

——— 1936, Foraminiferal, echinoid, molluscan correlations of the Santa Margarita and San Pablo Formations: Paleont. Soc., Pacific Coast Br., Program with Abs., April, p. 29-30.
Richmond, J. F., 1952, Geology of Burruel Ridge, northwestern Santa Ana Mountains, California: California Div. Mines Spec. Rept. 21, 16 p.
Roth, P. H., 1970, Oligocene calcareous nannoplankton biostratigraphy: Eclogae Geol. Helvetiae, v. 63, p. 802-879.
Rudel, C. H., 1968, *Pullenia moorei - Rotalia becki* (Pseudosaucesian) biofacies of the lower Mohnian, *in* S. E. Karp, ed., Geology and oilfields, west side southern San Joaquin Valley; Pacific Sec. AAPG-SEPM-SEG Guidebook.
Ruth, J., 1972, Panel discussion *of* E. H. Stinemeyer, ed., The Pacific Coast Miocene biostratigraphic symposium: Bakersfield, Calif., Pacific Sec. SEPM Symposium Proc., p. 352-353.
Savage, D. E., and L. Barnes, 1972, Miocene vertebrate geochronology of the West Coast of North America, *in* E. H. Stinemeyer, ed., The Pacific Coast Miocene biostratigraphy symposium: Bakersfield Calif., Pacific Sec. SEPM Symposium Proc., p. 124-145.
Schenck, H. G. 1935, What is the Vaqueros Formation of California and is it Oligocene?: AAPG Bull., v. 19, p. 521-536.
——— and T. S. Childs, 1942, Significance of *Lepidocyclina (Lepidocyclina) californica,* new species in the Vaqueros Formation (Tertiary), California: Stanford Univ. Pubs. Geol. Sci., v. 3, no. 2, 59 p.
——— and R. M. Kleinpell, 1936, Refugian Stage of Pacific Coast Tertiary: AAPG Bull., v. 20, p. 215-225.
——— and S. W. Muller, 1941, Stratigraphic terminology: Geol. Soc. America Bull., v. 52, p. 1419-1426.
Schmidt, R. R., 1970, Planktonic Foraminifera from the lower Tertiary of California: PhD thesis , Univ. California, Los Angeles, 302 p.
——— 1975, Upper Paleocene- middle Eocene planktonic biostratigraphy from the Great Valley of California and adjacent areas, and correlation to the West Coast microfaunal stages, *in* Future energy horizons of the Pacific Coast: Long Beach, Calif., Pacific Sec. AAPG-SEPM-SEG, p. 439-455.
Smith, H. P., Foraminifera from the Wagon Wheel Formation, Devils Den district, California: California Univ. Pubs. Geol. Sci., vol. 32, p. 65-126.
Smith, J. P., 1919, Climatic relations of Tertiary and Quaternary faunas of the California region: California Acad. Sci. Proc., 4th ser., v. 9, p. 123-173.
Smith, P. B, 1960, Foraminifera of the Monterey Shale and Puente Formation, Santa Ana Mountains and San Juan Capistrano area, California: U.S. Geol. Survey Prof. Paper 294-M, p. 463-495.
Smith, R. K., 1971, Foraminiferal studies in the lower and middle Tertiary of Soquel Creek, Santa Cruz County, California: California Univ. Pubs. Geol. Sci., v. 91, 111 p.
Smith, W. M., 1930, Some Foraminifera from the Elwood field, Santa Barbara County, California: Micropaleontology Bull., v. 2, no. 1, p. 5-7.
Snavely, P. D., Jr., et al, 1975, Alsea Formation—an Oligocene marine sedimentary sequence in the Oregon Coast Range: U. S. Geol. Survey Bull., 1395-F, p. F1-F21.
Snedden, L. B., 1932, Notes on the stratigraphy and micropaleontology of the Miocene formations in Los Sauces Creek, Ventura County, California: Micropaleontology Bull., v. 3, no. 2, p. 41-46.
Steineck, P. L. and J. M. Gibson, 1971, Age and correlation of the Eocene Ulatisian and Narizian Stages, California: Geol. Soc. America Bull., v. 82, p. 477-480.
——— ——— 1972, Age and correlation of the Eocene Ulatisian and Narizian Stages, California (reply): Geol. Soc. America Bull., v. 83, p. 535-536.
Stewart, R. E. and K. C. Stewart, 1930a, Post-Miocene Foraminifera from the Ventura quadrangle, Ventura County, California: Jour. Paleontology, v. 4, p. 60-72.
——— ——— 1930b, Lower Pliocene in eastern end of Puente Hills, San Bernardino, California: AAPG Bull., v. 14, p. 1445-1450.
Stinemeyer, E. H., 1972, ed., The Proceedings of the Pacific Coast Miocene biostratigraphic symposium: Pacific Sec. SEPM, March 9-10, Bakersfield, California.
Stirton, R. A., 1933, Critical review of the Mint Canyon mammalian fauna and its correlative significance: Am. Jour. Sci., 5th ser., v. 26, no. 156, p. 569-573.
——— 1960, A marine carnivore from the Clallam Miocene formation, Washington: California Univ. Pubs Geol. Sci., v. 36, p. 345-365.
Sullivan, F. R., 1962, Foraminifera from the type section of the San Lorenzo Formation, Santa Cruz County, California: California Univ. Pubs. Geol. Sci., v. 37, p. 233-252.
——— 1964, Lower Tertiary nannoplankton from the California Coast Ranges, I, Paleocene: California Univ. Pubs. Geol. Sci., v. 44, p. 163-228.
——— 1965, Lower Tertiary nannoplankton from the California Coast Ranges, II, Eocene: California Univ. Pubs. Geol. Sci., v. 53, no 210, p. 1-53.
Thorup, R. R., 1941, Vaqueros Formation (Tertiary) at its type locality, Junipero Serra quadrangle, Monterey County, California (abs): Geol. Soc. America Bull. v. 52 p. 1957-1958.
——— 1943, Type locality of the Vaqueros Formation: Calif. Div. Mines Bull. 118, p. 463-466.
Tipton, A., 1970, Oligocene faunas and biochronology in the subsurface southwestern San Joaquin Valley,

California: Master's thesis, Univ. California, Berkeley.
——— 1972, Zemorrian and Saucesian foraminiferal sequences in the subsurface, southwestern San Joaquin Valley, California, *in* The Pacific Coast Miocene biostratigraphic symposium: Bakersfield Calif., Pacific Sec. SEPM Symposium p. 111-122.
——— 1976a, Foraminiferal zonation of the Refugian Stage, uppermost Eocene of California: Bandy Mem. Vol., New York, Springer-Verlag (in press).
——— 1976b, The Refugian Stage of California—foraminiferal zonation, geologic history, and correlations to the Pacific Northwest: PhD Dissert., Univ. California, Santa Barbara.
——— R. M. Kleinpell, and D. W. Weaver, 1973, Oligocene biostratigraphy, San Joaquin Valley, California: California Univ. Pubs. Geol. Sci., v. 105, p. 1-72.
——— ——— ——— 1974, Oligocene correlations, geologic history, and lineages of Siphogenerine Foraminifera, west side, San Joauin Valley, California, *in* G. R. Hornaday, ed., The Paleogene of the Panoche Creek–Cantua Creek area: Pacific Sec. SEPM Fall 1974 Guidebook, p. 132-150.
Trask, P. D., 1922, The Briones Formation of middle California: California Univ. Pubs., Dept. Geol. Sci Bull., v. 13, p. 133-174.
Truex, J. N., 1976, Santa Monica and Santa Ana Mountains—relation to Oligocene, Santa Barbara basin: AAPG Bull., v. 60, p. 65-86.
Turner, D. L., 1970, Potassium-argon dating of Pacific Coast Miocene foraminferal stages: Geol. Soc. America Spec. Paper 124, p. 91-129.
Umbgrove, J. H. F., 1938, Geological history of the East Indies: AAPG Bull., v. 22, p. 1-20.
Vanderhoof, V. L., 1936, Nature and distribution of *Desmostylus,* a marine Tertiary mammal (abs.): Pan-American Geologist, v. 64, no. 1, p. 80; also Geol. Soc. America Proc. for 1935, p. 420.
——— 1937, A study of the Miocene sirenian *Desmostylus:* California Univ. Pubs., Dept. Geol. Sci. Bull., v. 24, no 8, p. 169-252.
——— 1941, Miocene sea-cow from Santa Cruz, California, and its bearing on intercontinental correlation (abs.): Geol. Soc. America Bull., v. 52, p. 1984-1985.
——— 1942, Bearing of sea-cows on the age of the Vaqueros: Stanford Univ. Pubs. Geol. Sci., v. 3, p. 40-42.
Vedder, J., 1972, Review of stratigraphic names and megafaunal correlations along the southeast edge of the Los Angeles basin, California, *in* E. H. Stinemeyer, ed., The Pacific Coast Miocene biostratigraphic symposium: Bakersfield, Calif., Pacific Sec. SEPM Symposium Proc., p. 158-172.
von Beyrich, E., 1854, Uber die Stellung der Hessischen Tertiurbildungen: Bericht Uber die zur Bekamntmachung geeigneben Verhandlungen der Konigl. Preuss. Akad. Wiss. Berlin, p. 640-666.
Wardle, W. C., 1957, Eocene foraminiferal fauna from the type Lucia Shale: Masters thesis, Univ. California, Berkeley.
Warren, A. D., 1972, Luisian and Mohnian biostratigraphy of the Monterey Shale at Newport Lagoon, Orange County, California, *in* E. H. Stinemeyer, ed., The Pacific Coast Miocene biostratigraphic symposium: Bakersfield, Calif., Pacific Sec. SEPM Symposium Proc., p. 27-36.
——— 1973, Luisian and Mohnian biostratigraphy of the Monterey Shale at Newport Lagoon, Orange County, California, *in* Miocene sedimentary environments and biofacies, southeastern Los Angeles basin: AAPG-SEPM-SEG Guidebook, SEPM Trip 1, p. 61-66.
——— and J. H. Newell, 1976a, Nannoplankton biostratigraphy of the upper Sacate and Gaviota Formations, Arroyo el Bulito, Santa Barbara County, California (abs.): San Francisco, Calif., Pacific Sec. AAPG-SEPM-SEG Ann. Mtg. Prog.
——— ——— 1976b, Plankton biostratigraphy of the Refugian and adjoining stages of the Pacific Coast Tertiary: Bandy Mem. Vol., New York, Springer-Verlag. (in press).
Waters, John, 1970. Foraminifera from the Church Creek Formation: Masters Thesis Univ. California, Berkeley.
Watkins, R., 1974, Molluscan paleobiology of the Miocene Wimer Formation, Del Norte County, California: Jour. Paleontology, v. 48, p. 1264-1282.
Weaver, C. E., 1909, Stratigraphy and paleontology of the San Pablo Formation in middle California: California Univ. Pubs., Dept. Geol. Sci. Bull., v. 5, p. 243-269.
Weaver, D. W., 1969, The limits of Lyellian series and epochs, *in* Colloque sur L'Eocene, Paris, 1968, vol. 3: Bur. Recherches Geol. et Minieres Mem. 69, p. 283-286.
——— and J. D. Frantz, 1967, Re-evaluation of the type Refugian Stage of the Pacific Coast Tertiary (abs.): Cordilleran Sec. Geol. Soc. America Ann. Mtg. Prog., Santa Barbara, Calif., p. 69-70.
——— and R. M. Kleinpell, 1972, Tertiary column of the Santa Barbara embayment, *in* D. W. Weaver, ed., Central Santa Ynez Mountains, Santa Barbara County, California: Pacific Sec. AAPG-SEPM Fall Guidebook.
——— and A. Tipton, 1972, Formations and age-subdivisions of the West Coast middle Tertiary, *in* E. H. Stinemeyer, ed., The Pacific Coast Miocene biostratigraphic symposium: Bakersfield, Calif., Pacific Sec. SEPM Symposium Proc., p. 52-62.
——— et al, 1969, Geology of the Northern Channel Islands: Pacific Sec. AAPG-SEPM Spec. Pub., p. 1-113.
Webster's New International Dictionary of the English Language, 1934, 2d edition: Springfield, Mass., G. & C. Merriam Co.
White, W. R., 1956, Pliocene and Miocene Foraminifera from the Capistrano Formation, Orange County,

California: Jour. Paleontology, v. 30, p. 237-260.
Wilcoxen, J. A., 1969, Tropical planktonic zones and calcareous nannoplankton correlations in part of the California Miocene: Nature, v. 221, no. 5184, p. 950-951.
Wilson, E. J., 1954, Foraminifera from the Gaviota Formation east of Gaviota Creek, California: California Univ. Pubs. Geol. Sci., v. 30, No. 2, p. 103-170.
Wise, S. W., Jr., 1973, Calcareous nannofossils from cores recovered during Leg 18, Deep Sea Drilling Project; biostratigraphy and observations of diagenesis, *in* L. D. Kulm et al, Initial reports of the Deep Sea Drilling Project, vol. 18: Washington, D.C., U.S. Govt. Printing Office, p. 569-615.
Wissler, S. G., 1943, Stratigraphic formations of the producing zones of the Los Angeles basin oil fields: California div. Mines Bull. 118, p. 209-233.
——— and F. E. Dreyer, 1943, Correlation of the oil fields of the Santa Maria district: California Div. Mines Bull. 118, p. 235-233.
Woodford, A. O., 1925, The San Onofre Breccia—its nature and origin: California Univ. Pubs. Dept. Geology Bull., v. 17, p. 265-304.
——— et al, 1954, Geology of the Los Angeles basin, *in* Geology of southern California: California Div. Mines Bull. 170, p. 65-82.
Woodring, W. P., 1970, Geology and paleontology of Canal Zone and adjoining parts of Panama—description of Tertiary mollusks (gastropods: Eulimidae, Marginellidae to Helminthoglyptidae): U.S. Geol. Survey Prof. Paper 306-D, p. 299-452.
——— and M. N. Bramlette, 1950, Geology and paleontology of the Santa Maria Distirct, California: U.S. Geol. Survey Prof. Paper 222, 185 p.
——— ——— and W. S. W. Kew, 1946, Geology and paleontology of Palos Verdes Hills, California: U.S. Geol. Survey Prof. Paper 207, 145 p.
——— ——— and R. M. Kleinpell, 1936, Miocene stratigraphy and paleontology of Palos Verdes Hills, California: AAPG Bull, v. 20, p. 125-149.
——— R. Stewart, and R. W. Richards, 1940, Geology of the Kettleman Hills oil field, California: U.S. Geol. Survey Prof. Paper 195, 170 p.
Wornardt, W. W., 1963, Stratigraphic distribution of diatom floras from the Mio-Pliocene of California: PhD thesis, Univ. California, Berkeley.
——— 1967, Miocene and Pliocene marine diatoms from California: California Acad. Sci. Occasional Paper 63, 108 p.
——— 1972, Late Miocene and early Pliocene correlations in the California province, *in* E. H. Stinemeyer, ed., The Pacific Coast Miocene biostratigraphic symposium: Bakersfield, Calif., Pacific Sec. SEPM Symposium Proc., p. 284-333.
——— 1973, Diatom, silicoflagellate, radiolarian, calcareous nannofossil and foraminiferal biostratigraphy of the middle and late Miocene and Pliocene of Newport Beach Bay, Newport Beach, California, *in* Miocene sedimentary environments and biofacies, southeastern Los Angeles basin: Pacific Sec. SEPM Guidebook, Trip 1, p. 39-53.

Pliocene Biostratigraphy of California[1]

C. R. HALLER[2]

Abstract Biostratigraphy of Neogene faunas from the Humboldt basin of northern California emphasizes six foraminifera aggregates encountered in the thick clastic sections. Other groups, notably mollusks, are included.

Detailed analysis of the chronologic and chorologic relations of the component elements (lineages, faunas, and stratal units) provide a framework for comparison and correlation of clastic sequences of roughly comparable age and thickness from the Los Angeles and Ventura basins.

Considerable discussion involves the two primary methods of defining the Pliocene-Pleistocene boundary—Reboul's 1833 concept implying evolutionary changes in fossil faunas gives way to Forbes' 1846 concept of paleoclimatic change reflected by fossil faunas. The writer herein places the boundary 4,000 ft (1, 220 m) lower than that cited by Ogle (1953), or Faustman (1964).

Paleoecologic interpretations and basin analyses rely on faunal variability and faunal dominance in both foraminiferal and molluscan assemblages. Sediment textures generally reflect basin filling and westward migration of the Humboldt basin shoreline. Turbidite sequences appear to be important, especially in sections containing bathyal assemblages. Upwelling possibly influenced the distribution of some key foraminiferal faunules in northern California.

More than 180 species of forams are documented in the systematics and distribution charts, and two species are new and previously undescribed. Important questions are raised concerning the nomenclature of *Bulimina subacuminata, Elphidium hughesi*, and other species. The use of the generic term *Islandiella* for certain species of *Cassidulina* is also questioned. The Neogene bolivinids, mainly *Bolivina, Bolivinoides,* and *Brizalina,* are regarded as the most promising for future studies on phylogeny through the extensive use of SEM pictures.

Introduction

Scope

The Humboldt basin is a triangular area roughly 26 by 28 mi (42 X 45 km) in its coastal onshore location (Fig. 1); it extends offshore into the northeast Pacific. Seismic

[1] Manuscript received, August 30, 1974; accepted, September 9, 1975; revised, November 3, 1976.

[2] Marathon Oil Company, Findlay, Ohio.

The writer owes a debt of gratitude to landowners for permission to enter posted property, including Keele's Ranch of Freshwater Creek, Barri Ranch south of Centerville, Rush Ranch south of Centerville and along the Bear River, and Lowry Ranch of the Bear River valley. The Pacific Lumber Company of Scotia granted permission to traverse extensive holdings on the North Fork of the Elk River as well as those at Larabee Ranch, Larabee Creek.

Information regarding fossil localities in Humboldt County was obtained from William Lowry of Ferndale and Percy Hollister of Eureka. C. H. Stevens of San Jose State College forwarded comparative Miocene and Pliocene material which he collected from outcrops in the Ventura and Los Angeles basins of southern California.

A. J. MacMillan, Jr., of Texaco Inc., Los Angeles, forwarded core samples of the Texaco No. 2 Eureka well (Humboldt County, California), and P. B. Harris, also of Texaco Inc., granted permission to publish foraminiferal data gained from the study of these samples. L. J. Simon, W. R. White, R. H. McGlasson, and others of the Texaco Paleontological Department at Los Angeles examined biotype and assemblage slides and provided pertinent suggestions regarding faunal analyses. They also permitted the writer to examine their extensive comparative material from both northern and southern California.

L. C. Pray, formerly of Marathon Oil Co., Denver Research Center, conveyed information about carbonate diagenesis, particularly with regard to organic carbonates.

R. W. Kopf of the U.S. Geological Survey, Menlo Park, provided information regarding use of formation names of the California Pliocene and Pleistocene.

The Department of Paleontology at the University of California, Berkeley, provided transportation funds for field trips to northern California in the summer of 1964, and for the field trip to southern California in the summer of 1965.

W. A. Berggren of the Woods Hole Oceanographic Institution has been most generous in supplying copies of many of his important publications. Numerous colleagues assisted in the work on the systematics, and their contribution is so credited in that section.

Finally, the writer is most indebted to R. M. Kleinpell, Z. M. Arnold, and C. M. Gilbert of the University of California, Berkeley, for reading the manuscript and providing critical comments.

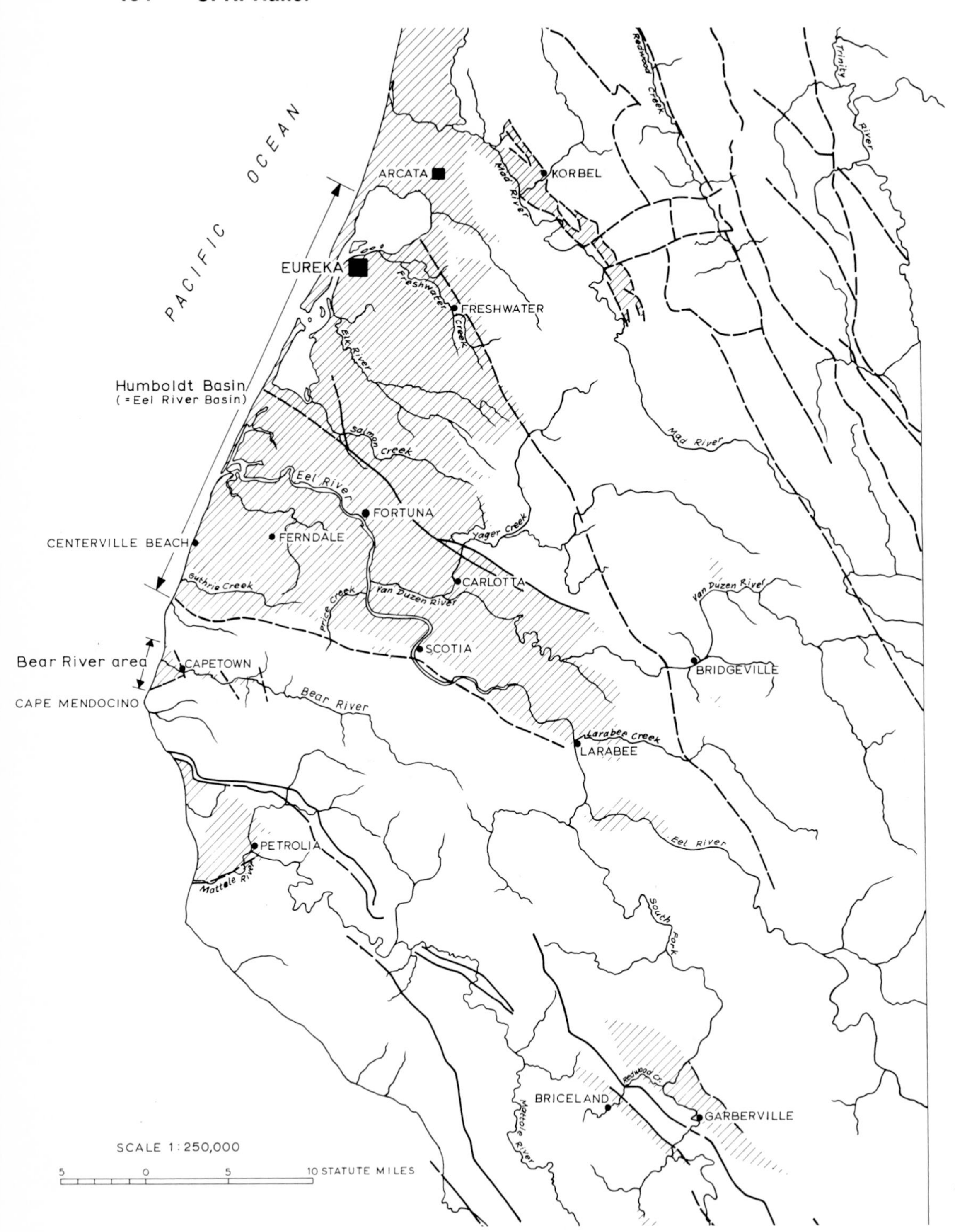

FIG. 1–Index map of a portion of Humboldt County showing major fault pattern and extent of Cenozoic outcrops.

investigations of the offshore area (Hoskins and Griffiths, 1971; Silver, 1971a, b) indicate a very large (4,000 sq mi or, 10,360 sq km), elongate, north-south sediment accumulation extending northward past the California-Oregon border. Kilkenny (1970, Fig. 2) relegated the Humboldt basin to a southern, landward extension of the much larger, offshore Eel River basin. Therefore, the Humboldt basin may be an embayment rather than a true basin. The geologic column, both onshore and offshore, includes 14,000± ft (4,200± m) of Neogene strata, much of which is interpreted to be Pliocene and Pleistocene in age.

During the summers of 1963 and 1964, the writer collected about 230 samples from outcrops in the Humboldt embayment and adjacent areas. In this paper, the superposition, faunal succession, and stratigraphic ranges of foram species in the three best-exposed, relatively least-disturbed, and most uniformly fossiliferous outcrop sections—Centerville coastal section, Scotia - Eel River section, and Price Creek section—are discussed (Figs. 1, 5, 6). Additional samples were taken from scattered outcrops for purposes of interpreting more broadly the age and ecologic relations of the various faunules. A few samples from the Bear River valley, directly south, were collected for purposes of Miocene age comparison.

Previous Work on Northern California Sequence

The outstanding exposures along the Eel River at Scotia and Rio Dell have long been regarded as classic collecting sites because of the abundance of well-preserved fossil mollusks. The first published work on the Humboldt basin area apparently was that by Gabb (1866-69), who described several molluscan species from the "Pliocene of Eagle Prairie [now the town of Rio Dell], Humboldt Co." Gabb included descriptions of mollusks and echinoderms from the "Miocene or Pliocene" of Bear River. In 1888, Cooper described other molluscan species, also from the "Pliocene of Eagle Prairie."

The term "Wildcat series"[3] was proposed by Lawson (1894, p. 256) primarily for outcrops exposed along the Wildcat Ridge Road (south of Ferndale). He regarded both the Ferndale section and the Scotia section as being of Pliocene age on the basis of mollusk identifications by J. C. Merriam. Lawson collected from the "Scotia beds"; evidently, his collection came largely from the upper part of the section along the Eel River just north of Scotia. Lawson (1894, p. 261) commented, ". . . Of these 36 (molluscan) species only 14, or 39%, are extinct. There are, moreover, 18 species which are not known in the Miocene." Diller (1902, p. 36) concluded that the "Wildcat series" was "probably Miocene," or "Upper Miocene," on the basis of W. H. Dall's analysis (*in* Diller, 1902, p. 39) of mollusks in the "lower portion of the [Ferndale] series."

The Scotia section of the Eel River valley was correlated with "latest Miocene or earliest Pliocene" formations in other parts of California, Oregon, and Washington by R. Arnold and Hannibal (1913, p. 592). They published a "partial list of molluscan species" collected from the Scotia section and based their age deductions to some extent on the percentage of living species (e.g., see Arnold and Hannibal, 1913, p. 574).

A. Harmon (1914, p. 456) considered the "Wildcat series" to be of Pliocene age, but suggested a Miocene age for the "lower nodular contorted shale" (i.e., the Pullen Formation in the terminology of Ogle, 1953, and the current usage). The term "Bear River formation" (and also, apparently synonymously, "Bear River series") was used by Stadler (1914, p. 447, 449) to denote isolated outcrops in the Bear River valley and along the coast north and south of the mouth of Bear River. He considered the Bear River formation to be upper Miocene. Stadler mentioned the "Wildcat series" but gave no age for it.

In two very significant publications, B. Martin (1914, 1916) proposed a twofold subdivision of the "Wildcat series" based on differences in lithology and molluscan faunas. He thought (1916, p. 238-239) that the upper division was "late Pliocene" and the lower division was "uppermost Miocene", and he also recognized a "Bear River

[3] The spelling "Wildcat" will be used throughout this paper, although "Wild-cat" and "Wild Cat" were used in some of the early references. *Editor*

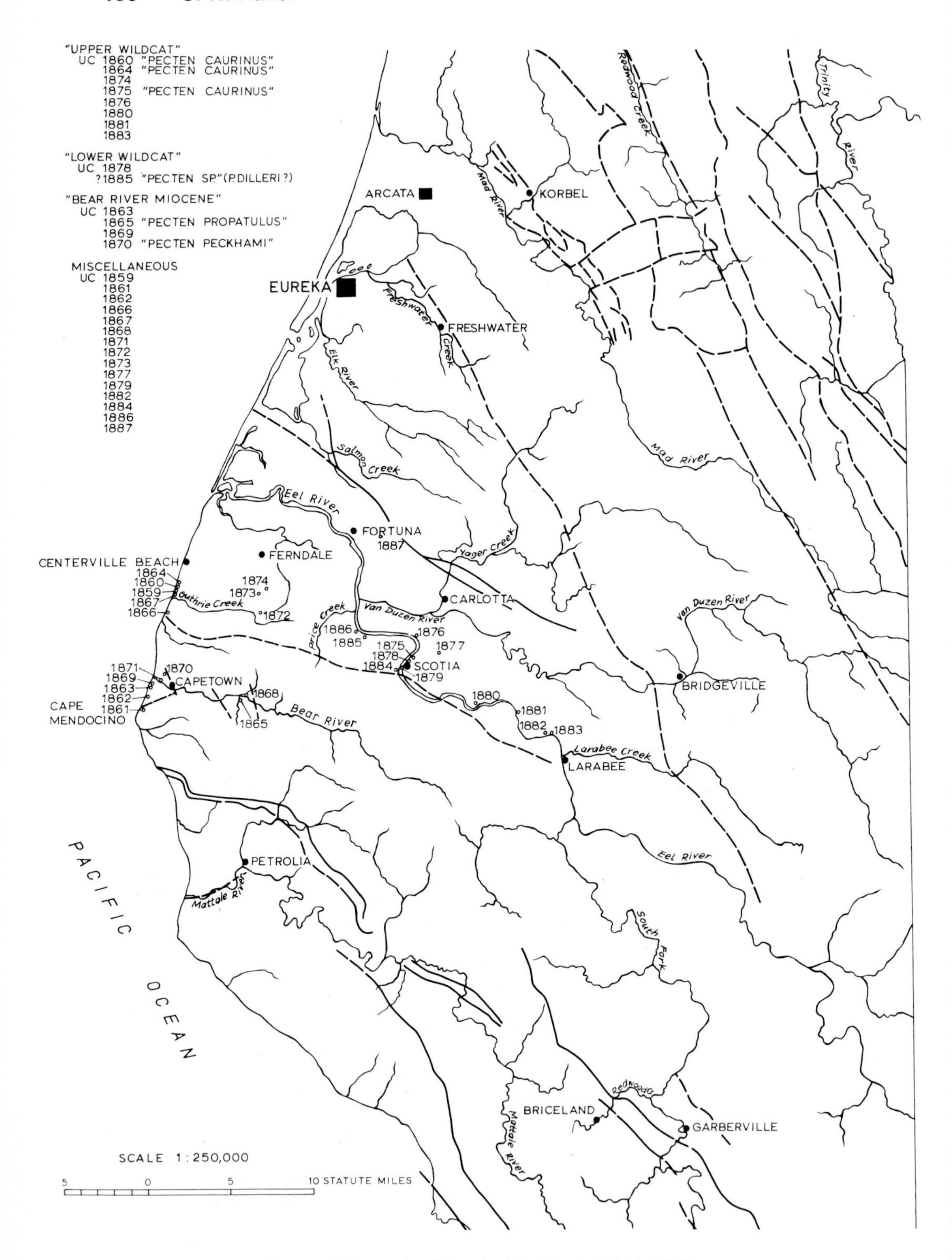

FIG. 2–Molluscan localities cited by Martin (1914; 1916).

Upper Miocene." The basis for these age determinations are (Martin, 1916, p. 248):

> . . . in general there are two methods on which faunal correlations are based: (1) the use of characteristic species; (2) the percentage method. The latter may include the percentage of species common among the formations themselves as well as the percentage of living species common to the formations.

Martin had extensive mollusk collections which he obtained from 31 localities, representing all the primary sections in the Humboldt basin and Bear River areas (see Fig. 2, this paper; comments in Faustman, 1964, p. 139). Martin's field map and field notes (dated 1914), his locality descriptions, and his "type molluscan specimens" are on file at the Museum of Paleontology, University of California, Berkeley.

In 1919, J. P. Smith concluded, on the basis of a study of individual collections by Gabb, Monroe, Cooper, Martin, and Hannibal, that the "Wildcat fauna" was the "upper part of the lower Pliocene" (p. 143), and living conditions were "about 50°F minimum." Evidently, he was referring largely to a few molluscan species from the upper portion of the "Wildcat formation" in interpreting the "cold-water" nature of the fauna. Clark (1921, 1929), basing his conclusion on a few molluscan species in the upper part of the "Wildcat formation," stated that the Wildcat was Pliocene in age.

Schenck et al (1925), in a survey of the literature, subscribed to a twofold division of the Wildcat, but regarded the upper Wildcat as upper Pliocene and placed the lower Wildcat as lower Pliocene.

Published analyses of foraminiferal faunas in the Humboldt basin and Bear River areas started with Cushman et al (1930). From a study of the foraminiferal faunas representing 48 localities, they concluded that most of the samples from the Humboldt basin were comparable to Pliocene faunas in the Ventura and Los Angeles basins. Foraminifera from the Bear River coastal area were diagnosed as Miocene (see Fig. 3 for locality data). However, problems regarding an apparent mixing and mislabeling of their samples were not entirely resolved (e.g., see comments in the section on systematics regarding *Elphidium hughesi* and *Bulimina subacuminata*).

Dorf (1930) and Axelrod (1944, 1956) referred to a flora in the "Wildcat formation" at Garberville, about 25 mi (40 km) south of Scotia. Both adhered to a Pliocene age for this flora by assuming that it overlies the molluscan faunas at Scotia. In fact, The Garberville outcrops are separated from the outcrops at Scotia by large blocks of Mesozoic rocks, and the former were shown by Kleinpell (1938) to be Miocene.

Gale (*in* Grant and Gale, 1931, p. 55-56) suggested that the Bear River Miocene and the Wildcat farther north represented but a "single Pliocene unit." Grant and Gale (p. 195) also stated that Martin's *Pecten propatulus* from the Bear River area is a synonym of *P. (Patinopecten) caurinus* Gould; evidently this is an error on their part in mixing locality data since *Pecten propatulus (s.l.)* occurs in the Miocene from Vallecitos and the Berkeley Hills north to Astoria (R. M. Kleinpell, personal commun.), and is not conspecific with *Pecten caurinus*, a younger, Plio-Pleistocene form.

In a significant paper, Reed (1937, p. 553) elucidated and restated his earlier work (1933, p. 228-233):

> . . . the Pliocene strata (of the Eel River basin) contain almost exactly the same (foraminifera) zones, in the same order, as the Pliocene of the Los Angeles Basin. The exceptions are particularly instructive: a. There is an omission of several subzones in the middle part of the Repetto, their position being marked by a striking disconformity . . . .

Reed (1937, Fig. 2, p. 552) shows another major unconformity at the "Repetto-Pico" contact, and he stated (p. 553) that, "the major folding of the Eel River Tertiary followed the deposition of this [Saugus] member." On the basis of microfaunal analyses by Texaco paleontologists (D. D. Hughes and others), Reed (1937, p. 552-553) recognized in Humboldt County the Miocene with its *Siphogenerina hughesi* "zone"; the "Repetto" with its Upper *Bolivinita*, Lower *Bolivinita*, and *Uvigerina hispida* var. *gigantea* "zones"; and the "Pico" with its *Uvigerina peregrina, Bolivina robusta,* and *Bulimina subacuminata* "zones."

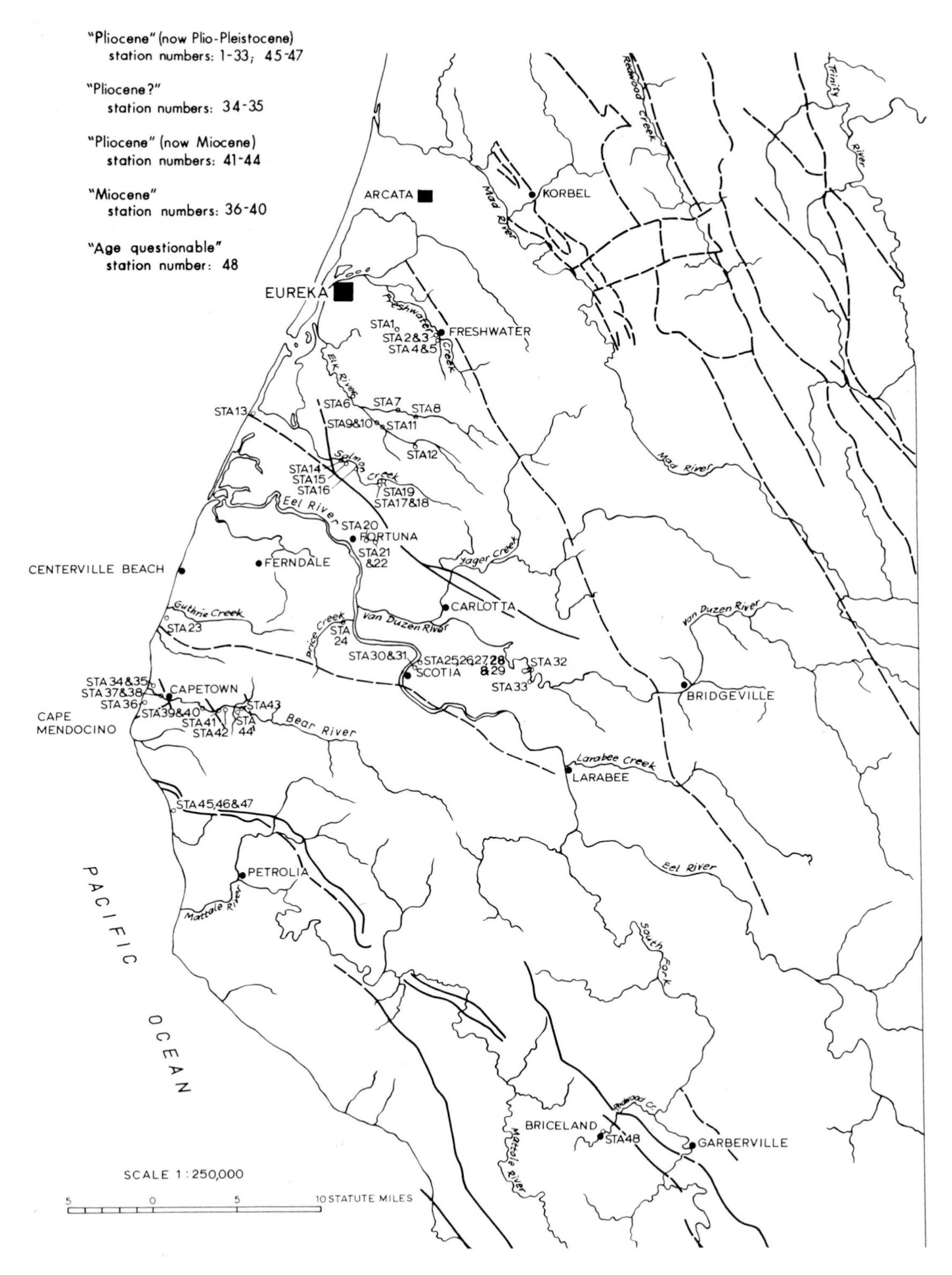

FIG. 3–Foraminiferal localities cited by Cushman, Stewart, and Stewart (1930).

In summary, Reed recognized three major unconformities in the Plio-Pleistocene of the Humboldt basin sequence: (1) one approximately at the base of the Eel River Formation (in the terminology of Ogle, 1953), (2) one at the approximate contact of the Eel River and Rio Dell Formations, and (3) another near the base of the Scotia Bluffs Sandstone. In his paper, Reed continued to use the terms "Repetto," "Pico," and "Saugus," but did not discriminate between their original rock-stratigraphic sense and their use in a time-stratigraphic sense.

The D. D. Hughes Collection, part of which is now in the Museum of Paleontology, University of California, Berkeley, contains a series of foraminiferal assemblage slides from samples collected by Hughes along the Centerville coastal section in 1934. Hughes did not live long enough to publish the results of his studies, but an informal analysis of his work was given by Reed (1937), as well as by Ogle (1953, p. 46).

Kleinpell (1938, Fig. 14) suggested that selected foraminiferal samples collected from the Bear River area by Cushman et al in 1930 were middle Miocene (upper and lower Relizian). Kleinpell tentatively correlated the isolated fault block at Briceland, west of the Garberville area, with the Miocene of southern and central California. He listed (p. 26) several foram species from Briceland in addition to those previously recorded by Cushman et al; this evidence led him to conclude that this locality apparently represents a "horizon" of late Relizian or early Luisian age.

A. I. Gregersen in 1940 collected microfaunal samples from the isolated outcrops of the Bear River area as well as along Price Creek. Gregersen (1942, *in* Keen and Bentson, 1944, p. 15) concluded that the foraminifera from the "Bear River beds" were of middle Miocene age, but differed from those in lithologically similar beds along Price Creek.

In their "Pliocene Correlation Chart," Grant and Hertlein, (1943, Fig. 85) recognized the upper and lower Wildcat (*s.l.*). MacGinitie (1943) used the terms "Wildcat," "Ferndale," "Upper Pico," "Middle Pico," "Lower Pico," and "Repetto" (the last four thereby extended northward from type localities in southern California) for strata in Humboldt County which he thought were of Pliocene age. MacGinitie also recognized the Santa Margarita Formation (perhaps correlative of the Wimer [or Wymer] beds of Del Norte County) which he listed as upper Miocene, and an unnamed Temblor Formation equivalent which he regarded as lower and middle Miocene.

Peculiarly, the Geological Society of America Correlation Committee (Weaver et al, 1944) dealt with the marine Cenozoic formations of western North America, but failed to mention the Humboldt basin or Bear River areas.

In a detailed analysis of the "ranges" of molluscan species in the "Wildcat series," Stewart and Stewart (1949) proposed 16 "lithologic zones" and "faunal assemblages" including two "faunal zones" with four "subzones." Their zones and subzones are, in fact, based on actual occurrences in one local section (the Centerville coastal section), not on ranges of species regionally under superpositional discipline (see Ogle's comments, 1953, p. 54). The lower part of the Centerville section was correlated (Stewart and Stewart, 1949, p. 170) with the lower Pliocene; the upper part of the Centerville section was related to upper and possibly middle Pliocene formations elsewhere. In addition, they included a "range chart" (p. 22) showing the most common foram species in that single section.

Durham (1950, p. 1257), in a general survey of Cenozoic marine climates, noted that the faunas of the Wildcat (of B. Martin, 1916) included the pelecypods *Anadara* and *Chione*, ". . . which would indicate a minimum temperature of about 12°C."

The basis for all recent geologic reports on the Humboldt basin is the work of Ogle (1951, 1953, 1960). His 1953 paper provided systematic *descriptions* of the lithology, foraminiferal occurrences, and molluscan occurrences (his pls. 3, 4, 5). Systematic *interpretation* of this paleontologic data for its age significance was necessarily omitted by Ogle (e.g., see Ogle, 1953, p. 45, bottom paragraph). The 1953 paper involved 106 microfossil localities with foram identification credited to D. Crawford, B. Jones, C. C. Church, and others. Macrofossils were taken from 45 localities, and mollusk identifications were determined in part by J. W. Durham. Lithologic descriptions were made of four key traverses: Centerville coastal, Ferndale - Wildcat Ridge Road, Price Creek, and Scotia - Eel River sections. Ogle (1953) proposed the terms "Pullen," "Eel River,"

"Rio Dell," "Scotia Bluffs," and "Carlotta" for formations within the Wildcat Group; he further recognized the younger Hookton and Rohnerville Formations. Ogle used the term "Eel River basin" (1953, p. 24-25; 1960, p. 32), apparently to define a structural basin which is superimposed on the sediments deposited in the Humboldt basin and which is limited by several major faults. He concluded (1953, p. 25) that upper Miocene, Pliocene, and lower Pleistocene rocks were represented, but he fixed only tentative age boundaries with respect to formations: the Miocene-Pliocene boundary falls within the Pullen Formation, and the Pliocene-Pleistocene boundary within the Carlotta Formation.

Evenson (1959), Irwin (1960), Valentine (1961, Fig. 15), Hawley (1961), Strand (1962), and Oakeshott (1964) more or less routinely followed the nomenclature and age relations of formations proposed by Ogle (1953).

Faustman (1964) dealt with the revision of molluscan systematics, and with the superpositional relations and occurrences of mollusk species in the Centerville coastal and Scotia - Eel River sections. Faustman recorded approximately 45 species each of pelecypods and gastropods, as well as a minor number of echinoids, scaphopods, and crustaceans; he noted that (p. 109) ". . . temperatures substantially cooler than those encountered at present off the northern California coast apparently prevailed at the time of Wildcat deposition." "Wildcat" referred to strata in the middle and upper Rio Dell Formation as determined by analysis of molluscan assemblages.

Recent publications by Kleinpell (1971, 1972) summarized the early history of paleontologic work on the West Coast; both papers are virtually required reading for all serious students of biostratigraphic studies in their broader scope.

Humboldt Basin: General Geology

Structural Versus Depositional-Basin Concepts

The general geology of the Humboldt basin strata has been comprehensively reviewed by Ogle (1953) in a paper devoted to the geology of the Eel River valley, Humboldt County, California (see Fig. 1, this paper, for areal details).

Ogle (1953, p. 24-25; 1960, p. 32) used the terms "Eel River basin" and "Eel River embayment" for much of the region in reference. His Eel River basin evidently refers to a mid-Pleistocene structural basin, the outline of which is marked by several major faults. The fault pattern, structural trends, and actual occurrence of outcrops within the basin are adequately shown by Ogle (1953) on his "Geologic Map of the Eel River Valley Area, California." Further, Strand (1962) showed the general pattern of faulting with reference to the Eel River, Bear River, and Briceland-Garberville areas and adjacent areas to the south.

Reed (1937, p. 552, Fig. 2; see also Woodring et al, 1941, Fig. 12) has previously referred to the "Eel River basin" in the same sense that he did the Los Angeles basin and stated that the ". . . remarkable parallelism of the Eel River and Los Angeles Basin faunas is still harder to account for except upon the assumption of closely similar depositional histories." The "parallelism" emphasized by Reed is indeed remarkable, in terms of both tectonic and faunal histories, yet as to detail, the distinction between zonal contemporaneities and parallel homotaxial sequences remains to be resolved.

R. Arnold (1909, Fig. 5) and Clark (1921, Fig. 11) were among the first to speculate on the outline of this sedimentary basin but did not name it. Kleinpell (1938, Fig. 2) applied the name "Humboldt Basin," evidently referring to a Neogene sedimentary basin, and postulated its limits very roughly. The writer prefers to use the term "Humboldt basin" in dealing with the sedimentary basin concept and late Cenozoic sediments deposited in the lower Eel River valley; in reality, the Humboldt basin is an embayment and a small part of the large offshore Eel River basin (Kilkenny, 1970, Fig. 2; Silver, 1974b, Fig. 3).

Axelrod's (1956, Fig. 16) paleogeographic interpretations of the Mio-Pliocene show a larger northern basin, embracing the Humboldt basin and Bear River areas, and a smaller southern basin in the Petrolia-Garberville area. Neither basin was named by him.

Natland and Kuenen (1951, Fig. 3), Bandy (1953a, Fig. 3; 1967), and Natland (1957) presented diagrammatic interpretations of the depositional history of the Pliocene sediments in the Ventura basin.

In summary, three prominent Neogene sedimentary basins of California–the Humboldt, Los Angeles, and Ventura basins–appear to have similar depositional histories. Formation of all three basins was initiated either during middle or late Miocene time (Silver, 1974b). Basin filling progressed throughout Pliocene and Pleistocene time with few relatively minor interruptions, and in the Humboldt basin, the shoreline regressed westward with sporadic transgressions. In the Humboldt basin, the Pleistocene orogeny (see Lawson, 1893; Reed, 1937; Bailey, 1943; Woodring, 1952; Izett et al, 1974) of the California coastal ranges is seemingly reflected, in part at least, by the massive conglomerates of the Hookton Formation. The present-day outline of the Humboldt basin is the result of middle and late Quaternary faulting.

Stratigraphy

A resume of lithologic and cartographic units recognized in the Humboldt basin and Bear River area follows (see Fig. 4). The physical stratigraphy is shown in greater detail in the columnar sections (Fig. 5). In general, terminology follows that of Ogle (1953), although primary *reference sections* are suggested to supplement the "type sections" of the two youngest formations.

Bear River Beds

This poorly defined series embraces a sequence of diverse lithologies in several fault blocks; one block contains early Miocene, others middle or late Miocene faunas. The term "Bear River formation" evidently was first used by Stadler (1914, p. 447) for exposures of unrecrystallized sedimentary rocks in the Bear River valley; unfortunately, he designated no type section. The term "Bear River beds" is used here in an informal sense; a reference section is that along the north flank of the Bear River coastal syncline (see Fig. 5). The isolated fault blocks of sedimentary rocks in the upper Bear River valley are here designated informally as the lower part of the Bear River beds, though beds of nearly the same age (i.e., early Miocene) would appear from the data of Cushman et al (1930, their station 36), to be represented also on the south flank of the Bear River coastal syncline; the south flank of the syncline was not sampled in the present study.

Pullen Formation

Ogle (1953, p. 26) stated, ". . . the thickest and most complete section of the Pullen formation is well exposed along Eel River near Scotia"; he had previously indicated (1951, p. 111) that this was to be considered the type section. He also stated, ". . . the name 'Pullen' is used because of excellent exposures and faunas at Pullen Ranch on Price Creek." Subsequently, however, much of the lower Pullen along Price Creek was obscured by massive landslides.

The Price Creek, Eel River, and Centerville coastal sections show similar lithologies of generally massive siltstones with intermittent sandy and glauconitic intervals. Microlithologic differences include high radiolarian content in the Eel River and coastal sections and low radiolarian content in the Price Creek section. The Eel River section has a stratigraphic thickness of about 1,012 ft (308 m), although an incomplete section was measured with the base obscured. In 1953, Ogle reported a thickness of 1,107 ft [337.4 m]. The base of the Pullen Formation is fractured to the point of pulverization and is obscured in the three key sections (Fig. 5; Appendix B). The False Cape shear zone forming the southern boundary of Humboldt basin (Strand, 1962) appears to have mixed the lower part of the Pullen Formation with rocks of Mesozoic age.

Macrofossils are scarce and poorly preserved in the type section along the Eel River. Microfossils, however, indicate that the formation is entirely of marine origin. Radiolarians and diatoms are common to abundant; but foram assemblages are generally impoverished with regard to number of species and occur in restricted intervals only.

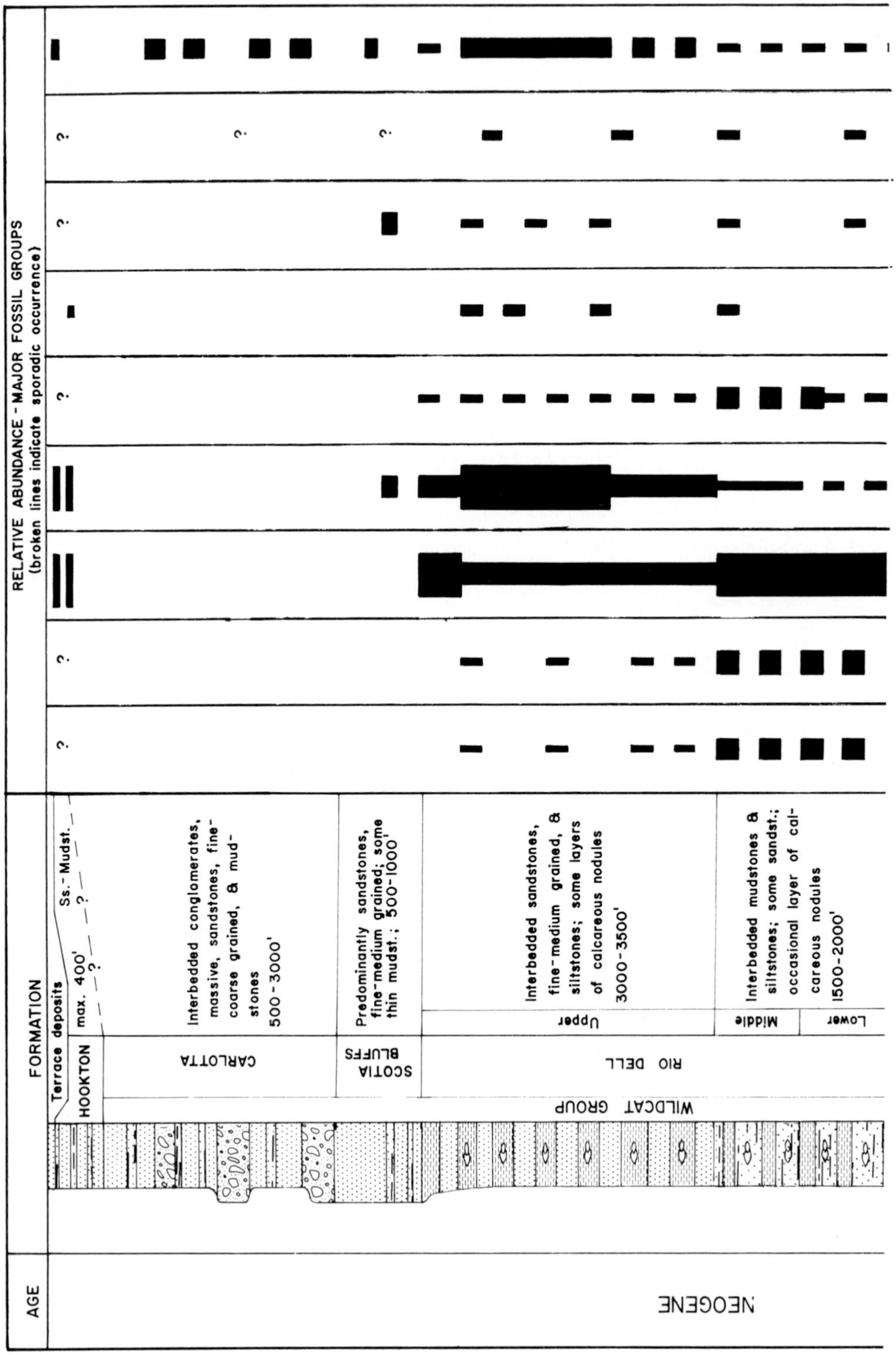

AGE

FORMATION

RELATIVE ABUNDANCE - MAJOR FOSSIL GROUPS
(broken lines indicate sporadic occurrence)

NEOGENE

WILDCAT GROUP

Terrace deposits

Ss.- Mudst.

HOOKTON

max. 400'

?

CARLOTTA

Interbedded conglomerates, massive, sandstones, fine-coarse grained, & mudstones
500 - 3000'

SCOTIA BLUFFS

Predominantly sandstones, fine-medium grained; some thin mudst.; 500-1000'

RIO DELL

Upper

Interbedded sandstones, fine-medium grained, & siltstones; some layers of calcareous nodules
3000-3500'

Middle

Lower

Interbedded mudstones & siltstones; some sandst.; occasional layer of calcareous nodules
1500-2000'

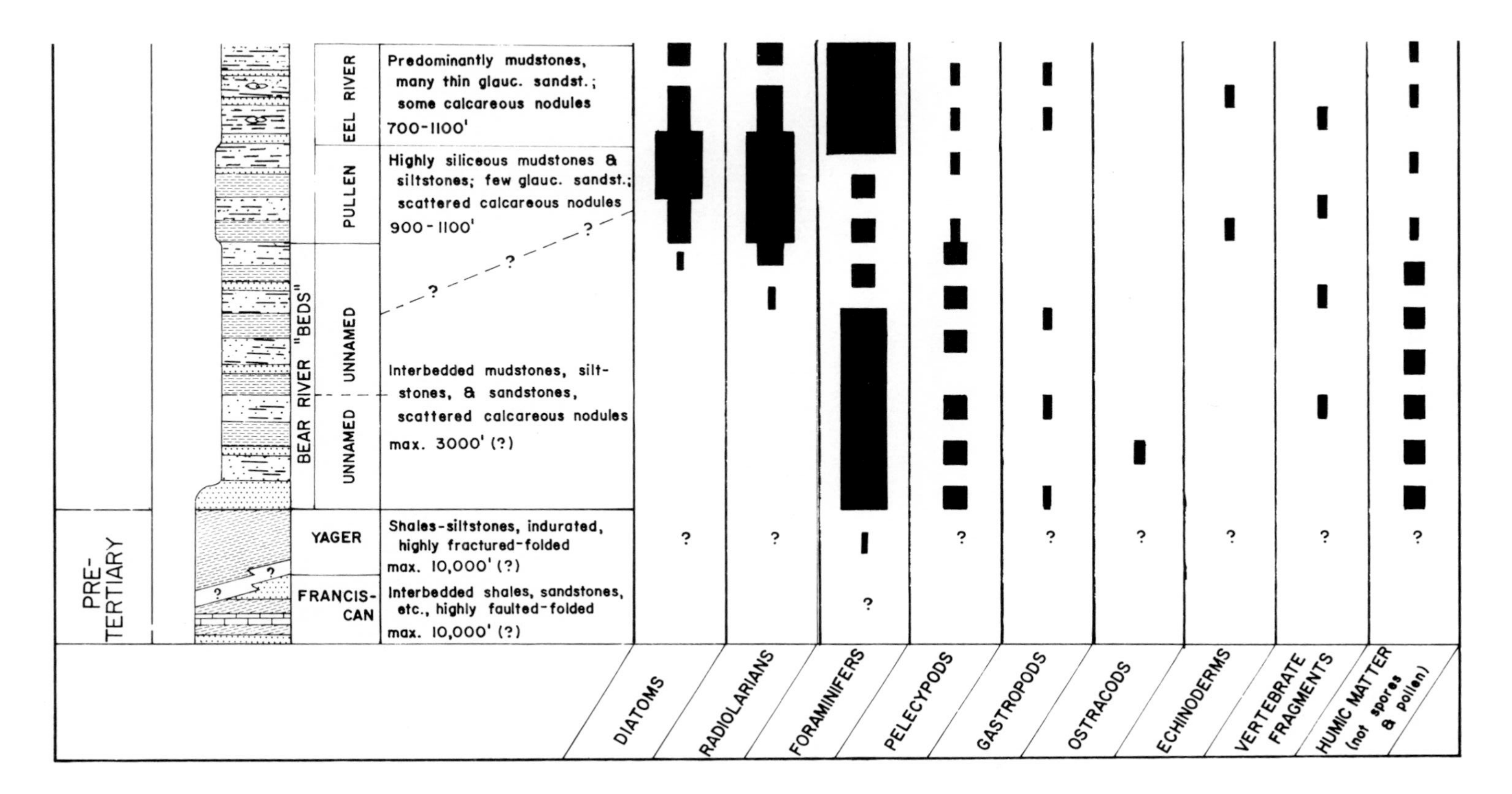

FIG. 4–Generalized biostratigraphic section, Humboldt basin and Bear River area.

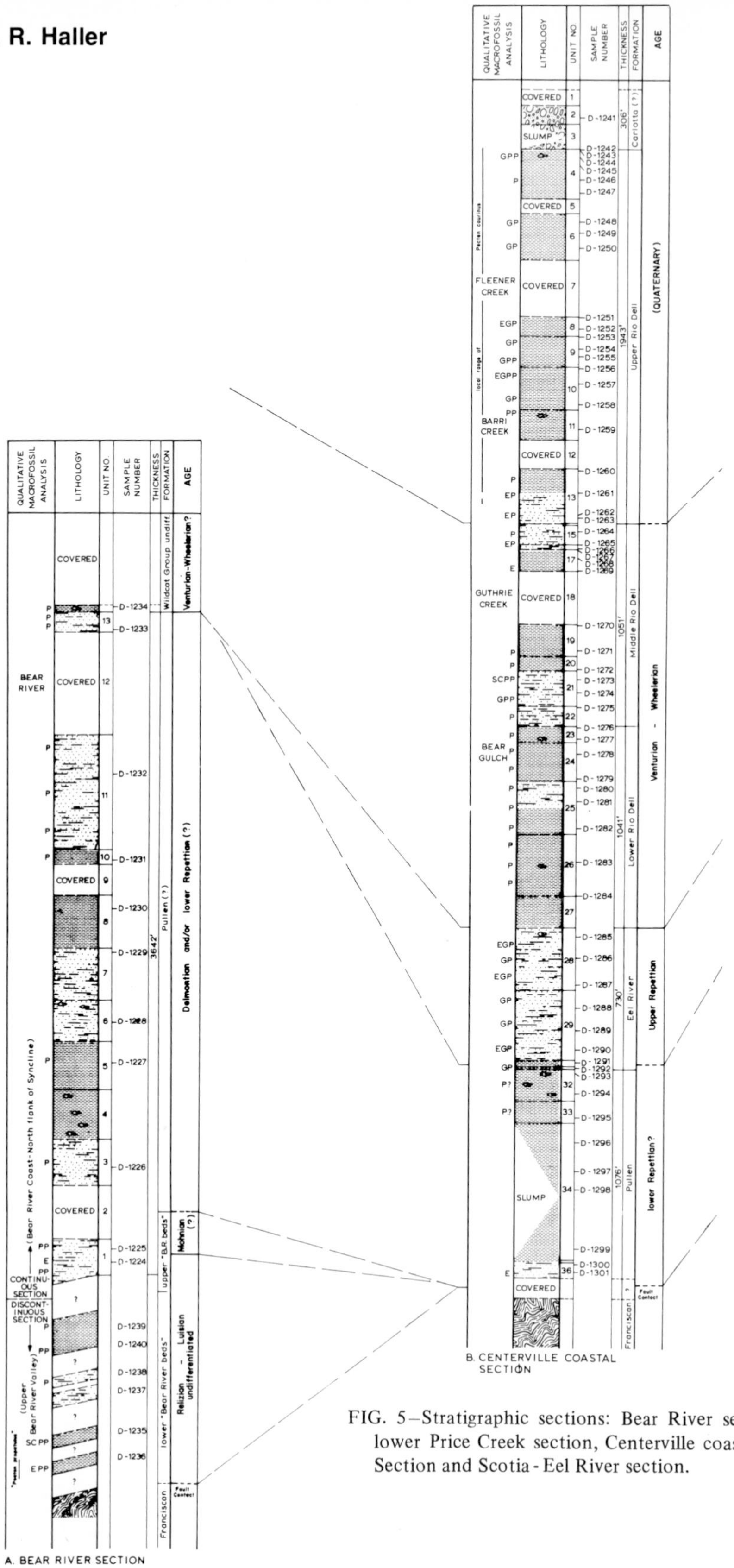

A. BEAR RIVER SECTION

B. CENTERVILLE COASTAL SECTION

FIG. 5–Stratigraphic sections: Bear River section, lower Price Creek section, Centerville coastal Section and Scotia - Eel River section.

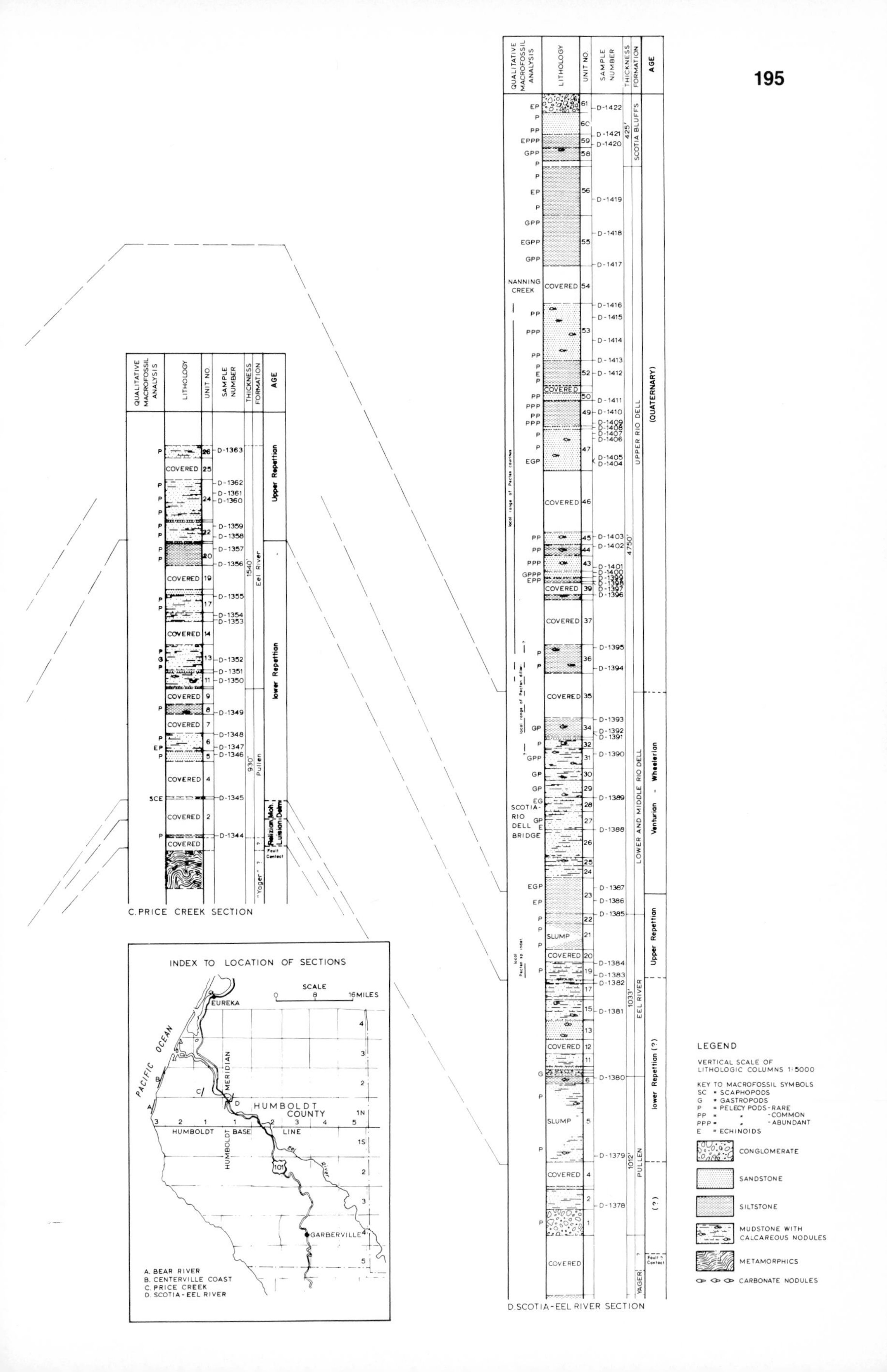
QUALITATIVE MACROFOSSIL ANALYSIS
LITHOLOGY
UNIT NO
SAMPLE NUMBER
THICKNESS
FORMATION
AGE
SCOTIA BLUFFS
425'
UPPER RIO DELL
(QUATERNARY)
4750'
NANNING CREEK
COVERED
SLUMP
SCOTIA-RIO DELL BRIDGE
LOWER AND MIDDLE RIO DELL
Venturian - Wheelerian
Upper Repetition
lower Repetition (?)
EEL RIVER
1033'
PULLEN
1012'
YAGER
Fault Contact
local range of Pecten caurinus
local range of Pecten dilleri
local Pecten sp. indet.
D. SCOTIA-EEL RIVER SECTION
Upper Repetition
lower Repetition
1540'
Eel River
930'
Pullen
"Yager"
Rueizian Moh Luisian Delmi
C. PRICE CREEK SECTION
INDEX TO LOCATION OF SECTIONS
SCALE
0 8 16 MILES
EUREKA
PACIFIC OCEAN
MERIDIAN
HUMBOLDT COUNTY
HUMBOLDT BASE LINE
HUMBOLDT
101
GARBERVILLE
A. BEAR RIVER
B. CENTERVILLE COAST
C. PRICE CREEK
D. SCOTIA-EEL RIVER
LEGEND
VERTICAL SCALE OF LITHOLOGIC COLUMNS 1:5000
KEY TO MACROFOSSIL SYMBOLS
SC = SCAPHOPODS
G = GASTROPODS
P = PELECY PODS - RARE
PP = " " - COMMON
PPP = " " - ABUNDANT
E = ECHINOIDS
CONGLOMERATE
SANDSTONE
SILTSTONE
MUDSTONE WITH CALCAREOUS NODULES
METAMORPHICS
CARBONATE NODULES

A thick radiolarian-bearing sequence on the north flank of the Bear River coastal syncline is analogous with the lithologies in the Humboldt basin and is therefore designated as Pullen(?) Formation (see Fig. 5).

Eel River Formation

Mudstones predominate, but many silty and sandy intervals occur in the type Eel River section at Scotia and also in the Centerville coastal section. Glauconite is a prominent constituent in many beds, especially toward the base of the formation.

Sandy intervals are peculiar in character: the beds are thin and "pseudocyclic" in nature, suggestive of turbidity current deposits, forming a turbidite sequence in the sense of Bouma (1961, 1964; Ingle, 1974, personal commun.). These pseudocyclic deposits however, are not as apparent as the classic turbidites exposed in Hall Canyon of southern California.

Forams are common to abundant, diatoms and radiolarians less so. Macrofaunas are represented by few species and specimens of pelecypods and by even scarcer gastropods.

In the type section, stratigraphic thickness is variously recorded: 1,033 ft or 314.9 m (this report), 940 ft or 286.5 m (Ogle, 1953), and 962 ft or 293.2 m (Faustman, 1964). Contact with the underlying Pullen Formation in the type section is highly irregular. Basal sands of the Eel River Formation fill depressions in the underlying siltstones of the Pullen Formation.

Rio Dell Formation

The type section is exposed along the Eel River at Scotia; another primary reference section is that seen in the Centerville coastal section. The formation is composed primarily of interbedded siltstones and fine-grained sandstones, with minor mudstones in the lower part. The coastal section was subdivided by Ogle into three members: the upper member is thickest, is distinguished as somewhat coarser grained, and usually contains mollusks; whereas the lower members contain fewer beds with mollusks. Comparison of the two sections shows that the upper part of the formation in the Eel River section contains even more abundant mollusks.

Ogle (1953, p. 48) assumed, as one might naturally conclude from a study of sediment textures, that there were considerable differences in water depths in the two areas, the Eel River section having been deposited in shallower waters and supposedly closer to the shoreline.

Forams are relatively abundant throughout the coastal section, becoming scarce or nonexistent in some parts—notably in the more clastic portions of the Eel River section.

In the type section, the thickness of the formation has been recorded as 4,642 ft or 1,414.9 m (this report), 4,259 ft or 1,298 m (Ogle, 1953), and 4,450 ft or 1,356.4 m (Faustman, 1964). The type section shows a minor unconformity at the contact with the underlying Eel River Formation which involves a general change in lithology from finer to coarser grained sediments.

Scotia Bluffs Sandstone

The type section along the Scotia Bluffs is typical of the formation—predominantly sandstone with some thin conglomeratic lenses. Other lenses are very rich in shallow-water mollusks and echinoids; these marine fossils are confined to the lower part of the formation. Marine microfossils occur very sparsely and only in the lower part of the section. Contact with the underlying Rio Dell Formation in the type section is, according to Ogle (1953) "gradational." The writer places the contact at the base of a 6-ft (2 m) sandy section containing well-rounded igneous pebbles scattered throughout. The thickness in the type section was estimated at 2,117 ft or 645.3 m (Ogle, 1953).

Carlotta Formation

In general, the Carlotta is a sequence of clay, sandstone, and conglomerate beds. Ogle (1953) stated, ". . . as the most important outcrops of the Carlotta formation are near the small town of Carlotta, this has been chosen as the type section." The writer visited this area in 1964 and found it to be lacking in good exposures, and suggests, therefore, that the primary reference section be that exposed in the bed and banks of the Van Duzen River at Riverside Park. There the lowermost exposures begin in Sec. 7, T1N, R2E, and the uppermost exposures end in Sec. 31, T2N, R2E (see Fig. 6). The nearly continuous outcrop at Riverside Park exposes about 2,500 ft (762 m) of section which illustrates the cyclic nature of deposition of the various lithologies in the formation. The formation in this section is evidently entirely nonmarine. Seven nonleached samples representative of the section were analyzed for microfaunas, with negative results. The only macrofossils noted were remains of land plants, including many logs and stumps throughout the formation. The larger logs are carbonized on the outside only and are still woody on the interior. Neither top nor bottom of the formation is visible at the Riverside Park location; the contact with the underlying Scotia Bluffs sandstone is best seen in the Ferndale-Wildcat Ridge Road section, where the lower part of the Carlotta is marked by a sequence of massive conglomeratic sands.

Hookton Formation

A series of clay, sandstone, and conglomeratic lenses make up this formation. In general, all beds are very poorly consolidated and form poor exposures. Ogle (1953, p. 58) stated, ". . . no adequate type section can be given because of the extreme variability of these beds. The name Hookton is used because there are numerous, more or less typical, random exposures of the sediments in the Hookton-Table Bluff area." A primary reference section embracing both marine (*Ostrea lurida* beds here included in the Hookton, contrary to Ogle) and nonmarine phases is suggested herein. This section is exposed along Ridgewood Drive—the lower portion of the outcrop begins at the junction of Ridgewood Drive and the main Elk River Road in Sec. 15, T4N, R1W. The fauna is represented by mollusks and forams at localities D-1328 and D-1329 (see Fig. 6). Another primary reference section is that exposed in a Highway 101 roadcut about 1.5 mi (2.4 km) south of Fields Landing (localities D-1331 and D-1332). The thickness of the Hookton is evidently less than 500 ft (150 m). Neither the contact with the underlying formations nor that with the overlying terrace deposits is known, and it is quite possible that the Hookton is a time equivalent of the upper part of the Carlotta Formation.

Paleontology

Biostratigraphy

In the introductory review of previous work, numerous isolated fossil foram localities were cited, two molluscan sequences were summarized, and prior biostratigraphic interpretations were outlined. In the Humboldt basin three key sections, roughly parallel with each other and with the coastline, are now exposed along the northward flank of Wildcat Ridge. From west to east these are the coastal sequences at Centerville, the Price Creek sequence, and the Eel River sequence at Scotia (Figs. 5, 6). By use of the prior reconnaissance studies as a reference, fossil forams were collected from the three biostratigraphic sequences mentioned and from isolated areas in the Bear River area south of Wildcat Ridge, to the southeast in Larabee syncline, and farther north on Freshwater Creek, on Salmon Creek, on Elk River, and at Fields Landing.

No new descriptive biostratigraphic terminology has been attempted here, since for the most part, the six foraminiferal aggregates may be referred to the formations of Ogle (1953), or to a discrete and stratigraphically restricted part of them, or to the "stage" names proposed by Natland (1952, 1957; Natland and Rothwell, 1954).

Foraminifera

More than 180 species of forams are recorded in the systematics and distribution charts. Less than 10% of the total species are planktonic. The distribution charts (Figs. 7, 8) record relative abundances of forams at the respective localities.

Two of the forams, *Cassidulina translucens* var. *natlandi* and *Elphidium humboldtensis*, are regarded as new and undescribed. Otherwise, the six (five post-Miocene)

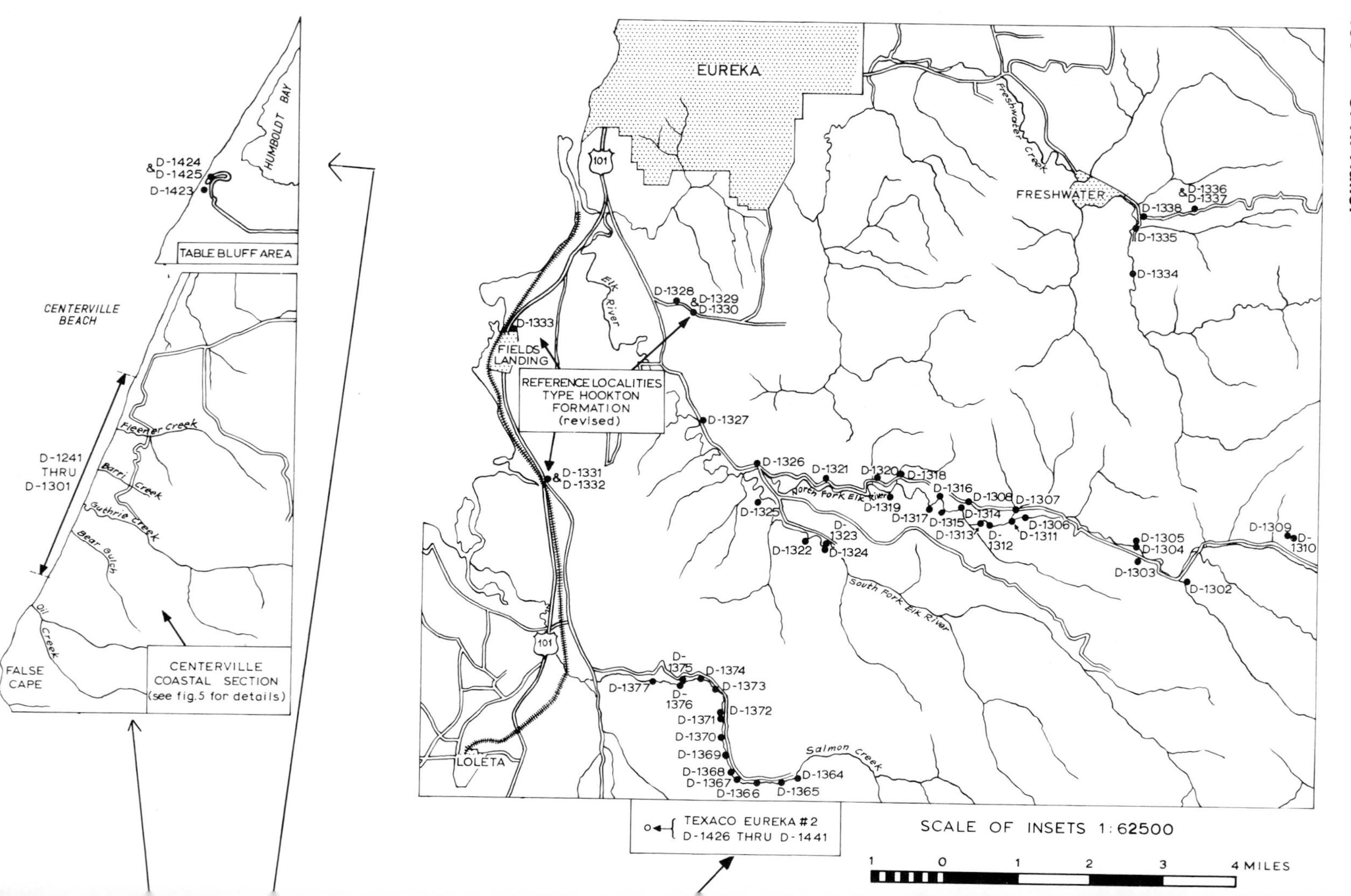
EUREKA
FRESHWATER
Freshwater Creek
Elk River
North Fork Elk River
South Fork Elk River
Salmon Creek
FIELDS LANDING
LOLETA
101
REFERENCE LOCALITIES TYPE HOOKTON FORMATION (revised)
TEXACO EUREKA #2 D-1426 THRU D-1441
SCALE OF INSETS 1:62500
1 0 1 2 3 4 MILES
D-1302
D-1303
D-1304
D-1305
D-1306
D-1307
D-1308
D-1309
D-1310
D-1311
D-1312
D-1313
D-1314
D-1315
D-1316
D-1317
D-1318
D-1319
D-1320
D-1321
D-1322
D-1323
D-1324
D-1325
D-1326
D-1327
D-1328
D-1329 & D-1330
D-1331 & D-1332
D-1333
D-1334
D-1335
D-1336 & D-1337
D-1338
D-1364
D-1365
D-1366
D-1367
D-1368
D-1369
D-1370
D-1371
D-1372
D-1373
D-1374
D-1375
D-1376
D-1377
HUMBOLDT BAY
TABLE BLUFF AREA
D-1424 & D-1425
D-1423
CENTERVILLE COASTAL SECTION (see fig.5 for details)
Fleener Creek
Barri Creek
Guthrie Creek
Bear Gulch
Oil Creek
CENTERVILLE BEACH
D-1241 THRU D-1301
FALSE CAPE

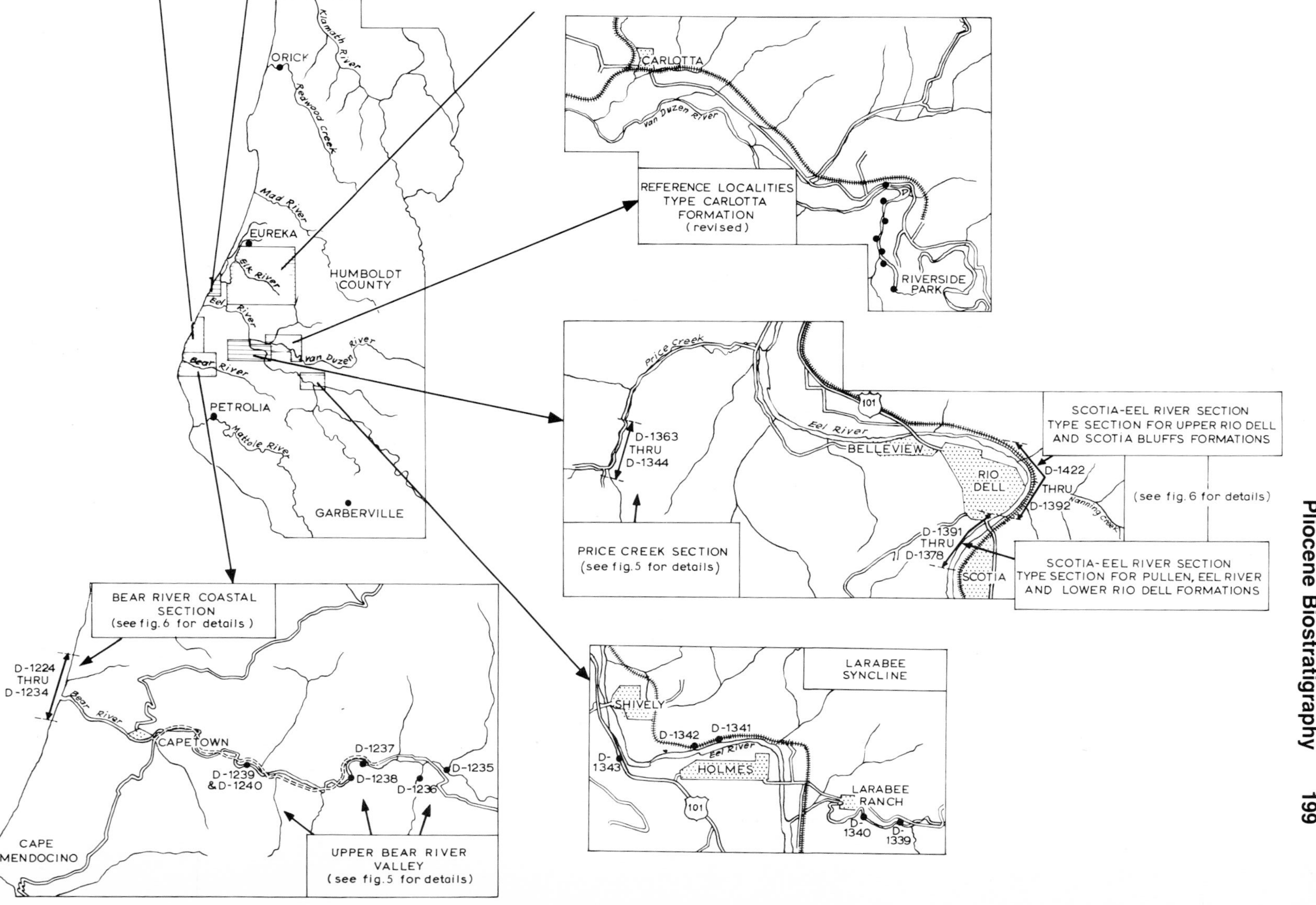

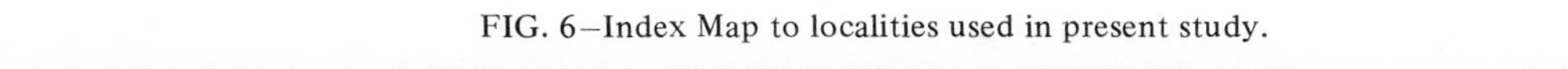

FIG. 6–Index Map to localities used in present study.

faunal aggregates (Berry, 1968, p. 135) contain species which are synonymous with those recorded in southern California and with those described in the extensive literature on species of Holocene forams off California. Of the four microfaunal aggregates and two partial assemblages recognized, the one from the Eel River Formation contains several lagenids and other species which are presumably invaders from the Indo-Pacific province. This assemblage is probably early Pliocene in age.

In addition to the lagenids mentioned, other major groups of forams include the plectofrondicularians, the bolivinids, and the buliminids, all conspicuously represented in the older assemblages. The cassidulinids and the elphidiids dominate the younger assemblages which, in outcrop sections, represent a more shoaly biofacies. Here they are accompanied by a few diverse species of rotalids, of which *Epistominella pacifica* is an abundant example.

Arenaceous forams are very sporadic; they apparently occupied ecologic niches where they occurred in abundance in a few samples. The miliolids are normally present only as a few individuals in any given sample.

Methods of Correlation

Biochronology

The Neogene of the Humboldt basin and Bear River areas is composed of a sequence of sedimentary formations whose total stratigraphic thickness is perhaps 14,000 ft (4,267 m). The late Tertiary and Quaternary age of these rocks has long been recognized although the boundaries between the series-epochs and smaller units have remained in dispute.

Prior studies, most notably those by mollusk workers, have used the following for age deductions: (1) percentage of living species (Merriam *in* Lawson, 1894, p. 261), (2) use of characteristic species (Martin, 1916, p. 248), and (3) percentage of species common among the formations (Martin, 1916, p. 248). Thus, for example, it has long been clear that the local fossil mollusks are represented by "Upper Wildcat" and "Lower Wildcat" faunas and come mainly from the Rio Dell Formation (Martin, 1916; Ogle, 1953, p. 44-45) of Pliocene and Pleistocene ages. Durham (*in* Ogle, 1953, p. 45) stated that all indications point to an older age than this for two species from the basal Pullen sandstone along the Eel River. One might suppose that the mollusks in the basal Pullen were correlative with Martin's "Bear River Miocene" macrofauna.

Among micropaleontologists, the use of guide fossils for correlation purposes between the microfaunas of the Humboldt basin of northern California and microfaunas of the Ventura and Los Angeles basins of southern California is a matter of long standing (Reed, 1933, p. 230, Table XV; 1937, Fig. 2; Stewart and Stewart, 1949). Through the years, intensive petroleum-industry activity has led to the recognition of three faunal units in the Pliocene and one (or two) faunal units in the Pleistocene of southern California. Correlation was made primarily on the basis of a large number of so-called guide fossils, all of which were benthonic forams (Driver, 1928; Wheeler, 1928; Woodring et al, 1932; Stewart and Stewart, 1930a, b; Reed, 1933, 1937; Wissler, 1937, 1943, 1958; Adams, 1939, 1940; Natland, 1933, 1952, 1953, 1957; Natland and Kuenen, 1951; Natland and Rothwell, 1954; Holman, 1958). Usage of these so-called guide or index forams is aptly summarized in Figure 11 and discussed in detail later. The extent to which these are biochronologic as distinguished from ecologic phenomena is not clearly shown in the available literature.

Neither the planktonic forams in southern California (see Ingle, 1967, Text-Fig. 35) nor those in northern California lend themselves readily to Oppelian zonation of the Neogene. The eight or so planktonic species of the Humboldt basin Neogene that are discussed here range through the Pliocene-Pleistocene and into the Holocene. As noted on Figure 12, planktonic forms occur in relatively insignificant numbers and thus are not useful for Oppelian zonation. Many of the planktonic specimens that do occur, especially in the younger sections, are difficult to assign to species and seem to be dwarfed forms or variants reacting to an unfavorable environment. For comparison, Ingle (1968) mentioned 10 planktonic foram species in the "Pliocene of the Centerville beach section," and Orr and Zaitzeff (1971) mentioned 14 species from the same section; neither of these works encountered the critical *Globorotalia tosaensis - G. trun-*

catulinoides lineage.

Table 1 shows the relative abundance of planktonic forams in the two most complete sections. Samples containing no forams may be relegated to either leached sandy sections or to diatomaceous and radiolarian siltstones. In the Centerville coastal section, an additional 23 samples contained common to abundant forams, but quantitative data are not available.

Table 1. Planktonic-Total Foraminifera Ratio, Humboldt Basin

	Centerville Coastal Section	*Scotia - Eel River Section*
20%	2 samples	1 sample
10-20%	5	2
5-10%	13	4
2½-5%	3	10
0-2½%	5	6
0	2	9
	—	—
	30 subtotal	32 all foram groups
	8 barren of forams	12 barren of forams
	—	—
	38 total samples	44 total samples

The present study emphasizes the benthonic foraminiferal faunas in the Humboldt basin in considerably more detail than previously was available. Under consideration are the various biochronologic principles (see Berry, 1968) and methods: superposition, faunal succession, percentage of living species, and "guide" or "index fossils" where used under adequate discipline. In addition, where the foraminiferal data permit, Oppelian zonation is applied to the smaller time-rock units (stages and zones). Those principles which best reflect the development of phylogenetic sequences in individual lineages and the evolution of whole faunas are of prime importance for the more refined correlations of stage and zone magnitude. Concerning diagnosis of these units, Berry (1964, p. 70; see also 1966) states:

> . . . The joint occurrence of certain species in a thickness of strata is interpreted to characterize the Zone. Because interpretation is needed to distinguish a Zone, the several species occurring together that characterized the rocks of the Zone may be termed a *congregation*. They constitute more than just an assemblage because an assemblage is a group of fossils *described* as occurring in a certain set of rocks. "Assemblage" is a descriptive term because it is used to merely describe the occurrence of a group of fossils. Congregation, on the other hand, applies to interpretive phenomena and interpretation is needed to discern it. . . .

and on page 71:

> . . . Units of fossiliferous rock characterized by and founded upon distinctive congregations are time-rock or time-stratigraphic units. Such units have a concrete basis in that they are founded upon specific sequences of fossiliferous strata through which the ranges of the several contained species have been ascertained, the overlap and the extent of the ranges may be interpreted under the discipline of the principles of superposition of strata and faunal succession to discern the characteristic congregation. . . .
>
> The vertical extent of the Zone is the vertical range through strata of the characteristic congregation. The horizontal or lateral extent of the Zone is also that of the characteristic congregation, which is one zoogeographic province.

In this sense, a congregation thus constitutes only a portion of the species and commonly only a very small, though biochronologically diagnostic, portion of any biostratigraphic assemblage zone.

Radioactive Dating

After a considerable period of instability regarding radioactive (and paleomagnetic) dating, biostratigraphers are beginning to receive valid assistance in interpreting dates of key sections. For instance, there is the classic case of radioactive dates of the "Bailey Ash bed" (Pico Formation) in Balcom Canyon, near Ventura, California (for location map, see Yeats and McLaughlin, 1970, Fig. 1), a key horizon mappable over several miles (Table 2).

Table 2. Radioactive Dates for Bailey Ash in Balcom Canyon

Yeats, 1965	Approx. 1.0 m.y.
Bandy, 1967	Approx. 5.1
Yeats et al, 1967	Approx. 8.7
Yeats and McLaughlin, 1970	Approx. 8.6
Izett et al, 1974	Approx. 1.2

Bandy and Wilcoxon (1970) discussed the biostratigraphy of the beds surrounding the Bailey Ash and referred the ash bed to a stratigraphic position near the top of the *Discoaster brouweri* Zone (=nannoplankton Zone NN18, Pliocene of Glaessner, 1970), just below the "*Globorotalia truncatulinoides* datum" (i.e., apparently below foram Zone N.22, Pleistocene of Glaessner, 1970) and within the Wheelerian Stage of Natland (1957). If Bandy and Wilcoxon's biostratigraphic allocation is correct (there is a strong possibility that the California plankton zones and "datum planes" have shifted in time in response to changing climates; Phillips, 1972), the age of the Bailey Ash may eventually prove to be close to 2.0 m.y. (i.e., within the uppermost Pliocene of Glaessner, 1970; Berggren, 1973; Sclater, et al, 1974). However, Berggren puts the base of the Pleistocene at about 1.8 m.y. (base of planktonic foram Zone N.22), whereas Lamb and Beard (1972, Table 1) place it somewhat below Zone N.22, at about 2.8 m.y.

A more complex problem involves radioactive dating of the base of the Pliocene (Fig. 9). Thus, Bandy (1973) assigned an age of 3.5 m.y., Berggren (1973) quoted 5.0 m.y., Glaessner (1970) used 5.5 m.y., and Addicott (1974) cited 10.0 m.y. Moreover, in the past, the age of the Miocene-Pliocene boundary ". . . itself has been estimated to be as young as 2.7-3.0 m.y. and as old as 15 m.y." (Berggren, 1973, p. 391). Paleomagnetic scales indicate that Berggren's figure of 5.0 m.y. is reasonably accurate (Sclater et al, 1974).

Unfortunately, no radioactive dates for the Neogene of northern California are currently available. Turner (1970) provided some analyses for the older Neogene stages of southern California (see Fig. 9). He indicated that the base of the Luisian Stage may be approximately 13.7 m.y., the base of the Relizian may be about 15.3 m.y., and the base of the Saucesian about 22.5 m.y. All of these dates reflect a Miocene age for the stages mentioned (Fig. 9), as Kleinpell inferred (Kleinpell, 1933, 1934b, 1935).

Phylogeny as a Key to Interpretation of Age Units

As a matter of convenience and necessity, in California *benthonic* forams were utilized for Oppelian zonation. There are only a few lineages recognizable at this time (monophyletic series, *s.l.*) which might be used for the purpose of naming zones; and, of these lineages, only a few are represented by species which are both common and morphologically distinctive (see Fig. 10). It is noted that Kleinpell's zones (1938; see also Tipton et al, 1973, Fig. 11) in the California Miocene were named after an evolutionary succession of species of *Siphogenerina* or closely related species of other genera (belonging to superfamily Buliminacea, families Bolivinitiidae and Uvigerinidae in the sense of Loeblich et al, 1964). Significantly, nearly all of Kleinpell's zonal species were adapted to relatively medium-depth benthonic environments–shelf-sea, central shelf to upper bathyal *s.l.* (see Bandy, 1960a, Fig. 9; Lutze, 1962; Harmon,

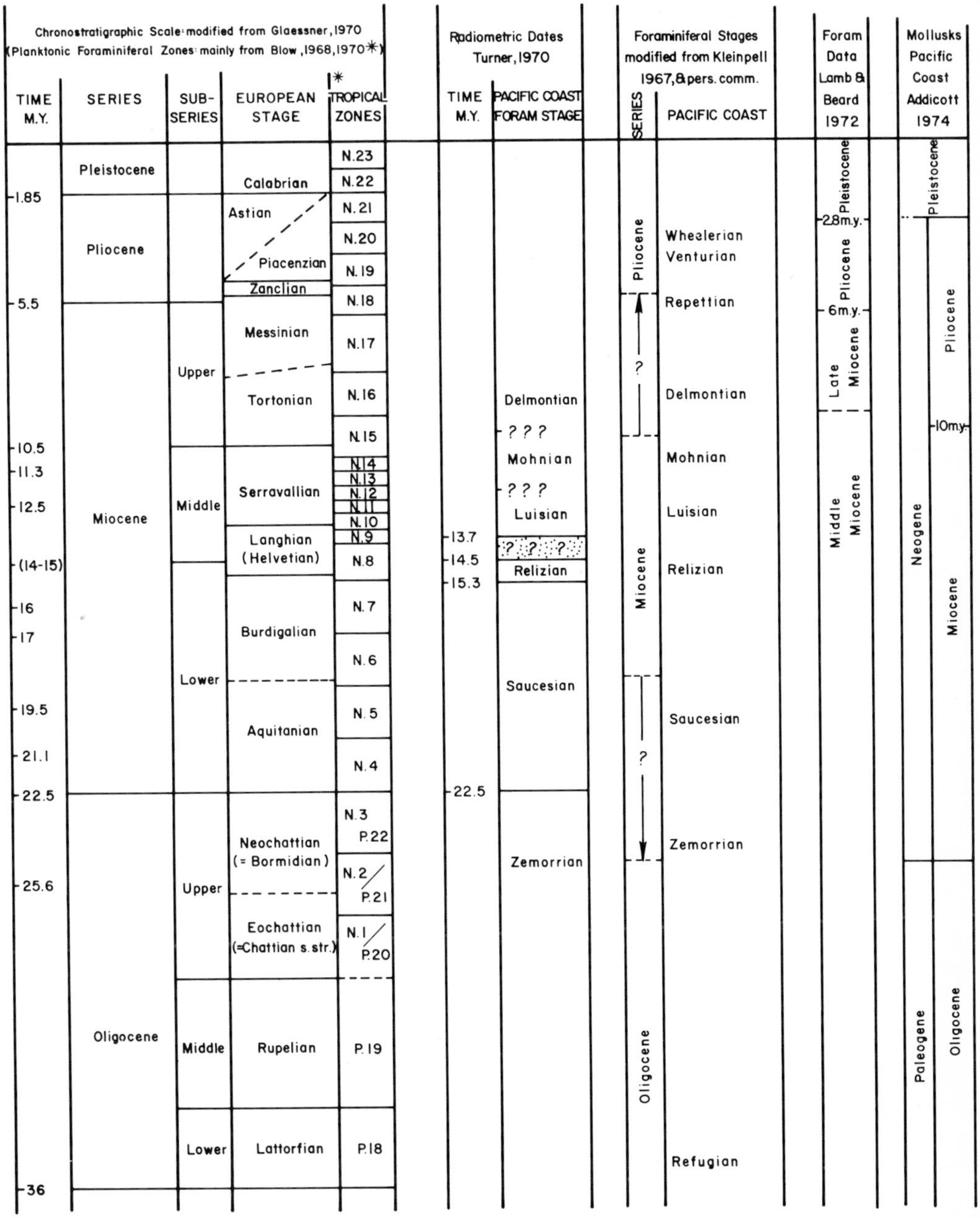

Chronostratigraphic Scale: modified from Glaessner, 1970 (Planktonic Foraminiferal Zones: mainly from Blow, 1968, 1970*)

TIME M.Y.	SERIES	SUB-SERIES	EUROPEAN STAGE	TROPICAL* ZONES
	Pleistocene		Calabrian	N.23
				N.22
1.85	Pliocene		Astian	N.21
				N.20
			Piacenzian	N.19
			Zanclian	N.18
5.5	Miocene	Upper	Messinian	N.17
			Tortonian	N.16
				N.15
10.5		Middle	Serravallian	N.14
11.3				N.13
				N.12
12.5				N.11
				N.10
			Langhian (Helvetian)	N.9
(14-15)				N.8
16		Lower	Burdigalian	N.7
17				N.6
19.5			Aquitanian	N.5
21.1				N.4
22.5	Oligocene	Upper	Neochattian (= Bormidian)	N.3 / P.22
25.6				N.2 / P.21
			Eochattian (=Chattian s.str.)	N.1 / P.20
		Middle	Rupelian	P.19
		Lower	Lattorfian	P.18
36				

Radiometric Dates Turner, 1970

- Time (M.Y.): 13.7, 14.5, 15.3, 22.5
- Pacific Coast foram stages (top to bottom): Delmontian; ? ? ?; Mohnian; ? ? ?; Luisian; ? ? ?; Relizian; Saucesian; Zemorrian

Foraminiferal Stages modified from Kleinpell 1967, & pers. comm.

- Series: Pliocene; ? (Pliocene/Miocene boundary); Miocene; ? (Miocene/Oligocene boundary); Oligocene
- Pacific Coast stages (top to bottom): Wheelerian; Venturian; Repettian; Delmontian; Mohnian; Luisian; Relizian; Saucesian; Zemorrian; Refugian

Foram Data Lamb & Beard 1972

- Pleistocene
- 2.8 m.y.
- Pliocene
- 6 m.y.
- Late Miocene
- Middle Miocene

Mollusks Pacific Coast Addicott 1974

- Neogene: Pleistocene; Pliocene; 10 m.y.; Miocene
- Paleogene: Oligocene

FIG. 9–Comparison of Neogene terminology.

		ELPHIDIUMS	PLECTOFRONDICULARIAS	BULIMINIDS
PLEISTOCENE		poeyanum; "tumidum"; oregonensis; hannai; crispum; foraminosum (= "hughesi" sensu lato)	advena	marginata; "rostrata"; "subacuminata"
PLIOCENE	Wheelerian	acutum of Natland; foraminosum (= "hughesi" sensu lato)	advena	deformata – denudata; "subacuminata"
	Venturian	foraminosum	advena	deformata – denudata; hebespinata; subcalva ("inflata"); "subacuminata"
	Repettian		advena; californica	fossa ("rostrata"); subcalva ("inflata"); "subacuminata"
MIOCENE (lower part ommited)	Delmontian	hughesi (sensu stricto)	"foliacea" (?); californica	fossa ("rostrata")
	Mohnian	hughesi (sensu stricto); granti	"foliacea" (?); californica	
	Luisian	hughesi (sensu stricto); granti	"foliacea" (?); californica; miocenica	
	Relizian	hughesi (sensu stricto); granti	miocenica; californica	

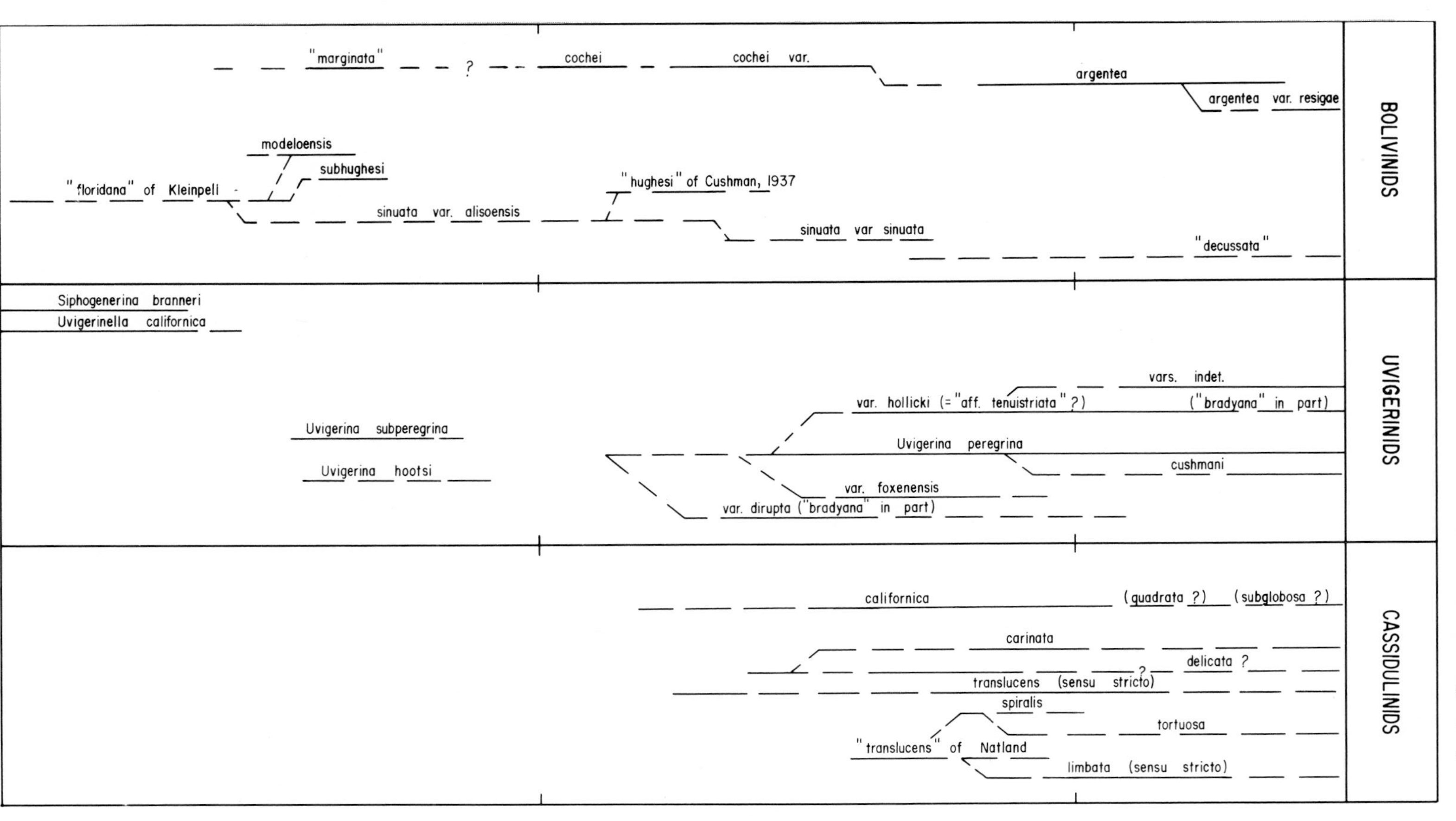

FIG. 10–Supposed phyletic relations of some key foram lineages in the California Neogene.

1964). From youngest to oldest, these foraminiferal zones and corresponding stages are:

Zone	*Stage*
Unnamed	Upper Delmontian
Bolivina obliqua	Lower Delmontian
Bolivina hughesi	Upper Mohnian
Bulimina uvigerinaformis *Bolivina modeloensis*	Lower Mohnian
Siphogenerina collomi *Siphogenerina nuciformis*	Upper Luisian
Siphogenerina reedi	Lower Luisian
Siphogenerina branneri	Upper Relizian
Siphogenerina hughesi	Lower Relizian
Uvigerinella obesa	Upper Saucesian

Woodring et al (1946, Fig. 4) subdivided the *Bolivina hughesi* Zone into an upper *Bolivina goudkoffi* subzone and a lower *Bolivina decurtata* subzone. Pierce (1956) recognized the following succession: upper Mohnian–*Bolivina granti* zone (= *B. goudkoffi*), *Bolivina benedictensis* zone (replaces *B. hughesi*), *Cassidulina renulinaformas* zone; lower Mohnian–*Bulimina uvigerinaformis* zone (? = *B. modeloensis* zone).

In addition to these lineages, Cushman (1937, p. 121) and Kleinpell (1938, p. 282) had indicated that other Neogene bolivinids might be utilized for stratigraphic purposes. The supposed evolutionary (time) sequence of eight foram groups is shown in Figure 10. One of the more prominent bioseries was thought to contain the following stratigraphically significant species: *Bolivina decussata* Brady (type from the Holocene; see Barker, 1960, pl. 53); *Bolivina sinuata* Galloway and Wissler, 1927b (type from the Pleistocene, California); *Bolivina sinuata* var. *alisoensis* Cushman and Adams, 1935 (type from the lower Mohnian, California); and *Bolivina floridana* Cushman, 1918 (type from the Miocene, Florida).

Although there is some question as to whether the first and last of these species are conspecific with specimens from California, nevertheless, there exists in the California Neogene a definite group of bolivinids which shows ancestral-descendant relationships (Table 7). This group was referred to the genus *Bolivinoides* by Hofker (1952, p. 381), returned to the genus *Bolivina* by Reyment (1959, p. 14), and apparently represents the true genus *Bolivina* (Loeblich et al, 1964) as distinct from the vast majority of species normally attributed to the genus *Bolivina* (Fig. 10; systematics herein); the group includes: *Bolivina decussata* of California authors (Holocene of California, range unknown); *Bolivina sinuata* Galloway and Wissler, 1927b (Pleistocene of California, range unknown); *Bolivina sinuata* var. *alisoensis* Cushman and Adams, 1935 (lower Mohnian of California, range unknown); *Bolivina hughesi* of Cushman, 1937 (pl. 14, fig. 7; not Cushman, 1926b; Miocene or Pliocene of California, range unknown); and *Bolivina floridana* of Kleinpell, 1938 (not Cushman, 1918; ranges from Saucesian through lower Mohnian in California).

In the Humboldt basin, *Bolivina costata* var. *interjuncta* may be used instead of *B. sinuata* as an indicator of the late Pliocene and Pleistocene, and *B. cochei* appears to be a substitute for *B. hughesi* in delineating early Pliocene units.

Advanced studies on the phylogeny of planktonic forams were published by Berggren (1968) and Cifelli (1969), among others mentioned by Riedel (1973, p. 254) and Jenkins (1971, p. 29-32). However, the usefulness of planktonic forams in California is sharply restricted, as already mentioned. Recent correspondence and manuscript data from N. de B. Hornibrook of the New Zealand Geological Survey indicates analogous worldwide correlation problems as they apply to New Zealand. Comparison of the 42 Tertiary planktonic foram zones designated by Blow (1970; see Fig. 9) from tropical areas with the mere 21 Tertiary planktonic zones defined by Jenkins (1971) in New Zealand shows little refined correlation of the two different climatic areas and pinpoints the inherent problems of using planktonic forams.

Age and Correlation

Major Biochronologic Subdivisions of Humboldt Basin Neogene

From the study of more than 8,000 molluscan species assembled from several localities in southern Europe, Lyell (1830-1833) deduced the following subdivisions according to percentage of species still living: newer Pliocene (90-100% living species), older Pliocene (35-50%), Miocene (17-18%), Eocene (about 3.5%). Revisions of Lyell's figures in the light of taxonomic changes and more precise documentation, as, for example, in Dunbar (1960, p. 353), indicates that modifications are in order: Pleistocene (90-100% living species), Pliocene (50-90%), Miocene (20-40%), Oligocene (10-15%), Eocene (1-5%), Paleocene (no living species).

Analysis of the relative percentage of living species in fossil foram aggregates in the Humbolt basin Neogene yields three divisions coinciding with the presence of three epochs (Miocene, Pliocene, and Pleistocene). The following percentages were determined (allowing for some inaccuracy in sampling techniques and identification): assemblages averaging 30-40% living species (see the qualifying discussion which follows) were considered to be Miocene from other lines of evidence. The richer assemblages ranging between 60 and 80% living species were designated as Pliocene, and those containing between 90 and 100% were considered Quaternary. These conclusions are in keeping with those derived from previous reconnaissance studies of both mollusks and forams in the area under study.

The oldest assemblages, from highly faulted Miocene sections along Bear River, show the lowest relative percentage of Holocene species. Fortunately, a few samples from the lower part of the Pullen Formation along Price Creek show correlative (in part) Miocene foraminifera faunules.

Overlying the Miocene faunas on Price Creek is an aggregate which occurs in the middle and upper portions of the Pullen Formation; this aggregate is considered tentatively to be the oldest of the Pliocene assemblages.

Overlying the assemblages from the Pullen Formation in the three key sections at Price Creek, Centerville coastal, and Eel River are three aggregates from the younger formations. The excellent exposures in the Centerville coastal and Eel River sections show an aggregate in the Eel River Formation, which is overlain by an aggregate in the lower and middle members of the Rio Dell Formation. The stratigraphically highest aggregate in these exposures is that from the upper Rio Dell and appears to belong to the Pleistocene (discussed in some detail in a later section).

Since the Scotia Bluffs and Carlotta formations are essentially barren, the sixth and youngest assemblages occur as scattered faunules from the Hookton Formation. Again, since all species are still living, this factor suggests a relative age of near recent.

Although the limits of the aggregate do not conform precisely to the designated boundaries of the formations in their type sections, the two features are sufficiently close to facilitate further discussion by reference to the corresponding formation name.

Of these six faunal units, the stratigraphically lowest, apparently both on Bear River and at Price Creek, has from 36 to 38% living species. This figure is somewhat vitiated by the small number of species involved and by the unsatisfactory nature of the exposures. The stratigraphic relations within the suite of beds containing these fossils are obscured by fault contact between fault blocks, and the various faunules are scattered over a rather wide terrane. The percentage of living species noted is nevertheless the lowest for the assemblages found here, and, although the percentages are too high for an average Miocene fauna, this unit is considered to be mainly of Miocene age. This Miocene unit contains several short-ranging species whose stratigraphic range is well known elsewhere, a factor which offsets the inherent difficulties of applying the Lyellian correlation principle to these scattered faunules. On the basis of the several short-ranging species, it is even possible to subdivide the Miocene assemblages into age differences of lesser magnitude. Further details bearing on these ages are given in the section devoted to correlations of infra-series-epoch magnitude.

Faunal aggregates from the middle and upper Pullen Formation with 63% living species, and those from the Eel River Formation with 76% living species, may be referred to the Pliocene Series and Epoch—a broad correlation in keeping with all other known lines of evidence.

The aggregate from the lower and middle members of the Rio Dell Formation contains at least 79 % living species, which seems well within a Pliocene age determination.

The stratigraphically highest faunal aggregate, which is from the upper Rio Dell Formation, and some isolated faunules from the Hookton Formation contain 90% or more living species, and consequently appears to be of Pleistocene rather than Pliocene age. This represents a younger date than has heretofore been attributed to the upper Rio Dell member.

Thus, both the lowest and the highest of the six faunal units are found in scattered outcrops and relatively discontinuous biostratigraphic sequences. Problems in these units are not so readily resolved as in the middle and main portion of the local Neogene sequence. Some conflicting evidence as to age and epoch boundaries is apparent.

Epoch Boundary Problems

Boundary problems at the series-epoch level have long been a matter of passionate dispute (see Williams, 1954).

As Kleinpell (1938) noted, in southern California the Miocene-Pliocene boundary might coincide with the termination of the Delmontian Stage. However, the matter is not settled (see Kleinpell, 1948; 1967; Kleinpell and Weaver, 1963, Fig. 2). Both Miocene and Pliocene faunas are recognized from the Humboldt basin, but sample collecting across the boundary has been too sketchy to permit an extensive discussion regarding this boundary. At present, the writer does not know of any uninterrupted sequence of well-exposed strata in the Humboldt basin which crosses the boundary. The Price Creek section, with its partially barren beds and intermittently covered intervals, presents the neareast approach to this desideratum. Recent evidence (Ingle, 1974, personal commun.) indicates that slump and fault blocks near the base of the Centerville coastal section are now paleontologically dated as probably Miocene.

The Pliocene-Pleistocene boundary problem continues to be a source of discussion and argument on a worldwide basis. Following a Commission Report for the International Geological Congress (1950), there was a divergence of opinion regarding techniques to mark the boundary: for example, Ericson (1959), Ericson and Wollin (1964), and Bandy (1960b) used the method of noting change in coiling direction (in response to temperatures) in certain planktonic forams to define the boundary in Atlantic and Pacific submarine cores. Van Voorthuysen (1952, 1953) suggested the cool-water benthonic foram *Elphidiella oregonensis* as a "guide marker" for the lower Pleistocene of northern Europe and North America; later, he (van Voorthuysen, 1957) cited the Commission's proposals and stated, ". . . The classic method to establish stratigraphical boundaries is based on changes in marine faunas. Pleistocene stratigraphy on the other hand rests on worldwide and broadly synchronous major climatic oscillations." Emiliani et al (1961) and Papani and Pelosio (1962) followed part of the Commission's specific recommendations regarding the influx of the "cold-marker" benthonic foram *Anomalina (Hyalinea) baltica* (Schroeter) in pointing out the Pliocene-Pleistocene break in southern Italy. Emiliani (1961) and Emiliani et al (1961) cited additional evidence based on oxygen-isotope temperatures derived from analyses of benthonic and pelagic foram species in the southern Italian sections. Arrhenius (1952) used a geochemical method based on distinction of rates of accumulation of organically formed calcium carbonate in submarine cores from the Pacific. Riedel (1957) suggested the upper limit of the common occurrence of a (specified) radiolarian species as a tentative marker for the end of the Pliocene in tropical Pacific submarine cores. Ericson et al (1963) defined the boundary at the extinction of all discoasters in Atlantic cores. Flint (1965) discussed the above arguments in somewhat greater detail, and Selli (1967) reaffirmed the traditional use of paleoclimatology in his discussion of the limits of the Pleistocene in the classic Italian marine sections. Hays and Berggren (1971) reviewed many of the preceding papers and concluded:

1. "The transition from Pliocene to Pleistocene was not marked by a major environmental change of world-wide significance."

2. "No significant stratigraphical or faunal discontinuity occurs between the Pliocene and Pleistocene as at present defined."

3. "The evolutionary transition from *G. tosaensis* to *G. truncatulinoides* is the most satisfactory paleontological criterion for recognizing the Pliocene-Pleistocene boundary [and] . . . occurs near the base of the holostratotype Calabrian at Santa Maria di Catanzaro, in southern Italy."

In summary, Emiliani (1961) stated, ". . . rapid temperature fluctuations were occurring already in the Pliocene, but became greater in Early Pleistocene. . . ." Much of the effort in solving the boundary problem, however, has been directed primarily toward recognizing in the fossil record a faunal change which correlates with changes in the molluscan and foraminiferal faunas at the base of the Calabrian in southern Italy—the appearance of the mollusk *Arctica islandica* and the forams *Hyalinea baltica* and *Globorotalia truncatulinoides*.

In the last decade or so, much work has been directed toward the establishment of stratotypes for the numerous European stages of the Pliocene and Pleistocene. In keeping with Lyell's fundamental work, most of the late Miocene, Pliocene, and Pleistocene stratotypes fall either within a limited area in northern Italy, on the toe of Italy, or on the island of Sicily. The primary Miocene-Pliocene contact in the stratotypes appears to be in Sicily, and the most relevant exposures of the Pliocene-Pleistocene contact are on the lower mainland (Hays and Berggren, 1971, Fig. 51.1; Lamb and Beard, 1972, Fig. 1; Fairbridge, 1973).

An intensive and continuing effort to document the macrofaunas and microfaunas of these stratotypes is seen in the publications of Barbieri (1967, 1969, 1971), Barbieri and Petrucci (1967), Bayliss (1969), Bertolino et al (1968), Carloni et al (1968), Cati et al (1968), Cita and Blow (1969), Cita and Premoli Silva (1968), Cita et al (1965), Colalongo (1968, 1970), Conato and Follador (1967), Dondi and Papetti (1968), D'Onofrio (1964, 1968), Follador (1967), Gianotti (1953), Gradstein (1970, 1973), Iaccarino (1967), Roda (1971), Ruggieri (1965), L. Smith (1969), Vervloet (1966), and Wezel (1968), among others.

Apt summaries of the results of many of these papers are given by Hays and Berggren (1971), Lamb and Beard (1972), and Fairbridge (1973). Although complete agreement does not exist, most workers place the Miocene-Pliocene boundary at or somewhat below the base of the Zanclian Stage and the Pliocene-Pleistocene boundary near the base of the Calabrian Stage (Fig. 9).

Particularly relevant to correlation of rocks in other parts of the world with the Italian stratotypes are certain species of planktonic forams and planktonic radiolarians. Sharp geographic limitations are placed on the distribution of these two groups of microfossils, as shown by Hays and Berggren (1971, Figs. 51.6, 51.7); and northern California falls outside the "areas of usefulness" of planktonic forams in regard to the Pliocene-Pleistocene boundary. Similar geographic problems in the usage of radiolarian assemblages were expressed by Casey (1972).

Much discussion has evolved around the occurrence of *Pecten caurinus* and other "cold-water" molluscan species (Ashley, 1895, p. 452; R. Arnold, 1903, p. 16, 20, 29, 69; J. Smith, 1919, p. 136, 137, 151; Hay, 1927, p. 166-173; Crickmay, 1929, p. 622-632; Grant and Gale, 1931, p. 22, 71-94, 195-196; Bailey, 1935; Grant, 1935, p. 349-350; Woodring et al, 1941, p. 109; Keen and Bentson, 1944; Schenck, 1945, p. 511; Durham, 1950, p. 1258-1260; Woodring, 1952; 1957b, p. 592, 595) in the Cenozoic of southern California. The key molluscan section at Santa Barbara has been described in some detail by Dibblee and Keen (*in* Keen and Bentson, 1944, Table 1, p. 12-14); they determined that *Pecten dilleri* occurs primarily below the Pliocene-Pleistocene boundary and *Pecten caurinus* occurs primarily above, although the ranges of the two species overlap across the supposed boundary.

In northern California, the Pliocene-Pleistocene boundary was tentatively placed by

Ogle (1953, Fig. 3) within the Carlotta Formation, apparently primarily on the basis of occurrence of the echinoid genus *Anorthoscutum* (i.e., *Scutellaster*) in the underlying Scotia Bluffs Sandstone. Durham (1955) cited this echinoid genus as "Middle and Upper Pliocene, central California to Oregon." Although Faustman (1964, p. 109) recognized the cool-water character of the molluscan fauna in the upper Rio Dell, he (p. 110) generally followed Ogle's placement of the boundary. In addition to *Scutellaster*, Faustman listed three molluscan species and one variety, and stated, ". . . a large part of the (Scotia) fauna is known only from Pliocene sediments."

As the Humboldt basin is 6 to 7° of latitude north of Santa Barbara (about 34-35° versus 40-41°), species with cold-temperature optima, such as *Pecten caurinus*, may have existed earlier in northern California than in southern California (see J. Smith, 1919). In any event, climatic criteria cannot be used in geochronology except where the two permanent gradients of environment—latitude and topographic elevation—are known constants in the areas being compared for correlation. In California, just as in the classic Italian region, data for comparison of the two areas are insufficient. Thus, biochronologic criteria were sought in the present study, only to find that, in the Humboldt basin, the evidence from the forams is in apparent conflict with the evidence from echinoids and mollusks. This conflicting evidence is discussed in detail in the following section on infra-epoch correlation.

Infra-Epoch Age Units in Coast Range Pliocene

Descriptive biostratigraphy of the Neogene foraminiferal sequences of southern California includes several publications which deserve some mention (see also Fig. 11).

Driver (1928) outlined the stratigraphy in Adams Canyon (Ventura County) and provided an empirical analysis of the foraminiferal species in that section. Driver did not name the foraminifera species encountered, but nonetheless distinguished three distinctive assemblages representing stratigraphic "Zones A, B, and C."

Wheeler (1928) more formally designated the forams at the generic, but not at the specific, level which characterized his "Alpha, Beta, Gamma, and Delta Zones" of Ventura County, in and along strike from the same Ex-Mission Hills sequence reported upon by Driver.

Subsequently, Woodring et al (1932) published a summary of the Neogene showing three divisions within the Pliocene, each division represented by two foraminiferal zones. In this classification, the foraminiferal guide fossils were identified at the specific level (Fig. 11). Reed (1933, p. 230, Table XV; p. 248, Table XX) published a slightly revised summary of Neogene subdivisions from southern California. In this revision, correlation with key mollusks was undertaken (Fig. 11). In 1933, Natland discussed five "faunal-, or life zones" (1933, p. 227) from a study of recent foraminifera collected in offshore southern California waters. Natland equated these life zones to biozones on the basis of assemblages in the strata of Hall Canyon and in the Ex-Mission Hills of Ventura County. This early work formed the basis for Natland's 1952 and 1957 publications (Fig. 11). Kleinpell (1938) provided a detailed analysis of foraminifera faunas from the Miocene (Fig. 11). As previously discussed, Kleinpell's zones were named after a sequence of closely related species.

Two papers by Adams (1939, 1940) concentrated on the local ranges and abundance of the foraminiferal species of *Bolivina, Bulimina,* and *Buliminella* in Canada de Aliso, adjacent to the Ex-Mission Hills sequences in Ventura County. Adams proposed the "zonal units" A, B, C, and D with some questions about their limits. His zonal units A, B, and C apparently represent Pliocene-Pleistocene assemblages stratigraphically high in these sequences. Adams' zonal unit D is of Miocene age and occurs stratigraphically just below the lowermost beds of Driver and of Wheeler (1928).

Woodring's (1938) paper documented the mollusks from 60 wells in the Los Angeles basin and integrated the molluscan data with work on forams by Wissler and other paleontologists. Wissler (1943, 1958) recognized 26 foraminiferal "zones" and a correspondingly larger number of "subzones" from the subsurface Pliocene of the Los Angeles basin (Fig. 11), a sequence which corresponds in most respects to the Ex-Mission Hills outcrop sequence of Ventura County. In his 1958 paper he grouped these zones and subzones into five "faunal divisions." Wissler's zones and subzones apparently are biostratigraphic reflections of ecologic phenomena, perhaps equivalent in magnitude to the zonule (see comments by Kleinpell, 1938, p. 98), whereas the larger faunal division ". . . must have been regarded as facies which follow environment and transgress much of Pliocene time . . ." (Holman, 1958, p. 191, also chart II, p. 2).

Natland's (1952, pl. IV) stratigraphic "Zone 2" became the type section for the "Upper Hallian Stage," and "Zone 3" became the type for the "Lower Hallian Stage." "Zone 4" formed the basis for the "Wheelerian Stage" and a type section was proposed from outcrops along the ridge east of Wheeler Canyon (Natland, 1952, pl. III). The "Venturian Stage" was named with the type section underlying the type Wheelerian at Wheeler Canyon (Natland, 1952, pl. III); Natland did not correlate the assemblage from the Venturian with a counterpart in Holocene sediments, and this interval became subject to the vagaries inherent in the designation of a "transition zone." "Zone 5" had been considered more or less synonymous with microfaunas in the Repetto Formation, a unit which was later clarified as the "Repettian Stage." The type section of both the formation and the "stage" is the same locality in the Repetto Hills, eastern Los Angeles (Reed, 1933, p. 239; Natland, 1952, pl. II; Natland and Rothwell, 1954, Fig. 3). The superpositional context of Natland's five "stages" can be viewed in two geographically close sequences: the Hall Canyon and the Wheeler Canyon sequences, both in Ventura County.

To summarize, four or five foraminiferal aggregates have been recognized as characterizing four or five stratigraphic intervals in the Pliocene and Pleistocene of southern California. Their superpositional relations are shown at four primary outcrop sequences in the Ventura basin: Hall Canyon, Wheeler Canyon, Adams Canyon, and Canada de Aliso. The outcrops at nearby Santa Paula Creek also contribute some information, as does the type Repettian outcrop section of the Los Angeles basin and many oil wells penetrating the subsurface of the Los Angeles and Ventura basins (Wissler, 1958). Apparently, these assemblages represent a succession of assemblage zones and a series of zonules, mainly of ecologic significance. Superpositional order of the many Pliocene zonules (zones in the sense of Wissler, 1937, and Natland, 1952; Fig. 11) may and does vary considerably between basins and even within a basin, as Bandy (1967) indicated in his many studies of the Los Angeles and Ventura basins. Bandy noted that the distribution of benthonic forams in these basins may be largely a function of bottom topography in association with sedimentation types and rates. Many of the "Pliocene" benthonic forams carried over into the Pleistocene because of suitable environmental conditions.

In Washington, Rau (1970, p. 12) encountered similar environmental problems. He stated:

> . . . Therefore, rather than to attempt conventional correlations based largely on evolution of species, it is perhaps more realistic to compare the dominating ecological conditions reflected by the foraminifers in each of these basins with those of the Quinault Formation. To this extent the foraminiferal assemblages of the various sections of the Quinault Formation are compared with those of other late Tertiary - early Quaternary basins of the West coast.

Certain elements, however undefined as such, do have biochronologic significance (Fig. 10).

Guide fossils were used for reconnaissance correlation purposes between the foraminiferal faunas of the Humboldt basin of northern California and the Ventura and Los Angeles basins of southern California by Reed (1937, Fig. 2), Stewart and Stewart (1949), and Ogle (1953; see also Fig. 11).

These workers attempted to distinguish in northern California the supposed equivalents of the four or five assemblages of southern California. Thus, Reed, and later Stewart and Stewart, in their studies of the Centerville coastal sequence in Humboldt County, referred to the guide species that were thought to be biochronologically the most diagnostic (Fig. 11).

The forams *Bolivinita quadrilatera* ("*Bolivina angelina*" of some authors), *Bulimina fossa, Bulimina subacuminata, Cassidulina translucens, Cassidulina limbata, Karreriella*

	Woodring, Hughes, & Wissler 1932		Reed, 1933 George H. Doane - Committee Southern California			Kleinpell, 1938 California Province		
PLEISTOCENE		Elphidium crispus						
PLIOCENE	Upper	Uvigerina aff. tenuistriata	PICO	Cibicides lobatus	Pecten caurinus	Plectofrondicularia californica and associated Foraminifera		
	Upper	Cibicides mckannai	PICO	Uvigerina n. sp. (U. aff. tenuistriata)	Pecten hemphilli			
	Middle	Uvigerina peregrina	PICO	Cibicides mckannai	Pecten hemphilli			
	Middle		PICO	Uvigerina peregrina	Pecten hemphilli			
	Middle	Bulimina subacuminata	PICO	Bulimina subacuminata	Trophoscyon nodiferum			
	Lower	Plectofrondicularia californica	REPETTO	Plectofrondicularia californica	Argobuccinum n. sp.			
	Lower	Arenaceous fauna	REPETTO	arenaceous forms	Argobuccinum n. sp.			
MIOCENE (lower part omitted)	Upper	Bolivina hughesi				DELMONTIAN	Upper	
						DELMONTIAN	Lower	Bolivina obliqua zone
						MOHNIAN	Upper	Bolivina hughesi zone
		Valvulineria californica Nodosaria koina				MOHNIAN	Lower	Bulimina uvigerinaformis zone Bolivina modeloensis zone
						LUISIAN	Upper	Siphogenerina collomi zone Siphogenerina nuciformis zone
	Middle					LUISIAN	Lower	Siphogenerina reedi zone
	Middle	Plectofrondicularia miocenica				RELIZIAN	Upper	Siphogenerina branneri zone
						RELIZIAN	Lower	Siphogenerina hughesi zone

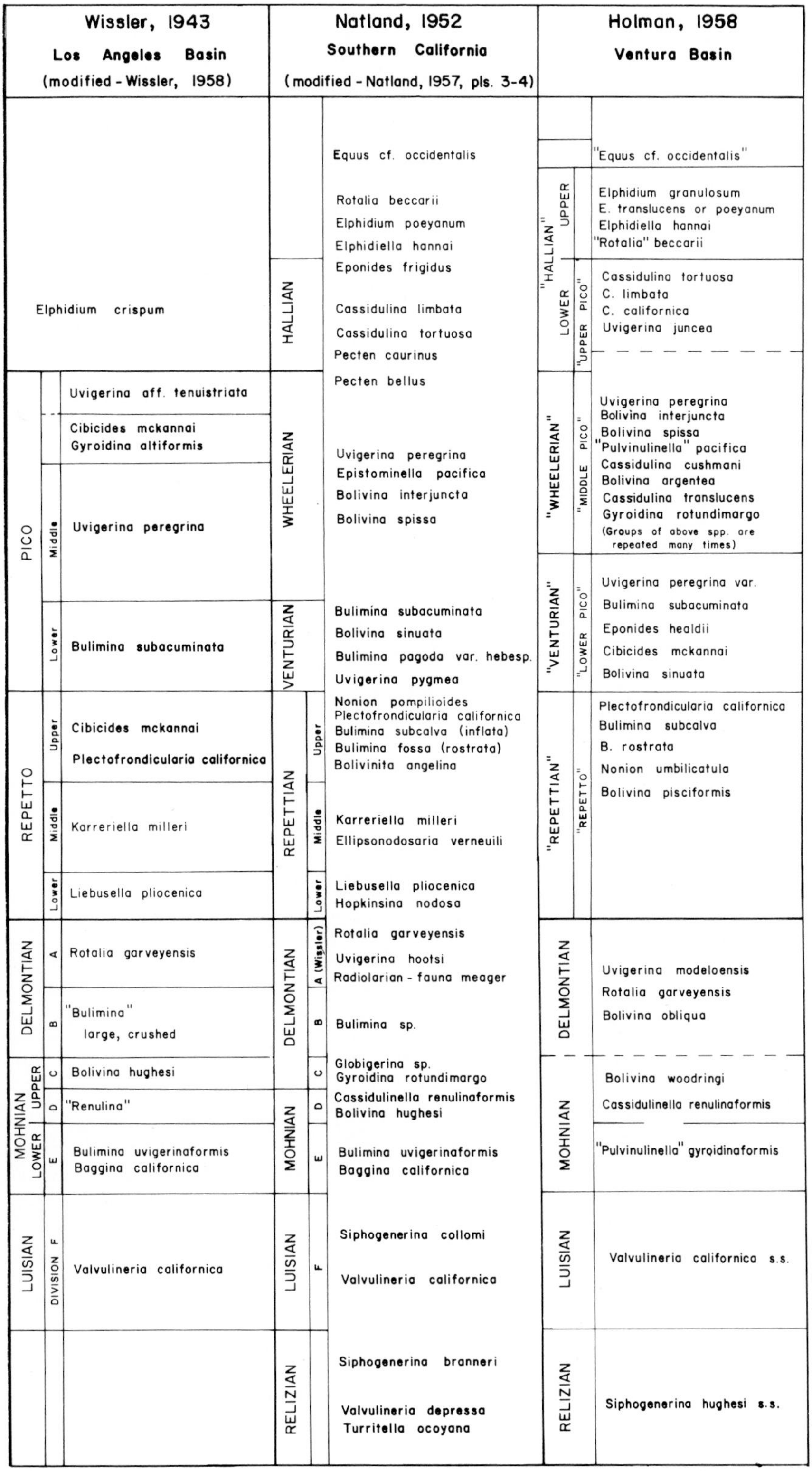

Wissler, 1943 Los Angeles Basin (modified - Wissler, 1958)			Natland, 1952 Southern California (modified - Natland, 1957, pls. 3-4)			Holman, 1958 Ventura Basin		
					Equus cf. occidentalis			"Equus cf. occidentalis"
		Elphidium crispum	HALLIAN		Rotalia beccarii Elphidium poeyanum Elphidiella hannai Eponides frigidus Cassidulina limbata Cassidulina tortuosa Pecten caurinus	"HALLIAN"	UPPER	Elphidium granulosum E. translucens or poeyanum Elphidiella hannai "Rotalia" beccarii
							LOWER "UPPER PICO"	Cassidulina tortuosa C. limbata C. californica Uvigerina juncea
PICO		Uvigerina aff. tenuistriata	WHEELERIAN		Pecten bellus	"WHEELERIAN"	"MIDDLE PICO"	Uvigerina peregrina Bolivina interjuncta Bolivina spissa "Pulvinulinella" pacifica Cassidulina cushmani Bolivina argentea Cassidulina translucens Gyroidina rotundimargo (Groups of above spp. are repeated many times)
		Cibicides mckannai Gyroidina altiformis			Uvigerina peregrina Epistominella pacifica Bolivina interjuncta Bolivina spissa			
	Middle	Uvigerina peregrina						
	Lower	Bulimina subacuminata	VENTURIAN		Bulimina subacuminata Bolivina sinuata Bulimina pagoda var. hebesp. Uvigerina pygmea	"VENTURIAN"	"LOWER PICO"	Uvigerina peregrina var. Bulimina subacuminata Eponides healdii Cibicides mckannai Bolivina sinuata
REPETTO	Upper	Cibicides mckannai Plectofrondicularia californica	REPETTIAN	Upper	Nonion pompilioides Plectofrondicularia californica Bulimina subcalva (inflata) Bulimina fossa (rostrata) Bolivinita angelina	"REPETTIAN"	"REPETTO"	Plectofrondicularia californica Bulimina subcalva B. rostrata Nonion umbilicatula Bolivina pisciformis
	Middle	Karreriella milleri		Middle	Karreriella milleri Ellipsonodosaria verneuili			
	Lower	Liebusella pliocenica		Lower	Liebusella pliocenica Hopkinsina nodosa			
DELMONTIAN	A	Rotalia garveyensis	DELMONTIAN	A (Wissler)	Rotalia garveyensis Uvigerina hootsi Radiolarian - fauna meager	DELMONTIAN		Uvigerina modeloensis Rotalia garveyensis Bolivina obliqua
	B	"Bulimina" large, crushed		B	Bulimina sp.			
MOHNIAN UPPER	C	Bolivina hughesi		C	Globigerina sp. Gyroidina rotundimargo	MOHNIAN		Bolivina woodringi
	D	"Renulina"	MOHNIAN	D	Cassidulinella renulinaformis Bolivina hughesi			Cassidulinella renulinaformis
MOHNIAN LOWER	E	Bulimina uvigerinaformis Baggina californica		E	Bulimina uvigerinaformis Baggina californica			"Pulvinulinella" gyroidinaformis
LUISIAN	DIVISION F	Valvulineria californica	LUISIAN	F	Siphogenerina collomi Valvulineria californica	LUISIAN		Valvulineria californica s.s.
			RELIZIAN		Siphogenerina branneri Valvulineria depressa Turritella ocoyana	RELIZIAN		Siphogenerina hughesi s.s.

FIG. 11—Prior usage of "guide, index, or zonal" foram names for units in the Neogene of southern California.

milleri, Liebusella pliocenica, Plectofrondicularia californica, and *Uvigerina peregrina* have loomed especially large in such correlations (Stewart and Stewart, 1949, p. 196-202, eight left-hand columns of pl. 22; also, Natland, 1957). For the most part these species have long geologic ranges. For example, *Plectofrondicularia californica* occurs in strata from upper Eocene (Narizian) through lower Pliocene in the California Coast Ranges (Kleinpell, personal commun.); *P. californica* apparently occupied an ecologic niche in deeper water, areas of high relief, and environmental optima (e.g., the Miocene of the southernmost San Joaquin Valley). Others, such as *Uvigerina peregrina*, are still living off the west coast; still others, such as *Bolivinita quadrilatera*, have a very short geologic range in the Coast Ranges, although in the Indo-Pacific faunal province this species ranges throughout most of the Neogene and into the Holocene.

Yet, despite all the time-rock terminology, stratigraphic typology, and meticulous biostratigraphy, no biochronologic diagnosis of the Pliocene and Pleistocene "stages" has been prepared or presented.

Published correlations using Natland's stages remain relatively few, and correlations actually stemming from the fossil foram species encompassed by them are even fewer. Fossil correlations of the upper Neogene still tend to rely on a few supposed guide fossils and are in general restricted to a reconnaissance correlation of whole formations or members.

The most comprehensive of such published studies is that of Ogle (1953). Ogle collected extensively in the Humboldt basin (see his pl. 6), both areally and in great stratigraphic detail (p. 4). His foram identifications and correlations depended upon Douglas Crawford of Union Oil Company, the records of the late Donald D. Hughes, (who was with Texaco), and those published by Stewart and Stewart. Discrepancies in identification, in the recorded local ranges of Pliocene foram species, and in the local ranges encountered by Ogle, were noted (Ogle, 1953, p. 46) though not specified. The evidence from invertebrate paleontology, paleobotany, and vertebrate paleontology was also taken into consideration. Ogle summarized his correlations in a single paragraph (1953, p. 47) worth quoting as the most succinct and definitive of all the previous summaries of such work in this field and this area:

> The abundance of fossils, especially foraminifera, makes it possible to establish correlation with other formations in California. Most of the Pullen formation may be equivalent to most of the Puente and Modelo formations of the Los Angeles basin, the lower part of the Purissima formation of the Santa Cruz basin, and the Tice shale (Monterey group) of the San Francisco Bay area. It is partly equivalent to the Wymer beds (Del Norte County). The upper part of the Pullen beds and the whole Eel River formation are probably equivalent to the Repetto formation of the Los Angeles and Ventura basins, the Jacalitos formation of the San Joaquin Valley, and part of the Purissima formation of the Santa Cruz basin. The Rio Dell formation is essentially equivalent to the Pico formation of the Los Angeles and Ventura basins, and the upper part of the Purissima beds (Santa Cruz basin). The upper part of the Rio Dell formation is equivalent to the lower part of the Empire formation at Coos Bay, Oregon. The upper Scotia Bluffs and the Carlotta formation are tentatively correlated with the San Pedro beds of southern California, the upper part of the Merced formation (near Santa Rosa), possibly with the beds at Battery Point, Crescent City, and part of the Pleistocene beds at the mouth of Elk River in southern Oregon.

Thus, correlations were based on mollusk species of demonstrated short range and upon comparably age-diagnostic Miocene foram species. No diagnostic Pliocene or Pleistocene foram species were so employed, nor are they in reference. Certain Pliocene foram species were noted as "characteristic" of certain "assemblages," but no evidence was given as to whether these are diagnostic of geologic age or simply of ecology. As in the Pliocene and Pleistocene foram assemblages in southern California, Bailey (1935) had shown that some fairly large-scale miscorrelations were involved within the classic Ex-Mission Hills area in sequences which were subsequently used as standards of comparison. The reason for not citing age-diagnostic foram species as a basis for correlation becomes apparent: the stages of the southern California Pliocene and Pleistocene were not interpreted as to demonstrable specific age-diagnostic, foraminiferal elements. In terms of relevant evidence, the Pliocene correlations determined between southern California and Humboldt County could be homotaxial (regarding facies) as well as geologically correlative (regarding age).

A problem therefore is posed as to which foram lineages, or associations of species, in the southern California Pliocene sequence are age diagnostic at an infra-epoch level, and which are not. Although this problem is beyond the scope of the present work, the published record of late Neogene foram species and their stratigraphic occurrences in southern California (Natland, 1933, 1953; Natland and Rothwell, 1954) is sufficient for at least a rudimentary analysis of age-diagnostic elements. This analysis is attempted in order to establish a foundation for comparison and biochronologic correlation.

Repettian

Of the so-called guide fossils, at least three appear to have had a relatively short geologic time span in the Coast Ranges: *Liebusella pliocenica, Karreriella milleri,* and *Bolivinita quadrilatera*. All are apparently limited to the biostratigraphic interval Natland defined as typical of his Repettian Stage—the first to its lowermost beds (it may actually appear in the underlying upper Delmontian, *sensu lato*, "uppermost Miocene"), the second to its lower and middle beds, and the third to thin beds within the upper half of the Repettian. The first two, being arenaceous forms, might seem to owe their stratigraphically restricted occurrences to ecologic causes involving local bottom-surface conditions; yet such suitable bottom conditions have persisted or recurred at higher horizons (horizons at which other arenaceous forms have continued to flourish) through the same stratigraphic profiles without the subsequent recurrence of these guide fossils. Such arenaceous species as *Cyclammina cancellata* and *Martinottiella communis* occur locally, seemingly restricted to the lower Pliocene, but are known from many local older Tertiary horizons and are recorded from the Holocene Pacific faunas. "*Bolivinita angelina*" Church, on the other hand, a calcareous and perforate species, has seemingly been a cryptogenetic species in the Coast Ranges; yet, since it is conspecific with *Bolivinita quadrilatera* (Schwager) of the Indo-Pacific Neogene and Holocene, its ancestry and origins become clear, and its presence in the Coast Ranges indicates a temporary, early Pliocene, warmer water invasion of the California province by elements from the Indo-Pacific province (Ingle, 1967; 1973b). Another calcareous species of seemingly short range is *Siphonodosaria insecta*, which appears also to be restricted to Natland's Repettian interval and to be an Indo-Pacific invader occurring temporarily in the California province. If, however, it should prove conspecific with *S. abyssorum*, it may actually represent a species still living in the Indo-Pacific province.

Several other calcareous perforate forms are also notably restricted in their stratigraphic ranges. Among these are four species which range upward from pre-Pliocene horizons in the Coast Range, but which apparently had become extinct by the close of Natland's Repettian time: *Bolivina cochei* (extinct by the mid-Repettian), *Plectofrondicularia californica, Stilostomella lepidula,* and *Uvigerina hootsi*. Of these, *P. californica* is the most conspicuous extinction; in the case of the other two, the term "extinct" should perhaps be applied with assurance only to these species *sensu stricto*. Two cassidulinids (*Cassidulina californica* and *C. translucens*) and three buliminids (*Bulimina fossa, Bulimina subacuminata*, and *Uvigerina peregrina*) appear apparently for the first time within the limits of Natland's Repettian. In *sensu stricto, B. fossa* and *B. subacuminata* appear in upper Repettian beds and persist into the directly overlying strata commonly referred to in southern California as a "Repetto-Pico transition." The Pliocene *B. subacuminata* may be conspecific with forms on record as "*B. subclava*" or "*B. subcalva*" (see Kleinpell, 1938, p. 259) from the Miocene, and other forms from the Holocene recorded under *B. inflata*. Some workers recognize these Miocene, Pliocene, and Holocene forms as three distinct varieties of *B. subacuminata*. Both *B. fossa* and *B. subacuminata* appear to be descended from species known in late Eocene, Oligocene, and Miocene horizons in the Coast Ranges (Kleinpell, personal commun.).

Thus, in the light of the current evidence, a joint occurrence of these species, or even of any two of those which overlap in time, would constitute a congregation diagnostic of the Repettian Stage and warrant correlation of the containing strata with its type.

Venturian

Ambiguities involving the so-called "transition zone," that is, approximate equivalents to Natland's Venturian Stage, have already been alluded to (see especially Natland's discussion of the Venturian in the Repetto Hills: Natland, 1957, p. 561). The vague stratigraphic interval in reference is that commonly designated as the "Repetto-Pico transition" beds (Reed, 1933, p. 248). The difference in usage and application of this designation by paleontologists is at least 400 ft (120 m) of the southern California Pliocene column (Kleinpell, personal commun.); in one case the "transition zone" is within the limits of the Repettian Stage of Natland; in another it is directly above the uppermost strata of the Repettian Stage and thus, by implication, within the Venturian. Conspicuously involved is systematics, especially within some of the more critical buliminid and uvigerinid lineages. The usage followed here is to include the highest beds bearing *Plectofrondicularia californica* as still within the Repettian.

Along with *Bulimina subacuminata* and *Uvigerina peregrina* var. *foxenensis* in their occurrences–not with, but just above the highest occurrence of *Plectofrondicularia californica*–three species of still-living cassidulinids enter the column: *Cassidulina cushmani, C. minuta* (perhaps even earlier?), and *Cassidulinoides bradyi*. During deposition of this supra-Repettian interval, the distinctive *foxenensis* variety of *Uvigerina peregrina* seems especially to have flourished where optimum environmental conditions were met. Thus, the joint occurrence of *Bulimina subacuminata* and *Uvigerina peregrina* var. *foxenensis*, neither as yet known from above the "Wheelerian," together with the introduction of the three cassidulinids, seemingly constitutes a congregation diagnostic of the Venturian Stage.

Wheelerian

In the Ex-Mission Hills of Ventura County, southern California, the progressive shallowing of the depositional basin, which is so noticeably reflected upward in the column from Repettian to Venturian strata, continued. The Wheelerian, which overlies the Venturian in the type sequence, is, however, still largely of moderately deep-water deposition, though the water was shallower than during Venturian deposition. In general, the Repettian and Venturian cassidulines as well as the Venturian bulimines persist. Other slight changes in the assemblages could be analyzed for biochronologic significance, of which the abundant occurrence of "*Uvigerina* aff. *tenuistriata* Reuss" is more apparent than real (see Reed, 1933, p. 230; Natland, 1933, Table 1; Kleinpell, 1938, p. 288-289; Wissler, 1943, p. 210, Table 1; Stewart and Stewart, 1949, pl. 22; Hughes *in* Ogle, 1953, p. 46). Unfortunately, not one of the cited references figures or describes *Uvigerina tenuistriata* or *U.* aff. *tenuistriata*. In all probability such a misnomer may refer to one or more varieties of *U. peregrina*, most likely *U. peregrina* var. *hollicki* herein (see systematics for synonymies of *U. juncea, U. cushmani, U. pygmaea*; and discussion in Barker, 1960, concerning *U. tenuistriata*). Later, however, Reed (1937, Fig. 2) referred not to *U.* aff. *tenuistriata*, but to *U. peregrina*. Moreover, "*Uvigerina* aff. *tenuistriata*," which is conspicuous in the "Mudpit Shale" of the Ventura area, occurs below as well as above the horizon of *Pecten bellus* in the inshore facies of the contemporaneous time-rock unit (see Natland, 1933, Table 1; Natland, 1957, pl. 4).

The biochronologic distinctions between Venturian and Wheelerian Stages, in southern California as well as in northern California, appear to be of such a minor nature as to suggest that these two stages might appropriately be regarded as two zones, or at most two substages of a single stage, of late Pliocene age. Moreover, the faunal differences that are so conspicuous between the assemblages in the type Venturian and Wheelerian sequences are of more ecologic than biochronologic significance (Natland, 1957, p. 561). It is noteworthy that the Wheelerian guide species (see Fig.

11) all range both below and above the Wheelerian (see Natland, 1933, Table 1; 1953; 1957).

"Hallian"

In the late Neogene of the Ex-Mission Hills of Ventura County, further shallowing appears to have continued, eventually involving strata of progressively coarser texture. Foraminiferal faunal changes are mainly of ecologic rather than biochronologic significance (see Bailey, 1935). At this geologic level in southern California, biochronology depends largely upon the evidence from the marine invertebrates of inshore habitat, notably such mollusks as *Pecten caurinus*. In the case of the "San Diego formation controversy" of the 1920s, correlations based on foraminiferal evidence proved purely empirical and unreliable owing to failures of distinction between the ecologic and biochronologic factors involved (see Bailey, 1935). In fact, the "Hallian Stage" of Natland appears to encompass a thick, mainly sandy section from beds overlying the Wheelerian to those as high and as young as the Palos Verdes Formation (see Natland, 1953). Consequently, the "Hallian" appears to be less useful as stage, substage, or zone, insofar as such an all-embracing application encompasses most of the Pleistocene Series and Epoch, which in California is represented mainly by sediments of inner-shelf and nonmarine origin.

Infra-Epoch Ages of Humboldt Basin Neogene

Lyellian percentages for the six foraminiferal aggregates (complete or partial) were noted as shown in Table 3.

With regard to the Miocene Bear River beds, much richer Miocene assemblages from the lower and middle members of the Monterey Formation in southern California showed percentages as low as 14 to 18.5 living species (Kleinpell, 1938, p. 171) in faunas of 122 Linnaean species.

In view of the thousand or so species traditionally requisite in distinguishing Lyellian subdivisions of the Tertiary, such small numbers of species hardly permit an assured Lyellian age determination of these microfaunas. Still, the change in percentage of living species is progressive and assists in relative age determination of bulk assemblages, which are then analyzed for age-diagnostic species.

Table 3. Percent of Foraminifera Still Living, Humboldt Basin

Assemblage in	*Total Foraminifera Species*	*Percent Still Living*
Hookton Formation	21	100
Scotia Bluffs - Carlotta	nil	–
Upper Rio Dell member	54	93
Lower-middle Rio Dell	72	79
Eel River Formation	115	76
Middle-upper Pullen Fm.	91	63
Upper Bear River beds[1]	26	38
Lower Bear River beds	47	36

[1] Includes lower Pullen of Price Creek section.

Faunas in Bear River Beds

Foraminiferal faunas from the lower part of the Bear River beds indicate an association of somewhat older Miocene age by the presence of such species as *Uvigerinella californica* var., *Buccella oregonensis, Lenticulina (Robulus) nikobarensis, Nonion costiferum,* and *Plectofrondicularia miocenica.*

Earlier workers (Cushman et al, 1930; Hughes, unpub.) reported *Siphogenerina branneri* (range upper Saucesian through Relizian into lower Luisian; Kleinpell, 1938,

p. 299) as an important local component of the Tertiary beds of Bear River valley. Cushman et al (1930, p. 70, pl. 5, fig. 13) and Kleinpell (1938, p. 301) indicated that *Siphogenerina hughesi* occurs in the lowermost strata of the south flank of the Bear River coastal syncline in association with *S. branneri*, a fact which suggests that the marine Miocene transgression of these beds in that area was early Relizian.

The upper part of the Bear River beds contains *Anomalina* cf. *A. salinasensis, Baggina californica, Bolivina advena* var. *striatella*, and other species (see Fig. 7), which suggests a correlation (see Kleinpell, 1938, p. 137) with the middle Miocene upper Relizian and/or Luisian Stage of southern and central California. Martin's (1916) commonly cited "*Pecten propatulus*" locality (U. C. locality 1865 and equivalent to D-1236 herein) contains *Nonion costiferum, Elphidium granti*, and other foram species and the mollusk *Pecten* sp. indet. (see Fig. 8) of Miocene (possibly as old as Relizian) age. The mollusks in the Bear River beds are relatively abundant at many localities (see Appendix A) and need a thorough restudy. *Pecten propatulus, sensu lato*, occurs with or just below forams of Relizian and Luisian age in the Vallecitos (Kleinpell, 1938, p. 119, 124, 258) and Berkeley Hills area of central California (Lutz, 1951, p. 383-387).

Ogle recorded a locality with Mohnian foraminifera from the Bear River coast, on the south flank of the syncline (Ogle, 1953, p. 47; his locality BA02, 1951, pl. IV). In the lowermost "Pullen formation" on Price Creek, Ogle (1953, p. 46-48, pl. 4) recognized a foraminiferal sequence of Mohnian age (late Miocene). To judge from the species Ogle listed (1953) from his four Mohnian samples from Price Creek, the entire stage from the *Bolivina modeloensis* Zone, below, to the "*Bolivina hughesi* Zone" (subdivided into *Bolivina benedictensis* Zone, below, and *Bolivina granti* Zone by Pierce, 1956), above, is represented. Ogle (p. 47) noted that he found no foraminiferal samples of Delmontian age, but that D. D. Hughes had reported *Bolivina obliqua*, a species restricted to that stage, from this section.

Another Delmontian species, "*Rotalia*" *garveyensis*, has been reported from the lower part of the section in Price Creek (L. J. Simon, personal commun.), from fault blocks at the base of the Centerville coastal section (J. C. Ingle, personal commun.), and from subsurface sequences in the Humboldt basin. Unfortunately, this critical part of the section in the two outcrop sections is no longer well exposed (see Fig. 5).

Thus, at Price Creek, the Bear River beds, together with the lowermost few hundred feet (less than 300 ft [90 m] stratigraphically) of the Pullen Formation, span, in age, the Coast Range Miocene interval from the Relizian Stage below, through the Luisian and Mohnian, to the Delmontian Stage above. The additional samples collected for study here serve to corroborate independently Ogle's conclusions as to the Miocene geologic sequence of events in this region.

Aggregate in Middle and Upper Pullen Formation

Samples in this study include those from the type Eel River sequence, those from stratigraphically above the Miocene beds in Price Creek, samples from the Pullen Formation of the Centerville coast, and samples from the Pullen(?) Formation of the Bear River syncline (see Fig. 5). In general, samples are dominated by abundant occurrences of radiolarians and diatoms; possibly because of rapid diagenesis of the silica in the organic state, recovery of the associated foraminifera in many, but not all, samples is poor. Those samples in which foraminiferal preservation is relatively good contain characteristic species such as *Cyclamina cancellata, Martinottiella communis, Uvigerina hootsi, Siphonodosaria insecta,* and locally *Karrierella milleri* and *Liebusella pliocenica.*

The unit may be partially equivalent to the upper Delmontian Stage (*s.l.*) or to the lower Repettian of southern California. At the present stage of investigation, the congregation includes *Karrierella milleri, Liebusella pliocenica,* and *Siphonodosaria insecta*, whose published ranges indicate that the congregation would be relegated to the lowermost Repettian.

Aggregate in Eel River Formation

The stratigraphically next-highest faunal aggregate contains many species in common with Natland's upper "Repettian Stage" of southern California. Characteristic species are *Bolivinita quadrilatera, Bulimina fossa, Bulimina subacuminata, Cibicides floridanus, Ehrenbergina compressa, Gyroidina soldanii* var. *altiformis, Gyroidina soldanii* var. *rotundimargo, Lagena melo, Lenticulina (Robulus) cushmani, Melonis pompilioides, Nodosaria arundinea, Nodosaria tosta, Oridorsalis umbonatus, Planulina wuellerstorfi, Plectofrondicularia californica, Stilostomella lepidula,* and *Uvigerina peregrina* var. *dirupta*.

The congregation is interpreted to include *Bolivinita quadrilatera, Bulimina fossa, Plectofrondicularia californica, Stilostomella lepidula,* and *Uvigerina peregrina* var. *dirupta*. Of these, *Plectofrondicularia californica* makes its last stand here as well as in Natland's "Repettian" of the type section and throughout southern California. The others appear apparently for the first time in the faunal column. In the California Coast Ranges, *Bolivinita quadrilatera* is not known from either above or below the range of the congregation; however, *B. quadrilatera* was part of a complex of species invading California from the Indo-Pacific.

Aggregate in Lower and Middle Rio Dell Formation

In the Centerville coastal and Eel River sections, the superpositional relation of the Rio Dell Formation to the underlying Eel River Formation is well exposed.

The aggregate in the lower and middle members of the Rio Dell Formation contains characteristic species such as *Bolivina spissa, Bulimina pagoda* vars., *Cassidulina californica, Cassidulina translucens* vars., *Cassidulinoides bradyi*, and *Uvigerina peregrina* var. *foxenensis*.

At these horizons, age-diagnostic foram congregations are becoming very poorly defined and are masked by abundant occurrences of such forams as *Epistominella pacifica* and *Uvigerina peregrina* vars. of more medium-depth biofacies.

The foram aggregate may be considered, in part at least, a correlative of Natland's "Venturian and Wheelerian Stages" on the basis of species in common. *Bulimina subacuminata*, a so-called guide species to the "Venturian" or "Repetto-Pico transition," represents a twofold problem in nomenclature (see Systematics) and in stratigraphic range, as it occurs abundantly in the underlying Repettian of both northern and southern California (see Natland and Kuenen, 1951, Fig. 6). Natland's (1957, pls. 3, 4, p. 567) Wheelerian guide fossils *Bolivina spissa, B. interjuncta, Epistominella pacifica,* and *Uvigerina peregrina* are present in the aggregate and similarly range into overlying and underlying assemblages.

However, because *Cassidulina cushmani, Cassidulina limbata s.s.,* and *Cassidulinoides bradyi* enter the local column here, in association with *Bulimina subacuminata s.s.* and *Uvigerina peregrina* var. *foxenensis*, a Venturian congregation is indicated. *Cassidulina translucens* var. *natlandi* Haller (see also Natland 1952, pl. 17, figs. 9-10) also does not appear to range below the congregation. "*Elphidium acutum*" of Natland (1952, pl. 7, fig. 5) appears to be characteristic of the upper part of the aggregate and may be restricted to it.

Aggregate in Upper Rio Dell Formation

Field relations in the Centerville coast and the Scotia-Eel River sections illustrate the superposition between this and underlying Pliocene assemblages.

The faunas involved are typically of relatively shallow-water origin and are characterized by *Cassidulina limbata s.s.* (not *C. limbata* of Kleinpell, 1938, pl. ix, fig. 21), *Buccella tenerrima, Elphidium foraminosum, Elphidiella hannai, Elphidiella oregonensis,* and *Gaudryina pliocenica.* Other forams, such as *Epistominella pacifica* and *Uvigerina peregrina* vars., are common in this as well as in the underlying aggregate.

Cassidulina limbata, a foram which Natland cited as a guide fossil to his Pleistocene "Hallian Stage" (1957, pl. 3-4), occurs in abundance in the upper member of the Rio Dell, but occurs also in the lower and middle members (i.e., Pliocene). Paratypes of this species apparently are not conspecific with *C. limbata* in Kleinpell (1938, pl. ix, fig. 21, Miocene), nor does *C. limbata* occur in favorable facies of Miocene age in the Humboldt basin.

Another "Hallian" foraminiferal guide fossil, *C. tortuosa*, occurs very rarely and in only a few samples of the present study, all of which are from the Wildcat Group undifferentiated, apparently of Pliocene age.

The foram *Elphidiella oregonensis*, a cool-water species, does not occur below the upper Rio Dell Formation; *E. hannai* ranges through the upper Rio Dell and, locally at least, does not occur below it (a few specimens in the uppermost part of the middle Rio Dell were questionably referred to *E. hannai* and may be ancestral). *Elphidiella oregonensis* usually occurs in association with *E. hannai* and the mollusk *Psephidia lordi*. This unique tripartite association was also found in the upper Merced of central California (see Glen, 1959; Stewart and Stewart, 1933) and in the Elk River beds of Oregon (Bandy, 1950).

In the two key sections–the Centerville coast and Scotia - Eel River–the pelecypods *Pecten caurinus, Psephidia lordi, Compsomyax subdiaphana* subsp. *gibbosus*, and *Thracia trapezoides* are locally abundant, and they apparently do not occur below the upper Rio Dell Formation in the Humboldt basin (see Faustman, 1964, Fig. 7).

However, *Pecten caurinus* and *Thracia trapezoides* both apparently range geologically well below and above the Pleistocene–*P. caurinus* from early Pliocene to Holocene (J. Smith, 1919, p. 130) and *T. trapezoides* in the Oligocene (J. Smith, 1919, p. 127). They are "cool-water species" characterizing the upper Santa Barbara Formation of the Ventura basin (Woodring, 1952, Fig. 1; 1957b, p. 592), a formation which is considered by most workers to be of early Pleistocene age.

Assemblage in Hookton Formation

Although Ogle (1953, p. 57) recorded a ". . . few *Elphidium* in mudstone members of . . . the Scotia Bluffs sediments . . . ," no forams were recovered by the writer from either the Scotia Bluffs or the Carlotta Formation (here revised). Consequently, the assemblage which is next-highest stratigraphically occurs in the youngest named formation, the Hookton.

The Hookton assemblage consists of only 21 foram species collected from three localities. Those species which are characteristic of the assemblage and not of the local, lower and older assemblages are *"Rotalia" beccarii, Elphidium poeyanum, Cassidulinoides cornuta*, and *Globorotalia inflata. Ostrea lurida* is a locally abundant molluscan species; other mollusks are abundant at locality D-1333 and have not yet been studied.

Both benthonic forams and mollusks indicate that most of the sediments from the limited outcrops were deposited in shallow-water habitats, perhaps in part even lagoonal (locs. D-1330, D-1329, D-1328, and D-1331), although locality D-1332 contains planktonic forams which indicate an open ocean environment.

As to the time at which the deposition of the Hookton Formation took place, the fact that 100% of the species are still living suggests a near-recent time. Furthermore, *Ostrea lurida* has been recorded as occurring abundantly in "Recent" and "near-Recent" sediments in San Francisco Bay; specific samples from San Francisco Bay were dated by radioactive analyses (see Story et al, 1966).

Faunas in Wildcat Group Undifferentiated

In those areas in the northeastern and southeastern part of the Humboldt basin (Elk River valley, Freshwater Creek, Larabee syncline, Salmon Creek; see Fig. 6 for sample locations) where superpositional control is lacking, dating of isolated outcrops on the basis of diagnostic species is possible in some, but not all, cases. Such species as *Plectofrondicularia californica* are occasionally present, indicating that the Repettian (lower Pliocene) as well as younger beds are represented at Freshwater Creek, Salmon Creek, and Elk River. At Larabee Creek, the only assemblage with abundant calcareous imperforate forams is present, but these forams add little to the knowledge of the age of the deposits that was not previously apparent from their areal geologic context. In many of the samples from the Wildcat Group undifferentiated, diagnostic species are absent. Most of the foraminiferal faunas in these areas are of apparent Pliocene age in general, as distinguished from Miocene or Pleistocene herin, and of relatively intermidiate- or shallow-water biofacies, or a combination thereof. A vast majority of localities in these areas contain common or abundant pelecypods and gastropods; these mollusks should furnish a clue to more precise dating when they are thoroughly studied.

Paleoecology

The bases for the present paleoecologic interpretations are three concepts which, although not necessarily original, are outlined in some detail and supported by a wealth of factual evidence in the papers cited. As many of the species studied here are still living, much importance was attached to the empirical distribution of species of Holocene foraminifera found along the West Coast (Anderson, 1963; Bandy, 1953a, b; Bradshaw, 1959; Crouch, 1952; Cushman and McCulloch, 1939, 1942a, b, 1948, 1950; Enbysk, 1960; Lalicker and McCulloch, 1940; Lankford, 1962; Lankford and Phleger, 1973; Lipps, 1965; Loeblich and Tappan, 1953; Lutze, 1962; Natland, 1938, 1957; Resig, 1958, 1962, 1964; Sliter, 1970; P. B. Smith, 1964; Uchio, 1960; and others). The published empirical data were used to support the inference of depth ranges shown in the right-hand column of Figure 12. Most of these more recent papers served merely to embellish the principles of geographic and depth distribution of planktonic and benthonic forams outlined by Cushman (1927), Norton (1930), and Natland (1933).

The reader is referred to Bandy's analysis of shell form in foraminifera as a key to environmental interpretations (Bandy, 1960a). This concept seems to have particular application to various species of cassidulinids recorded in the present paper. For purposes of communication, Bandy's generalized environmental classification is used (Table 4).

Bandy and Arnal's (1957, 1960) work on the West Coast and Walton's conclusions from studies of Gulf Coast foraminiferal ecology apply to assemblages in the Humboldt basin Pliocene and Pleistocene (Walton, 1964, p. 235):

1. "Faunal variability is inversely proportional to the variability of the environment .... numbers of species increase in an offshore direction, reach a maximum near the shelf edge, and decrease again deeper than 100 fathoms."
2. "Faunal dominance is a maximum in the intertidal zone marshes and decreases offshore to a minimum off the edge of the continental shelf."

In the following discussion, individual faunules are, in general, not distinguished from the whole assemblage; as such, environmental fluctuations within any given assemblage are noticeable mainly in the assemblages of the middle and upper members of the Rio Dell Formation.

A tabulation of the pertinent information from Figure 12 appears in Table 5. As seen from these data, the Centerville coastal section illustrates Walton's rules: the maximum number of foram species occurs in sediments from the Eel River Formation, whose environment was at the shelf edge. The samples from the Eel River section, which is 14 mi (22.4 km) nearer to the basin margin, reflect their relative geographic position by the lesser number of species; the anomalous situation with regard to the

Table 4. Classification of Marine Environments[1]

Bay
Shelf
inner shelf: 0 to 50 m
central and outer shelf: 50 to 150 m
Bathyal
upper bathyal: 150 to 610 m
middle bathyal: 610 to 2,438 m
lower bathyal: 2,438 to 4,000 m
Abyssal

[1] Depths vary from coast to coast, depending on bottom topography.

Table 5. Numbers of Foraminifera Species, Humboldt Basin

Formation	*Average (Maximum) Number of Foraminifera Species*			*Gross Environment*
	Centerville Section	*Eel River Section*	*Fields Landing*	
Hookton	–	–	19(19) 6(7)	marine and brackish
Scotia Bluffs-Carlotta	–	nil	–	inner shelf or nonmarine
Upper Rio Dell	17(26)	9(17)	–	central and inner shelf
Middle-lower Rio Dell	15(29)	25(39)	–	outer shelf and upper bathyal
Eel River	39(44)	24(42)	–	upper and middle bathyal
Pullen	32(46)	10(22)	–	middle bathyal

lower and middle members of the Rio Dell Formation indicates either inadequate sampling in the Centerville section or a mixing of ecologic niches in those samples that were collected from the Eel River section. Complications arise when attempts are made to interpret individual samples for the paleoenvironmental and paleoclimatic significance. The most important of these complications are: (a) displacement of faunas by turbidites and other physical means, and (b) shifting of environmental zones by upwelling currents, especially when considering the north-south alignment of basins along the northern West Coast (Jeletzky, 1973, p. 346).

A tabulation of molluscan species by Faustman (1964, Fig. 7) is shown in Table 6.

Comparison of Tables 5 and 6 upholds the rule that in marine formations the number of mollusk species shows an inverse relation to the number of foram species.

Faunules in Bear River Beds

The limited number of samples from the Miocene of Bear River, scattered over a long geologic time interval, precludes any overt discussion as to the paleoecology of the faunas. The samples uncovered all contained marine foram faunules whose depth distribution ranged probably from middle bathyal to inner shelf. With regard to the siphogenerids, Kleinpell (1938, p. 14) stated, ". . . the occasional abundance seems to indicate periods of warmest temperatures." More precise interpretations are hindered by the fact that the majority of species are no longer living and consequently the principle must be employed that homeomorphs had the same ecology that their Holocene counterparts have; as is readily apparent, interpretations become increasingly independent of this factor in the following assemblages.

Assemblages in Pullen Formation

The cyclamminids and the variety of other arenaceous forams which dominate most of the faunules indicate probable bathyal environments. Bottom waters, possibly stagnant, do not appear to have been favorable for many types of calcareous forams, but the occurrence of *Uvigerina hispida*(?) and *Bulimina subacuminata* in several faunules suggests middle bathyal depths (Bandy, 1967). Rare occurrences of *Melonis barleeanus* and *M. pompilioides* in the Centerville coastal section uphold that interpretation. The wealth of diatoms and radiolarians indicates surface waters highly populated by these organisms.

Assemblages in Eel River Formation

These assemblages show the maximum foraminiferal faunal diversity of any of

Table 6. Numbers of Molluscan Species, Humboldt Basin

	Average Number of Pelecypod/Gastropod Species	
Formation	*Centerville Section*	*Eel River Section*
Scotia Bluffs	–	8/11
Upper Rio Dell	19/25	34/28
Lower to middle Rio Dell	3/9	14/16
Eel River	2/2	2/0
Pullen	2/0	8/3

those encountered; that is, in any given sample, there are many species, but relatively few dominant ones. This condition characterizes faunas in all climates, and it is interesting to note that, in spite of the cool bottom temperatures indicated by the bathymetry, this fauna of maximum diversity is the one in which *Bolivinita quadrilatera* and a number of Schwager's lagenids (see McDonald, 1930) have been introduced from the Indo-Pacific.

Applying Bandy's rules to the cassidulinids, *Cassidulina californica* corresponds to a large, globose form indicative of outer-shelf - upper-bathyal environments, and *C. translucens* is a relatively small form with a sharp edge implying the same environment. However, the numbers of *Bulimina subacuminata, Melonis pompilioides,* and *Oridorsalis umbonatus* in the samples from the Centerville coastal section would suggest an upper- or middle-bathyal environment for that section.

Assemblages in Lower and Middle Rio Dell Formation

The dominant elements in these assemblages are *Epistominella pacifica* and *Uvigerina peregrina*; both occur in the Holocene and indicate an upper-bathyal to central-shelf environment. The coastal section contains a wealth of *Cassidulina translucens* vars., which suggests generally greater depth in relation to the Eel River section where *Cassidulina limbata* and *Buccella tenerrima* are present in abundance. The last two named species occur in the Holocene, mainly in inner-shelf environments, but *B. tenerrima* may also be found in considerable numbers in central-shelf environments. In the Humboldt basin, there appears to have been a strong lateral environmental distinction in the east-west trend at this time (see comments by Faustman, 1964, p. 104, regarding the pelecypod-gastropod ratio; Ogle, 1953, p. 48), a trend which continued through deposition of the upper Rio Dell.

Assemblages in Upper Rio Dell Formation

The cool-water nature of the molluscan and foraminiferal faunas has been discussed. At shallower depths, the uvigerinids are replaced by *Buccella tenerrima* (inner and central shelf, northern latitudes of the Holocene; Lankford, 1962, Fig. 6) as a dominant species, although the uvigerinids remain fairly common in the coastal section. *Epistominella pacifica* (upper-bathyal to central-shelf) is an important component in some faunules; here, its occurrence in a dominatly sandy section together with *Buccella, Cassidulina,* and especially the elphidiids would indicate the environment as being in the shallower range. Perhaps upwelling was an influencing factor at this time. *Cassidulina limbata*, which is representative of Bandy's "large, limbate cassidulinids" of the "inner shelf," is an important species in both sections. The elphidiids (*Elphidium foraminosum, Elphidiella hannai*) become more numerous. Walton's rule regarding faunal dominance applies to the *E. hannai* faunule whereby a few species, occurring in abundance, tend to dominate inner-shelf environments. Thus, at least six foraminiferal faunules may be distinguished in this assemblage–uvigerinid, rotalid (two), cassidulinid, and elphidiid (two). Each faunule occupies an environmental niche which overlaps others only in part. The total evidence argues for a general, but non-static, inner- to central-shelf environment for the eastern section, and a generally slightly deeper environment for the western section.

Assemblages in Scotia Bluffs and Carlotta Formations

The type sections apparently contain no forams other than the few elphidiids mentioned by Ogle (1953, p. 57). Only in the lower part of the Scotia Bluffs is there a molluscan fauna as well as an echinoid fauna. The coarse-grained sandstones, massive conglomerates, and large plant fragments support an analysis of shallow-water environments giving way to nonmarine deposits.

Assemblages in Hookton Formation

The assemblages are poorly known and probably involve a variety of environments. Nonmarine sediments occur, as do marine sediments with mollusks and forams that indicate brackish, marginal-marine, and open-ocean conditions. *Ostrea lurida* in association with "*Rotalia*" *beccarii* and *Elphidium* spp. suggests a brackish environment. Various species of mollusks are attributed to shelf environments, and abundant planktonic forams in at least one locality indicate temporary open-ocean conditions.

Faunules in Wildcat Group Undifferentiated

Beyond the regional ecologic diversity indicated by the Neogene faunas of the Humboldt basin, one interesting aspect of paleoecology derived from this study confirms certain conclusions given by Ogle (1953, p. 39, "Undifferentiated Wildcat"). This aspect concerns the thin veneer of Pliocene strata northeast of the Yager - Little Salmon Creek fault containing a basal glauconitic, boulder conglomerate (locs. D-1303, D-1309, D-1334) whereby a strong angular unconformity with underlying rocks of Mesozoic age is apparent. The boulder conglomerate is directly overlain by mudstones containing well-preserved early Pliocene foraminifera (i.e., an assemblage with abundant *Plectofrondicularia californica*). The indigenous foraminifera suggest a probable upper-bathyal environment with a mixture of outer-shelf species introduced by turbidity currents or by reworking. Locality D-1303 additionally contains several inner-shelf species, such as *Cibicides lobatulus* and *Rosalina columbiensis*, the adult forms of which live attached to kelp and other seaweeds (see Lankford, 1962, p. 78). The total evidence confirms Ogle's statement of "buried ridges" and further suggests the possible proximity of an offshore island adjacent to locality D-1303 during early Pliocene time.

For paleoecologic interpretations of the entire biostratigraphic column, reference is again made to the two key sections—the Centerville coastal and Scotia - Eel River sections. Indigenous assemblages from the Eel River section seem to be nearly intact, involving little reworking or contamination from turbidite sources. The physical evidence points to some occurrences of turbidites in the samples from bathyal environments. A small amount of stratigraphic section in the upper Rio Dell Formation shows massive slumping effects. Rock units and microfaunal assemblages from the Centerville coastal section show somewhat more evidence of displaced faunas and displaced sand bodies (Ingle, 1974, personal commun.), as might be expected from their apparent deposition in deeper waters. This trend should continue offshore where one might expect turbidity currents to have deposited reservoir beds comparable with those in the Los Angeles basin. The formations of Pliocene-Pleistocene age in the Ventura basin are also well known, not only because of extensive reworking of foraminiferal faunas in certain stratal intervals, but also for the many outstanding turbidite sequences containing displaced faunas (see Crowell et al, 1966; Natland, 1963; Natland and Kuenen, 1951).

Conclusions

The gross picture of the Humboldt basin suggests a deep-water basin in which late Miocene to early Pliocene sediments were deposited on highly faulted erosional remnants of older Miocene age. Ranging through the Pliocene and into the Pleistocene, the basin became shallower as infilling progressed. In the horizontal dimension, the Humboldt basin appears to have been an embayment open to the west with a generally westward-regressing shoreline. Grain size of the sediments generally supports the foraminiferal and molluscan evidence—the older sediments are essentially mudstones, and the stratigraphic sections become progressively coarser until coarse-grained sandstones and conglomerates predominate in the upper and eastern portions. Mollusks are fewer in numbers in the older and more central portions of the basin and become abundant in the younger and eastern, presumably marginal, portions of the basin. Conversely, foraminifera are relatively abundant in the shaly sections and become less so in the more sandy or conglomeratic basin-margin portions.

Six foraminiferal aggregates can be distinguished in the sediments from the Humboldt basin by reference to the following principles: (a) superpositional relation of assemblages and aggregates; (b) relative percentage of Holocene foraminifera species reflecting evolutionary development of aggregates; and (c) overlapping ranges of species, some of which are mutally exclusive (Oppelian zonation).

The molluscan and foraminiferal assemblages in the upper member of the Rio Dell Formation are suggestive of a Pleistocene age, perhaps the early Pleistocene of Woodring (1952, Fig. 1), on the basis of (a) superpositional relation to underlying Pliocene foram assemblages in two key sections; (b) relative percentage of Holocene foram species; (c) the molluscan species in common with those in the "Santa Barbara formation" of Woodring (1952); (d) the foram species, as well as molluscan species, in common with Natland's "Hallian Stage" of southern California, both groups of which reflect the influx of northern, cool-water species, although this may have occurred slightly earlier in northern California; and (e) the occurrence in the upper member of the Rio Dell of *Elphidiella oregonensis*, a species which apparently is restricted to the Pleistocene.

The writer prefers to place the Pliocene-Pleistocene boundary about 4,000 ft (1,220 m) stratigraphically lower than where Ogle (1953) and Faustman (1964) placed it. The transition is tentatively placed in the lowermost part of the upper Rio Dell member or at the approximate lowest occurrences, in the Humboldt basin, of the pelecypods *Pecten caurinus* and *Thracia trapezoides*.

Several foraminiferal lineages, notably the cassidulinids and uvigerinids, occur in abundance across the Pliocene-Pleistocene boundary, but the bolivinids are regarded as the most likely group for future studies on evolutionary development (Fig. 10, Appendix D). Well data from offshore portions of the basin are needed to provide better paleontologic continua.

Foraminifera of Humboldt County, California

Systematic Catalog

Type specimens of the following foraminifera are on deposit in the paleontologic collections of the University of California, Berkeley, and all types designated in the discussion are from outcrop samples collected by the writer from Humboldt County, California. Each fossiliferous sample is also represented by a microfaunal assemblage slide, also in the museum.

Through the courtesy of A. M. Keen and the late J. J. Graham, the writer was permitted to examine type specimens of cassidulinids on deposit at Stanford University.

J. H. van Voorthuysen of the Netherlands Geological Survey, Haarlem, permitted the writer to examine his collections of Neogene elphidiids.

I am also indebted to my colleagues, especially G. R. Hornaday and D. C. Steinker, at the University of California, Berkeley, for facilitating the examination of the extensive foraminiferal collections already in the university museum, and for discussion of certain taxonomic problems.

Ruth Todd and Dick Cifelli at the U.S. National Museum (now Museum of Natural History) kindly permitted the writer free access to the vast collections of primary and secondary types in the Cushman collection and otherwise assisted in locating pertinent reference material.

Systematic arrangement of the taxonomic notes generally follows that of Cushman (1950) with modifications at the generic level adapted from several sources. A determined attempt to synonymize all species cited in the literature proved futile; in many cases, only the original reference and primary types of the species are cited.

Phylum PROTOZOA
Class SARCODINA Bütschli, 1882
Order FORAMINIFERA d'Orbigny, 1826
Family ASTRORHIZIDAE
Genus RHABDAMMINA M. Sars, 1869
Rhabdammina abyssorum W. B. Carpenter?
Plate I, figure 1

?Rhabdammina abyssorum W. B. Carpenter, 1869, p. 60 (Holocene, Norway); Cushman and McCulloch, 1939, p. 31, pl. i, figs. 2-3 (Holocene, Pacific).
Hypotype 47782, loc. D-1301, Pullen Formation.

This hypotype is a small fragment comparing in general with numerous secondary types of nearly complete specimens from the Holocene now located in the U.S. Natl. Museum.

Genus BATHYSIPHON M. Sars, 1872
Bathysiphon arenacea Cushman
Plate I, figure 2

Bathysiphon arenacea Cushman, 1927, p. 129, pls. 1, 2 (Holocene, West Coast North America).
Bathysiphon arenaria Cushman; Natland, 1952, pl. 1-1 (Repettian, southern California).
Hypotype 47610, loc. D-1301, Pullen Formation.

This hypotype appears to be conspecific with the holotype and also similar to a paratype of *B. eocenica* in the U.S. Natl. Museum.

Family SACCAMMINIDAE
Genus SACCAMMINA M. Sars, 1869
Saccammina sphaerica M. Sars
Plate I, figure 3

Saccammina sphaerica M. Sars, 1869, p. 250 (Holocene, Norway); Cushman, 1910, p. 39, text-figs. 33-36 (Holocene, Pacific).
Hypotype 47786, loc. D-1301, Pullen Formation.

Family REOPHACIDAE
Genus REOPHAX Montfort, 1808
Reophax? sp.

Scarce specimens apparently belonging in this genus occur at loc. D-1349, Pullen Formation.

Genus HORMOSINA H. B. Brady, 1879
Hormosina globulifera H. B. Brady
Plate I, figure 4

Hormosina globulifera H. B. Brady, 1879, p. 60, pl. iv, figs. 4-5 (Holocene, loc. not designated); Cushman, 1910, 94 text-figs., p. 136-137 (Holocene, Pacific).
Hypotype 47718, loc. D-1301, Pullen Formation.

Family AMMODISCIDAE
Genus AMMODISCUS Reuss, 1861
Ammodiscus? sp.

Ammodiscus sp. (*minutum*) Natland, 1952, pl. 1-3 (Repettian, southern California).
Hypotype 47614, loc. D-1348, Pullen Formation.

This moderately well-preserved species occurs at two localities in the Price Creek section in association with other arenaceous forms.

Family LITUOLIDAE
Genus ALVEOLOPHRAGMIUM Stschedrina, 1936
Alveolophragmium scitulum (H. B. Brady)
Plate I, figures 7a, b

Alveolophragmium scitulum (H. B. Brady); Barker, 1960, pl. xxxiv, figs. 11-13 (Holocene, North Atlantic).
Haplophragmium scitulum H. B. Brady, 1881, p. 50 (Holocene, North Atlantic).
Haplophragmoides scitulum (H. B. Brady); Natland, 1952, pl. 1-2 (Repettian, southern California).
Hypotype 47601, loc. D-1349, Pullen Formation; hypotype 47602, loc. D-1381, Eel River Formation.

This hypotype is comparable to the secondary type of Cushman's *Haplophragmoides scitulum* from the Holocene (USNM 10277a).

Genus AMMOBACULITES Cushman, 1910
Ammobaculites cf. *A. americanus* Cushman

Ammobaculites americanus Cushman?, 1910, p. 117, text-figs. 184-185 (Holocene, Pacific).
Ammobaculites americanus Cushman; Bandy and Rodolfo, 1964, p. 828, fig. 5B (Holocene, Peru-Chile Trench).
Hypotype 47603, loc. D-1342, Wildcat Group undifferentiated.

A large number of specimens occurring at loc. D-1342 and rarely elsewhere in the Humboldt basin are comparable with those cited by Bandy and Rodolfo and also with Cushman's "cotypes" in the U.S. Natl. Museum.

Genus CYCLAMMINA H. B. Brady, 1876
Cyclammina cancellata H. B. Brady
Plate I, figures 5a, b

Cyclammina cancellata H. B. Brady, 1879, p. 62 (Holocene, loc. not designated); Cushman, Stewart, and Stewart, 1930, p. 50 (Pliocene, Humboldt County, California); Akers, 1954, p. 132-152; Barker, 1960, p. 76, pl. xxxvii, figs. 8-16 (Holocene, Atlantic and Pacific).
Cyclammina constrictimargo R. E. and K. C. Stewart, 1930b, p. 62, pl. vi, fig. 1 (Pliocene, southern Calif.).
Hypotype 47666, loc. D-1348, Pullen Formation; hypotype 47667, loc. D-1301, Pullen Formation.

This is a common species in certain deep-water (bathyal) sediments of Holocene age; often it occurs in basins of stagnant circulation in association with other arenaceous species and to the exclusion of calcareous benthonic and planktonic forms (see comments by Akers, 1954). The species has a very long range extending well back into the Tertiary—Eocene or older.

Cyclammina constrictimargo is regarded by Natland (1952, p. 93) as a flattened or crushed *C. cancellata* and thus an invalid species. Examination of primary types of *C. constrictimargo* confirm Natland's suggestion.

Cyclammina sp.
Plate I, figures 6a, b

Hypotype 47668, loc. D-1387, Rio Dell Formation.

This hypotype may or may not represent a species distinct from *C. cancellata*; the outer surfaces of the specimens in question show signs of abrasion.

Family TEXTULARIIDAE
Genus TEXTULARIA Defrance, 1824
Textularia? *flintii* Cushman
Plate I, figures 10a, b

Textularia flintii Cushman, 1911, p. 21, text-fig. 36 (Holocene, Pacific); Cushman, Stewart, and Stewart, 1930, p. 50, pl. i, fig. 1 (Pliocene, Humboldt County, California).
Siphotextularia flintii (Cushman); L. Martin, 1952, p. 117, pl. xvii, fig. 1 (Pliocene, southern California); White, 1956, p. 246, pl. xxvii, fig. 1 (Pliocene, Orange County, California).
Hypotype 47797, loc. D-1303, Wildcat Group undifferentiated.

This hypotype is comparable to the plesiotype of Cushman, Stewart, and Stewart in the U.S. Natl. Museum.

The species has a peculiar distribution in the Humboldt basin in that it occurs commonly in some Pliocene sections and is absent in other contemporaneous sequences. It is a readily distinguishable species characterized by the siphon-type aperture and therefore may belong in the genus *Siphotextularia*.

Family VERNEUILINIDAE
Genus VERNEUILINA d'Orbigny, 1840
Verneuilina scabra (Williamson)?
Plate I, figures 11a, b

Verneuilina scabra (Williamson); Cushman, Stewart, and Stewart, 1930, p. 51, pl. i, fig. 4 (Pliocene, Humboldt County, California).
Hypotype 47810, loc. D-1383, Eel River Formation.

The species as figured by Cushman, Stewart, and Stewart does not readily compare with that of Williamson; furthermore, the locality cited by Cushman, Stewart, and Stewart seems to be erroneous, as nearly duplicate samples produced an assemblage which differs in age and ecology.

Genus GAUDRYINA d'Orbigny, 1839
Gaudryina atlantica (Bailey) var.
Plate I, figures 12a, b

Gaudryina atlantica (Bailey); Barker, 1960, p. 96, pl. xlvii, fig. 18 (Holocene, West Indies).
Hypotypes 47699, 47700, loc. D-1303, Wildcat Group undifferentiated.

The rather distinctive specimens of a *Gaudryina* were found at only two localities; both localities contain *Plectofrondicularia californica* and thus are Repettian in age. Specimens recovered in this study are apparently conspecific with *Gaudryina atlantica* var. *pacifica* of McGlasson (R. H. McGlasson, personal commun., Holocene, off California).

Gaudryina pliocenica Cushman, Stewart, and Stewart
Plate I, figures 13a, b

Gaudryina atlantica (Bailey) var. *pacifica* Cushman and McCulloch; Cushman and Gray, 1946b, p. 2, pl. i, fig. 3 (Pliocene, California).
Gaudryina pliocenica Cushman, Stewart, and Stewart, 1949, p. 150, pl. xvii, fig. 2 (Pliocene, Washington).
Gaudryina triangularis Cushman; Cushman, Stewart, and Stewart, 1930, p. 51, pl. i, fig. 2 (Pliocene, Humboldt County, California).
Hypotype 47698, loc. D-1248, upper Rio Dell Formation.

This hypotype compares with the primary types in the U.S. Natl. Museum; the holotype of *G. triangularis* in the U.S. Natl. Museum represents a different species. Secondary types in the Museum, labeled *G. trinitatensis*, from the Miocene of California, are closely related and may be ancestral.

G. pliocenica is relatively common in certain samples and, where the optimum environment was found, appears to range through the entire section of Miocene, Pliocene, and Pleistocene.

Family VALVULINIDAE
Genus EGGERELLA Cushman, 1933
Eggerella? *bradyi* (Cushman)
Plate II, figures 1a-c

Eggerella bradyi (Cushman); Crouch, 1952, p. 823, pl. i, fig. 1 (Holocene, off California); Barker, 1960, p. 96, pl. xlvii, figs. 4-7 (Holocene, Pacific and West Indies).
Verneuilina bradyi Cushman, 1911, p. 54, text-fig. 87.
Hypotype 47673, loc. D-1355, Eel River Formation.

This hypotype is comparable to the primary types in the U.S. Natl. Museum. Although there is some question about the genus involved, specimens from several samples in the Humboldt basin Miocene(?) and Pliocene seem to be conspecific with those figured by Crouch and by Barker. As represented here, the species is very rare.

Genus CRIBROGOËSELLA Cushman, 1935
Cribrogoësella pacifica Cushman and McCulloch
Plate II, figures 2a, b

Cribrogoësella pacifica Cushman and McCulloch, 1939, p. 99, pl. xii, figs. 10-12 (Holocene, Pacific).
Hypotype 47665, loc. D-1303, Wildcat Group undifferentiated.

Paratype 35830 in the U.S. Natl. Museum has a similar chamber arrangement and size and differs only in the fact that the test is composed of somewhat larger sand grains.

The Humboldt basin specimens were found in only two samples, the one mentioned above and another from the Bear River Miocene(?).

Genus KARRERIELLA Cushman, 1933
Karreriella milleri Natland
Plate II, figures 3a, b

Karreriella milleri Natland, 1938, p. 140, pl. iii, figs. 11-12 (lower Pliocene, Los Angeles basin, California); Natland, 1952, pl. 2-6 (Repettian and Delmontian, southern California).
Hypotype 47719, loc. D-1357, Eel River Formation; hypotype 47220, loc. D-1301, Pullen Formation.

These hypotypes are identical to the primary types in the U.S. Natl. Museum. In southern California, the species characterizes the middle part of the "Repettian Stage" but according to Natland (1952) occurs also in the Delmontian Stage. In the Humboldt basin, the species is common in certain samples not younger than Repettian age and occurs also in samples which may be not older than Delmontian age.

Genus MARTINOTTIELLA Cushman, 1933
Martinottiella communis (d'Orbigny)
Plate I, figure 9

Martinottiella communis (d'Orbigny); Loeblich et al, 1964, p. C282, figs. 188-10, 188-11 (Holocene, Italy).

Clavulina communis d'Orbigny, 1826, p. 268, text-fig. 4 (level and loc. not designated); Cushman, Stewart, and Stewart, 1930, p. 51, pl. i, figs. 5-6 (Pliocene, Humboldt County, California).
Listerella pallida (Cushman); Cushman and McCulloch, 1939, p. 100, pl. x, fig. 17 (Holocene, Pacific).
Hypotype 47740, loc. D-1350, Eel River Formation.

The species is relatively common, occurring in many samples from the Humboldt basin.

Genus LIEBUSELLA Cushman, 1933
Liebusella pliocenica (Natland)
Plate I, figure 8

Liebusella pliocenica (Natland); Natland, 1952, pl. 2-8, fig. 9 (Lower Repettian, southern California).
Hypotypes 47738, 47739, loc. D-1348, Pullen Formation.

This species, which Natland cited as a marker for the lower Repettian, occurs only in the above cited sample in the Humboldt basin.

Family MILIOLIDAE
Genus QUINQUELOCULINA d'Orbigny, 1826
Quinqueloculina akneriana d'Orbigny
Plate II, figures 4a-c

Quinqueloculina akneriana d'Orbigny; Galloway and Wissler, 1927b, p. 38, pl. vii, fig. 3 (Pleistocene, southern California); Cushman, Stewart, and Stewart, 1930, p. 52, pl. ii, figs. 1-2 (Pliocene, Humboldt County, California); Cushman and Gray, 1946b, p. 3, pl. i, fig. 7 (Pliocene, Timms Point, California); Resig, 1958 (Holocene, California).
Hypotype 47779, loc. D-1383, Eel River Formation.

Specimens are scarce in any given sample, but occur at several Pliocene and Pleistocene localities in the Humboldt basin.

Quinqueloculina elongata Natland
Plate II, figures 6a-c

Quinqueloculina elongata Natland, 1938, p. 141, pl. iv, fig. 5 (Holocene, California); Cushman and Gray, 1946b, p. 4, pl. i, fig. 19 (Pliocene, Timms Point, California).
Hypotype 47780, loc. D-1379, Pullen Formation.

As Thalmann (1960) noted, the name is preoccupied and a new name should be designated. This species, as are most of the other quinqueloculines from the Humboldt basin, is a heavy-shelled form and occurs in many isolated samples.

Quinqueloculina vulgaris d'Orbigny
Plate II, figures 5a-c

Quinqueloculina vulgaris d'Orbigny, 1826, p. 302, no. 33 (Holocene, loc. not designated).
Hypotype 47781, loc. D-1254, upper Rio Dell Formation.

It is difficult to separate this species from *Q. akneriana* and the two species may be the same. In general, *Q. vulgaris* seems to be consistently larger and it retains a more distinctly triangular shape in cross section.

Quinqueloculina spp.

Isolated specimens representing probably several species cannot be referred to any of the quinqueloculines cited above; description of new species seems inadvisable in view of the few specimens per sample.

Genus SPIROLOCULINA d'Orbigny, 1826
Spiroloculina depressa d'Orbigny
Plate II, figures 7a, b

Spiroloculina depressa d'Orbigny, 1826, p. 298 (Plaisantin, Italy; Holocene, Mediterranean).
Hypotype 47792, loc. D-1383, Eel River Formation.

The species occurs rarely; its distribution apparently is limited to Pliocene formations in the Humboldt basin.

Genus SIGMOILINA Schlumberger, 1887
Sigmoilina celata (Costa)
Plate II, figures 8a, b

Sigmoilina celata (Costa); Cushman, Stewart, and Stewart, 1930, p. 52, pl. ii, fig. 3 (Pliocene, Humboldt County, California).
Hypotype 47787, loc. D-1384, Eel River Formation.

The species is a very distinct, arenaceous miliolid, usually occurring in association with calcareous, deep-water forams. It may be closely related to, but not conspecific with, *Sigmoilina schlumbergeri* in Phleger, Parker and Peirson (1953, Holocene, Atlantic).

"*Sigmoilina*" *tenuis* (Czjzek)
Plate II, figures 9a, b

"*Sigmoilina*" *tenuis* (Czjzek); Barker, 1960, p. 20, pl. x, figs. 7, 8, 11 (Holocene, South Pacific).
Sigmoilina tenuis Cjceki; Natland, 1952, pl. 3-4 (Venturian, southern California).
Sigmoilina elliptica Galloway and Wissler, 1927b, p. 39, pl. vii, fig. 2 (Pleistocene, southern California).
Hypotype 47788, loc. D-1385, Eel River Formation.

All species mentioned in the synonymy appear to be conspecific from comparison of published figures, but as Barker noted, the species may more properly be placed in another genus.

Genus PYRGO Defrance, 1824
Pyrgo oblonga (d'Orbigny)
Plate III, figures 1a, b

Pyrgo oblonga (d'Orbigny); Phleger, Parker, and Peirson, 1953, p. 29, pl. v, figs. 25-26 (Holocene, North Atlantic).
Hypotype 47775, loc. D-1383, Eel River Formation.

Pyrgo rotalaria Loeblich and Tappan
Plate III, figures 2a, b

Pyrgo rotalaria Loeblich and Tappan, 1953, p. 47, pl. vi, figs. 5-6 (Holocene, Bering Sea).
Pyrgo murrhina (Schwager); Cushman and Gray, 1946b, p. 10 (Pliocene, southern California).
Hypotype 47776, loc. D-1383, Eel River Formation.

Pyrgo vespertilio (Schlumberger)
Plate III, figures 3a, b

Pyrgo vespertilio (Schlumberger); Barker, 1960, p. 4, pl. ii, fig. 8 (Holocene, North Atlantic); Akers and Dorman, 1964, p. 49, pl. iii, figs. 29-32 (Pleistocene, Gulf of Mexico).
Hypotype 47777, loc. D-1254, upper Rio Dell formation.

Pyrgo sp.?

Hypotype 47778, loc. D-1249, upper Rio Dell Formation.

Several specimens occurring at the locality cited apparently belong to a species not previously described; comparison with type specimens currently unavailable is needed to verify this conclusion.

Pyrgo spp.

Isolated specimens, not referable to any known species, occur in several samples as shown in Figures 7 and 8 (distribution charts).

Family TROCHAMMINIDAE

Genus TROCHAMMINA Parker and Jones, 1859

Trochammina? sp.

Hypotype 47798, loc. D-1348, Pullen Formation.

Only one occurrence of the species in question was found, this being in late Miocene or early Pliocene sediments of the Price Creek section in an assembly also rich in other arenaceous forams.

Family LAGENIDAE

Genus LENTICULINA Lamarck, 1804

Lenticulina calcar (Linné)
Plate III, figures 4a, b

Lenticulina calcar (Linné); Barker, 1960, p. 146, pl. lxx, figs. 9-12 (Holocene, Philippines).
Hypotype 47733, loc. D-1303, Wildcat Group undifferentiated.

Lenticulina cushmani (Galloway and Wissler)
Plate III, figures 5a, b

Robulus cushmani Galloway and Wissler, 1927b, p. 51, pl. vii, fig. 11 (Pleistocene, southern California); Cushman and Gray, 1946b, p. 11, pl. ii, fig. 10 (Pliocene, southern California); White, 1956, p. 247, pl. xxvii, fig. 8 (Pliocene, Orange County, California).
Hypotype 47734, loc. D-1384, Eel River Formation.

The species is common throughout the Pliocene in the Humboldt basin; it usually occurs in association with other lagenids.

Lenticulina nikobarensis (Schwager)
Plate III, figures 6a, b

Robulus nikobarensis (Schwager); Cushman, Stewart, and Stewart, 1930, p. 53, pl. ii, figs. 5, 7 ("Pliocene," Humboldt County, California).
Hypotype 47735, loc. D-1237, "Bear River beds."

This hypotype apparently is conspecific with that figured by Cushman, Stewart, and Stewart (1930), although there is some question that they are the same as the type figured by Schwager. The species occurs only in the Miocene of the Bear River area, not in the younger sediments of the Humboldt basin. The secondary types of Cushman, Stewart, and Stewart in the U.S. Natl. Museum show a wide range in variation and may represent more than one species.

Lenticulina polita (Schwager)
Plate III, figures 7a, b

Cristellaria polita Schwager, 1866, pl. vi, fig. 86 (upper Tertiary?, Kar Nicobar).
Hypotype 47736, loc. D-1303, Wildcat Group undifferentiated.

This species is one of many lagenids which evidently are conspecific with those described by Schwager (1866; see also McDonald, 1930) from Kar Nicobar.

Lenticulina sp.?
Plate III, figures 8a, b

Hypotype 47737, loc. D-1237, "Bear River beds."

The species occurs in the Miocene of Bear River in association with *L. nikobarensis*; it cannot at present be attributed to any known species.

Lenticulina spp.

Various specimens which cannot be attributed to species figured in the literature occur in the Miocene and Pliocene of the Humboldt basin. Poor preservation accounts in part for the lack of species recognition.

Genus DENTALINA d'Orbigny, 1826
Dentalina quadrulata Cushman and Laiming?
Plate III, figure 11

?*Dentalina quadrulata* Cushman and Laiming, 1931, p. 99, pl. x, fig. 13 (Miocene, southern California); Sullivan, 1962, p. 264, pl. ix, figs. 13-14 (Zemorrian, Santa Cruz County, California).
Hypotype 47669, loc. D-1231, Pullen(?) Formation.

A few specimens from the late Miocene or early Pliocene of the Humboldt basin are recorded; these specimens appear to be related to the type figured by Cushman and Laiming; however, the primary types in the U.S. Natl. Museum are more strongly costate, and they may be ancestral.

Dentalina soluta Reuss
Plate IV, figures 4a, b; 5a, b

Dentalina soluta Reuss; L. Martin, 1952, p. 119, pl. xvii, fig. 8 (Pliocene, Los Angeles basin, California).
Hypotype 47670, loc. D-1303, Wildcat Group undifferentiated; hypotype 47671, loc. D-1262, upper Rio Dell Formation.

D. soluta is a rather large though otherwise nondescript, uniserial lagenid having no external ornamentation; the specimens from the Humboldt basin compare with Martin's figured specimens from southern California, but it is doubtful if they are the same species as that described by Reuss from the Oligocene of Germany.

Dentalina spp.

Poorly preserved specimens lacking definitive ornamentation are not identifiable as to species; their occurrence is shown in Figures 7 and 8 (distribution charts).

Genus NODOSARIA Lamarck, 1812
Nodosaria arundinea Schwager

Nodosaria arundinea Schwager, 1866, p. 211, pl. v, figs. 43-45 (upper Tertiary?, Kar Nicobar); Natland, 1952, pl. 4-13, 4-14, 4-15 (Repettian, southern California).
Hypotype 47745, loc. D-1383, Eel River Formation.

Specimens are composed only of fragments of very elongate chambers; chamber partitions are known in some cases, but the proloculus and apertural area are not.

Nodosaria moniliformis Ehrenberg
Plate IV, figure 1

Nodosaria moniliformis Ehrenberg, 1872, pl. vi, fig. 11 (Holocene, eastern Pacific);

Natland, 1952, pl. 4-1 - 4-4 (Wheelerian, southern California).
Hypotype 47746, loc. D-1272, middle Rio Dell Formation.

The species is distinctive in being composed of 4 or 5 nearly globular chambers. Although the type is from the Holocene, the species in the Humboldt basin apparently is confined to the Pliocene.

Nodosaria raphanistrum (Linné)
Plate IV, figures 2a, b

Nodosaria raphanistrum (Linné); Natland, 1952, pl. 4-5 - 4-7 (Repettian, southern California).
Hypotype 47747, loc. D-1383, Eel River Formation.

The species is very rare, and usually occurs in association with other lagenids. The primary type is from the Holocene.

Nodosaria tornata Schwager
Plate IV, figures 3a, b

Nodosaria tornata Schwager, 1866, p. 223, pl. v, fig. 51 (upper Tertiary?, Kar Nicobar).
Hypotype 47750, loc. D-1383, Eel River Formation.

Nodosaria tosta Schwager
Plate III, figure 9

Nodosaria tosta Schwager, 1866, p. 219, pl. v, fig. 42 (upper Tertiary?, Kar Nicobar); Cushman, Stewart, and Stewart, 1930, p. 55, pl. iii, fig. 1 (Pliocene, Humboldt County California).
Hypotype 47749, loc. D-1303, Wildcat Group undifferentiated.

Many exceedingly well-preserved specimens were obtained from the locality cited; there appears to be little doubt that the specimens herein are conspecific with the type from Kar Nicobar figured by Schwager.

Nodosaria tympaniplectriformis Schwager
Plate III, figure 10

Nodosaria tympaniplectriformis Schwager, 1866, p. 223, pl. v, fig. 34 (upper Tertiary?, Kar Nicobar).
Nodosaria parexilis Cushman and Stewart; Cushman, Stewart, and Stewart, 1930, p. 55, pl. ii, figs. 13-15 ("Pliocene," Humboldt basin, California).
Hypotype 47751, loc. D-1303, Wildcat Group undifferentiated.

The primary types of *N. parexilis* in the U.S. Natl. Museum appear to be immature specimens of one of Schwager's species, probably *N. tympaniplectriformis.*

Nodosaria spinosa? (d'Orbigny)

Dentalina spinosa d'Orbigny; Kleinpell, 1938, p. 215, pl. iv, fig. 3 (Zemorrian, southern California).
Hypotype 47748, loc. D-1287, Eel River Formation.

The species is very distinctive because of its ornamentation. It is related to *Dentalina spinosa* d'Orbigny of Kleinpell (1938) and to Kleinpell's plesiotype 497171 in the U.S. Natl. Museum. In the Humboldt basin, the species apparently is confined to the Pliocene.

Genus LAGENA Walker and Jacob, 1798
Lagena alcocki White
Plate IV, figure 6

Lagena alcocki White, 1956, p. 246, pl. xxvii, fig. 7 (lower Pliocene, Orange County, California).

Lagena williamsoni (Alcock); Cushman, 1929, p. 70, pl. xi, figs. 7-8 (Pliocene, southern California); Cushman, Stewart, and Stewart, 1930, p. 59, pl. vii, fig. 5 (Pliocene, Humboldt County, California); Cushman and McCulloch, 1950, p. 362, pl. xlviii, figs. 14-15 (Holocene, Pacific); L. Martin, 1952, p. 122, pl. xviii, fig. 10 (Pliocene, southern California).
Hypotype 47721, loc. D-1389, Rio Dell Formation.

This species is one of the more common lagenids, occurring in numerous Pliocene samples from the Humboldt basin. Specimens from the Miocene of Bear River have a slightly different ornamentation pattern and appear to be the ancestral form, perhaps at the subspecies level.

Lagena cf. *L. amphora* Reuss
Plate IV, figure 7

Lagena amphora Reuss; Cushman and McCulloch, 1950, p. 329, pl. xliii, figs. 11-14 (Holocene Pacific).
Hypotype 47722, loc. D-1303, Wildcat Group undifferentiated.

Lagena clypeata (Sidebottom)?
Plate IV, figure 10

?*Lagena clypeata* (Sidebottom); Cushman, 1929, p. 72, pl. xi, fig. 19 (Pliocene, southern California).
Hypotype 47723, loc. D-1383, Eel River Formation.

Lagena foveolata Reuss
Plate IV, figure 8

Lagena foveolata Reuss, 1862, p. 332, pl. v, fig. 65 (Oligocene, Germany).
Hypotype 47724, loc. D-1303, Wildcat Group undifferentiated.

The species is very rare, occurring in only a few samples. Comparison with type specimens is needed to verify its identification.

Lagena melo (d'Orbigny)
Plate IV, figure 11

Oolina melo d'Orbigny, 1839, p. 20, pl. v, fig. 9 (Holocene, Falkland Islands).
Lagena scalariformis (Williamson); L. Martin, 1952, p. 121, pl. xviii, fig. 5 (Pliocene, southern California).
Lagena foveolata Reuss; Cushman, Stewart, and Stewart, 1930, p. 58, pl. iii, fig. 11 ("Pliocene," Humboldt basin, California).
Hypotype 47725, loc. D-1383, Eel River Formation.

The Cushman, Stewart, and Stewart plesiotype in the U.S. Natl. Museum is conspecific with this hypotype.

Lagena pliocenica Cushman and Gray
Plate IV, figure 12

Lagena pliocenica Cushman and Gray, 1946a, p. 68, pl. xii, figs. 22-25 (Pliocene, southern California); Cushman and Todd, 1947, p. 11, pl. i, fig. 30 (Holocene, off Washington); Cushman & McCulloch, 1950, p. 344, pl. xlvi, figs. 6-8 (Holocene, Pacific).
Hypotype 47726, loc. D-1303, Wildcat Group undifferentiated.

This hypotype is comparable to the primary types in the U.S. Natl. Museum.

Lagena striata (d'Orbigny)
Plate IV, figure 9

Lagena striata (d'Orbigny); L. Martin, 1952, p. 121, pl. xviii, fig. 6 (Pliocene, Los Angeles County, California); Cushman and Gray, 1946b, p. 20, pl. iii, figs. 51-54 (Pliocene, southern California).

Lagena cf. *L. striata* (d'Orbigny); Cushman and McCulloch, 1950, p. 350, pl. xlvii, figs. 1-4 (Holocene, Pacific).
Hypotype 47727, loc. D-1303, Wildcat Group undifferentiated.

Lagena striata (d'Orbigny) var. *haidingeri* (Czjzek)

Lagena striata (d'Orbigny) var. *haidingeri* (Czjzek); Cushman, 1913, p. 19, pl. vii, fig. 6 (Holocene, Pacific).
Hypotype 47728, loc. D-1384, Eel River Formation.

The varietal form is sufficiently distinct from *L. striata* so that it probably should be raised to specific rank.

Lagena striata (d'Orbigny) var. *strumosa* (Reuss)

Lagena striata (d'Orbigny) var. *strumosa* (Reuss); Cushman, 1913, p. 20, pl. vii, figs. 7-10 (Holocene, Pacific); Cushman, 1929, p. 70, pl. xi, fig. 6 (Pliocene, southern California).
Hypotype 47729, loc. D-1303, Wildcat Group undifferentiated.

Lagena sulcata (Walker and Jacob)

Lagena sulcata (Walker and Jacob); Bagg, 1912, p. 52, pl. xiv, figs. 9-12 (Pliocene, California); Cushman, Stewart, and Stewart, 1930, p. 58, pl. iii, fig. 12 (Pliocene, Humboldt County, California); Cushman and Gray, 1946b, p. 19, pl. iii, fig. 45 (Pliocene, southern California).
Hypotype 47730, loc. D-1388, Rio Dell Formation.

The type is from the Holocene; presumably the figures listed in the synonymy represent the same species.

Lagena sp.?

Hypotype 47731, loc. D-1389, Rio Dell Formation.

The type cited may be a new species; comparison with other types is needed to verify an identification.

Lagena spp.

Poorly preserved specimens, usually few in number in any given sample, occur at several localities as shown on Figures 7 and 8 (distribution charts).

Family POLYMORPHINIDAE
Genus GLANDULINA d'Orbigny, 1826
Glandulina comatula (Cushman)
Plate IV, figure 14

Glandulina comatula (Cushman); Cushman and McCulloch, 1950, p. 325, pl. xlii, figs. 5-7 (Holocene, Pacific).
Glandulina tenuistrata (Bermudez); L. Martin, 1952, p. 118, pl. xvii, fig. 5 (Pliocene, Los Angeles County, California).
Hypotype 47701, loc. D-1303, Wildcat Group undifferentiated; hypotype 47702, loc. D-1383, Eel River Formation.

The species is comparatively rare, occurring in a limited number of samples from the Humboldt basin Pliocene.

Glandulina laevigata (d'Orbigny)
Plate IV, figure 13

Glandulina laevigata (d'Orbigny); Cushman, Stewart, and Stewart, 1930, p. 56, pl. iii, fig. 4 (Pliocene, Humboldt County, California); White, 1956, p. 246, pl. xxvii, figs. 4-5 (Pliocene, Orange County, California).

Pseudoglandulina laevigata (d'Orbigny); Cushman, Stewart, and Stewart, 1949, p. 151, pl. xvii, fig. 4 (Pliocene, Washington); Cushman and McCulloch, 1950, p. 325, pl. xlii, fig. 4 (Holocene, Pacific).
Hypotype 47703, loc. D-1389, Rio Dell Formation; hypotype 47704, loc. D-1303, Wildcat Group undifferentiated.

The species is common in the Pliocene of the Humboldt basin, less so in the Pleistocene and Miocene(?).

Genus PSEUDOPOLYMORPHINA Cushman and Ozawa, 1928
Pseudopolymorphina ovalis Cushman and Ozawa
Plate IV, figure 16

Pseudopolymorphina ovalis Cushman and Ozawa, 1930, p. 103, pl. xxvii, fig. 1, pl. xxix, fig. 6 (Miocene, Austria); Barker, 1960, p. 150, pl. lxxii, figs. 7-8 (Holocene, West Indies).
Hypotype 47770, loc. D-1389, Rio Dell Formation.

The species is comparatively scarce in the Neogene of the Humboldt basin.

Genus POLYMORPHINA d'Orbigny, 1826
Polymorphina charlottensis Cushman
Plate IV, figure 15

Polymorphina charlottensis Cushman, 1925a, p. 41, pl. vi, fig. 9 (Holocene, British Columbia); Cushman and Todd, 1947, p. 12, pl. ii, fig. 11 (Holocene, off Washington); Cushman, Stewart, and Stewart, 1930, p. 59, pl. iv, fig. 6 (Pliocene, Humboldt County, California).
Hypotype 47769, loc. D-1391, Rio Dell Formation.

This hypotype, although poorly preserved, is comparable to the primary types in the U.S. Natl. Museum.

Genus FISSURINA Reuss, 1850
Fissurina lucida (Williamson)
Plate V, figures 2a, b

Fissurina lucida (Williamson); Bandy, 1950, p. 274, pl. xli, fig. 12 (Pleistocene, Cape Blanco, Oregon).
Entosolenia lucida Williamson; Cushman and Gray, 1946b, p. 30, pl. v, figs. 16-18 (Pliocene, California); Cushman and Todd, 1947, p. 20, pl. iii, fig. 11 (Holocene, off Washington).
Hypotype 47696, loc. D-1387, Rio Dell Formation.

Fissurina semimarginata (Reuss)
Plate V, figures 1a, b

Fissurina semimarginata (Reuss); Loeblich and Tappan, 1953, p. 78, pl. xiv, fig. 3 (Holocene, off Alaska).
Hypotype 47697, loc. D-1303, Wildcat Group undifferentiated.

Fissurina spp.

Poorly preserved specimens, not comparable to any known species, occur in isolated samples as shown in Figures 7 and 8 (distribution charts).

Family NONIONIDAE
Genus NONION Montfort, 1808
Nonion costiferum (Cushman)
Plate V, figures 3a, b

Nonion costiferum (Cushman); Kleinpell, 1938, p. 229, pl. xv, fig. 13 (Miocene, California).

Nonionina costifera Cushman, 1926b, p. 90, pl. xiii, fig. 2 (Miocene, California); Cushman, Stewart, and Stewart, 1930, p. 60, pl. iii, fig. 13 ("Pliocene," Humboldt County, California).
Florilus costiferus (Cushman); Loeblich and Tappan, 1964, p. C747, fig. 612-6a, b (Miocene, California).
Hypotype 47752, loc. D-1240, "Bear River beds."

This hypotype is comparable to the primary types in the U.S. Natl. Museum.

In Humboldt County, the species apparently does not occur above the Miocene. The Pliocene report by Cushman, Stewart, and Stewart (1930) is apparently an erroneous age determination for the locality in reference.

Genus NONIONELLA Cushman, 1926
Nonionella cushmani R. E. and K. C. Stewart
Plate V, figures 6a-c

Nonionella cushmani R. E. and K. C. Stewart, 1933, p. 264, pl. xvii, fig. 1 (upper Pliocene, California).
Hypotype 47753, loc. D-1405, Rio Dell Formation; hypotype 47754, loc. D-1419, Rio Dell Formation.

These hypotypes are closely related, if not synonymous, to the primary types in the U.S. Natl. Museum.

N. cushmani is a common species in the shallow-water biofacies of the Humboldt basin Pliocene and Pleistocene.

Nonionella miocenica Cushman
Plate V, figures 4a-c

Nonionella miocenica Cushman, 1926a, p. 64 (for *Nonionina auris* Cushman, not d'Orbigny, Cushman, 1925b, p. 91, pl. xiii, fig. 4, Miocene, California); Cushman and McCulloch, 1950, p. 161, pl. xviii, fig. 1 (Holocene, Pacific); Bandy, 1950, p. 275, pl. xli, fig. 6 (Pliocene, Cape Blanco, Oregon); White, 1956, p. 247, pl. xxvii, fig. 10 (Pliocene, Orange County, California).
Hypotype 47755, loc. D-1243, upper Rio Dell Formation.

This hypotype is comparable to the primary types in the U.S. Natl. Museum.

Nonionella spp.

Specifically indeterminate specimens occur in several samples from both the Scotia-Eel River section and the Centerville coastal section as shown on Figures 7 and 8 (distribution charts).

Genus ELPHIDIUM Montfort, 1808
Elphidium acutum Natland
Plate V, figures 7a, b

Elphidium acutum Natland, 1952, pl. 7-5 (Wheelerian, southern California).
Hypotype 47675, loc. D-1389, lower Rio Dell Formation; hypotype 47676, loc. D-1276, lower Rio Dell Formation.

The Humboldt Basin specimens are similar to the unpublished type figures in thesis by Natland; the species is a distinctive one, possibly having affinities with *Elphidium foraminosum*. In the Humboldt basin, it is limited in range to the Pliocene.

Elphidium foraminosum Cushman
Plate V, figures 8a, b

Elphidium hughesi Cushman and Grant var., 1927, p. 75, pl. vii, fig. 5 (Pliocene, California).

Elphidium hughesi Cushman and Grant var. *foraminosum* Cushman, 1939a, p. 49, pl. xiii, fig. 8 (refig. as holotype; Pliocene, California).
Hypotype 47677, loc. D-1405, Rio Dell Formation; hypotype 47678, loc. D-1419, Rio Dell Formation.

These hypotypes differ from the primary type in the U.S. Natl. Museum in having latter chambers with curved sutures.

E. foraminosum is here raised to specific status; it is radically different morphologically from *E. hughesi (s.s.)*, not only in external shape, but also in the shape of individual chambers. *E. foraminosum* has umbilici with large irregular bosses whereas *E. hughesi* does not.

E. foraminosum is an abundant species in the Pleistocene of the Humboldt basin and occurs less commonly in the Pliocene. As far as known, its stratigraphic range does not overlap with that of *E. hughesi* (see below); the two species may not even be closely related. *E. hughesi* var. *obesum* (Cushman, 1939a, p. 49, pl. xiii, fig. 9) may be a varient of *E. foraminosum* which, in turn, could be related to *Elphidiella hannai*; more investigation is needed to verify this.

Although not known definitely from the late Pleistocene of the Humboldt basin, *E. foraminosum* is probably synonymous in part with *E. tumidum* of various authors (Holocene of California).

Elphidium granti Kleinpell
Plate V, figure 5

Elphidium granti Kleinpell, 1938, p. 238, pl. xix, figs. 1, 11 (Mohnian, southern California).
Hypotype 47679, loc. D-1236, "Bear River beds."

A few specimens from the locality cited appear to be conspecific with the type figured by Kleinpell and with the holotype in the U.S. Natl. Museum; the species may be ancestral to *E. crispum* from the Pleistocene and Holocene of southern California, but no intermediate link from the California Pliocene has been recorded.

Elphidium hughesi Cushman and Grant

Elphidium hughesi Cushman and Grant, 1927, p. 75, pl. vii, fig. 1, not pl. vii, figs. 4-5 ("lower Pliocene or upper Miocene(?)" California); Cushman, Stewart, and Stewart, 1930, p. 61, pl. iii, fig. 15 ("Pliocene," Humboldt County, California); Kleinpell, 1938, p. 238, not figured (upper Delmontian, California); Cushman, 1939a, pl. xiii, fig. 7, refigured holotype.

In the U.S. Natl. Museum, several paratypes of *E. hughesi* exist, two of which were later designated by Cushman as *E. hughesi* var. *foraminosum* and *E. hughesi* var. *obesum*. The type locality of *Elphidium hughesi* probably represents a horizon of late Miocene age (*fide* Kleinpell, 1938, p. 25). The figure of *E. hughesi* shown by Cushman, Stewart, and Stewart (1930), presumably from the Humboldt basin (note conflict in localities between the plate and the species distribution chart), could not be matched with specimens in the present study. Moreover, in the Cushman, Stewart, and Stewart collection in the U.S. Natl. Museum there exist specimens, not designated as types, but labeled as "*E. hughesi*," including two species of which neither is properly referred to *E. hughesi*. These two species show radically different types of preservation and obviously involve specimens of differing ages, indicating that a mixing of samples has occurred; one of their species is in place and is here designated as *E. humboldtensis* n. sp. (see below).

Elphidium humboldtensis Haller n. sp.
Plate V, figures 9a, b

Elphidium hughesi Cushman and Grant; Cushman, Stewart, and Stewart, 1930, p. 61, *not* pl. iii, fig. 15 (USNM 13805, loc. 44; "Pliocene," Humboldt County, California).

Holotype 47680, loc. D-1240, "Bear River beds." Bear River valley, Humboldt County, California, Miocene.
Paratypes 47681, 47682, loc. D-1240, as above.

Test planispiral, bilaterally symmetrical, involute; nine chambers in outer whorl, expands horizontally as well as vertically in final whorl; retral processes present, poorly defined; walls calcareous, finely perforate; aperture consists of row of very fine pores at base of septal face; periphery and chambers slightly lobate, umbilic depressed, partly filled with thin layers of retral processes; sutures arched slightly backward, depressed–Diameter, 0.42 mm; Thickness, 0.18 mm.

Discussion: *E. humboldtensis* resembles *E. tumidum* Natland (Holocene, California) from which it differs in the lateral chamber expansion in the final whorl, fewer chambers in the final whorl, and apparently fewer pores in the aperture. Specimens of *E. humboldtensis* were obtained from D-1240 and other localities along Bear River, as well as D-1344 on Price Creek; all localities are herein considered to be of Miocene age. Loc. D-1240 is identical with Station 44 as described by Cushman, Stewart, and Stewart (1930, p. 49) and also with Martin's "*Pecten propatulus*" locality (his loc. 1865, see Fig. 2 herein).

Elphidium poeyanum (d'Orbigny)
Plate VI, figures 1a, b

Elphidium poeyanum (d'Orbigny); Cushman, 1939a, p. 54, pl. xiv, figs. 25-26 (Holocene, Florida); Natland, 1952, pl. 6-13 - 6-17 (Holocene, southern California).
Cribroelphidium poeyanum (d'Orbigny); Loeblich et al, 1964, p. C635, figs. 508-3, 508-4 (Holocene, Cuba).
Hypotype 47683, loc. D-1329, Hookton Formation.

Very abundant in some samples from the Hookton Formation; scarce in a few samples from the uppermost Rio Dell Formation.

The primary types of *E. translucens* Natland in the U.S. Natl. Museum appear to be very closely related and may be synonymous.

Elphidium? sp.

Hypotype 47811, loc. D-1236, "Bear River beds." Miocene.

A few specimens, generally resembling *Elphidiella hannai* in gross morphology, occur in the Miocene of Bear River and also in the lowermost part (Miocene) of the Price Creek section. Preservation is poor so that analysis of wall structure and apertural characteristics is difficult. The species can not be referred to any known species figured in the literature, although *Elphidium hughesi* var. *obesum* may be related.

Genus ELPHIDIELLA Cushman, 1936

Elphidiella hannai (Cushman and Grant)
Plate VI, figures 3a, b

Elphidiella hannai (Cushman and Grant); Cushman and Todd, 1947, p. 15, pl. ii, fig. 15 (Holocene, off Washington); Bandy, 1950, p. 276, pl. xli, fig. 10 (Pliocene and Pleistocene, Cape Blanco, Oregon); Natland, 1952, pl. 7-6 (upper Pleistocene and Holocene, southern California); Resig, 1964, p. 393-396, text-fig. 1 (Holocene, California).
Elphidium hannai Cushman and Grant, 1927, p. 77, pl. viii, fig. 1 (Holocene, California).
Hypotype 47684, loc. D-1256, upper Rio Dell Formation; hypotype 47685, loc. D-1329, Hookton Formation.

The species occurs abundantly in some samples from the upper Rio Dell and Hookton Formations, both Pleistocene, of the Humboldt basin; a few specimens from the uppermost Pliocene were recorded; these may be ancestral.

The species is probably synonymous with *E.* cf. *E. arctica* of the Pleistocene from the Netherlands (e.g., see van Voorthuysen, 1953), although these widely separated occurrences may involve two subspecies.

Elphidiella hannai (Cushman and Grant) var.
Plate VI, figures 4a, b

Hypotype 47686, loc. D-1419, Rio Dell Formation.

This variety seems to be gradational between an unknown ancestral species and the smooth-surfaced form of *E. hannai.* Phylogenetic relations of the elphidiids in general need much more study as the status of terms such as *Elphidium, Elphidiella,* and *Cribroelphidium* is in dispute (e.g., see Ujiie, 1956; Wade, 1957).

Elphidiella oregonensis (Cushman and Grant)
Plate VI, figures 2a, b

Elphidiella oregonensis (Cushman and Grant); Bandy, 1950, p. 277, pl. xli, fig. 13 (Pleistocene, Cape Blanco, Oregon); Loeblich and Tappan, 1953, p. 103, pl. xviii, figs. 1-3 (Holocene, Bering Sea).
Elphidium oregonense Cushman and Grant, 1927, p. 79, pl. vii, fig. 3 (Pleistocene, Oregon); Cushman, Stewart, and Stewart, 1930, pl. iv, figs. 1-2 ("Pliocene," Humboldt County, California); Stewart and Stewart, 1933, p. 266 ("Pliocene or Pleistocene," San Mateo County, California); Cushman, 1941, p. 34, pl. ix, figs. 7-9 ("Pliocene or Pleistocene," Alaska); Anderson, 1963, pl. i, figs. 1-5 (Holocene–living, Bering Sea).
Hypotype 47687, loc. D-1256, upper Rio Dell Formation; hypotype 47688, loc. D-1419, Rio Dell Formation.

A paratype in the U.S. Natl. Museum is comparable to the two hypotypes.

In North America, this species ranges geologically from early Pleistocene to Holocene; the Holocene occurrences are recorded from the Bering Sea only. Geographically, the known fossil record is from San Mateo County, California (upper Merced Formation, Pleistocene) northward to Alaska. Van Voorthuysen (1952) cited *E. oregonensis* as a "possible marker for the Amstelian (lower Pleistocene) in North America and Northwestern Europe." The writer has viewed van Voorthuysen's specimens from the Netherlands subsurface; his specimens appear, at least surficially, to be conspecific with specimens from northern California.

E. oregonensis occurs sporadically, though in abundance, in the upper Rio Dell Formation of the Humboldt basin and is here taken as indicative of a Pleistocene age, contrary to the earlier Pliocene age interpretation reported by Cushman, Stewart, and Stewart (1930). Specimens from the Humboldt basin obtain a very large size; the maximum dimension is approximately 2.2 mm.

A very poor illustration of *Cribroelphidium oregonense* (Cushman and Grant) in Ujiie (1956; Holocene, Japan) does not appear to be conspecific with the North American forms. *E. sibirica* (Goes) in Tappan (1951; Pleistocene, Alaska) seems to be closely related and is probably synonymous at the specific level.

Family PLECTOFRONDICULARIIDAE
Genus PLECTOFRONDICULARIA Liebus, 1903
Plectofrondicularia advena (Cushman)
Plate VI, figure 6

Plectofrondicularia advena (Cushman); Barker, 1960, p. 138, pl. lxvi, figs. 8-12 (Holocene, Pacific).
Frondicularia advena Cushman, 1923, p. 141, pl. xx, figs. 1-2 (Holocene, Atlantic); Galloway and Wissler, 1927b, p. 47, pl. viii, figs. 7-8 (Pleistocene, southern California); Cushman and Gray, 1946b, p. 15, pl. iii, fig. 16 (Pliocene, southern California); L. Martin, 1952, p. 117, pl. xvii, fig. 2 (Pliocene, Los Angeles County, California); White, 1956, p. 252, pl. xxx, fig. 2 (Pliocene, Orange County, California).

Hypotype 47762, loc. D-1266, middle Rio Dell Formation; hypotype 47763, loc. D-1272, middle Rio Dell Formation.

The above hypotypes differ from the primary types in the U.S. Natl. Museum in being broader across the chambers.

In the Humboldt basin, the species is common in the Pliocene and rare in the Pleistocene.

Plectofrondicularia advena (Cushman) var.
Plate V, figure 7

Hypotype 47764, loc. D-1287, Eel River Formation.

This varietal form differs from the normal one in being more compressed in side view and less tapered in plan view; the distinction is sufficient so that it may represent a different species; it somewhat resembles *Frondicularia foliacea* Schwager. The variety occurs in only a few samples from the Humboldt basin.

Plectofrondicularia californica Cushman and Stewart
Plate VI, figure 9

Plectofrondicularia californica Cushman and Stewart, 1926, p. 39, pl. vi, figs. 9-11 (Pliocene, California); Cushman, Stewart, and Stewart, 1930, p. 63, pl. iv, figs. 3-4 (Pliocene, Humboldt County, California); Cushman and Gray, 1946b, p. 27, pl. v, figs. 3-6 (Pliocene, southern California); L. Martin, 1952, p. 130, pl. xxii, figs. 6-7 (Pliocene, Los Angeles County, California); Natland, 1952, pl. 7-8 - 7-11 (Repettian, southern California); White, 1956, p. 252, pl. xxx, fig. 3 (Pliocene, Orange County, California).

Hypotype 47765, loc. D-1383, Eel River Formation; hypotype 47766, loc. D-1303, Wildcat Group undifferentiated.

The above hypotypes are identical to the primary types in the U.S. Natl. Museum.

The species is very distinctive morphologically. Its apparent age range is from the lower Zemorrian (Kleinpell, 1938, p. 239) through the lower Pliocene (Repettian Stage of Natland, 1952). An upper Eocene (Narizian) form from California appears to be very closely related and though not conspecific, at least ancestral (G. R. Hornaday, personal commun.). The species has often been cited as especially characteristic of the upper Repettian Stage and age, a time interval in which it appears to have made its last stand.

Plectofrondicularia miocenica Cushman
Plate VI, figure 8

Plectofrondicularia miocenica Cushman, 1926a, p. 58, pl. vii, figs. 10-11, pl. viii, figs. 11-12 (Miocene, San Luis Obispo County, California); Cushman, Stewart, and Stewart, 1930, pl. iv, fig. 9 (Miocene, Humboldt County, California).

Hypotype 47767, loc. D-1238, "Bear River beds."

The above hypotype is comparable to the primary type in the U.S. Natl. Museum.

The species occurs in several samples from the Miocene of Bear River as well as one sample from the late Miocene or early Pliocene of the Bear River coast. The holotype of *P. billmani* Rau in the U.S. National Museum from the Miocene of Washington appears to be closely related and possibly synonymous.

Plectofrondicularia sp.?
Plate VI, figure 10

Hypotype 47768, loc. D-1238, "Bear River beds."

The species occurs in only one sample, that being from the Miocene of Bear River. It can not be shown to be related to any known species.

Plectofrondicularia spp.

A few specimens of specifically indeterminate *Plectofrondicularia* occur in the

Humboldt basin Miocene and Pliocene as shown on Figures 7 and 8 (distribution charts).

Family BULIMINIDAE
Genus BULIMINELLA Cushman, 1911
Buliminella elegantissima (d'Orbigny)
Plate VI, figures 12a, b

Buliminella elegantissima (d'Orbigny); Cushman, Stewart, and Stewart, 1930, p. 64, pl. iv, fig. 7 (Pliocene, Humboldt County, California); Cushman and Gray, 1946b, p. 28 (Pliocene, southern California); Bandy, 1950, p. 279, pl. xlii, fig. 10 (Pliocene, Cape Blanco, Oregon); Bandy, 1953a, p. 176, pl. xxiv, fig. 14 (Holocene, California).
Hypotype 47635, loc. D-1406, Rio Dell Formation.

In the Humboldt basin, the species ranges from the Miocene through Pleistocene.

Buliminella subfusiformis Cushman
Plate VI, figures 11, 13a, b

Buliminella subfusiformis Cushman, 1925c, p. 33, pl. v, fig. 12 (Miocene, California); L. Martin, 1952, p. 134, pl. xxiii, fig. 6 (Pliocene, Los Angeles County, California); White, 1956, p. 255, pl. xxx, fig. 13 (upper Miocene and lower Pliocene, Orange County, California).
Hypotype 47636, loc. D-1388, Rio Dell Formation; hypotype 47637, loc. D-1266, middle Rio Dell Formation.

Hypotype 47636 is comparable to the primary types in the U.S. Natl. Museum; hypotype 47637 is nearly identical to the primary type of *B. subfusiformis* var. *tenuata* Cushman from the Holocene.

The species does not occur above the Pliocene in the Humboldt basin. It may be closely related to *B. exilis* (Brady) of Bandy (1953a; Holocene, off California).

Genus BOLIVINITA Cushman, 1927
Bolivinita quadrilatera (Schwager)
Plate VI, figures 14a, b

Bolivinita quadrilatera (Schwager); Barker, 1960, p. 86, pl. xlii, figs. 8-12 (Holocene, Pacific).
Textilaria quadrilatera Schwager, 1866, p. 253, pl. vii, fig. 103 (Upper Tertiary?, Kar Nicobar).
Bolivinita angelina Church, 1927, p. 165, fig. 1 (Lower Pliocene, southern California).
Hypotype 47624, loc. D-1361, Eel River Formation.

The above hypotype is identical to topotypes (from Schwager's original Kar Nicobar material) now in the U.S. Natl. Museum.

In the Humboldt basin, the species is very rare and is known only from rocks of Pliocene age. Apparently it occurs also in the Holocene as indicated in the synonymy, but these Holocene occurrences are in the Indo-Pacific Province where the species ranges throughout the Neogene (see Kleinpell and Dalcon, pl. 35, "*Bolivinata* 1," *in* Corby et al, 1951).

Genus BULIMINA d'Orbigny, 1826
Bulimina fossa Cushman and Parker
Plate VII, figures 3a, b

Bulimina fossa Cushman and Parker, 1938, p. 56, pl. ix, fig. 10 (Pliocene, Ventura County, California); Natland, 1952, pl. 8-20 (Repettian, southern California).
Bulmina rostrata H. B. Brady; of Cushman, Stewart, and Stewart (*not* Brady), 1930, p. 65, pl. v, fig. 1 (Pliocene, Humboldt County, California).
Hypotype 47630, loc. D-1385, Eel River Formation.

Paratypes of *B. fossa* in the U.S. Natl. Museum compare with the above hypotype although the holotype of *B. fossa* is somewhat more finely striate. Plesiotypes of *B. rostrata* in the U.S. Natl. Museum from the Miocene, Pliocene, and Holocene appear to be conspecific with the primary types of *B. fossa*. The species is commonly cited by paleontologists in the petroleum industry as a "marker" for the Repettian of southern California.

Bulimina deformata Cushman and Parker
Plate VII, figures 4a, b

Bulimina pagoda Cushman var. *deformata* Cushman and Parker, 1938, p. 58, pl. x, fig. 3 (Pliocene, southern California).
Bulimina denudata deformata Cushman and Parker, 1946, p. 118, pl. xxvii, fig. 15 (Pliocene, southern California); White, 1956, p. 253, pl. xxx, fig. 5 (Pliocene, Orange County, California).
Hypotype 47628, loc. D-1391, Rio Dell Formation.

The above hypotype is comparable to the holotype in the U.S. Natl. Museum.

The species (including the following variety) apparently does not occur above the Pliocene in the Humboldt basin. The holotype of *B. pagoda* in the U.S. Natl. Museum does not appear to be closely related.

Bulimina denudata Cushman and Parker

Bulimina pagoda Cushman var. *denudata* Cushman and Parker, 1938, p. 57, pl. x, figs. 1-2 (Pliocene, Italy).
Bulimina denudata Cushman and Parker, 1946, p. 117, pl. xxvii, figs. 13-14 (Pliocene, southern California); Cushman and Gray, 1946b, p. 29, pl. v, figs. 13-15 (Pliocene, southern California); Bandy, 1953a, p. 176, pl. xxiv, fig. 11 (Holocene, California).
Hypotype 47629, loc. D-1393, Rio Dell Formation.

The above hypotype is comparable to the holotype in the U.S. Natl. Museum.

In the present study, separation of *B. deformata* and *B. denudata* was not possible in all samples in which they occur; consequently the two are treated as one on the range charts. The figures of *B. marginata* d'Orbigny of Galloway and Wissler (1927b; Pleistocene, southern California) and of Cushman and McCulloch (1948; Holocene, Pacific) may represent a synonymous species.

B. pagoda var. *hebespinata* Stewart and Stewart (1930b, p. 63, pl. viii, fig. 3), a southern California variant, was not encountered in the present study.

Bulimina subacuminata Cushman and Stewart, emend.
Plate VII, figures 1a, b, 2a, b

Bulimina subacuminata Cushman and Stewart; Cushman, Stewart, and Stewart, 1930, p. 65, pl. v, figs. 2-3 (Pliocene, Humboldt County, California); Cushman and Parker, 1938, p. 59, pl. x, fig. 9 (refigured holotype); L. Martin, 1952, p. 132, pl. xxii, fig. 12 (Pliocene, Los Angeles County, California); Natland, 1952, pls. 8-10, 8-11, 9-1 (Venturian, southern California); Bandy, 1953a, p. 176 (Holocene, California); White, 1956, p. 254, pl. xxx, fig. 8 (Pliocene, Orange County, California).
Bulimina subcalva Cushman and Stewart; Cushman, Stewart, and Stewart, 1930, p. 65, pl. iv, fig. 11 (Pliocene, Humboldt County, California); Cushman and Parker, 1938, p. 55, pl. ix, fig. 8 (Pliocene, Orange, County, California); Natland, 1952, pl. 8-13 - 8-16 (Repettian, southern California).
Hypotype 47631, loc. D-1284, lower Rio Dell Formation; hypotypes 47632, 47633, loc. D-1387, Rio Dell Formation.

The primary types of *B. subcalva* in the U.S. Natl. Museum (cf. hypotype 47633 here) show later chambers more bulbous and less spinose in relation to the primary types of *B. subacuminata* (cf. hypotype 47631 here).

The writer investigated the type localities for *B. subacuminata* and *B. subcalva* (Cushman, Stewart, and Stewart, 1930, locs. 42 and 48, Humboldt County, California). Neither species occurs in the site originally designated, the former being in a Miocene section (Bear River beds), and the latter in a Pleistocene section (upper Rio Dell Formation). Further investigation showed that the faunas associated with both types (Cushman, Stewart, and Stewart, 1930, "Species Distribution Chart") are typically Pliocene.

In the present study, specimens representative of both "species" occur together in several samples from the Pliocene (type Pullen in part, Eel River, and lower and middle Rio Dell Formations) of the Humboldt basin. The author has found it impossible to separate the two morphologic "types" and suggests that they represent variants of the same species and they are so treated herein.

It is proposed to retain the systematic description and holotype of *B. subacuminata* as originally published, and to treat the systematic description and holotype of *B. subcalva* as invalid. It is further proposed that a supplementary type locality be designated for the species as herein recognized; this new locality, exemplified by hypotype no. 47632, is designated as U.C. loc. D-1387 (lower or middle Rio Dell Formation).

Specimens occurring in the Pullen(?) Formation of the Bear River coast (see Fig. 8) possibly belong to *"B. cf. delreyensis"* of Adams (1940, chart I); comparison of type specimens is necessary to verify this conclusion. The holotype of *B. delreyensis* in the U. S. Natl. Museum is not conspecific with any of the Humboldt basin specimens.

Bulimina? sp.
Plate VII, figures 5a, b

Hypotype 47634, loc. D-1332, Hookton Formation.

A few specimens, belonging to an unknown species, were found only at the locality cited; they could perhaps be referred to *Virgulina* sp.

Genus GLOBOBULIMINA Cushman, 1927

Globobulimina affinis (d'Orbigny)
Plate VII, figures 6a, b

Globobulimina affinis (d'Orbigny) Phleger, Parker, and Peirson, 1953, p. 34, pl. vi, fig. 32 (Holocene, Mid-Atlantic).
Bulimina affinis d'Orbigny; Cushman and Parker, 1946, p. 122, pl. xxviii, figs. 23-25 (Holocene, western Atlantic); White, 1956, p. 253, pl. xxx, fig. 4 (Pliocene, Orange County California); P. Smith, 1964, p. B31, pl. ii, figs. 2-3 (Holocene, off Central America).
Hypotype 47708, D-1278, lower Rio Dell Formation.

This species, as well as the other globobuliminas, has an extremely fragile test; furthermore, the test is often filled with or replaced by iron sulfides. These circumstances make it unlikely that the genus is ever correctly represented in sample analyses with respect to other genera and species.

Globobulimina? *auriculata* (Bailey)
Plate VII, figures 7a, b

Bulimina auriculata Bailey, 1851, p. 12, figs. 25-27 (Holocene, Atlantic); P. Smith, 1964, p. B32, pl. ii, fig. 4 (Holocene, off Central America).
Bulimina (Desinobulimina) auriculata Bailey; Cushman and Parker, 1946, p. 129, pl. xxix, figs. 22-24 (Holocene, western Atlantic).
Hypotype 47709, loc. D-1278, upper Rio Dell Formation.

See remarks under *G. affinis.*

Globobulimina pacifica Cushman
Plate VII, figure 8

Globobulimina pacifica Cushman, 1927, p. 67, pl. xiv, fig. 12 (Holocene, eastern Pacific); Cushman, Stewart, and Stewart, 1930, p. 66, pl. v, fig. 4 (Pliocene, Hum-

boldt County, California); Cushman and Parker, 1946, p. 134, pl. xxiv, fig. 37 (Holocene, Pacific).
Hypotype 47710, loc. D-1278, upper Rio Dell Formation.

See remarks under *G. affinis*.

Genus FURSENKOINA Loeblich and Tappan, 1961

Fursenkoina? spp.

Rare specimens, occurring in a number of isolated samples from the Humboldt basin, can apparently be attributed to this genus, but not to a known species. Specimens usually show poor preservation which is caused in part by iron sulfide replacement.

Genus BOLIVINA d'Orbigny, 1839

Following the *Treatise on Invertebrate Paleontology* (Loeblich et al, 1964), the "genus" *Brizalina* is characterized as lacking basal chamber lobes, crenulations, or retral processes; the "genus" *Bolivina* is characterized as having basal margins of chambers with retral processes or backward directed chamber overlaps (also see discussion in Reyment, 1959, p. 11-14).

The "Key to Identification" (Appendix D) is presented as a temporary measure toward a reclassification of the California bolivinids; the group is in need of a thorough restudy as probably many species are synonyms, and the phylogeny of the others is poorly understood. For instance, *Bolivina spissa* and closely related subspecies have relatively large pores which distinguish them quite prominently from other bolivinids. Many of the finer details regarding pore size and distribution, chamber arrangement, and character of the sutures could best be studied by use of the scanning electron microscope, as Sliter (1970) has recently initiated. Other papers by Bandy (1960a), Lutze (1962), and Harmon (1964) contribute important environmental implications emphasizing the need for more study.

Bolivina advena Cushman var. *striatella* Cushman
Plate VIII, figure 17a, b

Bolivina advena Cushman var. *striatella* Cushman, 1925c, p. 29, pl. v, fig. 1 (Miocene, San Luis Obispo County, California).
Hypotype 47611, loc. D-1224, "Bear River beds."

The above hypotype is comparable to the holotype and many paratypes in the U.S. Natl. Museum.

Bolivina cochei Cushman and Adams
Plate VIII, figures 12a, b; 13a, b

Bolivina cochei Cushman and Adams, 1935, p. 19, pl. iii, figs. 6-7 (late Tertiary, Ventura County, California).
Hypotype 47612, loc. D-1231, Pullen(?) Formation; hypotype 47613, loc. D-1385, Eel River Formation.

The variant embraced by hypotype 47612 appears to be conspecific with paratypes of *B. cochei* in the U.S. Natl. Museum; in the Humboldt basin, it was not recorded in beds stratigraphically higher than the Eel River Formation. The hypotype 47613 is a variant close to some paratypes of *B. cochei* and may represent a form transitional between *B. cochei* and *B. argentea* although it is not conspecific with the holotype of *B. argentea* (Holocene).

Bolivina aff. *B. decussata* H. B. Brady
Plate VIII, figures 19a, b

Hypotype 47614, loc. D-1303, Wildcat Group undifferentiated.

Although Brady gave no type figure for *B. decussata*, the Humboldt basin specimens may be compared to *B. decussata* Brady in Barker (1960, p. 108, pl. liii, figs. 12-13; Holocene, eastern Pacific) and to plesiotype 26323 from the Holocene (Albatross D4839), now in the U.S. Natl. Museum.

Bolivina interjuncta (Cushman)
Plate VIII, figures 15a, b

Bolivina interjuncta (Cushman); Galloway and Wissler, 1927b, p. 70, pl. xi, figs. 10-13 (Pleistocene, southern California); Natland, 1952, pl. 10-6, 10-13 (Wheelerian, southern California); Natland and Rothwell, 1954, p. 34, fig. 2-5 (upper Pliocene, southern California).
Bolivina costata d'Orbigny var. *interjuncta* Cushman, 1926b, p. 41, pl. vi, fig. 3 (Pliocene, southern California).
Bolivina costata d'Orbigny var. *bicostata* Cushman; Cushman and McCulloch(?), 1942b, p. 195, pl. xxiii, figs. 9-11, 13-16 (Holocene, Pacific).
Hypotype 47615, loc. D-1248, upper Rio Dell Formation.

The hypotype above is comparable to the holotype of *B. costata* var. *interjuncta* in the U.S. Natl. Museum. Rare in the Humboldt basin, occurring in only a few samples from the Rio Dell Formation.

Bolivina pseudobeyrichi Cushman
Plate VIII, figures 11a, b

Bolivina alata Sequenza; Barker, 1960, p. 108, pl. liii, figs. 2-4 (Holocene, Pacific).
Bolivina beyrichi Reuss; Natland, 1952, pl. 11-1 - 11-3 (Pliocene, Pleistocene and Holocene, southern California).
Bolivina bramlettei Kleinpell, 1938, p. 267, pl. xxi, figs. 9-11 (upper Mohnian, California).
Bolivina pseudobeyrichi Cushman, 1926b, p. 45; Cushman, 1937.
Hypotypes 47616, 47617, loc. D-1303, Wildcat Group undifferentiated.

Bolivina bramlettei (holotype in the U.S. Natl. Museum) is more coarsely perforate and may be a valid, but closely related ancestral species. Topotypes of *B. beyrichi* from the Oligocene of Germany, now in the U.S. Natl. Museum, are also surficially comparable to the above hypotypes.

At the locality cited, many well-preserved specimens were found, the morphologic variations of which apparently embrace all the above cited "types." The species appears to have a high degree of phenotypic variation ranging from nearly nonkeeled to strongly keeled and from *Bolivina*-type arrangement of chambers and aperture to *Loxostomum*-type of chamber and aperture arrangement.

Bolivina sinuata Galloway and Wissler var. *alisoensis* Cushman and Adams
Plate VIII, figures 18a, b

Bolivina sinuata Galloway and Wissler var. *alisoensis* Cushman and Adams, 1935, p. 19, pl. iii, fig. 5 (Pliocene, California).
Hypotype 47618, loc. D-1379, Pullen Formation.

The holotype of *B. sinuata* var. *alisoensis* in the U.S. Natl. Museum is comparable to the above hypotype; the surface of the holotype of *B. floridana* is smoother; no type specimens of *B. sinuata* were available.

The species, which occurs in only a few samples in the Humboldt basin, is apparently extinct (Natland, 1957) and may be part of a bioseries containing the species *floridana, alisoensis, sinuata,* and *decussata* (Cushman, 1937; Kleinpell, 1938), of which the latter is living.

Bolivina subadvena Cushman var. *acuminata* Natland
Plate VIII, figures 14a, b

Bolivina subadvena Cushman var. *acuminata* Natland; Cushman and Gray, 1946b, p. 34, pl. v, fig. 46 (Pliocene, southern California).

Bolivina subadvena Cushman var. *serrata* Natland, 1938, p. 145, pl. v, figs. 8-9 (upper Pliocene, southern California).
Hypotype 47619, loc. D-1335, Wildcat Group undifferentiated.

It was difficult to separate specimens of this species from those of *B. spissa* and therefore both are shown in Figures 7 and 8 (distribution charts) under the latter term. It may be that the two are merely variants.

Bolivina subadvena Cushman var. *spissa* Cushman
Plate VIII, figures 9a, b; 10a, b

Bolivina spissa Cushman; Galloway and Wissler, 1927b, p. 72, pl. xi, figs. 14-16 (Pleistocene, southern California); Cushman, Stewart, and Stewart, 1930, p. 67, pl. v, fig. 7 (Pliocene, Humboldt County, California); Natland, 1952, pl. 10-1 - 10-5 (upper Pliocene, southern California); Bandy, 1953a, p. 176, pl. xxiv, fig. 5 (Holocene, California).
Bolivina subadvena Cushman var. *spissa* Cushman, 1926b, pl. vi, fig. 43 (Pliocene, California).
Hypotypes 47621, 47622, loc. D-1272, middle Rio Dell Formation; hypotype 47620, loc. D-1379, Pullen Formation.

The above hypotypes are comparable to numerous paratypes in the U.S. Natl. Museum. The species occurs relatively commonly in certain restricted biofacies in the Humboldt basin sediments.

Bolivina sulphurensis Cushman and Adams
Plate VIII, figures 16a, b

Bolivina subadvena Cushman var. *sulphurensis* Cushman and Adams, 1935, p. 20, pl. iii, figs. 8-9 (late Tertiary, Ventura County, California).
Hypotype 47623, loc. D-1303, Wildcat Group undifferentiated.

Paratypes in the U.S. Natl. Museum are comparable to the above hypotype. The species is probably extinct (R. M. Kleinpell, personal commun.).

Bolivina spp.

Poorly preserved specimens occur in the Miocene of Bear River; some of these might be identified by reference to type specimens, but can not definitely be referred to any published figures of known species of this genus.

Genus UVIGERINA d'Orbigny, 1826

Uvigerina hispida Schwager
Plate VIII, figure 1

Uvigerina hispida Schwager; Kleinpell(?), 1938, p. 295, pl. v, figs. 8, 16 (Zemorrian, California).
Hypotype 47799, loc. D-1383, Eel River Formation.

The species occurs relatively commonly in a number of samples from the Pullen and Eel River Formation; the identification of the above hypotype, which is coarsely spinose, is questionable.

Uvigerina hootsi Rankin?
Plate VII, figures 12a, b

?*Uvigerina hootsi* Rankin; Cushman and Kleinpell, 1934, p. 22, pl. iii, figs. 8-9 (Miocene, California).
Hypotype 47800, loc. D-1228, Pullen(?) Formation.

A holotype and a paratype in the U.S. Natl. Museum can be compared to the above hypotype as regards chamber arrangement and lack of ornamentation; the primary types differ in being slightly fatter.

Uvigerina peregrina Cushman

Uvigerina peregrina Cushman, 1923, p. 166 (Holocene, western Atlantic).
Euuvigerina peregrina (Cushman); Barker, 1960, p. 154, pl. lxxiv, figs. 11-12 (Holocene, West Indies).

Barker gives a résumé of the history of this species as cited in publications and notes that the species is highly variable. In the Humboldt basin, it ranges through the Pliocene and Pleistocene and comprises a high percentage of foraminiferal specimens in any given sample; here also, it exhibits a wide range of variation. The following varieties are thought to be subspecies; however, priority in nomenclature and the problem of phylogenetic relations in general in this group needs more study, a project recently initiated by Tipton et al, (1973, p. 37-39, fig. 11).

Uvigerina peregrina Cushman var. *dirupta* Todd
Plate VII, figures 13a, b; 14a, b

Uvigerina peregrina Cushman var. *dirupta* Todd; Cushman and McCulloch, 1948, p. 267, pl. xxxiv, fig. 3 (Holocene, California).
Hypotype 47801, loc. D-1285, Eel River Formation; hypotype 47802, loc. D-1386, Rio Dell Formation.

The holotype in the U.S. Natl. Museum differs from the above hypotype in being less spinose and in having costae somewhat more pronounced.

Uvigerina peregrina Cushman var. *foxenensis* Bramlette
Plate VIII, figure 3

Uvigerina foxenensis Bramlette, 1950, p. 59, pl. xxvi, fig. 16 (middle and late Pliocene, Santa Maria District, California).
Hypotype 47803, loc. D-1272, middle Rio Dell Formation.

Bramlette's *U. foxenensis* (holotype in U.S. Natl. Museum) evidently is closely related phyletically to the holotype of *U. peregrina*, differing mainly in the fact that the costae are somewhat more pronounced; it is therefore relegated to the status of subspecific rank. The Humboldt basin specimens are very common in Pliocene samples. *Uvigerina peregrina* var. *latalata* (Pliocene, southern California) has strongly pronounced costae.

Uvigerina peregrina Cushman var. *hollicki* Thalmann
Plate VIII, figures 4, 5

Uvigerina peregrina Cushman var. *bradyana* Cushman, 1923, p. 168, pl. xlii, fig. 12 (Holocene, Atlantic); preoccupied by *U. bradyana* Fornasini, 1900, and changed to *U. hollicki* Thalmann, 1950, nom. nov., p. 45.
Uvigerina peregrina Cushman; Natland, 1952, pl. 11-14 - 11-16 (Pliocene and Holocene, southern California).
Uvigerina hollicki Thalmann; Bandy, 1953a, p. 177, pl. xxv, fig. 8 (Holocene, California); White, 1956, p. 258, pl. xxxii, fig. 4 (lower Pliocene, Orange County, California).
Hypotype 47804, loc. D-1389, Rio Dell Formation; hypotype 47805, loc. D-1248, upper Rio Dell Formation.

The holotype of *U. peregrina* var. *bradyana* differs from the above hypotype in that its costae have a tendency to become broken (i.e., spinose) on the upper part of the latter chambers. The holotypes of *U. juncea* (Pliocene, California) and *U. cushmani* (Pliocene?, Alabama) are questionably placed in synonymy with this group, as is "*Uvigerina* aff. *tenuistriata*" of Stewart and Stewart (1949, p. 197). Note: Cushman (1923, p. 168) stated that *U. tenuistriata* of Brady is actually *U. peregrina* var. *bradyana.*

Uvigerina subperegrina Cushman and Kleinpell
Plate VIII, figure 2

Uvigerina subperegrina Cushman and Kleinpell, 1934, p. 12, pl. ii, figs. 9-11 (Miocene, California); Kleinpell, *in* Woodring, Stewart, and Richards, 1941, pl. 50, figs. 5-6 (Mohnian, Kettleman Hills, California).
Hypotype 47806, loc. D-1228, Pullen(?) Formation.

The above hypotype is comparable to the primary types in the U.S. Natl. Museum. The species apparently occurs only in a few samples from the Bear River coast, which are of Miocene age.

Uvigerina spp.

Various specimens not referable to any known species occur at a number of localities as shown in figures 7 and 8 (distribution charts).

Genus SIPHOGENERINA Schlumberger, 1882

Siphogenerina branneri (Bagg)

Siphogenerina branneri (Bagg); Kleinpell, 1938, p. 300, pl. vii, fig. 22, pl. xi, figs. 1, 5 (Miocene, California).

Although this species was not recorded in the present study, its stratigraphic importance is clarified by Tipton et al (1973, fig. 11). Prior records of localities from the Humboldt basin are herein repeated. Cushman, Stewart, and Stewart (1930, p. 70, pl. v, fig. 15) cited a locality from the lower part of the Miocene section on the south flank of the Bear River coastal syncline (5 specimens from their station 36 now exist in the U.S. Natl. Museum); Kleinpell (1938, p. 299) referred to this locality and designated it as "Lower and Upper Relizian." The D. W. Hughes Collection at the University of California, Berkeley, contains a foraminiferal assemblage slide with *Siphogenerina* cf. *S. branneri*; Hughes' slide is labeled "Briceland, Redwood Creek" (a locality about 25 mi [40 km] south of Scotia, see Fig. 1 herein); this locality is considered by Kleinpell (1938, Fig. 14) to be of early Luisian age.

Siphogenerina hughesi Cushman

Siphogenerina hughesi Cushman; Kleinpell, 1938, p. 301, pl. ix, fig. 19 (lower Relizian, California).

The species was not recorded in the present study, but was recorded by Cushman, Stewart, and Stewart (1930, p. 70, pl. v, fig. 13; one broken specimen, from their station 36, exists in the U.S. Natl. Museum) from the "Lower Relizian" (Kleinpell, 1938, p. 301) of the south flank of the Bear River coastal syncline. Reed (1937, Fig. 2) also noted *S. hughesi* as occurring in the "Eel River basin" Miocene, but gave no description, figures, or locality data for the species. The species appears to have a very short stratigraphic range as shown in Tipton et al (1973, Fig. 11).

Genus SIPHONODOSARIA Silvestri, 1924

Siphonodosaria insecta (Schwager)
Plate VIII, figure 6

Nodosaria insecta Schwager, 1866, p. 224, pl. v, figs. 53-54 (Pliocene?, Kar Nicobar).
Dentalina insecta (Schwager); Cushman, Stewart, and Stewart, 1930, p. 54, pl. ii, figs. 8-9 ("Pliocene," Humboldt County, California).
Ellipsonodosaria insecta (Schwager); Natland, 1952, pl. 13-5 - 13-7 (Repettian, southern California).
Siphonodosaria verneuili (d'Orbigny); Natland and Rothwell, 1954, p. 34, fig. 2-10 (lower Repettian, California).

Hypotype 45700, loc. D-1348, Pullen Formation; hypotype 45701, loc. D-1295, Pullen Formation.

This highly distinctive species evidently does not occur in California in rocks younger than Pliocene; however, it may be closely related to *S. abyssorum* (Brady) of Loeblich et al (1964; Holocene, south Pacific).

Genus STILOSTOMELLA Guppy, 1894
Stilostomella? *advena* (Cushman and Laiming)
Plate VIII, figure 8

Nodogenerina advena Cushman and Laiming, 1931, p. 106, pl. xi, fig. 9 (Miocene, California); Cushman, Stewart, and Stewart, 1947b, p. 16, pl. ii, figs. 5-6 (Miocene, Oregon).
Siphonodosaria advena (Cushman and Laiming); L. Martin, 1952, p. 138, pl. xxv, fig. 10 (Pliocene, Los Angeles County, California).
Hypotype 47793, loc. D-1303, Wildcat Group undifferentiated; hypotype 47794, loc. D-1266, middle Rio Dell Formation.

The above hypotypes have a slight tendency to become spinose at the base of the chambers and in this point, they differ from the primary types in the U.S. Natl. Museum.

The species apparently does not occur in rocks younger than Pliocene.

Stilostomella lepidula (Schwager)
Plate VIII, figure 7

Nodosaria lepidula Schwager, 1866, p. 210, pl. v, figs. 27-28 (upper Tertiary?, Kar Nicobar).
Siphonodosaria lepidula (Schwager); L. Martin, 1952, p. 139, pl. xxv, fig. 12 (Pliocene, Los Angeles County, California); Cushman, Todd, and Post, 1954, p. 356, pl. lxxx, figs. 17-18 (a single specimen from Holocene sediments containing reworked Pliocene species, Marshall Islands, Pacific).
Hypotype 47795, loc. D-1303, Wildcat Group undifferentiated.

The above hypotype differs from a topotype in the U.S. Natl. Museum by having spines more strongly accentuated and somewhat more numerous.

Evidently *S. lepidula* became extinct during the Pliocene; however, it has a close relative in *Stilostomella bradyi* (Cushman) in Loeblich et al, (1964; Holocene, Brazil) and also *Stilostomella antillea* in Bandy and Rodolfo (1964; Holocene, Chile-Peru Trench).

Genus UVIGERINELLA Cushman, 1926
Uvigerinella californica Cushman var. *ornata* Cushman?
Plate VII, figures 9a, b

Uvigerinella californica Cushman var. *ornata* Cushman, 1926a, p. 59, pl. viii, fig. 1 (Miocene, California); Cushman, Stewart, and Stewart, 1930, p. 68, pl. vi, fig. 8 (Miocene, Humboldt County, California).
Hypotype 47807, loc. D-1240, "Bear River beds."

The species is typical of the Miocene in Bear River valley and in Price Creek; it also occurs in the Miocene of the Bear River coastal section. Kleinpell (1938, p. 288) referred to Cushman, Stewart, and Stewart's specimen from their Bear River locality and gave its age as "Upper Relizian."

Genus ANGULOGERINA Cushman, 1927
Angulogerina aff. *A. hughesi* (Galloway and Wissler)
Plate VII, figures 11a, b

?*Uvigerina hughesi* Galloway and Wissler, 1927b, p. 76, pl. xii, fig. 5 (Pleistocene,

California).
Hypotype 47607, loc. D-1385, Eel River Formation.

The above hypotype differs from the figures of Galloway and Wissler in having bulbous final chambers.

The species is comparatively rare in the Humboldt basin and occurs in only a few samples.

Angulogerina angulosa (Williamson)
Plate VII, figures 10a, b

Angulogerina carinata Cushman; Cushman, Stewart, and Stewart, 1930, p. 71, pl. v, fig. 14 (Pliocene, Humboldt County, California).
Angulogerina angulosa (Williamson); Barker, 1960, p. 154, pl. lxxiv, figs. 15-16 (Holocene, Pacific).
Trifarina angulosa (Williamson); Loeblich et al, 1964, (p. C571, figs, 450, 1-3, paratype in part, Holocene, Great Britain).
Hypotype 47605, loc. D-1244, upper Rio Dell Formation; hypotype 47606 loc. D-1303, Wildcat Group undifferentiated.

A. angulosa is a very common species and occurs in many samples; it ranges through the Pliocene and lower Pleistocene in the Humboldt basin sediments. In the U.S. Natl. Museum, the plesiotypes (of Cushman, Stewart, and Stewart, 1930) *A. carinata* and *A. hughesi* from Humboldt County, California, appear to be synonymous.

Family ROTALIIDAE
Genus DISCORBIS Lamarck, 1804
Discorbis campanulata (Galloway and Wissler)
Plate IX, figures 1a, b, c

Globorotalia campanulata Galloway and Wissler, 1927b, p. 58, pl. ix, fig. 14 (Pleistocene, California).
Rotorbinella? *campanulata* (Galloway and Wissler); Lankford, 1962, p. 193, pl. v, fig. 13 (Holocene, Baja California).
Hypotype 47672, loc. D-1303, Wildcat Group undifferentiated.

Numerous specimens were found, but only at the locality cited.

The generic status of this species remains in doubt; the term *Discorbis* is used here in apparent agreement with the systematics of Loeblich et al (1964).

Genus BUCCELLA Andersen, 1952
Buccella oregonensis (Cushman, Stewart, and Stewart)
Plate IX, figures 3a, b, c

Eponides mansfieldi Cushman var. *oregonensis* Cushman, Stewart, and Stewart, 1947a, p. 48, pl. vi, fig. 4 (Miocene, Oregon).
Hypotype 47625, loc. D-1238, "Bear River beds."

The hypotype above is identical with the holotype in the U.S. Natl. Museum.

The species is relatively common in the Miocene of Bear River and Price Creek. Its geologic range in the Humboldt basin apparently does not overlap with that of *B. tenerrima* and consequently it may be ancestral to this later species.

Buccella tenerrima (Bandy)
Plate IX, figures 2a, b, c

Buccella tenerrima (Bandy); Lankford, 1962, p. 139, pl. iv, fig. 19 (Holocene, northern California).
Rotalia tenerrima Bandy, 1950, p. 278, pl. xlii, fig. 3 (Pleistocene, Cape Blanco, Oregon).

Buccella inusitata Andersen, 1952, p. 148, text-figs. 10-11 (Holocene, Washington); Loeblich and Tappan, 1953, p. 116, pl. xxii, fig. 1 (Holocene, Alaska).

Eponides cf. *E. frigidus* (Cushman); Natland, 1952, pl. 14-6 (Pliocene and Pleistocene, southern California).

Hypotype 47626, loc. D-1329, Hookton Formation; hypotype 47627, loc. D-1249, upper Rio Dell Formation.

The writer follows the usage of Lankford (1962), who compared near-topotypic material of *Rotalia tenerrima* Bandy and *Buccella inusitata* Andersen in synonymizing the forms listed above. The holotype of *B. inusitata* in the U.S. Natl. Museum is synonymous with the above hypotypes. As Lankford notes, it is a typical "turbulent zone" species ranging along the greater part of the modern-day shoreline of northwestern North America. It has a very wide range in variation within any given sample or population. In the Humboldt basin, the species is common in many Pleistocene samples and occurs in limited numbers in the Pliocene.

Genus EPISTOMINELLA Husezima and Maruhasi, 1944

Epistominella exigua (H. B. Brady)
Plate X, figures 3a, b, c

Epistominella exigua (H. B. Brady); Barker, 1960, p. 212, pl. ciii, figs. 13-14 (Holocene, south Atlantic).

Pulvinulina exigua H. B. Brady, 1884, p. 696, pl. ciii, figs. 13-14 (Holocene, south Atlantic).

Pseudoparrella exigua (H. B. Brady); Lipps, 1965, p. 133, pl. i, fig. 9 (Holocene, San Diego; holotype of *E. sandiegoensis* Uchio).

Epistominella sandiegoensis Uchio, 1960, p. 68, pl. ix, figs. 6-7 (Holocene, southern California).

Hypotype 47689, loc. D-1249, upper Rio Dell Formation.

Epistominella pacifica (Cushman)
Plate X, figures 5a, b, c; 6a, b, c

Epistominella pacifica (Cushman); L. Martin, 1952, p. 136, pl. xxiv, fig. 8 (Pliocene, Los Angeles County, California); Bandy, 1953a, p. 177, pl. xxiii, fig. 2 (Holocene, California); Natland and Rothwell, 1954, p. 34, fig. 2-7 (Pliocene and upper Miocene, California).

Pulvinulinella pacifica Cushman, 1927, p. 165, pl. v, figs. 14-15 (Holocene, Pacific); Cushman, Stewart, and Stewart, 1930, p. 73, pl. vi, fig. 5 (Pliocene, Humboldt County, California); Natland, 1952, pl. 16-1 (Pliocene and Holocene, southern California).

Epistominella smithi (Stewart and Stewart); Lipps, 1965, p. 129, pl. ii, fig. 6 (Pliocene, Ventura County, California; Holotype of *Pulvinulinella smithi* Stewart and Stewart).

Hypotype 47690, loc. D-1389, Rio Dell Formation; hypotype 47691, loc. D-1332, Hookton Formation.

Hypotype 47690 is comparable to the holotype and some of the paratypes in the U.S. Natl. Museum; hypotype 47691 represents an extreme, flattened variation, comparable to at least one of the paratypes in the U.S. Natl. Museum. *Pulvinulinella parva* (primary types in U.S. Natl. Museum) from the Miocene of California appears to be closely related and may be ancestral; specimens from the Bear River (Miocene) localities perhaps could be referred to this species. The primary types of *P. smithi*, now in the U.S. Natl. Museum, may be a subspecies of *E. pacifica.*

E. pacifica is one of the most abundant species in the Pleistocene of the Humboldt basin; it occurs also in the Pliocene and Miocene.

Epistominella subperuviana (Cushman)
Plate X, figures 7a, b, c

Pulvinulinella subperuviana Cushman, 1926a, p. 63, pl. ix, fig. 9 (Miocene, California).

Pseudoparrella subperuviana (Cushman); Lipps, 1965, p. 135, pl. i, figs. 1-2 (Miocene, California).
Hypotype 47692, loc. D-1303, Wildcat Group undifferentiated.

The primary types in the U.S. Natl. Museum compare with the above hypotype. Lipps (1965) included the holotype of *Pulvinulinella bradyana* Cushman (Holocene, eastern Pacific) within this species although there seems to be some question whether this should be done from the figures he illustrated.

Epistominella spp.

Poorly preserved specimens occurring at loc. D-1231 can not be identified at the specific level.

Genus LATICARININA Galloway and Wissler, 1927
Laticarinina halophora (Stache)
Plate X, figures 1a, b

Laticarinina halophora Stache); Barker, 1960, p. 214, pl. civ, figs. 3-11 (Holocene, Atlantic).
Laticarinina pauperata (Parker and Jones); Natland, 1952, pl. 20-4 (Repettian and Holocene, southern California).
Hypotype 47732, loc. D-1357, Eel River Formation.

The species occurs very rarely in the Pliocene of the Humboldt basin; in addition, a single specimen, apparently of the same species, was found in the Miocene(?) of the Bear River area.

Genus ROSALINA d'Orbigny, 1826
Rosalina columbiensis (Cushman)
Plate X, figures 2a, b, c; 4a, b, c

Rosalina columbiensis (Cushman); Lankford, 1962, p. 191, pl. v, figs. 10-12 (Holocene, Baja California).
Discorbis columbiensis Cushman, 1925a, p. 43, pl. vi, fig. 13 (Holocene, British Columbia).
Hypotypes 47783, 47784, loc. D-1303, Wildcat Group undifferentiated.

A limited number of specimens comparable to figures of Holocene specimens (Cushman, 1925a) were found in a single sample from the Pliocene of the Humboldt basin; Lankford (1962) included "*Tretomphalus bulloides*" as a reproductive stage in the life cycle of *R. columbiensis*, but this stage was not present in the Humboldt basin samples.

Genus BAGGINA Cushman, 1926
Baggina californica Cushman
Plate IX, figures 4a, b, c

Baggina californica Cushman, 1926a, p. 64, pl. ix, fig. 8 (Miocene, California); Kleinpell, 1938, p. 324, pl. xiii, fig. 3 (Miocene , California).
Hypotype 47609, loc. D-1224, "Bear River beds."

The holotype in the U.S. Natl. Museum appears to be synonymous with the above hypotype.

Genus VALVULINERIA Cushman, 1926
Valvulineria araucana (d'Orbigny)
Plate IX, figures 5a, b, c; 6a, b, c

Valvulineria araucana (d'Orbigny); Cushman, 1927, p. 160, pl. iv, figs. 7-8 (Holocene,

California); Cushman, Stewart, and Stewart, 1930, p. 71, pl. vi, fig. 4 (Pliocene, Humboldt County, California).
Hypotype 47808, loc. D-1272, middle Rio Dell Formation; hypotype 47809, loc. D-1231, "Bear River beds."

Valvulineria? spp.

Moderately well preserved specimens from locs. D-1244 and D-1245 (Bear River coastal section) can not be compared with any known published figures.

Genus GYROIDINA d'Orbigny, 1826

Gyroidina soldanii d'Orbigny var. *altiformis* R. E. and K. C. Stewart
Plate XI, figures 2a, b, c

Gyroidina soldanii d'Orbigny var. *altiformis* R. E. and K. C. Stewart, 1930b, p. 67, pl. ix, fig. 2 (lower Pliocene, California); L. Martin, 1952, p. 125, pl. xix, fig. 8 (Pliocene, Los Angeles County, California); White, 1956, p. 248, pl. xxviii, fig. 4 (Pliocene, Orange County, California).
Gyroidina altiformis R. E. and K. C. Stewart; Resig, 1958, p. 306 (reworked? Pliocene or Holocene, California).
Hypotype 47715, loc. D-1386, Rio Dell Formation.

The above hypotype compares with the primary types in the U.S. Natl. Museum.

Gyroidina soldanii d'Orbigny var. *rotundimargo* R. E. and K. C. Stewart
Plate XI, figures 1a, b, c

Gyroidina soldanii d'Orbigny var. *rotundimargo* R. E. and K. C. Stewart, 1930b, p. 68, pl. ix, fig. 3 (lower Pliocene, California).
Hypotype 47716, loc. D-1384, Eel River Formation.

The above hypotype compares with the primary types in the U.S. Natl. Museum. The species, including both varieties, is common in the older Pliocene sediments of the Humboldt basin.

Genus EPONIDES Montfort, 1808

Eponides healdi Stewart and Stewart
Plate XI, figures 4a, b, c

Eponides healdi Stewart and Stewart, 1930b, p. 70, pl. viii, fig. 8 (lower Pliocene, California); Cushman, Stewart, and Stewart, 1947b, p. 21, pl. iii, fig. 4 (Miocene, Oregon).
Hypotype 47693, loc. D-1285, Eel River Formation; hypotype 47694, loc. D-1228, Pullen(?) Formation.

The above hypotypes are comparable to the primary types in the U.S. Natl. Museum.

Eponides repandus (Fichtel and Moll)
Plate XI, figures 5a, b, c

Eponides repandus (Fichtel and Moll); Resig, 1962, p. 55-57, pl. xiv, (Pleistocene and Holocene, California); Todd, 1965, p. 20, pl. vii, figs, 3-4 (Holocene, Pacific).
Poroeponides cribrorepandus Asano and Uchio, 1951, p. 18, text-figs. 134-135; Bandy, 1953a, p. 177, pl. xxiv, fig. 1 (Holocene, southern California); Lankford, 1962, p. 180, pl. iv, fig. 25 (Holocene, Baja California).
Sestronophora arnoldi Loeblich and Tappan, 1957, p. 229; Loeblich et al, 1964, p. C683, Fig. 546-6 (Pleistocene, California).
Hypotype 47695, loc. D-1303, Wildcat Group undifferentiated.

This paper follows the usage of Resig (1962) in recognizing *Poroeponides cribrorepandus* as a transition stage and *Sestronophora arnoldi* as an adult stage in the life cycle of *Eponides repandus*, this latter term having priority. A limited number of specimens were found in the Pliocene of the Humboldt basin, representing the first two stages (juvenile and transition) only.

Genus ORIDORSALIS Andersen, 1961
Oridorsalis umbonatus (Reuss)
Plate XI, figures 3a, b, c

Oridorsalis umbonatus (Reuss); Todd, 1965, p. 23, pl. vi, fig. 2 (Holocene, Pacific).
Pulvinulina umbonata Reuss; Cushman, Stewart, and Stewart, 1947b, p. 21, pl. iii, figs. 2-3 (Miocene, Oregon); Natland, 1952, pl. 14-4 (Venturian and Holocene, southern California).
Eponides tenera (Brady); L. Martin, 1952, p. 124, pl. xix, fig. 7 (Pliocene, Los Angeles County, California).
Eponides? *tenera* (Brady); Barker, 1960, p. 196, pl. xcv, fig. 11 (Holocene, Pacific).
Pseudoeponides umbonatus (Reuss); P. Smith, 1964, p. B43, pl. iv, fig. 8 (Holocene, eastern Pacific).
Hypotype 47757, loc. D-1287, Eel River Formation; hypotype 47758, loc. D-1237, "Bear River beds."

The species occurs in the Miocene and Pliocene of the Humboldt basin, but was not recorded from the Pleistocene.

Genus ROTALIA Lamarck, 1804
'Rotalia' becarrii (Linné)
Plate XI, figures 6a, b, c

Rotalia beccarii (Linné); H. B. Schenck, 1940, p. 1761, fig. 4 (Holocene, San Francisco Bay); Natland, 1952, pl. 15-6, 15-7 (upper Pleistocene, southern California); Bandy, 1953a, p. 177, pl. xxii, fig. 8 (Holocene, California); Natland and Rothwell, 1954, p. 34, fig. 2-1 (Pleistocene, California).
Hypotype 47785, loc. D-1329, Hookton Formation.

The systematics of this species at the generic level has undergone several changes in recent years; the long used term "*Rotalia*" is retained herein while the reader is referred to the following for pertinent discussion regarding the status of this term: Frizzell and Keen (1949), Cifelli (1962), and Wood et al (1963).

The published occurrence of this species in the geologic record of California is from the upper Pleistocene only; however, it evidently occurs elsewhere in sediments as old as Miocene (Kleinpell, personal commun.). In the Humboldt basin, it occurs only in the upper Pleistocene.

Genus HÖGLUNDINA Brotzen, 1948
Höglundina elegans (d'Orbigny)
Plate XII, figures 1a, b, c

Höglundina elegans (d'Orbigny); Phleger, Parker, and Peirson, 1953, p. 43, pl. ix, figs. 24-25 (Holocene, Mid-Atlantic); Loeblich et al, 1964, p. C775, fig. 636-3 (Holocene, Caribbean).
Hypotype 47717, loc. D-1303, Wildcat Group undifferentiated.

Family CASSIDULINIDAE
Genus CASSIDULINA d'Orbigny, 1826

Generally speaking, among the hyaline foraminifera, radiate refers to the fact that the calcite grains (=fibers?, or crystals?) comprising the test wall are oriented with their C-axes perpendicular to the outer surface of the test wall. Granular wall structure refers either to randomly oriented grains or to C-axes oriented tangential to the test wall surface.

Cassidulinid foraminifera were shown by Wood (1949) to have two types of wall structure. This dichotomy was further investigated by Nørvang (1958), who erected the new genus *Islandiella* and distinguished it from *Cassidulina*. In *Islandiella*, Nørvang cited an "unconditional association of a radiate fibrous wall and an aperture with an internal tooth;" he characterized *Cassidulina* by its "granular wall structure and the lack of an internal tooth." Nørvang recognized the following species of each group:

Islandiella
- *I. islandica* (Nørvang)–type species of *Islandiella*
- *I. californica* (Cushman and Hughes)
- *I. limbata* (Cushman and Hughes)
- *I. tortuosa* (Cushman and Hughes)
- *I. quadrata* (Cushman and Hughes)
- *I. norcrossi* (Cushman)
- *I. vejlensis* Nørvang

Cassidulina
- *C. laevigata* d'Orbigny–type species of *Cassidulina*
- *C. oblonga* Reuss
- *C. subglobosa* Brady
- *C. crassa* d'Orbigny
- *C. cruysi* Marks

In 1958, through the courtesy of M. N. Bramlette of the Scripps Institution of Oceanography, the writer was permitted to determine wall structure petrographically in certain cassidulinid foraminifera as follows:

- *Cassidulinoides* sp. (Holocene, California)–radial
- *Cassidulina californica* (Plio-Pleistocene, California)–radial
- *C.* cf. *C. californica* (Holocene, California)–radial?
- *C. limbata* (Plio-Pleistocene, California)–radial
- *C. modeloensis* (Miocene, California)–radial?
- *C. pulchella* (Plio-Pleistocene, California)–granular
- *C. globosa*? (Holocene, Gulf of Mexico)–granular
- *C. tortuosa* (Plio-Pleistocene, California)–granular

In 1965, the writer examined (surficially only) the following paratypes in paleontological collections at the University of California, Berkeley, and at Stanford University:

- *Cassidulina californica* Cushman and Hughes–16 specimens
- *C. limbata* Cushman and Hughes–7 specimens
- *C. subglobosa quadrata* Cushman and Hughes–5 specimens
- *C. translucens* Cushman and Hughes–7 specimens
- *C. tortuosa* Cushman and Hughes–8 specimens

Of these, *C. translucens* and *C. subglobosa quadrata* are from the Pleistocene, Lomita Quarry, Palos Verdes Hills, California; the other three species are from the "Pliocene" (now considered to be Pleistocene by Woodring *in* Woodring et al, 1946, p. 96), Timm's Point, California. The cited specimens from the Berkeley and Stanford collections show poor preservation in that all are translucent to opaque (not transparent), some are iron-stained, many show pit solution cavities on the surface, some have gastropod(?) borings through the wall, and many show mechanical abrasion as well as chemical weathering. In summary, doubt remains as to whether or not the original wall structure of the types of these five species can accurately be determined.

The results given by Nørvang (1958) are similar to those obtained by my unpublished 1958 study of the Bramlette slides except for the discrepancy in *Cassidulina* (or *Islandiella*) *tortuosa*. Several possibilities accounting for this discrepancy exist: (a) the Bramlette or the Nørvang specimen was recrystallized resulting in a pseudo-wall texture; (b) the Nørvang "topotype" or the Bramlette specimen was misidentified; (c) different types of wall structure exist in the same species or the same population.

Z. M. Arnold (1964, p. 22, fig. 3) has illustrated the hazard of placing overemphasis on aperture characteristics as a key to systematics in the forams. The present paper recognizes the general dichotomy (trichotomy?) in the wall structure of cassidulinid forams, but refrains from subjugating the various types to different superfamilies as Loeblich et al (1964) have done; this necessitates presupposing an undemonstrated and unlikely amount of convergent evolution in unrelated, morphologically similar forms. The adoption of the generic term *Islandiella* is questioned until opportunity is granted for a concentrated study of cassidulinid wall structure, using well-preserved specimens, in either fossil or living populations (preferably both), and until the phylogenetic relations of Tertiary, Quaternary, and Holocene species are demonstrated.

Cassidulina californica Cushman and Hughes?
Plate XII, figures 2a, b, c

?*Cassidulina californica* Cushman and Hughes, 1925, p. 12, pl. ii, fig. (Pleistocene, southern California); Cushman, Stewart, and Stewart, 1949, p. 154, pl. xviii, fig. 3 (Pliocene, Washington); L. Martin, 1952, p. 134, pl. xxiv, fig. 2 (Pliocene, Los Angeles County, California); Bandy, 1953a, p. 176, pl. xxv, fig. 1 (Holocene, California); White, 1956, p. 255, pl. xxxi, fig. 1 (Pliocene, Orange County, California).
Hypotype 47638, loc. D-1357, Eel River Formation; hypotype 47639, loc. D-1389, Rio Dell Formation.

The types herein are somewhat more flattened than the holotype and paratypes at the U.S. Natl. Museum, at Berkeley, and at Stanford and possibly represent a form intermediate between *C. californica* and *C. lomitensis*. *Islandiella californica* (Cushman and Hughes) in Loeblich et al (1964, Fig. 439-4a, b) does not appear to be conspecific with the primary types since the types indicate a much smoother outline, sutures which are only slightly depressed, tests which are less compressed and more spheroid, and a lip on the ventral side of the aperture, but not a pronounced rim on the dorsal side of the aperture as the figures by Loeblich et al show.

Cassidulina carinata Silvestri
Plate XII, figures 5a, b, c

Cassidulina carinata Silvestri; Todd, 1965, p. 40, pl. xvii, fig. 4 (Holocene, Pacific).
Hypotype 47640, loc. D-1272, middle Rio Dell Formation.

The above hypotype is comparable to Todd's plesiotype in the U.S. Natl. Museum.

Cassidulina crassa d'Orbigny?
Pl. XII, figures 4a, b, c

?*Cassidulina crassa* d'Orbigny, 1839, p. 56, pl. vii, figs. 18-20 (Holocene, loc. not designated); Cushman, 1926a, p. 56, pl. vii, fig. 4.
Hypotype 47641, loc. D-1237, "Bear River beds."

The specimens from the Miocene of Bear River are questionably placed in synonymy with this species pending investigation of type specimens.

Cassidulina delicata? Cushman
Plate XII, figures 3a, b, c

Cassidulina delicata Cushman?; Cushman, 1927, p. 168, pl. vi, fig. 5 (Holocene, Pacific).
Hypotype 47642, loc. D-1250, upper Rio Dell Formation.

Cassidulina delicata Cushman, 1927, may be synonymous with the above hypotype although the holotype in the U.S. Natl. Museum shows a much more elongate aperture.

Cassidulina limbata Cushman and Hughes
Plate XIII, figures 1a, b, c

Cassidulina limbata Cushman and Hughes, 1925, p. 12, pl. ii, fig. 2 (Pleistocene,

southern California); Cushman, Stewart, and Stewart, 1949, p. 155, pl. xviii, fig. 2 (Pliocene, Washington); Bandy, 1950, p. 280, pl. xlii, fig. 4 (Pleistocene, Cape Blanco, Oregon); Natland, 1952, pl. 17-1 (upper Pliocene, Pleistocene, and Holocene, southern California); Bandy, 1953a, p. 176, pl. xxv, fig. 2 (Holocene, California).
Hypotype 47643, loc. D-1262, middle Rio Dell Formation; hypotype 47644, loc. D-1389, Rio Dell Formation.

The hypotypes herein compare favorably with the holotype and paratypes at the U.S. Natl. Museum, at Berkeley, and at Stanford. The species is very common in the Pleistocene.

Cassidulina limbata Cushman and Hughes vars.

Specimens occurring in the Pliocene and Pleistocene of the Humboldt basin seem to be varietal forms of *C. limbata* and possibly represent a new subspecies. Specimens in the U.S. Natl. Museum labeled *C. corbyi* by Cushman, Stewart, and Stewart (1930) fall within this group.

Cassidulina minuta Cushman
Plate XII, figures 6a, b, c

Cassidulina minuta Cushman, 1933, p. 92, pl. x, fig. 3 (Holocene, Pacific).
Hypotype 47645, loc. D-1389, Rio Dell Formation.

The holotype in the U.S. Natl. Museum may be synonymous or closely related to the above hypotype.

Cassidulina tortuosa Cushman and Hughes
Plate XIII, figures 4a, b, c

Cassidulina tortuosa Cushman and Hughes, 1925, p. 14, pl. ii, fig. 4 (Pliocene, southern California); Natland, 1952, pl. 17-2 (Pleistocene and Holocene, southern California); Bandy, 1953a, p. 176, pl. xxv, fig. 3 (Holocene, California).
Hypotype 47646, loc. D-1337, Wildcat Group undifferentiated.

A limited number of specimens, all of which apparently occur in sediments of Pliocene age, were encountered. The above cited hypotype appears to be conspecific with the holotype and paratypes at the U.S. Natl. Museum, at Berkeley, and at Stanford.

Cassidulina translucens Cushman and Hughes
Plate XIII, figures 3a, b, c

Cassidulina translucens Cushman and Hughes, 1925, p. 15, pl. ii, fig. 5 (Pleistocene, California)
Hypotype 47647, loc. D-1272, middle Rio Dell Formation.

The original publication of this species contains only a series of line drawings; consequently, the author found it impossible to identify the Humboldt basin specimens without recourse to paratypes at Berkeley and at Stanford. The identity of *C. translucens* in Martin (1952), Bandy (1953a), Natland (1952), and White (1956) is questioned; the holotype and paratypes are strongly biumbonate with a sharply defined keel; part or all of the hypotypes figured in the papers cited might be more readily compared to hypotype 47649 below.

Cassidulina translucens Cushman and Hughes var. *natlandi* Haller n. var.
Plate XIII, figures 2a, b, c

Cassidulina translucens Cushman and Hughes; Bandy, 1953a, p. 176, pl. xxv, fig. 6 (Holocene, southern California).

Holotype 47649, loc. D-1249, upper Rio Dell Formation, Centerville coastal section. Humboldt County, California, Quaternary.
Paratypes: 47650, 47651, loc. D-1249, as above.

Test free, lenticular, biumbonate, with clear central bosses; chambers biserially arranged in a coil, alternating on each side of periphery; wall calcareous, hyaline, very finely perforate; surface smooth, aperture an elongate slit, with bordering lip on outer margin; sutures very slightly depressed in part; keel distinct, periphery smooth– Diameter, 0.50 mm; Thickness, 0.28 mm.

Discussion: The new variety differs from the species in being less sharply keeled, in having smoother surfaces, and in being less strongly umbonate. It differs from *C. limbata* Cushman and Hughes in being less sharply keeled, having a smoother periphery, and in the lesser indentation of the sutures. It is common in a number of samples in the Rio Dell Formation.

Cassidulina translucens Cushman and Hughes vars.

Hypotype 47648, loc. D-1387, Rio Dell Formation.

A number of specimens occurring in the Pliocene and Pleistocene of the Humboldt basin appear to be varietal forms of *C. translucens* and possibly some represent new species or subspecies. The hypotypes herein are somewhat flattened and lack a sharply defined keel in contrast to *C. translucens (s.s.).*

Cassidulina spp.

Hypotype 47652, loc. D-1237, "Bear River beds."

Specimens occurring in the Miocene of Bear River, probably embracing several species, cannot be attributed to any known species. *C. pulchella* of Kleinpell (1938, p. 335) and Cushman, Stewart, and Stewart (1930, p. 74) falls within this group.

Genus CASSIDULINOIDES Cushman, 1927

Cassidulinoides bradyi (Norman)
Plate XIII, figures 6a, b, c

Cassidulinoides bradyi (Norman); Natland, 1952, pl. 19-3 (Wheelerian, southern California); Barker, 1960, p. 112, pl. liv, figs. 6-9 (Holocene, Atlantic).
Hypotype 47653, loc. D-1283, lower Rio Dell Formation.

The species is comparatively rare in the Humboldt basin, occurring in only a few samples from the Pliocene of the Centerville Coastal section.

Cassidulinoides cornuta (Cushman)
Plate XIII, figures 5a, b, c

Cassidulinoides cornuta (Cushman); Bandy, 1961, p. 15, pl. iv, fig. 12 (Holocene, Gulf of California); White, 1956, p. 256, pl. xxxi, fig. 6 (lower Pliocene, Orange County, California).
Virgulina cornuta Cushman, 1913, p. 637, pl. lxxx, fig. 1 (Holocene, eastern Pacific).
Hypotype 47654, loc. D-1332, Hookton Formation.

The holotype in the U.S. Natl. Museum is comparable to the above mentioned hypotype. The species was found only at the late Pleistocene locality listed above.

Genus EHRENBERGINA Reuss, 1850

Ehrenbergina compressa Cushman
Plate XIV, figures 1a, b, c

Ehrenbergina compressa Cushman, 1927, p. 168, pl. vi, fig. 7 (Holocene, eastern Pacific); Cushman, Stewart, and Stewart, 1930, p. 75, pl. vi, fig. 9 (Pliocene, Humboldt County, California); White, 1956, p. 256, pl. xxxi, fig. 8 (Pliocene, Orange County, California).
Hypotype 47674, loc. D-1384, Eel River Formation.

Family CHILOSTOMELLIDAE
Genus CHILOSTOMELLA Reuss, 1850
Chilostomella czizeki Reuss
Plate XIV, figures 2a, b, c

Chilostomella czizeki Reuss; Cushman, Stewart, and Stewart, 1949, p. 156, pl. xviii, fig. 5 (Pliocene, Washington); L. Martin, 1952, p. 126, pl. xx, fig. 7 (Pliocene, Los Angeles County, California); White, 1956, p. 250, pl. xxix, fig. 2 (Pliocene, Orange County, California); Bandy and Rodolfo, 1964, p. 829, fig. 5C (Holocene, Peru-Chile Trench).
Hypotype 47655, loc. D-1383, Eel River Formation.

Genus PULLENIA Parker and Jones, 1862
Pullenia bulloides (d'Orbigny)
Plate XIV, figures 4a, b

Pullenia bulloides (d'Orbigny); Cushman, Stewart, and Stewart, 1930, p. 76, pl. vii, fig. 3 (Pliocene, Humboldt County, California); Barker, 1960, p. 174, pl. lxxxiv, figs. 12-13 (Holocene, Pacific and Atlantic).
Hypotype 47771, loc. D-1389, Rio Dell Formation.

Pullenia miocenica Kleinpell?
Plate XIV, figures 3a, b

?*Pullenia miocenica* Kleinpell, 1938, p. 338, pl. xiv, fig. 6 (Miocene, California).
Hypotype 47772, loc. D-1231, Pullen(?) Formation.

The species is comparatively rare, occurring in only a few samples from the Miocene of the Humboldt basin; it is questionably placed in synonymy with Kleinpell's type figure pending comparison of specimens. Plesiotypes in the U.S. Natl. Museum are similar to the above hypotype.

Pullenia salisburyi R. E. and K. C. Stewart
Plate XIV, figures 6a, b

Pullenia salisburyi R. E. and K. C. Stewart, 1930b, p. 72, pl. vii, fig. 2 (lower Pliocene, California).
Hypotype 47773, loc. D-1237, "Bear River beds."

Pullenia subcarinata (d'Orbigny)
Plate XIV, figures 5a, b

Pullenia subcarinata (d'Orbigny); Barker, 1960, p. 174, pl. lxxxiv, figs. 14-15 (Holocene, Pacific).
Pullenia quinqueloba (Reuss); Natland, 1952, pl. 19-8 (Pliocene, southern California).
Hypotype 47774, loc. D-1387, Rio Dell Formation.

As Barker (1960) noted, *P. quinqueloba* (of some authors) is probably a junior synonym of *P. subcarinata*.

Genus SPHAEROIDINA d'Orbigny, 1826
Sphaeroidina bulloides d'Orbigny
Plate XIV, figure 7

Sphaeroidina bulloides d'Orbigny; Cushman, Stewart, and Stewart, 1930, p. 76, pl. vii, fig. 2 (Pliocene, Humboldt County, California); Cushman, Stewart, and Stewart, 1949, p. 156, pl. xviii, fig. 7 (Pliocene, Washington); Barker, 1960, p. 174, pl. lxxxiv, figs. 1-7 (Holocene, Atlantic and Pacific).
Sphaeroidina chilostomata Galloway and Morrey; L. Martin, 1952, p. 127, pl. xx, fig. 6 (Pliocene, Los Angeles County, California).
Hypotype 47791, loc. D-1287, Eel River Formation.

Family GLOBIGERINIDAE

Genus GLOBIGERINA d'Orbigny, 1826

Globigerina bulloides d'Orbigny
Plate XV, figures 1a, b, c

Globigerina bulloides d'Orbigny; Galloway and Wissler, 1927b, p. 40, pl. vii, fig. 4 (Pleistocene, southern California); Bandy, 1950, p. 279, pl. xlii, fig. 2 (Pleistocene, Cape Blanco, Oregon); Barker, 1960, p. 164, pl. lxxix, figs. 3-7 (Holocene, loc. not designated).
Hypotype 47705, loc. D-1388, Rio Dell Formation.

The species is a relatively common one in the Pliocene and Pleistocene of the Humboldt basin, occurring in most of the open-ocean biofacies; its occurrence in the Miocene is questionable because of the generally poor preservation of globigerinids therein. Some of the specimens here included under *G. bulloides* probably belong to the new species *Globigerina umbilicata* Orr and Zaitzeff (1971).

Globigerina pachyderma (Ehrenberg)
Plate XV, figures 3a, b, c

Globigerina pachyderma (Ehrenberg); Galloway and Wissler, 1927b, p. 43, pl. vii, fig. 13 (Pleistocene, southern California); Parker, 1962, p. 224, pl. i, figs. 26-35, pl. ii, figs. 1-6 (Holocene, Pacific).
Hypotype 47706, loc. D-1393, Rio Dell Formation.

No attempt was made to determine the direction of coiling in this species as Ericson (1959), Bandy (1960b), and others have done. Although the species is relatively common in the Humboldt basin Pliocene and Pleistocene, many specimens occur which may or may not be attributed to this species; these indeterminate specimens seem to be variants reacting to an unfavorable environment. The prerequisite paleontological continuum, or lack of it, does not appear to support extensive work on coiling ratios in the present study.

Globigerina spp.
Plate XV, figures 2a, b, c

Hypotype 47707, loc. D-1419, Rio Dell Formation.

A number of specimens occur, as shown on Figures 7 and 8, which cannot be identified at the specific level.

"*Globigerina quinqueloba*" recorded from the "Lower Pliocene, Rio Dell fm., Centerville Beach, Calif." by Asano et al (1968, p. 232), was not identified in the present study and apparently occurs in few samples in the Humboldt basin; its occurrence in the "Pliocene of Centerville Beach" was mentioned also by Orr and Zaitzeff (1971).

Genus ORBULINA d'Orbigny, 1839

Orbulina universa d'Orbigny
Plate XVI, figure 1

Orbulina universa d'Orbigny; Galloway and Wissler, 1927b, p. 45, pl. vii, fig. 3 (Pleistocene, southern California); White, 1956, p. 250, pl. xxix, fig. 1 (Pliocene, Orange County, California); Barker, 1960, p. 168, pl. lxxxi, figs. 8-26 (Holocene, Pacific and Atlantic).
Hypotype 47756, loc. D-1303, Wildcat Group undifferentiated.

Family GLOBOROTALIIDAE

Genus GLOBOROTALIA Cushman, 1927

Globorotalia crassaformis (Galloway and Wissler)
Plate XV, figures 5a, b, c

Globorotalia crassaformis (Galloway and Wissler); Parker, 1962, p. 235, pl. iv, figs.

17-18, 20-21 (Holocene, Pacific).
Globorotalia punctulata (d'Orbigny); Barker, 1960, p. 212, pl. ciii, figs. 11-12 (Holocene, Atlantic).
Globigerina crassaformis Galloway and Wissler, 1927b, p. 41, pl. vii, fig. 12 (Pleistocene, southern California).
Hypotype 47711, loc. D-1389, Rio Dell Formation; hypotype 47712, loc. D-1402, Rio Dell Formation.

The species occurs in several samples from the Humboldt basin although specimens are relatively few. It ranges through the Pliocene and Pleistocene.

Hypotype 47711 above appears to be close to *Globorotalia (Turborotalia) crassaformis ronda* in Blow (1969, p. 388-390, pl. 4, figs. 4-6), and hypotype 47712 is similar to Blow's (1969, pl. 4, figs. 7-9) *Globorotalia (Turborotalia) crassaformis oceanica*. Lidz (1972) discussed the variability of this species.

Globorotalia crassula Cushman and R. E. Stewart

The Cushman, Stewart, and Stewart (1930) locality 7 could not be duplicated and consequently topotypic specimens of *G. crassula* could not be obtained, nor were specimens from other samples in this study referred to this species.

In the U.S. Natl. Museum, at least one paratype of *G. crassula* is close to *G. scitula*, whereas the holotype and one other paratype represent a different species in being somewhat more inflated and lacking a well-defined apertural lip. Blow (1969, p. 361, pl. 9, figs. 1-3) has refigured the holotype of *G. crassula* Cushman and Stewart and clarified the earlier figures which represent three separate specimens.

Globorotalia inflata (d'Orbigny)
Plate XV, figures 4a, b, c

Globorotalia inflata (d'Orbigny); Parker, 1962, p. 236, pl. v, figs. 6-9 (Holocene, Pacific).
Globigerina inflata d'Orbigny; Barker, 1960, p. 164, pl. lxxix, figs. 8-10 (Holocene, Atlantic and Pacific).
Hypotype 47713, loc. D-1332, Hookton formation.

In the Humboldt basin, the species occurs only at the locality listed, although Ingle (1968) stated that it comprises more than 40% of the total (foram) population "in two intervals within the early Pliocene" of the Centerville beach section.

Globorotalia scitula (Brady)
Plate XVI, figures 2a, b, c

Globorotalia scitula (Brady); Parker, 1962, p. 238, pl. vi, figs. 4-6 (Holocene, Pacific).
Globorotalia scitula scitula (Brady); Parker, 1964, p. 631, pl. cii, figs. 18-19 (late Miocene, Baja California).
Hypotype 47714, loc. D-1287, Eel River Formation.

The species is comparatively rare in the Humboldt basin, occurring in isolated samples from the Pliocene and Pleistocene.

Family ANOMALINIDAE
Genus ANOMALINA d'Orbigny, 1826
Anomalina salinasensis? Kleinpell
Plate XVI, figures 3a, b, c

?*Anomalina salinasensis* Kleinpell, 1938, p. 347, pl. xii, fig. 1 (Relizian and Luisian, southern California).
Hypotype 47608, loc. D-1224, "Bear River beds."

The hypotype herein is tentatively referred to the figure of Kleinpell's type pending further investigation and comparison of actual specimens.

Genus MELONIS Montfort, 1808
Melonis barleeanus (Williamson)
Plate XVI, figures 4a, b; 6

Nonion barleeanus (Williamson); Bandy, 1953a, p. 177, pl. xxi, fig. 8 (Holocene, California).
Gavelinonion barleeanum (Williamson); Barker, 1960, p. 224, pl. cix, figs. 8-9 (Holocene, Pacific).
Hypotype 47741, loc. D-1303, Wildcat Group undifferentiated; hypotype 47742, loc. 1287, Eel River Formation.

Loeblich et al's (1964) use of the term *Melonis* is followed in denoting the distinction between this species and the true nonionids; this species, as are those following, evidently is related phyletically to the anomalinids as indicated by wall structure characteristics.

Melonis pompilioides (Fichtel and Moll)
Plate XVI, figures 5a, b

Melonis pompilioides (Fichtel and Moll); Loeblich et al, 1964, p. C761, fig. 627-1 (Pliocene, Albania).
Nonion pompilioides (Fichtel and Moll); Natland, 1952, (Repettian, southern California).
Nonion umbilicatula (Montagu); Cushman, Stewart, and Stewart, 1930, p. 60, pl. iii, fig. 14 (Pliocene, Humboldt County, California).
Nonion? *pompilioides* (Fichtel and Moll); Barker, 1960, p. 224, pl. cix, figs. 10-11 (Holocene, eastern Atlantic).
Hypotype 47743, loc. D-1303, Wildcat Group undifferentiated.

Melonis? sp.

Hypotype 47744, loc. D-1231, Pullen(?) Formation.

A number of specimens occurring in the Miocene and possibly also the Pliocene of the Humboldt basin were not identified as to species; they appear to belong either to the genus *Melonis* or to the genus *Anomalina.*

Genus PLANULINA d'Orbigny, 1826
Planulina ariminensis d'Orbigny
Plate XVI, figures 7a, b, c

Planulina ariminensis d'Orbigny; Galloway and Wissler, 1927b, p. 66, pl. xi, fig. 2 (Pleistocene, southern California); Resig, 1958, p. 307, unfigured (Holocene, California).
Planulina astoriensis Cushman, Stewart, and Stewart, 1947b, p. 23, pl. iv, figs. 4-5 (Miocene, western Oregon).
Hypotype 47759, loc. D-1379, Pullen Formation.

Planulina mexicana Cushman
Plate XVII, figures 1a, b, c

Planulina mexicana Cushman, 1927, p. 113, pl. xxii, fig. 5 (upper Eocene, Mexico); Cushman, Stewart, and Stewart, 1947b, p. 23, pl. iv, fig. 6 (Miocene, Oregon).
Hypotype 47760, loc. D-1266, middle Rio Dell Formation.

Planulina wuellerstorfi (Schwager)
Plate XVII, figures 2a, b, c

Planulina wuellerstorfi (Schwager); Barker, 1960, p. 192, pl. xciii, fig. 9 (Holocene, Pacific).
Anomalina wuellerstorfi Schwager, 1866, p. 258, pl. vii, figs. 105, 107 (upper Ter-

tiary?, Kar Nicobar).
Hypotype 47761, loc. D-1285, Eel River Formation.

Genus CIBICIDINA Bandy, 1949

Cibicidina? *washingtonensis* (Cushman, Stewart, and Stewart)
Plate XVII, figures 3a, b, c; 4a, b, c

Cibicides concentricus (Cushman) var. *washingtonensis* Cushman, Stewart, and Stewart, 1949, p. 157, pl. xviii, fig. 8 (Pliocene, western Washington).
Valvulineria menloensis Rau, 1951, p. 446, pl. lxvi, figs. 17-22 (Miocene, Washington).
Hypotype 47663, loc. D-1385, Eel River Formation; hypotype 47664, loc. D-1240, "Bear River beds."

The holotype in the U.S. Natl. Museum compares with the above hypotypes.

The species occurs at several Miocene localities in the Humboldt basin and also at a few Pliocene localities. The generic status of this species is uncertain; perhaps the term *Hanzawaia* should have priority. Moreover, controversy exists as to its relation to other foraminifera; this paper prefers to group the species (and genus) with the anomalinids rather than the rotalids.

Genus CIBICIDES Montfort, 1808

Cibicides floridanus (Cushman)
Plate XVII, figures 5a, b, c

Cibicides floridanus (Cushman); Cushman, Stewart, and Stewart, 1947b, p. 23, p. iv, pl. 7 (Miocene, Oregon); P. Smith, 1964, p. B46, pl. vi, fig. 6 (Holocene, eastern Pacific).
Hypotype 47656, loc. D-1387, Rio Dell Formation; hypotype 47657, loc. D-1237, "Bear River beds."

Cibicides lobatulus (Walker and Jacob)
Plate XVIII, figures 3a, b, c; 4a, b, c; 5a, b, c

Cibicides lobatulus (Walker and Jacob); Barker, 1960, p. 192, pl. xciii, fig. 1, 4, 5 (Holocene, Pacific and Atlantic).
Cibicides lobatulus (d'Orbigny); Galloway and Wissler, 1927b, p. 64, pl. xi, fig. 1 (Pleistocene, southern California).
Hypotypes 47658, 47659, 47660, loc. D-1303, Wildcat Group undifferentiated.

The species apparently exhibits a very wide range in variation, part of which is due to the fact that specimens often live firmly attached to algae or other organisms and consequently the test varies in relation to the configuration of the substratum.

Cibicides mckannai Galloway and Wissler
Plate XVIII, figures 2a, b, c

Cibicides mckannai Galloway and Wissler, 1927b, p. 65, pl. x, figs. 5-6 (Pleistocene, southern California); White, 1956, p. 249, pl. xxviii, fig. 6 (Pliocene, Orange County, California).
Hypotype 47661, loc. D-1357, Eel River Formation.

It is impossible to tell from the figures given by P. Smith (1964; Holocene, eastern Pacific) if her *C. mckannai* is synonymous.

Cibicides mckannai Galloway and Wissler var. *spiralis* Natland
Plate XVIII, figures 1a, b, c

Cibicides spiralis Natland, 1938, p. 151, pl. vii, fig. 7 (Repettian, southern California); Natland, 1952, pl. 20-7 (Repettian and Holocene, southern California); Crouch, 1952, p. 842, pl. vii, figs. 13-14 (Holocene, California).

Hypotype 47662, loc. D-1357, Eel River Formation.

Populations involving both *Cibicides mckannai* and *C. spiralis* in the Humboldt basin samples seem to be transitional and show phenotypic rather than genotypic variation; therefore, *C. spiralis* is here relegated to the status of subspecies.

PLATES I – XVIII

Plate I

FIG. 1– *Rhabdammina abyssorum*? W. B. Carpenter. Hypotype 47782, U.C. (University of California, Berkeley) loc. D-1301. ×22.
FIG. 2– *Bathysiphon arenacea* Cushman. Hypotype 47610, U.C. loc. D-1301. ×20.
FIG. 3– *Saccammina sphaerica* M. Sars. Hypotype 47786, U.C. loc. D-1301. ×50.
FIG. 4– *Hormosina globulifera* H. B. Brady. Hypotype 47718, U.C. loc. D-1301. ×33.
FIG. 5– *Cyclammina cancellata* H. B. Brady. **a**, lateral view; **b**, apertural view. Hypotype 47666, U.C. loc. D-1348. ×13.
FIG. 6– *Cyclammina* sp. **a**, lateral view; **b**, apertural view. Hypotype 47668, U.C. loc. D-1387. ×50.
FIG. 7– *Alveolophragmium scitulum* (H. B. Brady). **a**, lateral view; **b**, apertural view. Hypotype 47601, U.C. loc. D-1349. ×39.
FIG. 8– *Liebusella pliocenica* (Natland). Hypotype 47738, U.C. loc. D-1348. ×11.
FIG. 9– *Martinottiella communis* (d'Orbigny). Hypotype 47740, U.C. loc. D-1350. ×25.
FIG. 10– *Textularia*? *flintii* Cushman. **a**, lateral view; **b**, apertural view. Hypotype 47797, U.C. loc. D-1303. ×55.
FIG. 11– *Verneuilina scabra*? (Williamson). **a**, lateral view; **b**, apertural view. Hypotype 47810, U.C. loc. D-1383. ×45.
FIG. 12– *Gaudryina atlantica* (Bailey) var. **a**, lateral view; **b**, apertural view. Hypotype 47699, U.C. loc. D-1303. ×31.
FIG. 13– *Gaudryina pliocenica* Cushman, Stewart, and Stewart. **a**, lateral view; **b**, apertural view. Hypotype 47698, U.C. loc. D-1248. ×24.

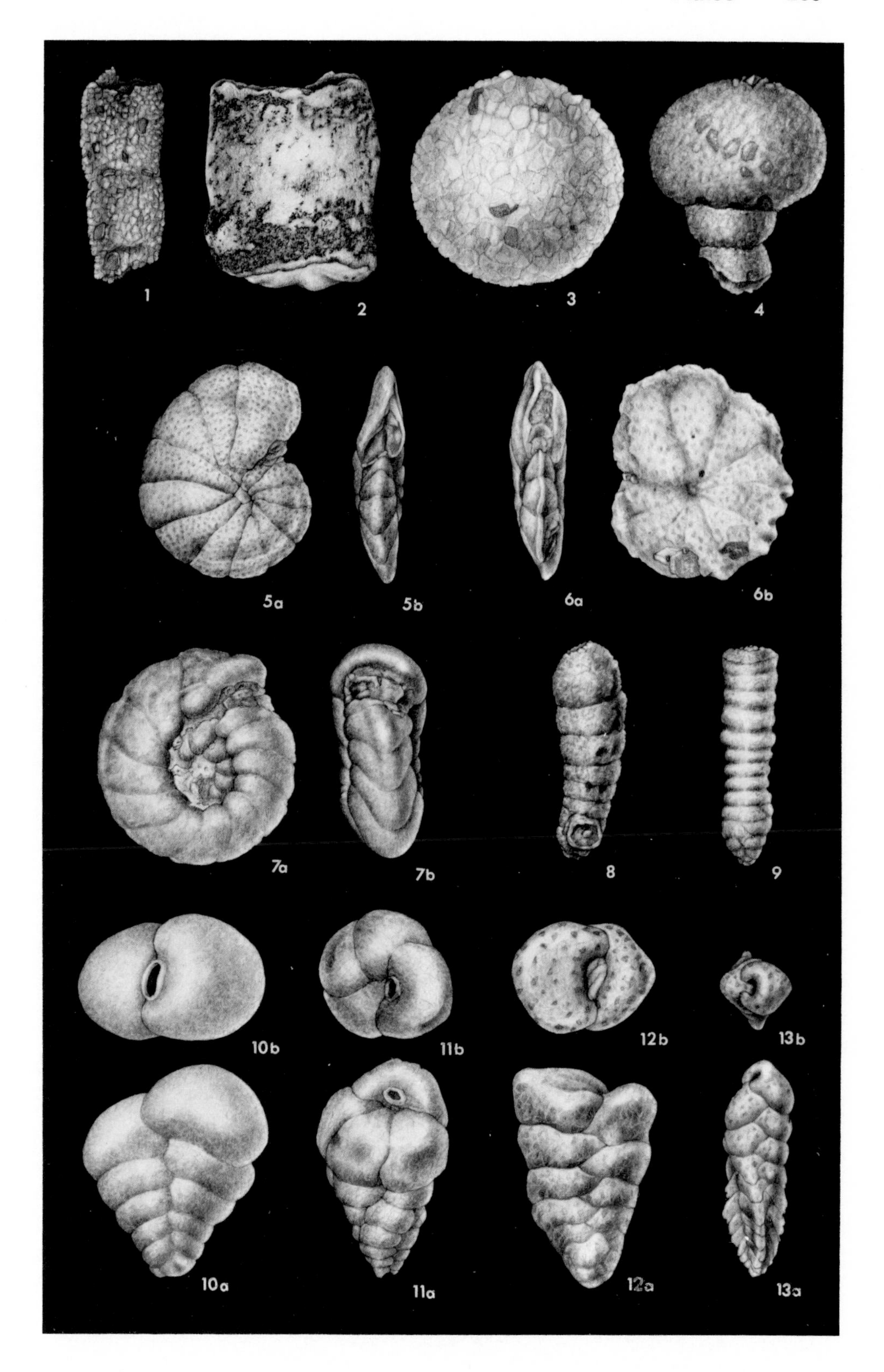
1
2
3
4
5a
5b
6a
6b
7a
7b
8
9
10b
11b
12b
13b
10a
11a
12a
13a

Plate II

FIG. 1–*Eggerella*? *bradyi* (Cushman). **a**, lateral view; **b**, apertural view; **c**, lateral view. Hypotype 47673, U.C. loc. D-1355. ×71.

FIG. 2–*Cribrogoësella pacifica* Cushman and McCulloch. **a**, lateral view; **b**, apertural view. Hypotype 47665, U.C. loc. D-1303. ×15.

FIG. 3–*Karreriella milleri* Natland. **a**, lateral view; **b**, apertural view. Hypotype 47719, U.C. loc. D-1357. ×25.

FIG. 4–*Quinqueloculina akneriania* d'Orbigny. **a**, lateral view; **b**, apertural view; **c**, lateral view. Hypotype 47779, U.C. loc. D-1383. ×71.

FIG. 5–*Quinqueloculina vulgaris* d'Orbigny. **a**, lateral view; **b**, apertural view; **c**, lateral view. Hypotype 47781, U.C. loc. D-1254. ×26.

FIG. 6–*Quinqueloculina elongata* Natland. **a**, lateral view; **b**, apertural view; **c**, lateral view. Hypotype 47780, U.C. loc. D-1379. ×35.

FIG. 7–*Spiroloculina depressa* d'Orbigny. **a**, lateral view; **b**, apertural view. Hypotype 47792, U.C. loc. D-1383. ×71.

FIG. 8–*Sigmoilina celata* (Costa). **a**, lateral view; **b**, apertural view. Hypotype 47787, U.C. loc. D-1384. ×63.

FIG. 9–*Sigmoilina tenuis* (Czjzek). **a**, lateral view; **b**, apertural view. Hypotype 47788, U.C. loc. D-1385. ×83.

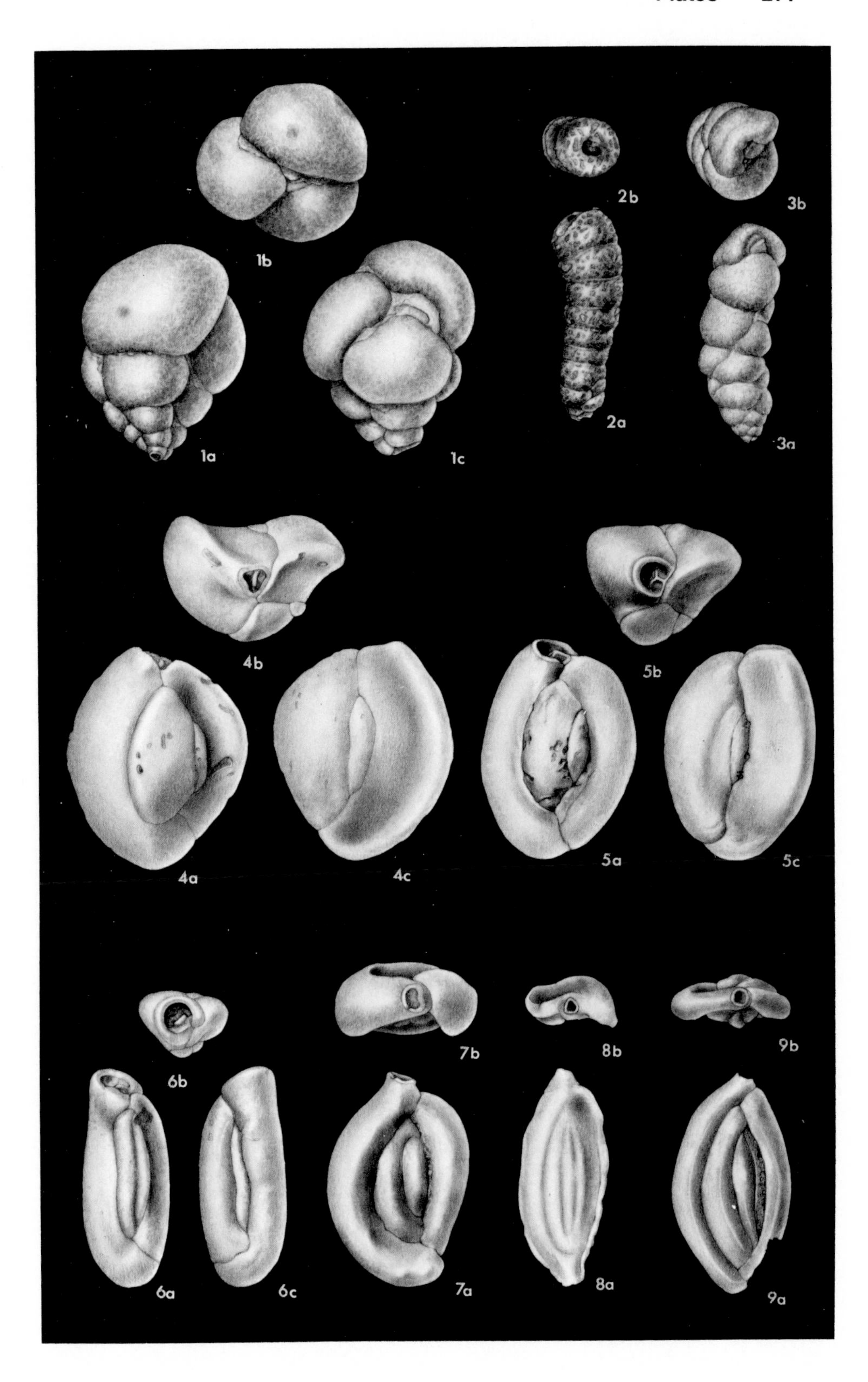
1b
1a
1c
2b
2a
3b
3a
4b
4a
4c
5b
5a
5c
6b
6a
6c
7b
7a
8b
8a
9b
9a

Plate III

FIG. 1– *Pyrgo oblonga* (d'Orbigny). **a,** lateral view; **b,** apertural view. Hypotype 47775, U.C. loc. D-1385, ×100.

FIG. 2– *Pyrgo rotalaria* Loeblich and Tappan. **a,** lateral view; **b,** apertural view. Hypotype 47776, U.C. loc. D-1385. ×71.

FIG. 3– *Pyrgo vespertilio* (Schlumberger). **a,** lateral view; **b,** apertural view. Hypotype 47777, U.C. loc. D-1254. ×29.

FIG. 4– *Lenticulina calcar* (Linné). **a,** lateral view; **b,** apertural view. Hypotype 47733, U.C. loc. D-1303. ×39.

FIG. 5– *Lenticulina cushmani* (Galloway and Wissler). **a,** lateral view; **b,** apertural view. Hypotype 47734, U.C. loc. D-1384. ×55.

FIG. 6– *Lenticulina nikobarensis* (Schwager). **a,** lateral view; **b,** apertural view. Hypotype 47735, U.C. loc. D-1237. ×40.

FIG. 7– *Lenticulina polita* (Schwager). **a,** lateral view; **b,** apertural view. Hypotype 47736, U.C. loc. D-1303. ×33.

FIG. 8– *Lenticulina* sp.? **a,** lateral view; **b,** apertural view. Hypotype 47737, U.C. loc. D-1237. ×30.

FIG. 9– *Nodosaria tosta* Schwager. Hypotype 47749, U.C. loc. D-1303. ×12.

FIG. 10– *Nodosaria tympaniplectriformis* Schwager. Hypotype 47751, U.C. loc. D-1303. ×20.

FIG. 11– *Dentalina quadrulata*? Cushman and Laiming. Hypotype 47669, U.C. loc. D-1231. ×53.

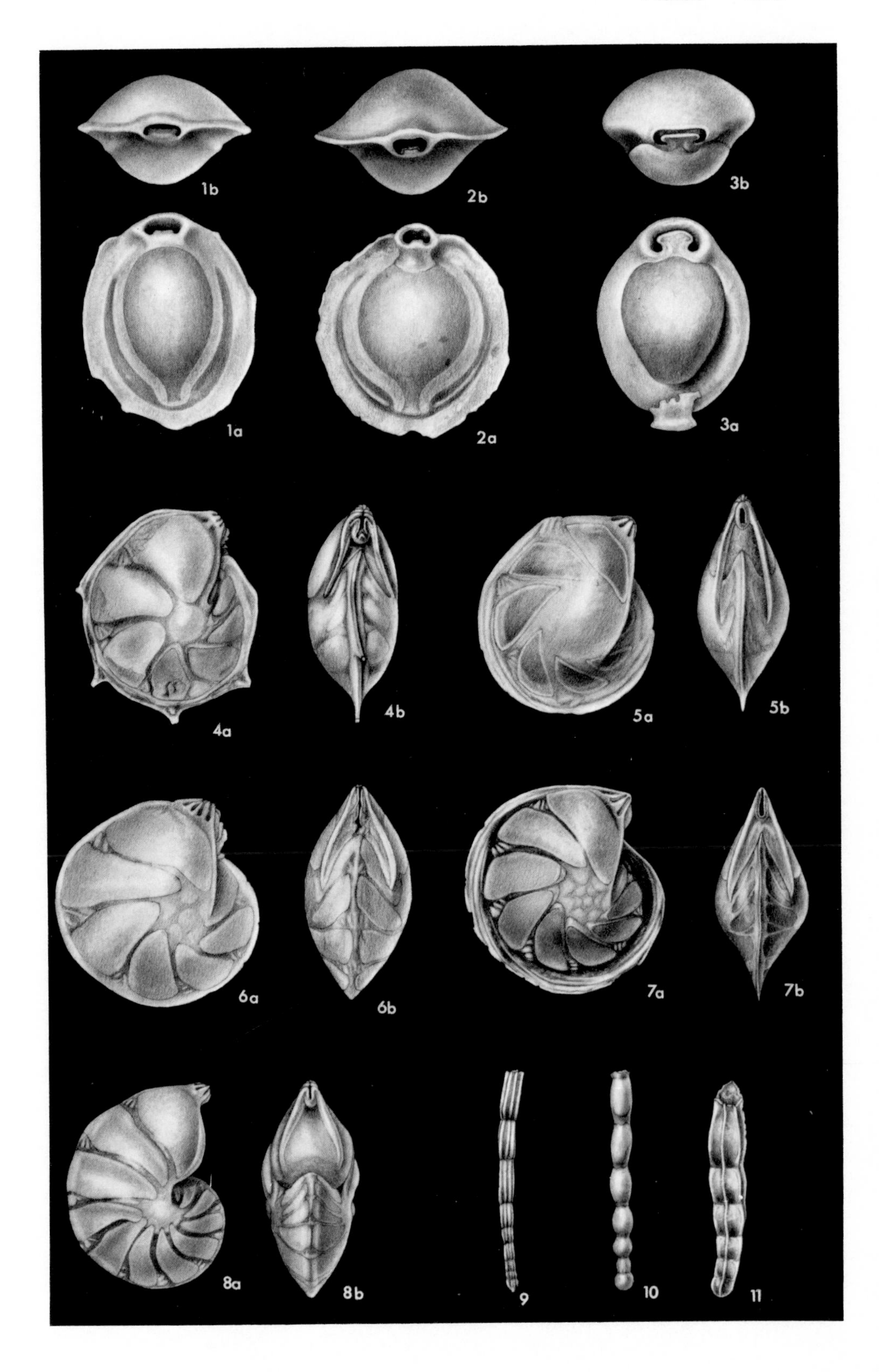
1b
2b
3b
1a
2a
3a
4a
4b
5a
5b
6a
6b
7a
7b
8a
8b
9
10
11

Plate IV

FIG. 1– *Nodosaria moniliformis* Ehrenberg. **a**, lateral view; **b**, apertural view. Hypotype 47746, U.C. loc. D-1272. ×20.
FIG. 2– *Nodosaria raphanistrum* (Linné). **a**, lateral view; **b**, apertural view. Hypotype 47747, U.C. loc. D-1383. ×17.
FIG. 3– *Nodosaria tornata* Schwager. **a**, lateral view; **b**, apertural view. Hypotype 47750, U.C. loc. D-1383. ×18.
FIG. 4– *Dentalina soluta* Reuss (microspheric form). **a**, lateral view; **b**, apertural view. Hypotype 47670, U.C. loc. D-1303. ×21.
FIG. 5– *Dentalina soluta* Reuss (macrospheric form). **a**, lateral view; **b**, apertural view. Hypotype 47671, U.C. loc. D-1262. ×21.
FIG. 6– *Lagena alcocki* White. Hypotype 47721, U.C. loc. D-1389. ×100.
FIG. 7– *Lagena* cf. *L. amphora* Reuss. Hypotype 47722, U.C. loc. D-1303. ×91.
FIG. 8– *Lagena foveolata* Reuss. Hypotype 47724, U.C. loc. D-1303. ×100.
FIG. 9– *Lagena striata* (d'Orbigny). Hypotype 47727, U.C. loc. D-1303. ×71.
FIG. 10– *Lagena clypeata* (Sidebottom). Hypotype 47723, U.C. loc. D-1383. ×71.
FIG. 11– *Lagena melo* (d'Orbigny). Hypotype 47725, U.C. loc. D-1383. ×111.
FIG. 12– *Lagena pliocenica* Cushman and Gray. Hypotype 47726, U.C. loc. D-1303. ×67.
FIG. 13– *Glandulina laevigata* (d'Orbigny). Hypotype 47703, U.C. loc. D-1303. ×55.
FIG. 14– *Glandulina comatula* (Cushman). Hypotype 47701, U.C. loc. D-1303. ×53.
FIG. 15– *Polymorphina charlottensis* Cushman. Hypotype 47769, U.C. loc. D-1391. ×26.
FIG. 16– *Pseudopolymorphina ovalis* Cushman and Ozawa. Hypotype 47770, U.C. loc. D-1389. ×77.

1b
2b
3b
4b
5b
1a
2a
3a
4a
5a
6
7
8
9
10
11
12
13
14
15
16

Plate V

FIG. 1–*Fissurina semimarginata* (Reuss). **a,** lateral view; **b,** apertural view. Hypotype 47697, U.C. loc. D-1303. ×111.

FIG. 2–*Fissurina lucida* (Williamson). **a,** lateral view; **b,** apertural view. Hypotype 47696, U.C. loc. D-1387. ×111.

FIG. 3–*Nonion costiferum* (Cushman). **a,** lateral view; **b,** apertural view. Hypotype 47752, U.C. loc. D-1240. ×63.

FIG. 4–*Nonionella miocenica* Cushman. **a,** lateral view; **b,** apertural view; **c,** lateral view. Hypotype 47755, U.C. loc. D-1243. ×77.

FIG. 5–*Elphidium granti* Kleinpell. Hypotype 47679, U.C. loc. D-1236. ×36.

FIG. 6–*Nonionella cushmani* R. E. and K. C. Stewart. **a,** lateral view; **b,** apertural view; **c,** lateral view. Hypotype 47753, U.C. loc. D-1405. ×50.

FIG. 7–Elphidium acutum Natland **a,** lateral view; **b,** apertural view. Hypotype 47675, U. C. loc. D-1276. X55.

FIG. 8–*Elphidium foraminosum* Cushman. **a,** lateral view; **b,** apertural view. Hypotype 47677, U.C. loc. D-1405. ×67.

FIG. 9–*Elphidium humboldtensis* Haller n. sp. **a,** lateral view; **b,** apertural view. Holotype 47680, U.C. loc. D-1240. ×67.

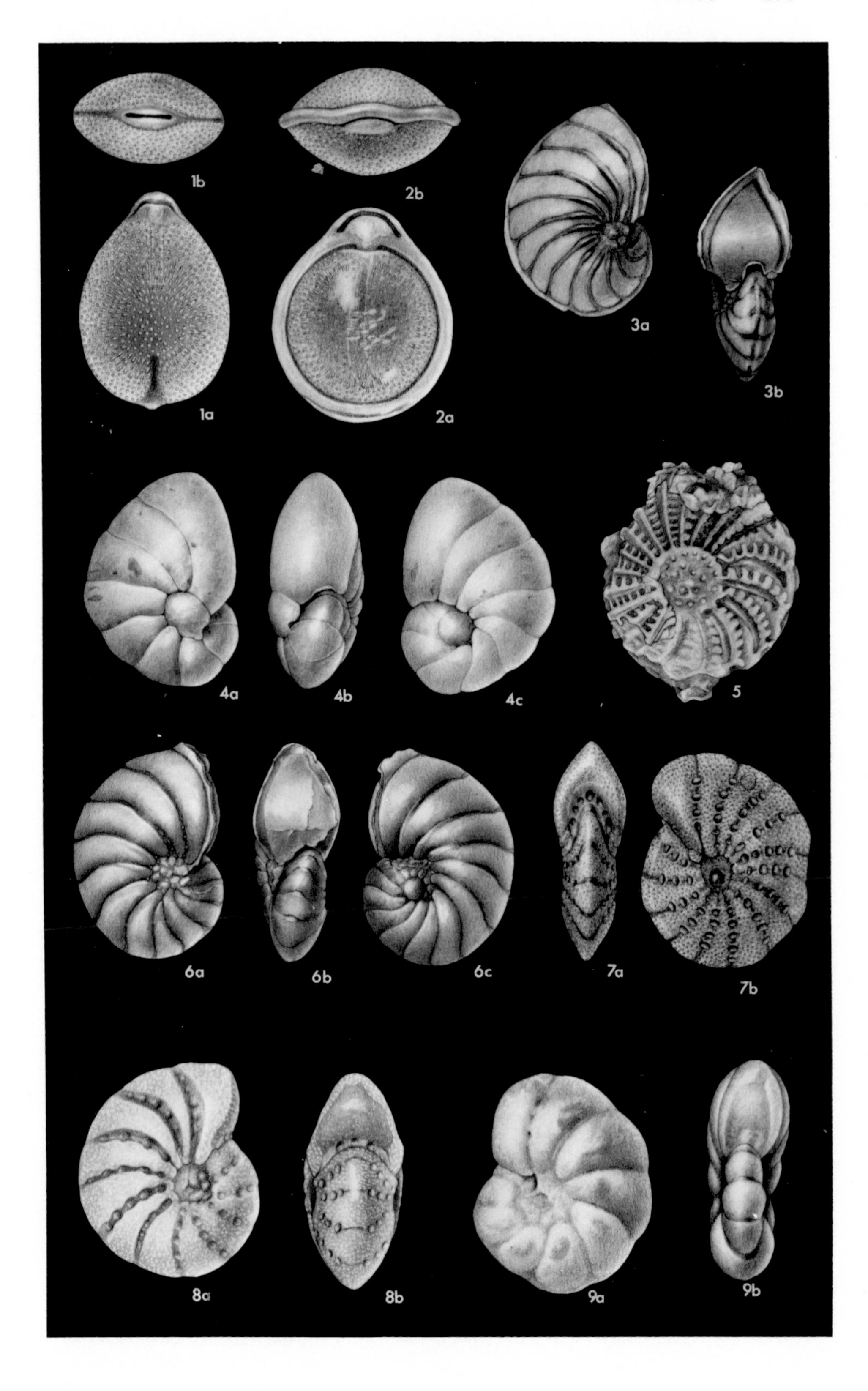
1b
2b
3a
1a
2a
3b
4a
4b
4c
5
6a
6b
6c
7a
7b
8a
8b
9a
9b

Plate VI

FIG. 1– *Elphidium poeyanum* (d'Orbigny). **a,** lateral view; **b,** apertural view. Hypotype 47683, U.C. loc. D-1329. ×50.

FIG. 2– *Elphidiella oregonensis* (Cushman and Grant). **a,** lateral view; **b,** apertural view. Hypotype 47687, U.C. loc. D-1256. ×14.

FIG. 3– *Elphidiella hannai* (Cushman and Grant). **a,** lateral view; **b,** apertural view. Hypotype 47684, U.C. loc. D-1256. ×33.

FIG. 4– *Elphidiella hannai* (Cushman and Grant) var. **a,** lateral view; **b,** apertural view. Hypotype 47686, U.C. loc. D-1419. ×55.

FIG. 5– *Plectofrondicularia advena* (Cushman). Hypotype 47763, U.C. loc. D-1272. ×25.

FIG. 6– *Plectofrondicularia advena* (Cushman). Hypotype 47762, U.C. loc. D-1266. ×23.

FIG. 7– *Plectofrondicularia advena* (Cushman) var. Hypotype 47764, U.C. loc. D-1287. ×37.

FIG. 8– *Plectofrondicularia miocenica* Cushman. Hypotype 47767, U.C. loc. D-1238. ×35.

FIG. 9– *Plectofrondicularia californica* Cushman and Stewart. Hypotype 47766, U.C. loc. D-1303. ×13.

FIG. 10– *Plectofrondicularia* sp.? Hypotype 47768, U.C. loc. D-1238. ×45.

FIG. 11– *Buliminella subfusiformis* Cushman var. Hypotype 47637, U.C. loc. D-1266. ×32.

FIG. 12– *Buliminella elegantissima* (d'Orbigny). **a,** lateral view; **b,** apertural view. Hypotype 47635, U.C. loc. D-1406. ×100.

FIG. 13– *Buliminella subfusiformis* Cushman. **a,** lateral view; **b,** apertural view. Hypotype 47636, U.C. loc. D-1388. ×63.

FIG. 14– *Bolivinita quadrilatera* (Schwager). **a,** lateral view; **b,** apertural view. Hypotype 47624, U.C. loc. D-1361. ×42.

1a
1b
2a
2b
3a
3b
4a
4b
5
6
7
8
9
12b
13b
14b
10
11
12a
13a
14a

Plate VII

FIG. 1– *Bulimina subacuminata* Cushman and Stewart emend. (*B. subcalva s.l.*). **a**, lateral view; **b**, apertural view. Hypotype 47633, U.C. loc. D-1387. ×67.

FIG. 2– *Bulimina subacuminata* Cushman and Stewart (*B. subacuminata s.s.*). **a**, lateral view; **b**, apertural view. Hypotype 47631, U.C. loc. D-1284. ×67.

FIG. 3– *Bulimina fossa* Cushman and Parker. **a**, lateral view; **b**, apertural view. Hypotype 47630, U.C. loc. D-1385. ×100.

FIG. 4– *Bulimina deformata* Cushman and Parker. **a**, lateral view; **b**, apertural view. Hypotype 47628, U.C. loc. D-1391. ×59.

FIG. 5– *Bulimina*? sp. **a**, lateral view; **b**, apertural view. Hypotype 47634, U.C. loc. D-1332. ×53.

FIG. 6– *Globobulimina affinis* (d'Orbigny). **a**, lateral view; **b**, apertural view. Hypotype 47708, U.C. loc. D-1278. ×50.

FIG. 7– *Globobulimina*? *auriculata* Bailey. **a**, lateral view; **b**, apertural view. Hypotype 47709, U.C. loc. D-1278. ×39.

FIG. 8– *Globobulimina pacifica* Cushman. Hypotype 47710, U.C. loc. D-1278. ×63.

FIG. 9– *Uvigerinella californica* Cushman var. *ornata*? Cushman. **a**, lateral view; **b**, apertural view. Hypotype 47807, U.C. loc. D-1240. ×77.

FIG. 10– *Angulogerina angulosa* (Williamson). **a**, lateral view; **b**, apertural view. Hypotype 47605, U.C. loc. D-1244, ×59.

FIG. 11– *Angulogerina* aff. *A. hughesi* (Galloway and Wissler). **a**, lateral view; **b**, apertural view. Hypotype 47607, U.C. loc. D-1385. ×91.

FIG. 12– *Uvigerina hootsi*? Rankin. **a**, lateral view; **b**, apertural view. Hypotype 47800, U.C. loc. D-1228. ×63.

FIG. 13– *Uvigerina peregrina* Cushman var. *dirupta* Todd. **a**, lateral view; **b**, apertural view. Hypotype 47801, U.C. loc. D-1285. ×50.

FIG. 14– *Uvigerina peregrina* Cushman var. *dirupta* Todd. **a**, lateral view; **b**, apertural view. Hypotype 47802, U.C. loc. D-1386. ×55.

1b
2b
3b
4b
5b
1a
2a
3a
4a
5a
6b
7b
9b
8
6a
7a
9a
10b
11b
12b
13b
14b
10a
11a
12a
13a
14a

Plate VIII

FIG. 1– *Uvigerina hispida* Schwager. Hypotype 47799, U.C. loc. D-1383. ×50.
FIG. 2– *Uvigerina subperegrina* Cushman and Kleinpell. Hypotype 47806, U.C. loc. D-1228. ×71.
FIG. 3– *Uvigerina peregrina* Cushman var. *foxensis* Bramlette. Hypotype 47803, U.C. loc. D-1272. ×32.
FIG. 4– *Uvigerina peregrina* Cushman var. *hollicki* Thalmann. Hypotype 47805, U.C. loc. D-1248. ×33.
FIG. 5– *Uvigerina peregrina* Cushman var. *hollicki* Thalmann. Hypotype 47804, U.C. loc. D-1389. ×40.
FIG. 6– *Siphonodosaria insecta* (Schwager). Hypotype 47790, U.C. loc. D-1295. ×22.
FIG. 7– *Stilostomella lepidula* (Schwager). Hypotype 47795, U.C. loc. D-1303. ×26.
FIG. 8– *Stilostomella*? *advena* (Cushman and Laiming). Hypotype 47793, U.C. loc. D-1303. ×31.
FIG. 9– *Bolivina subadvena* Cushman var. *spissa* Cushman (microspheric form). **a**, lateral view; **b**, apertural view. Hypotype 47621, U.C. loc. D-1272. ×39.
FIG. 10– *Bolivina subadvena* Cushman var. *spissa* Cushman (macrospheric form). **a**, lateral view; **b**, apertural view. Hypotype 47622, U.C. loc. D-1272. ×42.
FIG. 11– *Bolivina pseudobeyrichi* (Cushman). **a**, lateral view; **b**, apertural view. Hypotype 47616, U.C. loc. D-1303. ×28.
FIG. 12– *Bolivina cochei* Cushman and Adams. **a**, lateral view; **b**, apertural view. Hypotype 47612, U.C. loc. D-1231. ×53.
FIG. 13– *Bolivina cochei* Cushman and Adams var. **a**, lateral view; **b**, apertural view. Hypotype 47613, U.C. loc. D-1385. ×47.
FIG. 14– *Bolivina subadvena* Cushman var. *acuminata* Natland. **a**, lateral view; **b**, apertural view. Hypotype 47619, U.C. loc. D-1335. ×59.
FIG. 15– *Bolivina interjuncta* (Cushman). **a**, lateral view. **b**, apertural view. Hypotype 47615, U.C. loc. D-1248. ×45.
FIG. 16– *Bolivina sulphurensis* Cushman and Adams. **a**, lateral view; **b**, apertural view. Hypotype 47623, U.C. loc. D-1303. ×59.
FIG. 17– *Bolivina advena* Cushman var. *striatella* Cushman. **a**, lateral view; **b**, apertural view. Hypotype 47611, U.C. loc. D-1224. ×63.
FIG. 18– *Bolivina sinuata* Galloway and Wissler var. *alisoensis* Cushman and Adams. **a**, lateral view; **b**, apertural view. Hypotype 47618, U.C. loc. D-1379. ×50.
FIG. 19– *Bolivina* aff. *B. decussata* H. B. Brady. **a**, lateral view, **b**, apertural view. Hypotype 47614, U.C. loc. D-1303. ×77.

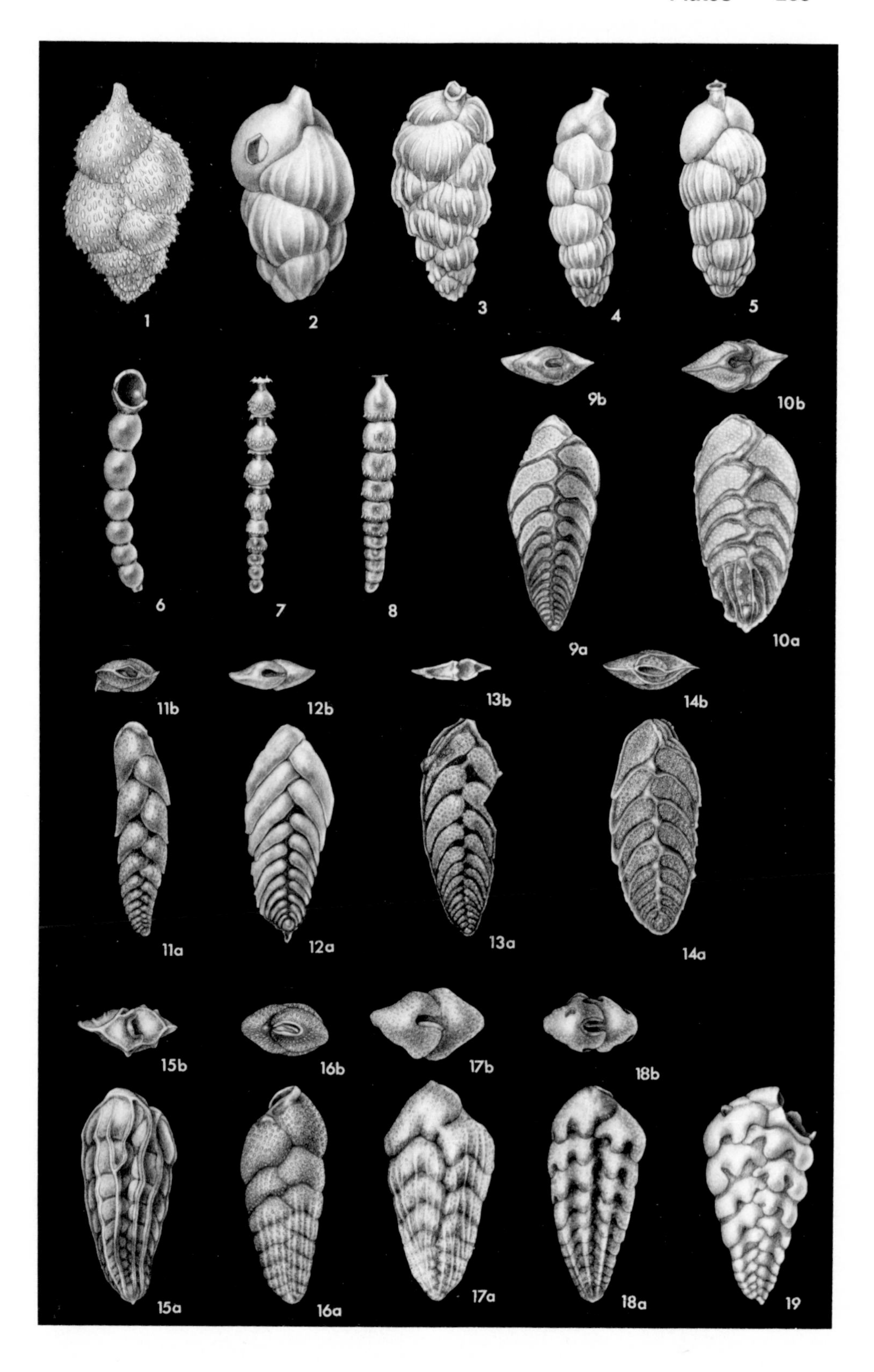
1
2
3
4
5
9b
10b
6
7
8
9a
10a
11b
12b
13b
14b
11a
12a
13a
14a
15b
16b
17b
18b
15a
16a
17a
18a
19

Plate IX

FIG. 1–*Discorbis campanulata* (Galloway and Wissler). **a,** dorsal view; **b,** apertural view; **c,** ventral view. Hypotype 47672, U.C. loc. D-1303. ×83.

FIG. 2–*Buccella tenerrima* (Bandy). **a,** dorsal view; **b,** apertural view; **c,** ventral view. Hypotype 47627, U.C. loc. D-1249. ×55.

FIG. 3–*Buccella oregonensis* (Cushman, Stewart, and Stewart). **a,** dorsal view; **b,** apertural view; **c,** ventral view. Hypotype 47625, U.C. loc. D-1238. ×53.

FIG. 4–*Baggina californica* Cushman. **a,** dorsal view; **b,** apertural view; **c,** ventral view. Hypotype 47609, U.C. loc. D-1224, ×77.

FIG. 5–*Valvulineria araucana* (d'Orbigny). **a,** dorsal view; **b,** apertural view; **c,** ventral view. Hypotype 47808, U.C. loc. D-1272. ×55.

FIG. 6–*Valvulineria araucana* (d'Orbigny). **a,** dorsal view; **b,** apertural view; **c,** ventral view. Hypotype 47809, U.C. loc. D-1321. ×71.

1a
1b
1c
2a
3c
3b
2b
2c
3a
4c
4b
5c
5b
4a
5a
6a
6b
6c

Plate X

FIG. 1–*Laticarinina halophora* (Stache). **a**, lateral view; **b**, apertural view. Hypotype 47732, U.C. loc. D-1351. ×64.

FIG. 2–*Rosalina columbiensis* (Cushman). **a**, dorsal view; **b**, apertural view; **c**, ventral view. Hypotype 47784, U.C. loc. D-1303. ×36.

FIG. 3–*Epistominella exigua* (H. B. Brady). **a**, dorsal view; **b**, apertural view; **c**, ventral view. Hypotype 47689, U.C. loc. D-1249. ×143.

FIG. 4–*Rosalina columbiensis* (Cushman). **a**, dorsal view; **b**, apertural view; **c**, ventral view. Hypotype 47783, U.C. loc. D-1303. ×53.

FIG. 5–*Epistominella pacifica* (Cushman) var. **a**, dorsal view; **b**, apertural view; **c**, ventral view. Hypotype 47691, U.C. loc. D-1332. ×108.

FIG. 6–*Epistominella pacifica* (Cushman). **a**, dorsal view; **b**, apertural view; **c**, ventral view. Hypotype 47690, U.C. loc. D-1389. ×100.

FIG. 7–*Epistominella subperuviana* (Cushman). **a**, dorsal view; **b**, apertural view; **c**, ventral view. Hypotype 47692, U.C. loc. D-1303. ×111.

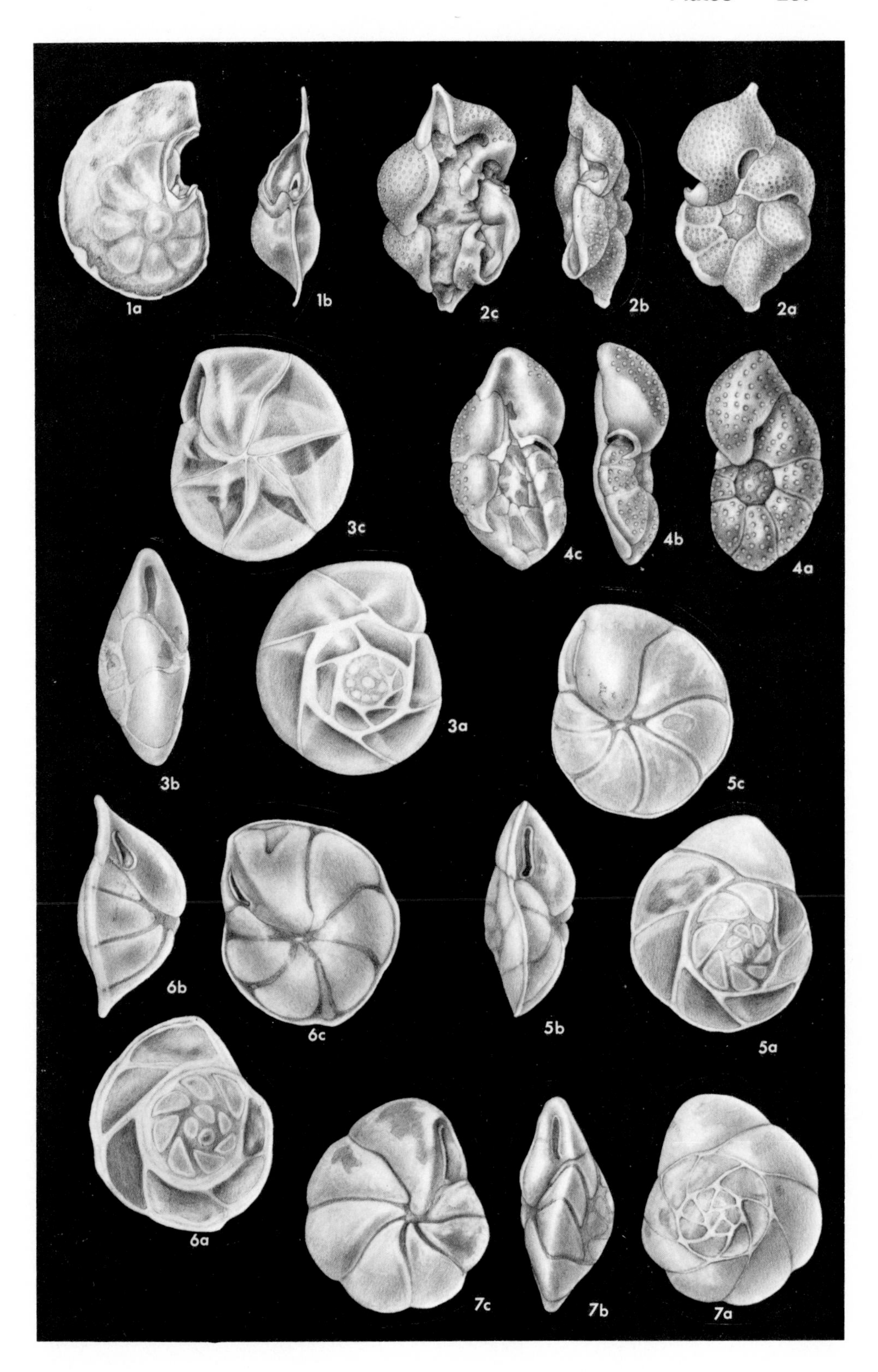
1a
1b
2c
2b
2a
3c
4c
4b
4a
3b
3a
5c
6b
6c
5b
5a
6a
7c
7b
7a

Plate XI

FIG. 1–*Gyroidina soldanii* d'Orbigny var. *rotundimargo* R. E. and K. C. Stewart. **a**, dorsal view; **b**, apertural view; **c**, ventral view. Hypotype 47716, U.C. loc. D-1384. ×77.

FIG. 2–*Gyroidina soldanii* d'Orbigny var. *altiformis* R. E. and K. C. Stewart. **a**, dorsal view; **b**, apertural view; **c**, ventral view. Hypotype 47715, U.C. loc. D-1386. ×71.

FIG. 3–*Oridorsalis umbonatus* (Reuss). **a**, dorsal view; **b**, apertural view; **c**, ventral view. Hypotype 47758, U.C. loc. D-1237. ×67.

FIG. 4–*Eponides healdi* R. E. and K. C. Stewart. **a**, dorsal view; **b**, apertural view; **c**, ventral view. Hypotype 47693, U.C. loc. D-1285. ×67.

FIG. 5–*Eponides repandus* (Fichtel and Moll). **a**, dorsal view; **b**, apertural view; **c**, ventral view. Hypotype 47695, U.C. loc. D-1303. ×29.

FIG. 6–"*Rotalia*" *beccarii* (Linné). **a**, dorsal view; **b**, apertural view; **c**, ventral view. Hypotype 47785, U.C. loc. D-1329. ×83.

1a
1b
1c
2a
3b
3c
2c
2b
3a
4a
4c
5b
5c
4b
5a
6c
6b
6a

Plate XII

FIG. 1–*Höglundina elegans* (d'Orbigny). **a**, dorsal view; **b**, apertural view; **c**, ventral view. Hypotype 47717, U.C. loc. D-1303. ×42.

FIG. 2–?*Cassidulina californica* Cushman and Hughes. **a**, dorsal view; **b**, apertural view; **c**, ventral view. Hypotype 47639, U.C. loc. D-1389. ×83.

FIG. 3–*Cassidulina delicata*? Cushman. **a**, dorsal view; **b**, apertural view; **c**, ventral view. Hypotype 47642, U.C. loc. D-1250. ×91.

FIG. 4–*Cassidulina crassa*? d'Orbigny. **a**, dorsal view; **b**, apertural view; **c**, ventral view. Hypotype 47641, U.C. loc. D-1237. ×63.

FIG. 5–*Cassidulina carinata* Silvestri. **a**, dorsal view; **b**, apertural view; **c**, ventral view. Hypotype 47640, U.C. loc. D-1272. ×100.

FIG. 6–*Cassidulina minuta* Cushman. **a**, dorsal view; **b**, apertural view; **c**, ventral view. Hypotype 47645, U.C. loc. D-1389. ×143.

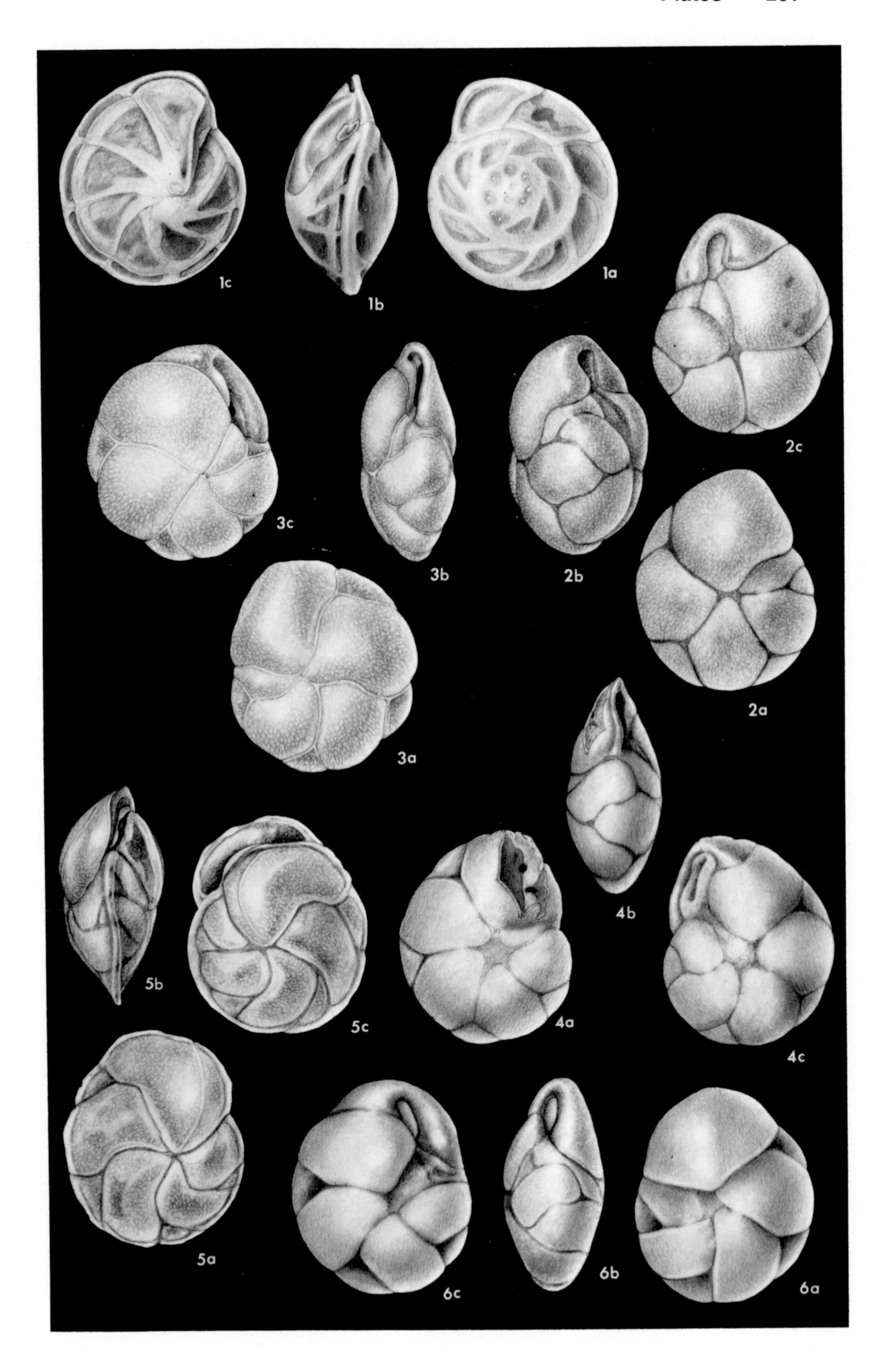
1c
1b
1a
2c
3c
3b
2b
2a
3a
4b
5b
5c
4a
4c
5a
6c
6b
6a

Plate XIII

FIG. 1—*Cassidulina limbata* Cushman and Hughes. **a**, dorsal view; **b**, apertural view; **c**, ventral view. Hypotype 47643, U.C. loc. D-1262. ×47.

FIG. 2—*Cassidulina translucens* Cushman and Hughes var. *natlandi* Haller n. var. **a**, dorsal view; **b**, apertural view; **c**, ventral view. Holotype 47649, U.C. loc. D-1249. ×59.

FIG. 3—*Cassidulina translucens* Cushman and Hughes. **a**, dorsal view; **b**, apertural view; **c**, ventral view. Hypotype 47647, U.C. loc. D-1272. ×55.

FIG. 4—*Cassidulina tortuosa* Cushman and Hughes. **a**, dorsal view; **b**, apertural view; **c**, ventral view. Hypotype 47646, U.C. loc. D-1337. ×43.

FIG. 5—*Cassidulinoides cornuta* (Cushman). **a**, dorsal view; **b**, apertural view; **c**, ventral view. Hypotype 47654, U.C. loc. D-1332. ×91.

FIG. 6—*Cassidulinoides bradyi* (Norman). **a**, dorsal view; **b**, apertural view; **c**, ventral view. Hypotype 47653, U.C. loc. D-1283. ×67.

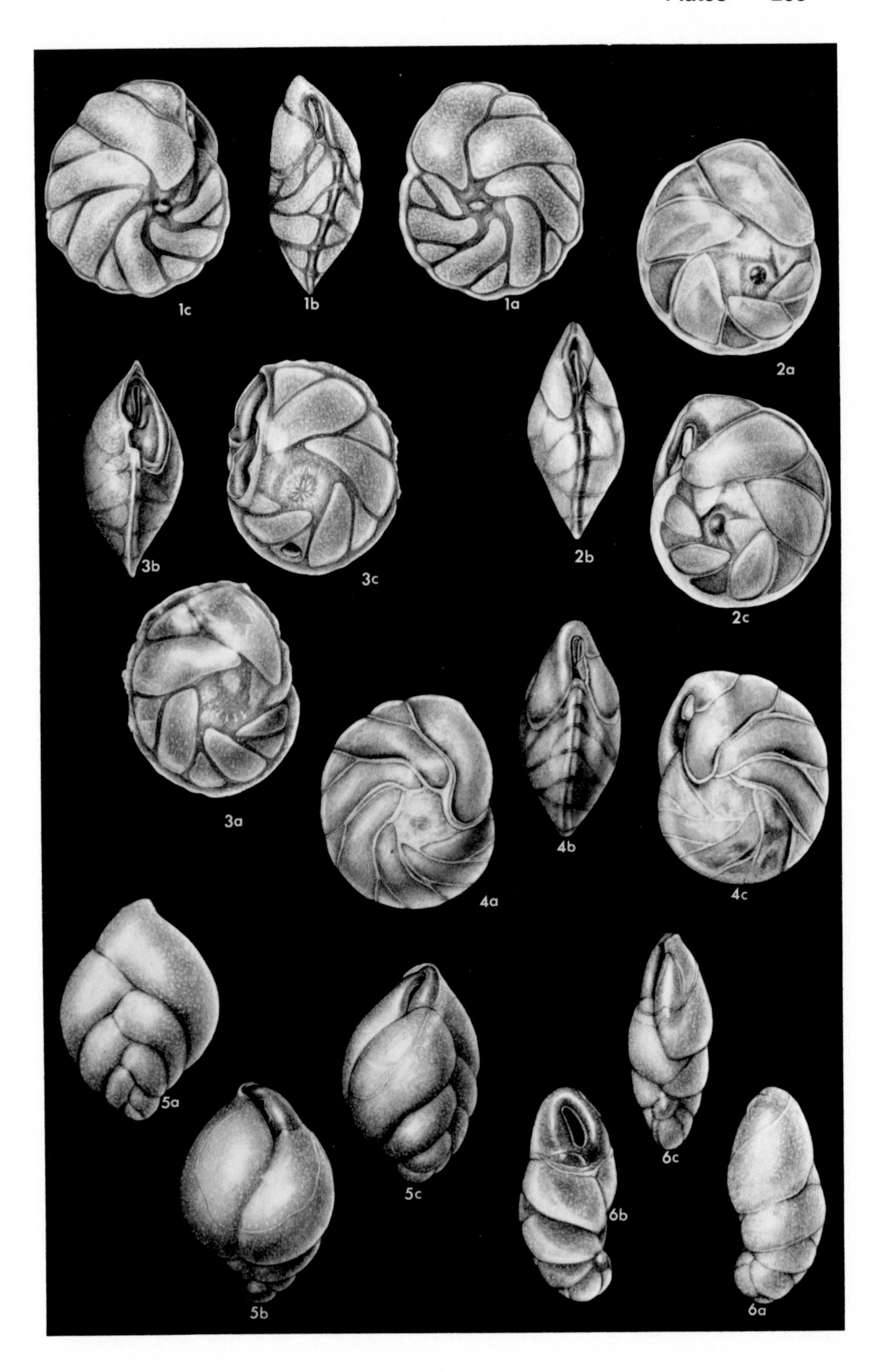
1c
1b
1a
2a
3b
3c
2b
2c
3a
4b
4a
4c
5a
5c
5b
6c
6b
6a

Plate XIV

FIG. 1–*Ehrenbergina compressa* Cushman. **a**, lateral view; **b**, apertural view; **c**, lateral view. Hypotype 47674, U.C. loc. D-1384. ×71.

FIG. 2–*Chilostomella czizeki* Reuss. **a**, lateral view; **b**, apertural view; **c**, lateral view. Hypotype 47655, U.C. loc. D-1383. ×50.

FIG. 3–*Pullenia miocenica*? Kleinpell. **a**, lateral view; **b**, apertural view. Hypotype 47772, U.C. loc. D-1231. ×100.

FIG. 4–*Pullenia bulloides* (d'Orbigny). **a**, lateral view; **b**, apertural view. Hypotype 47771, U.C. loc. D-1389. ×100.

FIG. 5–*Pullenia subcarinata* (d'Orbigny). **a**, lateral view; **b**, apertural view. Hypotype 47774, U.C. Loc. D-1387. ×100.

FIG. 6–*Pullenia salisburyi* R. E. and K. C. Stewart. **a**, lateral view; **b**, apertural view. Hypotype 47773, U.C. loc. D-1237. ×100.

FIG. 7–*Sphaeroidina bulloides* d'Orbigny. **a**, lateral view; **b**, apertural view. Hypotype 47791, U.C. loc. D-1287. ×67.

1b
1c
1a
2b
2a
2c
3a
3b
4a
4b
5a
5b
6a
6b
7a
7b

Plate XV

FIG. 1–*Globigerina bulloides* d'Orbigny. **a**, dorsal view; **b**, apertural view; **c**, ventral view. Hypotype 47705, U.C. loc. D-1388. ×100.

FIG. 2–*Globigerina* sp. **a**, dorsal view; **b**, apertural view; **c**, ventral view. Hypotype 47707, U.C. loc. D-1419. ×100.

FIG. 3–*Globigerina pachyderma* (Ehrenberg). **a**, dorsal view; **b**, apertural view; **c**, ventral view. Hypotype 47706, U.C. loc. D-1393. ×91.

FIG. 4–*Globorotalia inflata* (d'Orbigny). **a**, dorsal view; **b**, apertural view; **c**, ventral view. Hypotype 47713, U.C. loc. D-1332. ×67.

FIG. 5–*Globorotalia crassaformis* (Galloway and Wissler). **a**, dorsal view; **b**, apertural view; **c**, ventral view. Hypotype 47711, U.C. loc. D-1389. ×100.

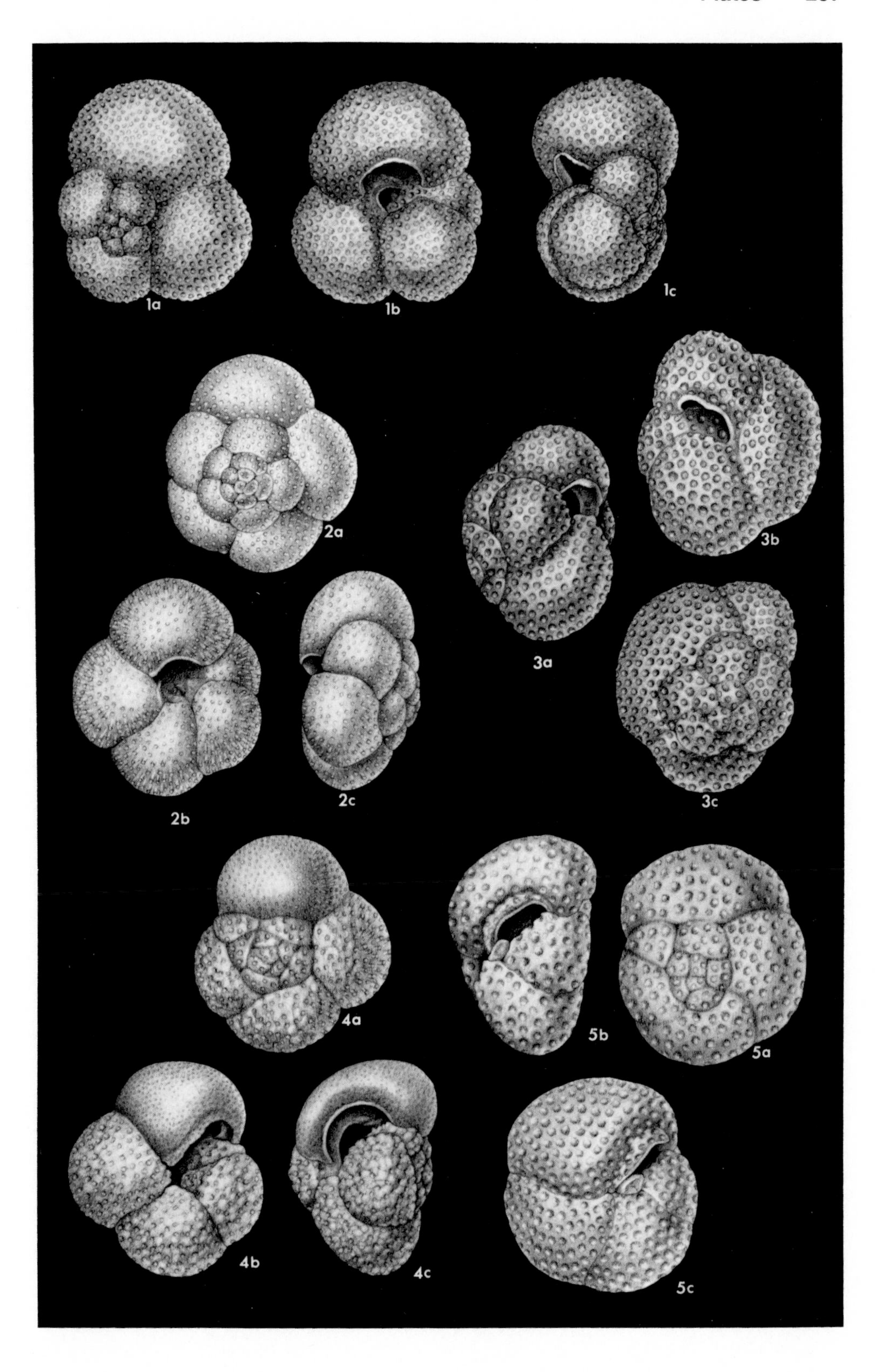
1a
1b
1c
2a
3b
3a
2b
2c
3c
4a
5b
5a
4b
4c
5c

Plate XVI

FIG. 1–*Orbulina universa* d'Orbigny. Hypotype 47756, U.C. loc. D-1303. ×71.

FIG. 2–*Globorotalia scitula* (H. B. Brady). **a,** dorsal view; **b,** apertural view; **c,** ventral view. Hypotype 47714, U.C. loc. D-1287. ×83.

FIG. 3–*Anomalina* cf. *A. salinasensis*? Kleinpell. **a,** dorsal view; **b,** apertural view; **c,** ventral view. Hypotype 47608, U.C. loc. D-1224. ×67.

FIG. 4–*Melonis barleeanus* (Williamson). **a,** lateral view; **b,** apertural view. Hypotype 47742, U.C. loc. D-1287. ×100.

FIG. 5–*Melonis pompilioides* (Fichtel and Moll). **a,** lateral view; **b,** apertural view. Hypotype 47743, U.C. loc. D-1303. ×59.

FIG. 6–*Melonis barleeanus* (Williamson). **a,** lateral view; **b,** apertural view. Hypotype 47741, U.C. loc. D-1303. ×67.

FIG. 7–*Planulina ariminensis* d'Orbigny. **a,** dorsal view; **b,** apertural view; **c,** ventral view. Hypotype 47759, U.C. loc. D-1379. ×53.

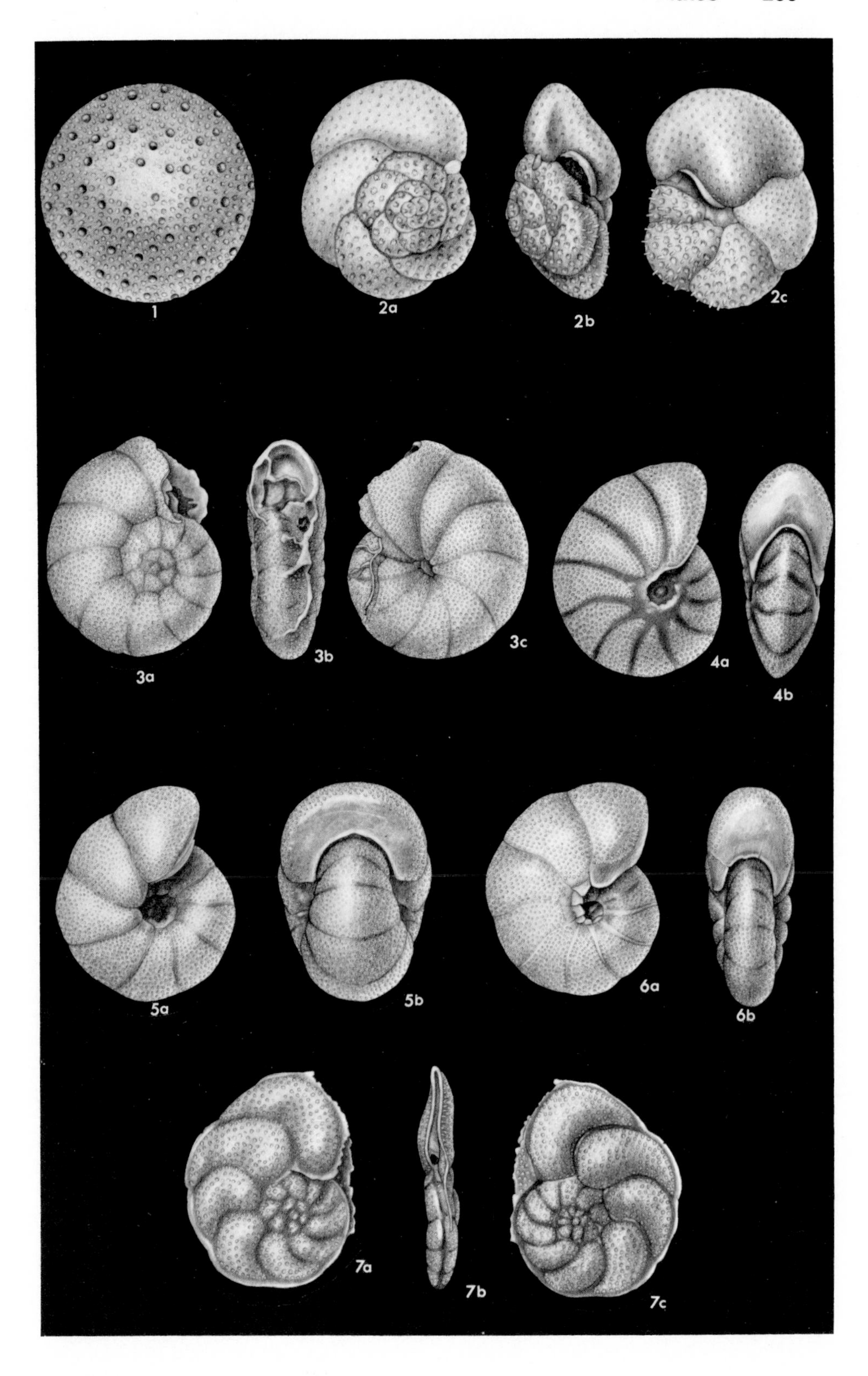
1
2a
2b
2c
3a
3b
3c
4a
4b
5a
5b
6a
6b
7a
7b
7c

Plate XVII

FIG. 1–*Planulina mexicana* Cushman. **a**, dorsal view; **b**, apertural view; **c**, ventral view. Hypotype 47760, U.C. loc. D-1266. ×33.

FIG. 2–*Planulina wuellerstorfi* (Schwager). **a**, dorsal view; **b**, apertural view; **c**, ventral view. Hypotype 47761, U.C. loc. D-1285. ×39.

FIG. 3–*Cibicidina*? *washingtonensis* (Cushman, Stewart, and Stewart). **a**, dorsal view; **b**, apertural view; **c**, ventral view. Hypotype 47663, U.C. loc. D-1385. ×33.

FIG. 4–*Cibicidina*? *washingtonensis* (Cushman, Stewart, and Stewart). **a**, dorsal view; **b**, apertural view; **c**, ventral view. Hypotype 47664, U.C. loc. D-1240. ×45.

FIG. 5–*Cibicides floridanus* (Cushman). **a**, dorsal view; **b**, apertural view; **c**, ventral view. Hypotype 47657, U.C. loc. D-1237. ×55.

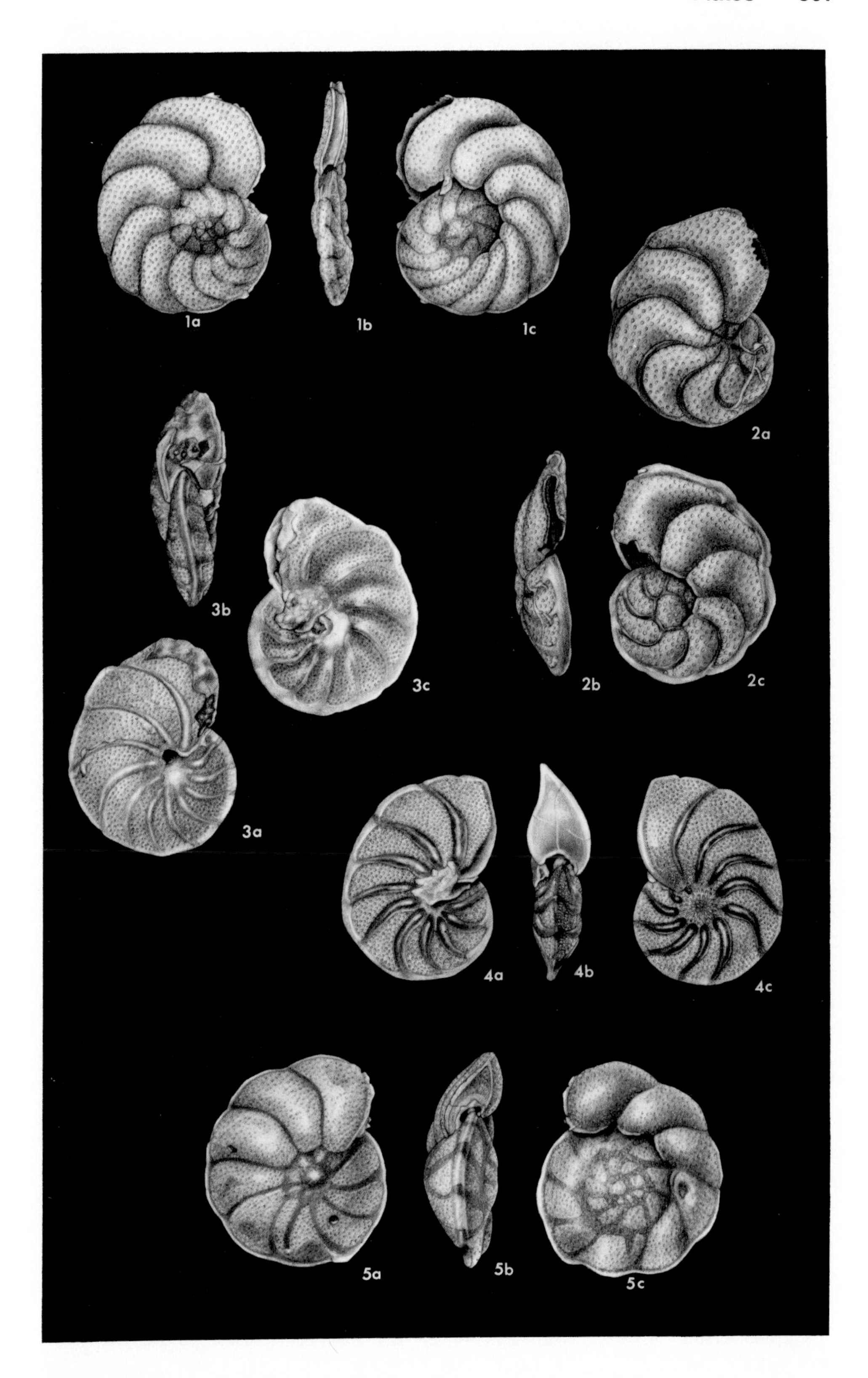
1a
1b
1c
2a
3b
2b
2c
3c
3a
4a
4b
4c
5a
5b
5c

Plate XVIII

FIG. 1–*Cibicides mckannai* Galloway and Wissler var. *spiralis* Natland. **a**, dorsal view; **b**, apertural view; **c**, ventral view. Hypotype 47662, U.C. loc. D-1357. ×83.

FIG. 2–*Cibicides mckannai* Galloway and Wissler. **a**, dorsal view; **b**, apertural view; **c**, ventral view. Hypotype 47661, U.C. location D-1351. ×67.

FIG. 3–*Cibicides lobatulus* (Walker and Jacob). **a**, dorsal view; **b**, apertural view; **c**, ventral view. Hypotype 47658, U.C. loc. D-1303. ×35.

FIG. 4–*Cibicides lobatulus* (Walker and Jacob). **a**, dorsal view; **b**, apertural view; **c**, ventral view. Hypotype 47659, U.C. loc. D-1303. ×31.

FIG. 5–*Cibicides lobatulus* (Walker and Jacob). **a**, dorsal view; **b**, apertural view; **c**, ventral view. Hypotype 47660, U.C. loc. D-1303. ×33.

1a
1b
1c
2a
2b
2c
3a
3b
3c
4a
4b
4c
5a
5b
5c

Appendix A: Register of Localities

Index of Appendix A

Appendix A

Register of Localities

The following 15-minute series (topographic) maps published by the U.S. Geological Survey and the U.S. Army Corps of Engineers were used to determine field locations and to plot localities: Cape Mendocino, California (1950), Eureka, California (1959), Ferndale, California (1959), Fortuna, California (1959), Iaqua Buttes, California (1950), Scotia, California (1950), and Weott, California (1949).

Primary assistance in locating fault and formation contacts and determining areal geology was obtained by reference to geologic maps of the Eureka (Evenson, 1959), Eel River (Ogle, 1953), and Redding (Strand, 1962) areas of California.

Appendix Table 1. Current Localities, Humboldt County, California

UC Loc.[1] No.	Field No.	Area	Material on File UC[1] Dept. Paleo.		
			Assemblage Slides	Washed Vials	Bulk
D-1224	Bcl-1a	Bear River coast	*	*	
D-1225	Bcl-1	"	*	*	*
D-1226	Bcl-2	"	*	*	*
D-1227	Bcl-3	"	*	*	*
D-1228	Bcl-4	"	*	*	*
D-1229	Bcl-5	"		*	*
D-1230	Bcl-6	"	*	*	*
D-1231	Bcl-7	"	*	*	*
D-1232	Bcl-8	"		*	*
D-1233	Bcl-9	"		*	*
D-1234	Bcl-10	"	*	*	*
D-1235	Brl-1	Bear River valley	*	*	*
D-1236	Brl-2	"	*	*	*
D-1237	Brl-3	"	*	*	*
D-1238	Brl-4	"	*	*	*
D-1239	Brl-5	"	*	*	*
D-1240	Brl-6	"	*	*	*
D-1241	Cel-1	Centerville coast		*	*
D-1242	Cel-1a	"		*	*
D-1243	Cel-2	"	*	*	*
D-1244	Cel-3	"		*	*
D-1245	Cel-4	"		*	*
D-1246	Cel-5	"	*	*	*
D-1247	Cel-6	"		*	*
D-1248	Cel-7	"	*	*	*
D-1249	Cel-8	"	*	*	*
D-1250	Cel-9	"	*	*	*
D-1251	Cel-10	"	*	*	*

[1] University of California at Berkeley.

UC Loc.[1] No.	Field No.	Area	Material on File UC[1] Dept. Paleo.		
			Assemblage Slides	Washed Vials	Bulk
D-1252	Cel-11	"		*	*
D-1253	Cel-12	"		*	*
D-1254	Cel-13	"		*	*
D-1255	Cel-14	"		*	*
D-1256	Cel-15	"	*	*	*
D-1257	Cel-16	"		*	*
D-1258	Cel-17	"	*	*	*
D-1259	Cel-18	"		*	*
D-1260	Cel-19	"	*	*	*
D-1261	Cel-20	"	*	*	*
D-1262	Cel-21	"	*	*	*
D-1263	Cel-22	"	*	*	*
D-1264	Cel-23	"	*	*	*
D-1265	Cel-24	"	*	*	*
D-1266	Cel-25a	"	*	*	*
D-1267	Cel-25	"		*	*
D-1268	Cel-26	Centerville coast		*	*
D-1269	Cel-27	"	*	*	*
D-1270	Cel-28	"		*	*
D-1271	Cel-29	"	*	*	*
D-1272	Cel-30	"	*	*	*
D-1273	Cel-31	"		*	*
D-1274	Cel-32	"		*	*
D-1275	Cel-33	"	*	*	*
D-1276	Cel-34	"	*	*	*
D-1277	Cel-35	"		*	*
D-1278	Cel-36	"		*	*
D-1279	Cel-37	"		*	*
D-1280	Cel-38	"	*	*	*
D-1281	Cel-39	"		*	*

UC Loc.[1] No.	Field No.	Area	Material on File UC[1] Dept. Paleo.		
			Assemblage Slides	Washed Vials	Bulk
D-1282	Cel-40	”		*	*
D-1283	Cel-41	”	*	*	*
D-1284	Cel-42	”	*	*	*
D-1285	Cel-43	”	*	*	*
D-1286	Cel-44	”		*	*
D-1287	Cel-45	”	*	*	*
D-1288	Cel-46	”		*	*
D-1289	Cel-47	”	*	*	*
D-1290	Cel-48	”		*	*
D-1291	Cel-49	”	*	*	*
D-1292	Cel-50	”		*	*
D-1293	Cel-51	”		*	*
D-1294	Cel-52	”		*	*
D-1295	Cel-53	”	*	*	*
D-1296	Cel-54	”		*	*
D-1297	Cel-55	”	*	*	*
D-1298	Cel-56	”		*	*
D-1299	Cel-57	”		*	*
D-1300	Cel-58	”		*	*
D-1301	Cel-59	”	*	*	*
D-1302	Ekl-1a	Elk River valley	*	*	*
D-1303	Ekl-4	”	*	*	*
D-1304	Ekl-5	”		*	*
D-1305	Ekl-6	”		*	*
D-1306	Ekl-7	”		*	*
D-1307	Ekl-8	”		*	*
D-1308	Ekl-9	”	*	*	*
D-1309	Ekl-10	”		*	*
D-1310	Ekl-11	”		*	*
D-1311	Ekl-12	”		*	*
D-1312	Ekl-13	”		*	*
D-1313	Ekl-14	”	*	*	*
D-1314	Ekl-15	”		*	*
D-1315	Ekl-16	”	*	*	*
D-1316	Ekl-17	”		*	*
D-1317	Ekl-18	”	*	*	*
D-1318	Ekl-19	”		*	*
D-1319	Ekl-20	”		*	*
D-1320	Ekl-21	”	*	*	*
D-1321	Ekl-22	”	*	*	*
D-1322	Ekl-23	”		*	*
D-1323	Ekl-24	”	*	*	*
D-1324	Ekl-25	”		*	*
D-1325	Ekl-26	”		*	*
D-1326	Ekl-27	”		*	*
D-1327	Ekl-28	”		*	*
D-1328	Ekl-29	”		*	*
D-1329	Ekl-30	”	*	*	*
D-1330	Ekl-31	”		*	*
D-1331	Fll-1	Fields landing	*	*	*
D-1332	Fll-2	”	*	*	*
D-1333	Fll-3	”		*	*
D-1334	Frl-1	Freshwater Creek	*	*	*
D-1335	Frl-3	”	*	*	*
D-1336	Knl-2	”		*	*
D-1337	Knl-3	”	*	*	*
D-1338	Knl-4	”	*	*	*
D-1339	Lbl-2	Larabee ranch		*	*
D-1340	Lbl-3	”	*	*	*
D-1341	Lbl-6	”		*	*

UC Loc.[1] *No.*	*Field No.*	*Area*	*Material on File UC*[1] *Dept. Paleo.*		
			Assemblage Slides	*Washed Vials*	*Bulk*
D-1342	Lbl-7	”	*	*	*
D-1343	Lbl-8	”	*	*	*
D-1344	Pcl-1	Price Creek	*	*	*
D-1345	Pcl-2	”	*	*	*
D-1346	Pcl-3	”	*	*	*
D-1347	Pcl-4	”	*	*	*
D-1348	Pcl-5	”	*	*	*
D-1349	Pcl-6	”	*	*	*
D-1350	Pcl-7	”		*	*
D-1351	Pcl-8	”		*	*
D-1352	Pcl-9	”	*	*	*
D-1353	Pcl-10	”		*	*
D-1354	Pcl-11	”		*	*
D-1355	Pcl-12	”	*	*	*
D-1356	Pcl-13	”		*	*
D-1357	Pcl-14	”	*	*	*
D-1358	Pcl-15	”		*	*
D-1359	Pcl-16	”		*	*
D-1360	Pcl-17	”		*	*
D-1361	Pcl-18	”	*	*	*
D-1362	Pcl-19	”		*	*
D-1363	Pcl-20	”	*	*	*
D-1364	Sal-1	Salmon Creek		*	*
D-1365	Sal-2	”	*	*	*
D-1366	Sal-3	”	*	*	*
D-1367	Sal-4	”		*	*
D-1368	Sal-5	”		*	*
D-1369	Sal-6	”	*	*	*
D-1370	Sal-7	”		*	*
D-1371	Sal-8	”		*	*
D-1372	Sal-9	”	*	*	*
D-1373	Sal-10	”		*	*
D-1374	Sal-11	”		*	*
D-1375	Sal-12	”		*	*
D-1376	Sal-13	”	*	*	*
D-1377	Sal-14	”		*	*
D-1378	Sc2-15	Scotia - Eel River	*	*	*
D-1379	Sc2-14	”	*	*	*
D-1380	Sc2-12	”	*	*	*
D-1381	Sc2-11	”	*	*	*
D-1382	Sc2-10	”	*	*	*
D-1383	Sc2-9	”	*	*	*
D-1384	Sc2-8	”	*	*	*
D-1385	Sc2-7	”	*	*	*
D-1386	Sc2-6	”	*	*	*
D-1387	Sc2-5	”	*	*	*
D-1388	Sc2-4	”	*	*	*
D-1389	Sc2-3	”	*	*	*
D-1390	Sc2-2	”	*	*	*
D-1391	Sc2-1	”	*	*	*
D-1392	Scl-1	”		*	*
D-1393	Scl-2	”	*	*	*
D-1394	Scl-3	”	*	*	*
D-1395	Scl-4	”	*	*	*
D-1396	Scl-5	”	*	*	*
D-1397	Scl-6	”	*	*	*
D-1398	Scl-7	”	*	*	*
D-1399	Scl-8	”	*	*	*
D-1400	Scl-9	”		*	*
D-1401	Scl-10	”		*	*

UC Loc.[1] No.	Field No.	Area	Material on File UC[1] Dept. Paleo.		
			Assemblage Slides	Washed Vials	Bulk
D-1402	Scl-11	”	*	*	*
D-1403	Scl-12	”		*	*
D-1404	Scl-13	”		*	*
D-1405	Scl-14	”	*	*	*
D-1406	Scl-15	”	*	*	*
D-1407	Scl-16	”	*	*	*
D-1408	Scl-17	”		*	*
D-1409	Scl-18	”	*	*	*
D-1410	Scl-19	”	*	*	*
D-1411	Scl-20	”		*	*
D-1412	Scl-21	”		*	*
D-1413	Scl-22	”		*	*
D-1414	Scl-23	”		*	*
D-1415	Scl-24	”	*	*	*
D-1416	Scl-25	”	*	*	*
D-1417	Scl-26	”	*	*	*
D-1418	Scl-27	”	*	*	*
D-1419	Scl-28	”	*	*	*
D-1420	Scl-29	”		*	*
D-1421	Scl-30	”		*	*
D-1422	Scl-32	”		*	*
D-1423	Tbl-2	Table Bluff		*	*
D-1424	Tbl-3	”		*	*
D-1425	Tbl-4	”		*	*
D-1426	Txl-1	Texaco-Eureka No. 2	*	*	
D-1427	Txl-2	”	*	*	
D-1428	Txl-3	”		*	
D-1429	Txl-4	”		*	
D-1430	Txl-5	”	*	*	
D-1431	Txl-6	”	*	*	
D-1432	Txl-7	”		*	
D-1433	Txl-8	”		*	
D-1434	Txl-9	”		*	
D-1435	Txl-10	”		*	
D-1436	Txl-11	”	*	*	
D-1437	Txl-12	”		*	
D-1438	Txl-13	”		*	
D-1439	Txl-14	”	*	*	
D-1440	Txl-15	”	*	*	
D-1441	Txl-16	”		*	

Appendix Table 2. Localities in Bear River Valley—Description

D-1235	NE¼ NW¼ Sec. 22, T1N, R2W. South bank of main Bear River; approximately 300 ft (90 m) upstream from Lowry cabin (abandoned school). Mudstone and siltstone, fractured but not pulverized; many scattered pelecypods, rare scaphopods; several calcareous nodule layers. Dip appears to be low SE.
D-1236	NE¼ NE¼ Sec. 21, T1N, R2W. East bank of South Fork of Bear River. Siltstone and fine to medium-grained sandstone; many pelecypods including "*Pecten propatulus*" of Martin.
D-1237	NW¼ NW¼ Sec. 21, T1N, R2W. South bank of main Bear River. Mudstone; scattered pelecypods.
D-1238	NE¼ NE¼ Sec. 20, T1N, R2W. South bank of main Bear River. Mudstone; many scattered pelecypods. Steep SE dip may be slump(?).
D-1239	NE¼ NW¼ Sec. 19, T1N, R2W. Northwest bank of main Bear River. Siltstone and sandstone with much humic matter in some layers, several calcareous nodule layers. Dip 60° N12°W; many fractures with slight displacement. Many scattered pelecypods.
D-1240	Same location, approximately 75 ft downstream from D-1239 and apparently stratigraphically below.

Remarks: All samples belong to "Bear River beds" (Informal term).

Appendix Table 3. Localities in Bear River Coastal Section

Locality No.	*Above Base Stratigraphically (feet)*	*Above Base Horizontally (feet)*	*Dip*	
"Bear River beds"				
D-1224	70	85		
D-1225	143	175	78°	S58°W
Pullen(?) Formation				
D-1226	569	675	67°	S47°W
D-1227	1,104	1,275	66°	S38°W
D-1228	1,319	1,495	64°	S33°W
D-1229	1,677	1,890	60°	S34°W
D-1230	1,881	2,110		
D-1231	2,163	2,420	56°	S36°W
D-1232	2,619	3,085	44°	S45°W
D-1233	3,361	4,600	31°	S55°W
Wildcat Group undiff.				
D-1234	3,497	5,185	18°	S64°W

Remarks: The section above begins at the north end of the Tertiary exposures starting in SW¼ NW¼ Sec. 11, T1N, R3W; see Figure 5 for lithologic details, and Appendix B for description of sections.

Appendix Table 4. Localities in Centerville Coastal Section

Locality No.	Below Fm. Top Stratigraphically (feet)	Below Fm. Top Horizontally (feet)	Dip
"Hookton Formation"			
D-1241	147	900	12° N22°W
Upper Member Rio Dell Formation			
D-1242	1	1	
D-1243	10	50	
D-1244	48	250	
D-1245	89	475	
D-1246	161	850	
D-1247	227	1,200	
D-1248	376	2,000	11° due N
D-1249	436	2,235	
D-1250	515	2,700	
D-1251	871	4,035	
D-1252	933	4,325	
D-1253	972	4,595	
D-1254	1,037	4,875	
D-1255	1,075	5,060	
D-1256	1,139	5,365	13° N17°W
D-1257	1,217	5,715	
D-1258	1,326	6,075	20° N3°W
D-1259	1,444	6,430	
D-1260	1,667	7,150	
D-1261	1,787	7,550	
D-1262	1,904	7,960	
D-1263	1,941	8,095	18° N16°W
Middle Member Rio Dell Formation			
D-1264	36	90	
D-1265	109	274	22° N15°E
D-1266	132	349	
D-1267	135	354	18° N13°E
D-1268	175	475	
D-1269	244	680	
D-1270	535	1,505	
D-1271	661	1,830	22° N3°W
D-1272	762	2,100	
D-1273	811	2,200	
D-1274	879	2,350	27° due N
D-1275	955	2,515	
Lower Member Rio Dell Formation			
D-1276	1	1	27° N8°E
D-1277	56	120	
D-1278	137	295	28° N3°W
D-1279	265	550	
D-1280	314	650	32° N5°W
D-1281	383	775	34° N4°W
D-1282	527	1,070	32° N5°E
D-1283	702	1,385	33° N4°E
D-1284	873	1,675	
Eel River Formation			
D-1285	50	80	37° N10°E
D-1286	161	255	35° N4°E
D-1287	296	495	
D-1288	414	705	31° N5°E
D-1289	531	910	
D-1290	629	1,075	36° due N
D-1291	699	1,190	
D-1292	728	1,244	
Pullen Formation			
D-1293	23	35	40° due N

Locality No.	*Below Fm. Top Stratigraphically (feet)*	*Below Fm. Top Horizontally (feet)*	*Dip*
D-1294	125	190	39° N5°E
D-1295	248	380	
D-1296	378	580	
D-1297	528	825	
D-1298	629	980	
D-1299	933	1,430	
D-1300	1,010	1,565	
D-1301	1,052	1,620	

Remarks: Section measurements start at north end of Tertiary exposures at Centerville Beach in SE¼ SE¼ NE¼ Sec. 12, T2N, R3W. All samples were taken just above beach level except for D-1242, which is from a road cut on hilltop overlooking beach. The following checkpoints are noted:

1. Fleener Creek–center of valley approximately 2,825 ft horizontally below top of upper member Rio Dell.
2. Barri Gulch–center of valley approximately 6,295 ft horizontally below top of upper member Rio Dell.
3. Guthrie Creek–center of valley approximately 1,225 ft horizontally below top of middle member Rio Dell.

Appendix Table 5. Localities in Elk River Valley—Description

North Fork Elk River

D-1309	SE¼ NE¼ NE¼ Sec. 36, T4N, R1E. Interbedded siltstone and fine-grained sandstones. Worm tubes(?), a few small pelecypods. Sample about 5 ft (1.5 m) stratigraphically above Wildcat Group - Yager Formation contact. Basal Wildcat contains some boulders and many well rounded pebbles. Dip of Wildcat apparently a few degrees due N, underlying Yager dip 27° N8°W.
D-1310	Locality about 300 ft (90 m) west of D-1309 and perhaps about 30 ft (9 m) stratigraphically above. Only fossil evidence is poorly preserved fragments of an echinoid(?).
D-1302	SW¼ SE¼ Sec. 35, T4N, R1E. About 600 ft (180 m) east of road junction in road cut along new logging road, road being parallel with South Branch Elk River. Sample from Wildcat Group apparently less than 50 ft (15 m) stratigraphically above Yager Formation. Mudstone with scattered mollusks.
D-1303	NE¼ SE¼ Sec. 34, T4N, R1E. Approximately 5 ft (1.5 m) stratigraphically above Wildcat Group - Yager Formation contact, south side of road. Mudstone with abundant forams, common pelecypods, and a few solitary corals.
D-1304	SE¼ NE¼ Sec. 34, T4N, R1E. South bank of North Fork of Elk River, just above Wildcat Group - Yager formation contact. Basal part of Wildcat contains boulders 6 to 8 ft (1.8 to 2.4 m) in diameter and many pebbles, also common glauconite. Yager Formation dips steeply; Wildcat appears to have a low angle dip to northwest, but bedding is very obscure other than a lens-shaped calcareous nodule.
D-1305	Same location as D-1304, approximately 180 ft (55 m) west and downstream, and apparently about 15 ft (4.6 m) higher stratigraphically. Mudstone with scattered pelecypods.
D-1306	SE¼ SW¼ Sec. 28, T4N, R1E. Southwest bank of North Fork of Elk River. Apparently less than 50 ft (15 m) stratigraphically above Wildcat Group - Yager Formation contact. Scattered gastropods, rare pelecypods.
D-1307	SW¼ SW¼ Sec. 28, T4N, R1E. North side of road in roadcut. Partly weathered siltstone.
D-1311	SW¼ SW¼ Sec. 28, T4N, R1E. South bank of North Fork of Elk River. Approximately 300 ft (90 m) west of D-1306. Scattered pelecypods and gastropods.
D-1312	NE¼ NE¼ Sec. 32, T4N, R1E. Southwest bank of North Fork of Elk River. Mudstone.
D-1313	NW¼ NE¼ Sec. 32, T4N, R1E. North bank of North Fork of Elk River. Approximately 900 ft (270 m) west of D-1312. Scattered gastropods in mudstone.
D-1308	SW¼ SE¼ Sec. 29, T4N, R1E. North side of road in roadcut. Siltstone.
D-1314	SE¼ SW¼ Sec. 29, T4N, R1E. North bank of North Fork of Elk River. Mudstone.
D-1315	SW¼ SW¼ Sec. 29, T4N, R1E. South bank of North Fork of Elk River. Mudstone.
D-1316	NW¼ SW¼ Sec. 29, T4N, R1E. North bank of North Fork of Elk River. Scattered pelecypods in mudstone.

D-1317	SW¼ SW¼ Sec. 29, T4N, R1E. South bank of North Fork of Elk River. Scattered pelecypods in mudstone.
D-1318	SW¼ NE¼ Sec. 30, T4N, R1E. North side of road in road cut, approximately midway along road between Browns Gulch and Boy Scout camp.
D-1319	NE¼ SW¼ Sec. 30, T4N, R1E. South bank of North Fork of Elk River. Leached(?) siltstone.
D-1320	NE¼ SW¼ Sec. 30, T4N, R1W. North side of road in roadcut. Partly leached siltstone.
D-1321	SW¼ NE¼ Sec. 25, T4N, R1W. North side of road in roadcut. Scattered mollusks including ornamented gastropods in mudstone.
D-1326	SW¼ NE¼ Sec. 26, T4N, R1W. North bank of North Fork of Elk River. Partly leached siltstone.
South Fork Elk River	
D-1323	SW¼ NE¼ Sec. 36, T4N, R1W. North bank of South Fork of Elk River. Mudstone.
D-1324	Same location as D-1323, about 450 ft (140 m) downstream on south bank. Scattered pelecypods and gastropods in mudstone.
D-1322	SE¼ NW¼ Sec. 36, T4N, R1W. South bank of South Fork of Elk River. Scattered mollusks in mudstone.
D-1325	SW¼ SE¼ Sec. 26, T4N, R1W. South bank of South Fork of Elk River. Scattered mollusks in mudstone.
Main Elk River valley	
D-1327	NE¼ SE¼ Sec. 22, T4N, R1W. North side of road in roadcut. Carlotta(?) Formation.
D-1330	NW¼ NE¼ Sec. 15, T4N, R1W. North side of Ridgewood Drive in road cut, approximately 0.5 mi (0.8 km) uphill from intersection with main Elk River road. Uppermost *Ostrea lurida* bed in series of 3 or 4 oyster beds. Mudstone. Bedding apparently nearly horizontal. Hookton Formation.
D-1329	Same location as D-1330, 2 ft (0.6 m) stratigraphically below. Partly leached *Ostrea lurida* bed. Hookton Formation.
D-1328	NE¼ NW¼ Sec. 15, T4N, R1W. North side of Ridgewood Drive in road cut, approximately 0.3 mi (0.5 km) uphill from intersection with main Elk River road. Siltstone. Hookton Formation.

Remarks: All localities except for last four are from Wildcat Group undifferentiated.

Appendix Table 6. Localities in Fields Landing—Description

D-1331	NW¼ SE¼ Sec. 29, T4N, R1W. East side of Highway 101 in roadcut, approximately 1.5 mi (2.4 km) south of Fields Landing. About 5 ft (1.5 m) above road level in siltstone with *Ostrea lurida*. Total outcrop includes lenses of conglomerate, sandstone, siltstone, and mudstone with much humic matter in thin layers. Hookton Formation.
D-1332	Same location as D-1331, upper part of road cut and approximately 30 ft (9 m) stratigraphically above D-1331. Dip apparently a few degrees to south.
D-1333	SW¼ NW¼ Sec. 17, T4N, R1W. East side of Highway 101 in road cut, approximately 0.25 mi (0.4 km) north of Fields Landing, about 50 ft (15 m) above highway level. Mudstone, silty, with abundant pelecypods. *Ostrea lurida* elsewhere at this same general locality. Hookton formation.

Appendix Table 7. Localities in Freshwater Creek—Description

Freshwater Creek valley	
D-1334	NE¼ SE¼ Sec. 10, T4N, R1E. Southeast bank of Freshwater Creek. Siltstone, just above basal conglomeratic sandstone which overlies Franciscan(?) Formation. Scattered pelecypods.
D-1335	SE¼ SE¼ Sec. 3, T4N R1E. Southeast side of road in road cut along southeast bank of Freshwater Creek. Siltstone; scattered pelecypods. Apparently less than 50 ft (15 m) stratigraphically above underlying Franciscan(?) Formation and forming strong angular unconformity with it.
Kneeland Road	
D-1336	NW¼ SE¼ Sec. 2, T4N, R1E. Road cut in north side of road. Scattered pelecypods in siltstone.
D-1337	Same location as D-1336, approximately 300 ft (90 m) downhill on north side of road and stratigraphically below. Scattered pelecypods in siltstone.
D-1338	NE¼ SE¼ Sec. 3, T4N, R1E. East side of road in roadcut. Scattered pelecypods in siltstone.

Remarks: All localities listed are from Wildcat Group undifferentiated.

Appendix Table 8. Localities in Larabee Creek—Description

D-1339	NE¼ SW¼ Sec. 1, T1S, R2E. South bank of Larabee Creek. Siltstone interbedded with mudstones; scattered gastropods and pelecypods.
D-1340	SE¼ NE¼ Sec. 2, T1S, R2E. South bank of Larabee Creek. Interbedded mudstones and siltstones; scattered gastropods and a few gastropod concentrations.
D-1341	SE¼ NE¼ Sec. 33, T1N, R2E. Railroad cut, north bank of Eel River. Mudstone. Dip 40° N29°E.
D-1342	SE¼ NW¼ Sec. 33, T1N, R2E. Railroad cut, north bank of Eel River. Sample from mudstone in interbedded mudstone and sandstone sequence. Mudstone units range up to 6 ft (1.8 m) thick; sandstones are less than 1 ft (0.3 m). Scattered mollusks. Dip 36° N10°E.
D-1343	NE¼ SW¼ Sec. 32, T1N, R2E. Roadcut along Highway 101, west bank of Eel River. Scattered mollusks in siltstones. Dip 57° N38°E.

Remarks: All localities are from Wildcat Group undifferentiated.

Appendix Table 9. Localities in Price Creek

Locality No.	*Above Fm. Base Stratigraphically (feet)*	*Dip*	
Pullen Formation			
D-1344	12		
D-1345	255		
D-1346	531		
D-1347	574		
D-1348	656	49°	N20°W
D-1349	792		
Eel River Formation			
D-1350	50		
D-1351	111		
D-1352	178	46°	N15°W
D-1353	433	54°	N11°W
D-1354	467	47°	N10°W
D-1355	583	47°	N8°W
D-1356	787		
D-1357	888		
D-1358	960		
D-1359	1,034	42°	N4°W
D-1360	1,159		
D-1361	1,211		
D-1362	1,296		
D-1363	1,511		

Remarks: Section begins in stream bed of south branch of Price Creek upstream from main Pullen farmhouse. Measurements start in SW¼ NW¼ NE¼ Sec. 5, T1N, R1W, presumed base of Pullen Formation in lieu of good exposures.

Appendix Table 10. Localities in Salmon Creek—Description

D-1364	NW¼ SW¼ Sec. 13, T3N, R1W. South side of road along south bank of Salmon Creek. Mudstone, highly fractured; local concentrations of mollusks.
D-1365	SE¼ SE¼ Sec. 14, T3N, R1W. South bank of Salmon Creek. Mudstone, highly fractured; local concentrations of mollusks.
D-1366	SW¼ SE¼ Sec. 14, T3N, R1W. South bank of Salmon Creek. Mudstone, highly fractured; scattered mollusks.
D-1367	SE¼ SW¼ Sec. 14, T3N, R1W. Southwest bank of Salmon Creek. Mudstone, highly fractured; scattered mollusks.
D-1368	NE¼ SW¼ Sec. 14, T3N, R1W. West bank of Salmon Creek. Mudstone, highly fractured; local concentrations of mollusks.
D-1369	SW¼ NW¼ Sec. 14, T3N, R1W. West bank of Salmon Creek. Mudstone, highly fractured; scattered mollusks.
D-1370	NW¼ NW¼ Sec. 14, T3N, R1W. East bank of Salmon Creek. Mudstone; scattered mollusks. Dip 37° N5°E.
D-1371	SW¼ SW½ Sec. 11, T3N, R1W. West bank of Salmon Creek. Mudstone; scattered mollusks.
D-1372	SW¼ SW¼ Sec. 11, T3N, R1W. East bank of Salmon Creek. Approximately 700 ft (210 m) downstream from D-1371. Mudstone; scattered mollusks.
D-1373	SW¼ NW¼ Sec. 11, T3N, R1W. East bank of Salmon Creek. Mudstone; scattered mollusks.
D-1374	SE¼ NE¼ Sec. 10, T3N, R1W. North bank of Salmon Creek. Mudstone; scattered mollusks.
D-1375	SW¼ NE¼ Sec. 10, T3N, R1W. North bank of Salmon Creek. Mudstone; scattered mollusks.
D-1376	SW¼ NE¼ Sec. 10, T3N, R1W. South bank of Salmon Creek. Approximately 600 ft (180 m) downstream from D-1375. Mudstone with scattered mollusks.
D-1377	SW¼ NW¼ Sec. 10, T3N, R1W. South bank of Salmon Creek. Mudstone with common mollusks.

Remarks: All localities are from Wildcat Group undifferentiated.

Appendix Table 11. Localities in Scotia - Eel River Section

Locality No.	*Above Fm. Base Stratigraphically (feet)*	*Above Fm. Base Horizontally (feet)*	*Dip*
Pullen Formation			
D-1378	181	223	
D-1379	505	725	
D-1380	1,001	1,289	63° N13°E
Eel River Formation			
D-1381	416	648	
D-1382	591	939	52° N6°E
D-1383	645	1,023	
D-1384	722	1,151	
D-1385	1,030	1,676	53° N13°E
Rio Dell Formation			
D-1386	88	145	52° N12°E
D-1387	172	282	
D-1388	542	883	52° N4°E
D-1389	745	1,218	
D-1390	1,025	1,679	
D-1391	1,155	1,906	62° N14°E
D-1392	1,163	1,929	
D-1393	1,218	2,014	
D-1394	1,571	2,527	
D-1395	1,706	2,731	
D-1396	2,044	3,246	53° N15°E
D-1397	2,121	3,362	
D-1398	2,124	3,367	
D-1399	2,155	3,408	
D-1400	2,176	3,434	
D-1401	2,214	3,479	
D-1402	2,349	3,633	57° N19°E
D-1403	2,413	3,712	
D-1404	2,882	4,283	
D-1405	2,889	4,293	
D-1406	3,039	4,474	
D-1407	3,071	4,513	
D-1408	3,111	4,561	
D-1409	3,138	4,594	
D-1410	3,209	4,680	
D-1411	3,281	4,767	
D-1412	3,456	4,980	49° N4°W
D-1413	3,536	5,077	
D-1414	3,665	5,233	
D-1415	3,809	5,408	
D-1416	3,889	5,511	
D-1417	4,144	5,820	
D-1418	4,340	6,143	36° N13°E
D-1419	4,557	6,679	
Scotia Bluffs Sandstone			
D-1420	138	299	
D-1421	197	560	
D-1422	364	–	

Remarks: Measurements begin along west bank and low water

level of main Eel River with samples D-1378 through D-1391 continuing northward (downstream). Samples D-1391 through D-1422 were taken along east bank just above railroad bed level. Presumed base of Pullen Formation is taken as that in SW¼ SW¼ NE¼ Sec. 7, T1N, R1E.

The following checkpoints are noted:

1. Scotia-Rio Dell bridge, center of west abutment 1,019 ft (311 m) horizontally above base of Rio Dell Formation.
2. Nanning Creek bridge, beginning (south end) of bridge 5,649 ft (1722 m) horizontally above base of Rio Dell Formation.

Appendix Table 12. Localities in Table Bluff Coast—Description

D-1423	SE¼ SE¼ Sec. 27, T4N, R2W. Just below Table Bluff Lighthouse in top of exposures. Mudstone, silty, with scattered pelecypods. Dip 53° S15°E.
D-1424	SW¼ SW¼ Sec. 26, T4N, R2W. South side of Table Bluff, road leading to beach. Mudstone.
D-1425	Same location as D-1424; approximately 100 ft (30 m) northeast and uphill. Approximately 15 ft (4.6 m) vertically below Hookton Formation-Wildcat Group contact—contact is represented by a strong angular unconformity. Scattered pelecypods in siltstone.

Remarks: All localities represent Wildcat Group undifferentiated.

Appendix Table 13. Localities from Well Texaco-Eureka (NCT-1) No. 2

Locality No.	*Depth in Well (feet)*	*Formation*
D-1426	3,242	Rio Dell
D-1427	3,518	"
D-1428	3,690	"
D-1429	3,795	"
D-1430	3,909-17	"
D-1431	4,189	"
D-1432	4,356	"
D-1433	4,703	"
D-1434	5,086	"
D-1435	5,404	"
D-1436	5,670	"
D-1437	5,936	"
D-1438	5,993	"
D-1439	6,197	Eel River
D-1440	6,643	"
D-1441	7,322	"

Remarks: This well is from Tompkins Hill Gas Field, Sec. 22, T3N, R1W, H. B. & M., 2978 ft (908 m) N and 696 ft (212 m) E of SW corner of section. For stratigraphic relations, see California Division of Oil and Gas (1960, p. 478-479).

Appendix B
Description of Sections

Bear River Coastal Syncline

The section begins at the north end of the Tertiary exposures and covers the north flank of the Bear River syncline in the coastal section. Measurements start at the bottom of the section in the SW¼, NW¼, Sec. 11, T1N, R3W and indicate stratigraphic thicknesses.

	Bear River Beds Undifferentiated	Feet
1.	Interbedded mudstone and siltstone; beds mostly 1-6 in. thick, but some thicker; highly fractured; mollusks and forams common	183
2.	Massive slump (possible fault sliver) of Franciscan-type lithology, highly fractured	285
	Pullen(?) Formation	
3.	Mudstone with many 1 to 2 in. sandstone layers near top; highly fractured	240
4.	Siltstone with several layers of large (6 to 9 ft diameter) calcareous modules; intensely fractured throughout	253
5.	Interbedded siltstone and sandstone; beds in lower part not over 6 in thick, becoming thicker at top; base of interval marked by a 10-ft glauconitic layer; intensely fractured	255
6.	Mudstone with 1 to 2-ft sandy and silty layers and a few thin calcareous nodule layers near top; intensely fractured	210
7.	Mudstone; large *Cyclammina* in some intervals; lower part intensely fractured; upper part highly fractured and faulted(?)	276
8.	Siltstone, lower 20 ft highly glauconitic; highly fractured throughout; large *Cyclammina* in places	276
9.	Covered (small gulch)	159
10.	Interbedded siltstone and fine-grained sandstone becoming mudstone at top; sandstones less than 4 in. thick, siltstones up to 4 ft thick; scattered pelecypods	80
11.	Mudstone; most beds with pyritic "streaks"; apparently high diatom and radiolarian content; scattered pelecypods throughout	603
12.	Covered (valley floor of main Bear River)	537
13.	Mudstone as in unit 11; apparent high diatom and radiolarian content; scattered pelecypods	105
	Total	3,462
	Wildcat Group Undifferentiated	
14.	Interbedded siltstones and sandstones with a few calcareous nodule layers	36
15.	Covered. (The next good outcrop occurs 3,200 ft horizontally south in a very steep ravine with about 100 ft of excellent stratigraphic exposure including at least one 10-ft glauconitic interval.) Dips in this interval are reversed from SW to NW	

Remarks: The section south of Bear River covers more than mi horizontally, but contains only the two limited exposures described above, and a few lesser ones of questionable nature. Contact with rocks of Franciscan-type lithology at the south end of the syncline is obscured by a wide slump area; contact with rocks of Franciscan-type lithology at the north end of the syncline occurs in a zone of highly fractured rocks, probably a fault zone.

Ogle (1951) records Relizian, Luisian, and Mohnian Foraminifera from the steep ravine mentioned in unit 15 above (his locality BA02, pl. IV, p. 236; south end of syncline).

The Bear River beds in unit 1 are of Miocene age according to analyses of the foraminiferal content; contact with the somewhat younger radiolarian beds of the Pullen(?) Formation is obscured as indicated.

One sample taken in the undifferentiated Wildcat near the center of the syncline (unit 14) indicates a general correlation with the Pliocene of the Humboldt basin and appears to be much younger than the underlying Pullen(?) Formation.

Centerville Coastal Section

This section is measured from top to bottom. It begins at the north end of Tertiary exposures at Centerville Beach in the SE¼, SE¼, NE¼, Sec. 12, T2N, R3W. Measurements begin at winter berm level on the beach just below the point on the hill which contains California State Historical Marker no. 173. Measurements are stratigraphic thicknesses.

	"Hookton Formation" of Ogle (see remarks)	Feet
1.	Covered	80
2.	Sandstone, medium to coarse-grained, with abundant well-rounded pebbles and	

Unit	Description	Thickness
	cobbles—some as lenses and near layers; cobbles have a maximum 6-in. diameter; formation reddish-brown on surface, cross bedded in part; no apparent macrofauna. .	99
3.	Massive slump area—appears to be same lithology as described above	127
	Total	306
	Upper Member Rio Dell Formation	
4.	Siltstone, gray, massive or thick bedding, some bedding planes due to apparent change in lithology; a few calcareous nodules in upper part; upper part contains many *Pecten*; also small, high-spired gastropods; macrofossils present throughout.	263
5.	Covered and slump	70
6.	Siltstone, gray, some massive bedding due to apparent slight changes in lithology; scattered pelecypods and gastropods throughout	245
7.	Covered and slump	291
8.	Series of small slump blocks; apparent lithology is siltstone, massive bedding, gray; scattered pelecypods and gastropods	94
9.	Siltstone, gray, little apparent bedding; scattered pelecypods and gastropods in upper part; many pelecypods and large *Pecten (P. caurinus)* in lower part; *Pecten* normally contain both valves together, other pelecypods generally contain both valves together	159
10.	Siltstone, gray, massive bedding; many scattered pelecypods and gastropods with concentrations of mollusks in upper part	221
11.	Siltstone, gray, massive bedding in general with upper part broken by a series of thin sandstone nodules; upper part contains many large *Pecten* and other pelecypods. . . .	160
12.	Massive slump and covered area	152
13.	Siltstone, gray, massive bedding becoming interbedded with thin mudstones in lower part; scattered pelecypods throughout; last occurrence of *Pecten caurinus* in place 140 ft below top of interval with float somewhat lower	282
14.	Silty mudstone, gray; many large worm borings, large fragments of humic matter, and some gastropods; contact with unit below appears to be a minor erosional unconformity.	6
	Total	1,943
	Middle Member Rio Dell Formation	
15.	Mudstone, gray, with some silty intervals and a 1-ft sandy interval in lower part; beds in general not more than 8 to 10 ft thick; scattered small pelecypods throughout . . .	110
16.	Mudstone, gray, massive bedding; scattered thin diatom streaks in lower part; small pelecypods.	26
17.	Interbedded sandstone and siltstone, gray; beds 1 in. to 1 ft thick; no apparent macrofauna other than worm borings(?); also a few calcareous nodule layers.	109
18.	Covered and slump area	288
19.	Interbedded sandstone and siltstones, as above; noticeable layers of humic matter in lower part; small, scattered pelecypods in lower part.	164
20.	Interbedded sandstones and siltstones with 2 to 3-in. limestone layers in upper part; beds not over 1 ft thick; lower 6 ft a unit of contorted bedding with many scattered, small pelecypods; most pelecypods double-valved	80
21.	Mudstone, gray, massive 6 to 8-ft beds; bedding marked by apparent slight changes in lithology; three 6-in. to 1-ft sandstones in middle interval; many small pelecypods in upper part; many scattered small pelecypods and gastropods in lower part; pelecypods mostly double-valved; oriented more or less along bedding planes.	176
22.	Interbedded mudstones and siltstones, gray; beds generally between 1 and 6 in thick; scattered small pelecypods; diatom streaks near top	98
	Total	1,051
	Lower Member Rio Dell Formation	
23.	Siltstone, gray; massive 6 to 8-ft beds; bedding marked by slight apparent changes in lithology; scattered, irregular layers of calcareous nodules in lower part; scattered small pelecypods throughout; pelecypods generally single-valved with convex side upwards	82
24.	Interbedded mudstones and siltstones, gray; a few irregular sandstone layers and calcareous nodule layers in lower part; bed thickness highly variable from a few inches to a few feet; scattered small pelecypods throughout.	195
25.	Interbedded mudstones and siltstones becoming interbedded siltstones and sandstones at base; gray; thick-bedded at top becoming thin-bedded in lower part; scattered single- and double-valved pelecypods throughout	281
26.	Interbedded siltstones and sandstones, gray, beds in general a few inches thick;	

several calcareous nodule layers at middle of interval; scattered pelecypods throughout, mostly single-valved, convex upward 318

27. Interbedded siltstones and sandstones becoming mostly siltstones toward base; base marked by thin glauconitic stringers in lower 3 ft with concentrations of glauconite and scattered small calcareous nodules at contact with unit below; macrofauna scarce 165

Total 1,041

Eel River Formation

28. Interbedded mudstones and siltstones, gray, thin-bedded; sandy lenses and layers scattered throughout; several layers of thin calcareous nodules near top; scattered small pelecypods and gastropods throughout 327
29. Interbedded mudstones and siltstones; gray; mudstones predominate; thin-bedded in upper part becoming thick-bedded in lower part; becoming highly fractured with minor displacement; scattered small pelecypods and gastropods 363
30. Siltstone, sandy, glauconitic, gray becoming green on weathered surface; much fracturing and minor displacement; pelecypods and gastropods scarce to common. . . 28
31. Siltstone, gray; two 1 to 3-ft glauconitic sandstones in interval—one at base; lower sandstone contains scattered, well-rounded pebbles; one layer of calcareous nodules at middle interval; highly fractured with minor displacement; contact with unit below appears to be a slight erosional unconformity 12

Total 730

Pullen Formation

32. Siltstone, gray, weathers buff-brown; several layers of calcareous nodules and several 1 to 3-in glauconitic layers; possibly some poorly preserved pelecypods; diatoms and radiolarians irregularly common; highly fractured with minor displacement along fractures 166
33. Siltstone, gray, weathers buff-brown; bedding apparently massive; possibly some poorly preserved pelecypods; diatoms and radiolarians irregularly common; highly fractured with minor displacement along fractures 114
34. Massive slump area; apparent lithology: siltstone, gray, weathers buff-brown; no apparent macrofauna; diatoms and radiolarians sporadic and common; highly fractured in upper part becoming pulverized in lower part. 719
35. Massive slump area; apparent lithology: siltstone, glauconitic, gray, weathers brownish; pulverized. 5
36. Massive slump area; apparent lithology: mudstone, dark gray; no apparent macrofauna 82

Total 1,086

Remarks: Franciscan-type lithologies occur about 125 ft horizontally below unit 36. The intermediate interval is obscured by slump and cover.

Geology above follows that of Ogle (1953); formation terms and contact locations are synonymous with his except that it is noted here that his "Hookton Formation" compares lithologically with his "Carlotta Formation" of the Scotia-Eel River section.

Price Creek Section

The section begins in the stream bed of the south branch of Price Creek just upstream from the main Pullen farmhouse and traverses downstream (north). Measurements begin at the bottom of the section in the SW¼, NW¼, NE¼, Sec. 5, T1N, R1W and represent stratigraphic thicknesses.

Pullen Formation — Feet

1. Siltstone, sandy, gray, weathers brown; poorly preserved pelecypods and pelecypod casts 21
2. Covered and slumped area 224
3. Mudstone, gray; scattered scaphopods; highly fractured 13
4. Covered and slumped area 232
5. Sandstone, fine- to medium-grained, gray, weathers brown; highly fractured; indeterminate molluscan fragments 65
6. Interbedded sandstone and mudstones, gray; sandstone fine-grained with beds less than 4 in. thick; mudstone beds less than 2 ft thick; small pelecypods. 112
7. Covered and slumped area 119

8.	Siltstone, gray; calcareous nodule layers and a 1 ft glauconitic sandstone near base; scattered mollusks	63
9.	Covered and slumped area	83
	Total	932

Eel River Formation

10.	Sandstone, glauconitic, medium-grained, gray-green, weathers brownish; fractured with slight displacement; base apparently forming a slight erosional unconformity with underlying unit (unit 9)	16
11.	Mudstone, silty in part, gray; at least three calcareous nodule layers	87
12.	Siltstone, glauconitic, gray-green, weathers brownish.	14
13.	Mudstone, gray; at least two 1-ft glauconitic siltstones; lower part fractured with minor displacement; upper part slumped and distorted; scattered gastropods and pelecypods.	165
14.	Covered and slumped area	143
15.	Interbedded mudstones and siltstones, thin-bedded, gray	59
16.	Sandstone, fine-grained, glauconitic stringers in lower part; gray-green, weathers brown	3
17.	Mudstone, gray; scattered pelecypods	110
18.	Sandstone, gray, fine- to medium-grained; lower 6 in. contains concentrations of glauconite, humic matter and molluscan fragments.	23
19.	Covered and slumped area	155
20.	Interbedded mudstones, siltstones, and fine-grained sandstones; thin-bedded, gray; scattered pelecypods	138
21.	Sandstone, fine-grained, gray	12
22.	Mudstone, gray; a few 1-in. to 1-ft silty layers; scattered pelecypods.	124
23.	Sandstone, fine- to medium-grained, gray; some glauconite	8
24.	Interbedded mudstones and siltstones where mudstones predominate; thin-bedded; scattered pelecypods (contact with unit 23 not an apparent unconformity).	251
25.	Covered and slumped area	147
26.	Mudstone, gray; scattered pelecypods; section discontinued in Rio Dell Formation approximately 450 ft upstream from Brookdale ranch house	75
	Total	1,530

Remarks: The contact of the Pullen Formation with underlying rocks is obscure although these underlying rocks apparently are predominantly of metamorphic origin with some possible small fault slivers of Yager-type lithology.

Formation terms and contact locations are synymous with those described by Ogle (1953).

Scotia-Eel River Section

The section begins along the west bank and bed of the main Eel River at low water level and continues northward (downstream). Units 1-34 are along the west bank; the remaining units (35-61) were measured along the east bank just above railroad level. Measurements begin at the bottom of the section in the SW¼, SW¼, NE¼, Sec. 7, T1N, R1E and represent stratigraphic thicknesses.

	Pullen Formation (base not exposed; see remarks)	Feet
1.	Sandstone, medium to coarse-grained, glauconitic, gray, weathers brownish; scattered well-rounded igneous pebbles throughout, locally conglomeratic; scattered pelecypods at mid-interval; highly fractured	166
2.	Interbedded medium-grained sandstone and silty mudstone, gray, weathers brownish; sandstone tends to be lenticular, glauconitic; scattered calcareous nodule layers; no apparent macrofauna	126
3.	Siltstone, sandy, gray; becoming increasingly coarser grained and glauconitic toward base.	23
4.	Covered and slumped area	156
5.	Massive slump area; apparent lithology: mudstone, silty, gray; scattered calcareous nodule layers; some thin sandstone layers in lower part; scattered pelecypods	491
6.	Siltstone, sandy, gray; a 1-ft-thick limestone in upper portion and calcareous nodules at base	50
	Total	1,012

Eel River Formation

7.	Siltstone, sandy, massive, gray; base of unit marked by a 1 to 2-in. coarse-grained sandstone and highly irregular contact with unit 6	18
8.	Sandstone, coarse-grained, gray; many well-rounded igneous pebbles; intermittent limestone-cobble conglomerate with associated gastropods; locally thickens to 3 ft	1
9.	Sandstone, silty, gray-brown, fine-grained, massive bedding; scattered siltstone "boulders," some having calcareous nodule as a nucleus	21
10.	Sandstone, medium-grained, green and gray, highly glauconitic in lower 7 ft; apparently a fracture zone with displacement minor	20
11.	Mudstone, silty, gray, massive bedding; highly fractured	83
12.	Covered and slumped area	91
13.	Sandstone, silty, fine grained, gray, weathers brown; scattered "limestone and siltstone boulders" most of which are more or less elongate parallel with bedding plane, but are not aligned in a bed themselves–maximum diameter 6 ft; scattered well-rounded igneous pebbles	116
14.	Sandstone, medium-grained, highly glauconitic, green and gray	6
15.	Mudstone, silty, gray; massive bedding; several calcareous nodule layers	126
16.	Sandstone, fine-grained, silty, glauconitic in lower part; green and gray	21
17.	Mudstone, silty, gray	78
18.	Siltstone, sandy, gray-brown, massive bedding; several calcareous nodule layers; some nodules show evidence of reworking and weathering	21
19.	Mudstone, silty, gray, massive bedding except for a few scattered glauconitic stringers; scattered pelecypods	117
20.	Covered and slumped area	85
21.	Massive slumped area; apparent lithology: interbedded mudstones, siltstones, and sandstones, thin-bedded; scattered pelecypods	157
22.	Interbedded mudstones and sandy siltstones, thin-bedded, gray	72
	Total	1,033

Lower and Middle Members Rio Dell Formation

23.	Interbedded siltstones and fine-grained sandstones, thin-bedded, gray, scattered mollusks; base of unit marked by irregular layer of calcareous-cemented sandstone containing many well-rounded igneous pebbles, pelecypod fragments (including *Pecten*) and forming irregular contact with unit 22	227
24.	Mudstone, silty, gray; massive bedding; no apparent macrofossils	75
25.	Siltstone, gray; many layers of sandy nodules	56
26.	Mudstone, silty, gray; massive bedding except for thin sandy layers at top and bottom; no apparent macrofauna	178
27.	Mudstone, interbedded with siltstone; thin-bedded, gray; scattered mollusks	114
28.	Mudstone, gray; massive bedding; scattered gastropods	94
29.	Mudstone, interbedded with layers of thin calcareous nodules, gray; scattered mollusks	107
30.	Mudstone, silty, gray; massive beds; some small calcareous nodules in lower part; scattered mollusks	78
31.	Mudstone, silty, gray; thin-bedded in upper and lower parts; thick-bedded in middle portion; scattered calcareous nodules throughout; lenses of mollusks including *Pecten* in upper part	133
32.	Mudstone, silty, gray, massive beds; scattered calcareous nodules and mollusks	53
33.	Sandstone, fine-grained, gray and brown, massive beds	13
34.	Siltstone, gray; massive bedding; scattered calcareous nodules; scattered mollusks including *Pecten*	124
	Total	1,252

Upper Member Rio Dell Formation

35.	Covered and slumped area (contact of Middle and Upper Members thought to be not exposed)	287
36.	Siltstone, gray, massive bedding; intermittent calcareous ironstone concretions; scattered pelecypods	181
37.	Covered and slumped area	283
38.	Siltstone, gray; massive bedding, some thin calcareous lenses near top	35
39.	Covered and slumped area	68
40.	Siltstone, gray, massive bedding, scattered pelecypods and humic matter	9

Unit	Description	Thickness
41.	Sandstone, silty, gray; calcareous lenses in lower part; weathers yellow brown in lower 2 ft; contains siltstone fragments in lower 6 in. weathered from unit 40; abundant pelecypods, some echinoids in lower part	25
42.	Sandstone, fine-grained, gray; massive bedding; abundant pelecypods including *Pecten*, scattered gastropods.	58
43.	Sandstone, fine-grained, gray with some yellow-brown weathering; thin-bedded; several layers of calcareous lenses; abundant pelecypods.	96
44.	Interbedded fine-grained sandstones and silty mudstones; thin-bedded; gray; high humic matter content; several layers of calcareous lenses; common pelecypods.	72
45.	Sandstone, fine-grained, gray, massive bedding; several layers of calcareous lenses; pelecypods common	71
46.	Covered and slumped area	414
47.	Sandstone, fine-grained, gray; some thin layers of siltstone and calcareous lenses; scattered pelecypods including *Pecten*; high humic matter content.	211
48.	Sandstone, fine-grained, gray; some calcareous lenses; abundant pelecypods in upper part; strong contorted bedding	20
49.	Interbedded fine-grained sandstones and siltstones, thin-bedded; gray; calcareous nodule layers in lower part; pelecypods becoming abundant and concentrated in layers in upper part; high humic matter content	162
50.	Sandstone, fine-grained, gray, massive bedding; pelecypods common—mostly single valved with concave side downward; high humic matter content	48
51.	Covered and slumped area	41
52.	Interbedded fine-grained sandstones and siltstones; thin-bedded, gray; bedding at middle of interval apparently cross-bedded; scattered pelecypods; several streaks of ophiuriods at middle of interval; high humic matter content throughout	167
53.	Sandstone, fine-grained, gray and brown; massive bedding marked by slight apparent variation in grain size; several layers of calcareous nodules and lenses; pelecypods common (including *Pecten*), locally abundant in layers; larger pelecypods generally concave side downward; smaller forms have random orientation	362
54.	Covered and slumped area	238
55.	Siltstone, sandy, gray, massive beds in general; gastropods and pelecypods common; locally abundant	321
56.	Siltstone, sandy, gray; thin-bedded in lower part becoming thick-bedded in upper part; scattered pelecypods	329
	Total	3,498

Scotia Bluffs Sandstone

Unit	Description	Thickness
57.	Sandstone, medium-grained, gray and brown; scattered well-rounded igneous pebbles in lower 6 ft; high humic matter content in middle of interval; scattered pelecypods	32
58.	Siltstone, sandy, gray; massive bedding; upper part contains many calcareous lenses crowded with pelecypods and gastropods	87
59.	Siltstone, sandy, gray; becoming fine-grained sandstone in upper part; many layers of pelecypod concentrations; many pelecypods have both valves together; little or no apparent orientation of single valves	88
60.	Sandstone, medium-grained, gray and brown, massive bedding; pelecypods common in lower part and less so in upper part	132
61.	Interbedded medium-grained sandstones and conglomerate lenses and layers; several lenses of sand dollars (*Scutellaster oregonensis*) and pelecypods in lower part	86
	Total	425

Remarks: The base of the Pullen Formation is not exposed; however, the Yager Formation is exposed approximately 300 ft upriver (south) of the lowest visible Pullen outcrops, the intervening area being covered by a massive landslide and heavily wooded area.

Appendix C
Treatment of Samples

Approximately one-half quart of loosely packed, dry sample was crushed to about ½-in. size. The sample was soaked overnight and normally allowed to disintegrate of its own accord with a minimum of wet crushing. A few samples in the lower part (Pullen Formation) required treatment by kerosene and NaOH methods (see, for instance, Bolli, 1952).

Residue is that retained on sieve number 115 (opening .0049 in. or .124 mm), but passing through sieve number 20 (opening .0328 in. or .833 mm) i.e., roughly coarse, medium, and fine-grained sands of the Wentworth grade scale.

Washed material is stored in a standard plastic container (approximately 25 cc) used by the University of California, Berkeley, Paleontology Department.

241 samples (= localities) were collected for study. Of this group, 116 were selected for semiquantitative microfaunal analysis. Generally, approximately 500 specimens were picked from a fraction of the sample and mounted on an assemblage slide. Analysis of assemblage slides is shown on Figures 7, 8, and 12.

The faunal abundance grade scale shown on the figures is a modified geometric grade scale whose grade limits follow those used by Walton (1964, Fig. 3). These grades are: greater than 20% (abundant), 10-20% (frequent), 5-10% (common), 2½-5% (few) and less than 2½% (rare). The scientific basis for this grade scale is unknown; empirically, it seems to fit the ecologic distribution pattern rather well and generally reflects population changes through the section.

Appendix D

Key to Identification of California Neogene Bolivinids[1]

Genus *Bolivina*, subgenus *Brizalina* (type *B. aenariensis* Costa, 1856)

- A. Test compressed
  - 1. Costae present
    - *B. aenariensis* of some California authors
    - *B. argentea* Cushman, 1926
    - *B. argentea* var. *resigae* Zalesny, 1959 (originally var. *monica* Zalesny, 1959)
    - *B. costata* var. *bicostata* Cushman, 1926
    - *B. costata* var. *interjuncta* Cushman, 1926
    - *B. imbricata* Cushman, 1925
    - *B. imbricata* var. *inflata* Kleinpell, 1938
    - *B. interjuncta* Cushman, 1937
  - 2. Costae not present; margins carinate or serrate.
    - *B. bramlettei* Kleinpell, 1938
    - *B. marginata* of some California authors
    - *B. marginata* var. *adelaidana* Cushman and Kleinpell, 1934
    - *B. marginata* var. *monica* Pierce, 1956
    - *B. pseudobeyrichi* Cushman, 1926
    - *B. subadvena* var. *acuminata* Natland, 1946 (originally var. *serrata* Natland, 1946)
    - *B. subadvena* var. *spissa* Cushman, 1926
    - *B. subargentea* Uchio, 1960
  - 3. Lacking ornamentation
    - *B. barbarana* Cushman and Kleinpell, 1934
    - *B. cochei* Cushman and Adams, 1935
    - *B. cuneiformis* Kleinpell, 1938
    - *B. girardensis* Rankin, 1934
    - *B. pseudospissa* Kleinpell, 1938
- B. Test becoming inflated, but not fusiform.
  - 1. Costae present
    - *B. advena* var. *acutula* Bandy, 1953
    - *B. advena* var. *ornata* Cushman, 1925
  - 2. Lacking ornamentation
    - *B. brevior* Cushman, 1925
    - *B. chehalisensis* Rau, 1949 (originally *B. astoriensis* Rau, 1948)
    - *B. decurtata* Cushman, 1926
    - *B. lomitensis* Galloway and Wissler, 1927
    - *B. malagensis* Kleinpell, 1938
    - *B. modesta* Galloway and Wissler, 1927
    - *B. pocheensis* White, 1956
    - *B. subadvena* Cushman, 1926
    - *B. subadvena* var. *sulphurensis* Cushman and Adams, 1935
- C. Test subfusiform or fusiform
  - 1. Costae present
    - *B. californica* Cushman, 1925
    - *B. salinasensis* Kleinpell, 1938
  - 2. Lacking ornamentation
    - *B. acerosa* var. *pacifica* Cushman and McCulloch, 1942
    - *B. benedictensis* Pierce, 1956
    - *B. foxenensis* Bramlette, 1951
    - *B. goudkoffi* Rankin, 1934
    - *B. granti* Rankin, 1934
    - *B. hootsi* Rankin, 1934
    - *B. hughesi* of Cushman, 1926
    - *B. hughesi* var. *parva* Cushman and Galliher, 1934
    - *B. obliqua* Barbat and Johnson, 1934

---

[1] Dates refer to date of creation of species name as cited in the Ellis and Messina "Catalogue of Foraminifera" (1940ff.); see also Sherborn (1955) and Thalmann (1960).

B. quadrata Cushman and McCulloch, 1942
B. rankini Kleinpell, 1938
B. seminuda Cushman, 1911
B. seminuda var. *foraminata* R. E. and K. C. Stewart, 1930
B. seminuda var. *humilis* Cushman and McCulloch, 1942
B. semiperforata Martin, 1952
B. ticensis Kleinpell, 1938
B. tumida Cushman, 1925
B. tumida var. *cuneata* Kleinpell, 1938

Genus *Bolivina*, subgenus *Bolivinoides* (type not here designated), forms apparently transitional between subgenera *Brizalina* and *Bolivina*.

1. Costae present
   B. advena var. *striatella* Cushman, 1925
2. Lacking ornamentation
   B. advena Cushman, 1925
   B. astoriensis Cushman, Stewart, and Stewart, 1948
   B. marginata Cushman var. gracillima Cushman, 1938
   B. modeloensis Cushman and Kleinpell, 1934
   B. perrini Kleinpell, 1938
   B. torqueata Cushman and McCulloch, 1942
   B. vaughani Natland, 1938
   B. woodringi Kleinpell, 1938

Genus *Bolivina*, subgenus *Bolivina* (type *B. plicata* d'Orbigny, 1839)

A. Test compressed
   Examples unknown
B. Test inflated
   1. Costae present
      B. conica Cushman, 1925
   2. Lacking costae
      B. decussata Brady, 1881
      B. floridana of some California authors
      B. hughesi of Cushman, 1937
      B. sinuata Galloway and Wissler, 1927
      B. sinuata var. *alisoensis* Cushman and Adams, 1935

Selected References

Adams, B. C., 1939, Distribution of Foraminifera of the genus *Bolivina* in Canada de Aliso, Ventura Country, California: Am. Jour. Sci., v. 237, p. 500-511.

—— 1940, Foraminifera in zonal paleontology: 6th Pacific Sci. Cong. Proc., p. 665-670.

Addicott, W. O., 1969, Tertiary climatic change in the marginal northeastern Pacific Ocean: Science, v. 165, p. 583-586, 3 figs.

—— 1970, Latitudinal gradients in Tertiary molluscan faunas of the Pacific Coast: Palaeogeography, Palaeoclimatology, Palaeoecology, v. 8, p. 287-312, 7 figs.

—— 1974, Giant pectinids of the eastern North Pacific margin: significance in Neogene zoogeography and chronostratigraphy: Jour. Paleontology, v. 48, p. 180-194, 7 figs.

Akers, W. H., 1954, Ecologic aspects and stratigraphic significance of the Foraminifera *Cyclammina cancellata* Brady: Jour. Paleontology, v. 28, p. 132-152, 10 figs.

—— 1972, Planktonic Foraminifera and biostratigraphy of some Neogene formations, northern Florida and Atlantic coastal plain: Tulane Studies Geology and Paleontology, v. 9, nos. 1-4, 139 p., 60 pls.

—— and J. H. Dorman, 1964, Pleistocene Foraminifera of the Gulf Coast: Tulane Studies Geology and Paleontology, v. 3, no. 1, 93 p., 15 pls.

Andersen, H. V., 1952, *Buccella*, a new genus of rotalid Foraminifera: Washington Acad. Sci. Jour., v. 42, no. 5, p. 143-151.

—— 1961, Genesis and paleontology of the Mississippi River mudlumps: Pt. 2, Foraminifera of the mudlumps, lower Mississippi River delta: Louisiana Geol. Survey Geol. Bull. 35, p. 1-208, 29 pls.

Anderson, G. J., 1963, Distribution patterns of recent Foraminifera of the Bering Sea: Micropaleontology, v. 9, no. 3, p. 305-317, 12 figs., 1 pl.

Arnold, R., 1903, The paleontology and stratigraphy of the marine Pliocene and Pleistocene of San Pedro, California: California Acad. Sci. Mem. 3, 419 p., 37 pls.

—— 1906, The Tertiary and Quaternary pectens of California: U.S. Geol. Survey Prof. Paper 47, 264 p., 57 pls.

—— 1909, Environment of the Tertiary faunas of the Pacific Coast of the United States: Jour. Geology, v. 17, p. 509-533, 5 figs.

—— and H. Hannibal, 1913, The marine Tertiary stratigraphy of the north Pacific Coast of America: Am. Philos. Soc. Proc., v. 52, p. 559-605.

Arnold, Z. M., 1964, Biological observations on the foraminifer *Spiroloculina hyalina* Schulze: California Univ. Pub. Zool., v. 72, 93 p., 7 pls.

Arrhenius, G., 1952, Sediment cores from the east Pacific: Swedish Deep-Sea Exped. Repts., v. 5, pts. 1-4, 227 p., 29 pls., 197 figs.

Asano, K., and T. Uchio, 1951, *Poroeponides cribrorepandus*, in K. Asano, Illustrated catalog of Japanese Tertiary smaller Foraminifera, pt. 14: p. 18., Tokyo, Maruzen Co. Ltd.

—— J. C. Ingle, and Y. Takayanagi, 1968, *Globigerina quinqueloba* Natland; origin and distribution in late Cenozoic of the North Pacific: Gior. Geologia, ser. 2a, v. 35, pt. 2, p. 217-246, figs. 1-15.

Ashley, G. H., 1895, Studies in the Neocene of California: Jour. Geology, v. 3, p. 434-460.

Axelrod, D. I., 1944, The Sonoma flora (California): Carnegie Inst. Washington Pub. 553, p. 167-206.

—— 1956, Mio-Pliocene floras from west-central Nevada: California Univ. Pubs. Geol. Sci., v. 33, 322 p., 32 pls., 18 figs.

Bagg, R. M., 1905, Miocene Foraminifera from Monterey Shale of California with a few species from the Tejon Formation: U.S. Geol. Survey Bull. 268, 78 p., 11 pls.

—— 1912, Pliocene and Pleistocene Foraminifera from southern California: U.S. Geol. Survey Bull. 513, 153 p., 28 pls.

Bailey, J. W., 1851, Microscopial examination of soundings made by the United States Coast Survey of the Atlantic Coast of the United States: Smithsonian Contrib. Knowledge, v. 11, art. 3, p. 1-15, 1 pl.

Bailey, T. L., 1935, Lateral change of fauna in the lower Pleistocene: Geol. Soc. America Bull., v. 46, p. 489-502, 1 fig., 44 pls.

—— 1943, Late Pleistocene Coast Range orogenesis in southern California: Geol. Soc. America Bull., v. 54, p. 1549-1567, 2 figs., 2 pls.

Bandy, O. L., 1950, Some later Cenozoic Foraminifera from Cape Blanco, Oregon: Jour. Paleontology, v. 24, p. 269-281, 2 figs., pls. 41-42.
--- 1953a, Ecology and paleoecology of some California Foraminifera; Pt. 1, Frequency distribution of recent Foraminifera off California: Jour. Paleontology, v. 27, p. 161-182, 4 figs., pls. 21-25.
--- 1953b, Ecology and paleocology of some California Foraminifera; Pt. 2, Foraminiferal evidence on subsidence rates in Ventura basin: Jour. Paleontology, v. 27, p. 200-213.
—— 1960a, General correlation of foraminiferal structure with environment: 21st Internat. Geol. Cong. Rept., pt. 22, p. 7-19.
—— 1960b, The geologic significance of coiling ratios in the foraminifer *Globigerina pachyderma* (Ehrenberg): Jour. Paleontology, v. 34, p. 671-681.
—— 1961, Distribution of Foraminifera, Radiolaria and diatoms in sediments of the Gulf of California: Micropaleontology, v. 7, p. 1-26, pls. 1-5.
—— 1967, Foraminiferal definition of the boundaries of the Pleistocene in southern California, U.S.A.: Progress Oceanography, v. 4, p. 27-49, 7 figs.
—— 1972a, Neogene planktonic foraminiferal zones, California, and some geologic implications: Palaeogeography, Palaeoclimatology, Palaeoecology, v. 12, p. 131-150, 5 figs.
—— 1972b, Late Paleogene-Neogene planktonic biostratigraphy and its geologic implications, California, *in* E. H. Steinmeyer, ed., Pacific Coast Miocene Biostratigraphic Symposium: SEPM Pacific Sec., p. 37-51.
—— 1973, Paleontology: Antarctic Jour. U.S., May-June, p. 86-92, 2 figs.
—— and R. E. Arnal, 1957, Distribution of recent Foraminifera off west coast of Central America: AAPG Bull., v. 41, p. 2037-2053, 3 figs.
—— —— 1960, Concepts of foraminiferal paleoecology: AAPG Bull., v. 44, p. 1921-1932, 14 figs.
—— R. E. Arnal, and F. Theyer, 1972, Planktonic events, upper Gilbert-Gauss magnetic epochs: Geol. Soc. America Abs., p. 442-443.
—— and R. E. Casey, 1969, Major late Cenozoic planktonic datum planes, Antarctica to the tropics: Antarctic Jour. U.S., Sept.-Oct., p. 170-171, 1 fig.
—— and J. C. Ingle, Jr., 1970, Neogene planktonic events and radiometric scale, California: Geol. Soc. America Spec. Paper 124, p. 131-172, 7 figs.
—— and K. S. Rodolfo, 1964, Distribution of Foraminifera and sediments, Peru-Chile Trench area: Deep-Sea Research, v. 11, p. 817-837.
—— and J. A. Wilcoxon, 1970, The Pliocene-Pleistocene boundary, Italy and California: Geol. Soc. America Bull., v. 81, p. 2939-2948, 7 figs.
—— —— and R. C. Wright, 1971, Late Neogene planktonic zonation, magnetic reversals, and radiometric dates, Antarctic to the tropics: Am. Geophys. Union, Antarctic Res. Ser., v. 15, p. 1-26.
Barbieri, Francesco, 1967, The Foraminifera in the Pliocene section Vernasca-Castell'Arquato, including the "Piacenzian stratotype" (Piacenza Province): Soc. Italiana Sci. Nat. e Museo Civico Storia Nat. Milano Mem., v. 15, no. 3, p. 145-163, 1 table, 10 figs.
—— 1969, Planktonic Foraminifera in western Emily Pliocene (North Italy): 1st Internat. Conf. Planktonic Microfossils Proc. (Geneva), v. 1, p. 58-65.
—— 1971, Comments on some Pliocene stages and on the taxonomy of a few species of *Globorotalia*: Ateneo Parmense, Acta Nat., v. 7, p. 1-24.
—— and F. Petrucci, 1967, La série stratigraphique du messinien au calabrien dans la vallée du Crostolo (Reggio Emilia, Italie sept.): Soc. Italiana Sci. Nat. e Museo Civico Storia Nat. Milano Mem., v. 15, no. 3, p. 181-188, 1 fig., 1 pl.
Barker, R. W., 1960, Taxonomic notes on the species figured by H. B. Brady in his report on the Foraminifera dredged by H.M.S. *Challenger* during the years 1873-1876: SEPM Spec. Pub. 9, 238 p., 114 pls.
Bayliss, D. D., 1969, The distribution of *Hyalinea balthica* and *Globorotalia truncatulinoides* in the type Calabrian: Lethaia, v. 2, no. 2, p. 133-143, 5 figs.
Berggren, W. A., 1968, Phylogenetic and taxonomic problems of some Tertiary planktonic foraminiferal lineages: Tulane Studies Geology, v. 6, no. 1, p. 1-22, 3 figs.
—— 1969, Rates of evolution in some Cenozoic planktonic Foraminifera: Micropaleontology, v. 15, p. 351-365, 13 figs.
—— 1969, Cenozoic chronostratigraphy, planktonic foraminiferal zonation and the radiometric time scale: Nature, v. 224, no. 5224, p. 1072-1075.
—— 1972, A Cenozoic time-scale—some implications for regional geology and paleobiogeography: Lethaia, v. 5, p. 195-215, 10 figs.
—— 1973, The Pliocene time scale: calibration of planktonic foraminiferal and calcareous nannoplankton zones: Nature, v. 243, no. 5407, p. 391-397, 5 tables, 1 fig.
Berry, W. B. N., 1964, The Middle Ordovician of the Oslo region, Norway, No. 16, Graptolites of

the *Ogygiocaris* series: Norsk Geol. Tidsskr., v. 44, pt. 1, p. 61-170, 16 pls.
—— 1966, Zones and zones–with exemplification from the Ordovician: AAPG Bull., v. 50, p. 1487-1500, 2 tables.
—— 1968, Growth of a prehistoric time scale: San Francisco, W. H. Freeman Co., 158 p., 16 figs.
—— 1972, Criteria for biostratigraphic correlation, *in* E. H. Stinemeyer, ed., Pacific Coast Miocene Biostratigraphic Symposium: SEPM Pacific Sec., p. 214-225.
Bertolino, Vera, et al, 1968, Proposal for a biostratigraphy of the Neogene in Italy based on planktonic Foraminifera: Gior. Geologia, ser. 2, v. 35, pt. 2, p. 23-30, 1 fig.
Blow, W. H., 1969, Late middle Eocene to recent planktonic foraminiferal biostratigraphy: 1st Internat. Conf. Planktonic Microfossils Proc., v. 1, p. 199-421.
—— 1970, Validity of biostratigraphic correlations based on the Globigerinacea: Micropaleontology, v. 16, p. 257-268.
Bolli, Hans, 1950, The direction of coiling in the evolution of some Globorotaliidea: Cushman Found. Foram. Research Contr., v. 1, p. 82-89.
—— 1952, Note on the disintegration of indurated rocks: Micropaleontologist, v. 6, p. 46-48.
Bouma, A. H., 1961, Sedimentology of some flysch deposits: a graphic approach to facies interpretation: Amsterdam, Elsevier, 168 p.
—— 1964, Ancient and recent turbidites: Geologie en Mijnbouw, v. 43, no. 8, p. 375-379, 3 figs.
Bradshaw, J. S., 1959, Ecology of living planktonic Foraminifera in the north and equatorial Pacific Ocean: Cushman Found. Foram. Research Contr., v. 10, p. 25-64, pls. 6-8.
Brady, H. B., 1879, Notes on some of the reticularian Rhizopoda of the *Challenger* expedition: Quart. Jour. Micros. Sci., v. 19, p. 20-63, pls. 3-5, p. 261-299, pl. 8.
—— 1881, Notes on some of the reticularian Rhizopoda of the *Challenger* expedition: Quart. Jour. Micros. Sci., v. 21, p. 37-71.
—— 1884, Report on the Foraminifera dredged by H.M.S. *Challenger*, during the years 1873-1876: Rept. Voyage Challenger, Zool., v. 9, 814 p., 115 pls.
Bramlette, M. N., 1950, Foraminifera of Santa Maria District, *in* W. P. Woodring and M. N. Bramlette, Geology and paleontology of Santa Maria District, California: U.S. Geol. Survey Prof. Paper 222, p. 58-61, pls. 22-23.
California Division of Oil and Gas, 1960, San Joaquin-Sacramento Valleys and northern coastal regions, Pt. 1 *of* California oil and gas fields, maps and data sheets: San Francisco, California Div. Oil and Gas, 493 p.
Carloni, G. C., Franco Cati, and A. M. Borsetti, 1968, Stratigrafia del Miocene Marchigiano in facies di "Schlier": Gior. Geologia, ser. 2, v. 35, no. 2, p. 341-368, 1 fig., pls. 8-10.
Carpenter, W. B., 1869, On the rhizopod fauna of the deep sea: Royal Soc. [London] Proc., v. 18, no. 114, p. 59-62.
Casey, R. E., 1972, Neogene radiolarian biostratigraphy and paleotemperatures–southern California, the experimental Mohole, Antarctic Core E 14-8: Palaeogeography, Palaeoclimatology, Palaeoecology, v. 12, p. 115-130, 4 figs., 5 tables.
Cati, Franco, et al, 1968, Biostratigraphia del Neogene mediterraneo basata sui foraminiferi planctonici: Soc. Geol. Italiana Boll., v. 87, no. 3, p. 491-503, 2 figs.
Chapman, Frederick, 1900, Foraminifera from the Tertiary of California: California Acad. Sci. Proc., 3d Ser., Geology, v. 1, no. 8, p. 241-260, 2 pls.
Church, C. C., 1927, A new species of *Bolivinita* from lower Pliocene of California: Jour. Paleontology, v. 1, p. 265-268, 1 fig.
Cifelli, R., 1962, The morphology and structure of *Ammonia beccarii* (Linné): Cushman Found. Foram. Research Contr., v. 13, p. 119-126.
—— 1969, Radiation of Cenozoic planktonic Foraminifera: Systematic Zoology, v. 18, no. 2, p. 154-168, 8 figs.
Cita, M. B., and W. H. Blow, 1969, The biostratigraphy of the Langhian, Serravallian and Tortonian Stages in the type-sections in Italy: Riv. Italiana Paleontologia e Stratigrafia, v. 75, no. 3, p. 549-603, 10 figs.
—— and I. Premoli Silva, 1968, Evolution of the planktonic foraminiferal assemblages in the stratigraphical interval between the type-Langhian and the type-Tortonian and biozonation of the Miocene of Piedmont: Gior. Geologia, ser. 2, v. 35, no. 3, p. 1-23, pls. 1-2.
—— —— and R. C. Rossi, 1965, Foraminiferi planctonici del tortoniano-tipo: Riv. Italiana Paleontologia Stratigrafia, v. 71, no. 1, p. 217-308, pls. 18-31.
Clark, B. L., 1921, The marine Tertiary of the west coast of the United States–its sequence, paleogeography, and the problems of correlation: Jour. Geology, v. 29, p. 583-614, 12 figs.
—— 1929, Stratigraphy and faunal horizons of the Coast Ranges of California: privately published, 132 p., 50 pls., correlation charts.
Colalongo, M. L., 1968, Cenozone a foraminiferi ed ostracodi nel Pliocene e basso Pleistocene della serie del Santerno e dell'Appennino Romagnolo: Gior. Geologia, ser. 2, v. 35, no. 3, p. 29-61, 2 figs.

—— 1970, Appunti biostratigrafici sul Messiniano: Gior. Geologia, ser. 2, v. 36, p. 515-542, 1 fig., 2 pls.
—— and Samuele Sartoni, 1967, *Globorotalia hirsuta aemiliana* nuova sottospecie cronologica del Pliocene in Italia: Gior. Geologia, ser. 2, v. 34, p. 255-284, 2 figs., 2 pls.
Conato, Vittorio, and Umberto Follador, 1967, *Globorotalia crotonensis* e *Globorotalia crassacrotonensis* nuove species del Pliocene Italiano: Soc. Geol. Italiana Boll., v. 84, p. 555-563, 6 figs.
Conrey, B. L., 1967, Early Pliocene sedimentary history of the Los Angeles basin, California: California Div. Mines and Geology, Spec. Rept. 93, 63 p.
Cooper, J. G., 1888, Catalogue of California fossils: 7th Ann. Rept. State Mineralogist (1886-87), p. 223-308.
Corby, G. W., et al, 1951, Geology and oil possibilities of the Philippines: Philippines Bur. Mines Rept. Inv., Tech. Bull. 21, 363 p., 30 text-figs., 57 pls.
Crickmay, C. H., 1929, The anomalous stratigraphy of Deadman's Island, California: Jour. Geology, v. 37, p. 617-638.
Crouch, R. W., 1951, *Nodosarella verneuili* (d'Orbigny) from the Pliocene of the Los Angeles basin: Cushman Found. Foram. Research Contr., v. 2, p. 9-10.
—— 1952, Significance of temperature on Foraminifera from deep basins off southern California: AAPG Bull., v. 36, p. 807-843, pls. 1-7, figs. 1-5.
Crowell, J. C., et al, 1966, Deep-water sedimentary structures, Pliocene Pico Formation, Santa Paula Creek, Ventura basin, California: California Div. Mines and Geology Spec. Rept. 89, 40 p., 20 figs., 4 tables.
Cushman, J. A., 1910, A monograph of the Foraminifera of the north Pacific Ocean; Pt. 1, Astrorhizidae and Lituolidae: U.S. Natl. Mus. Bull., v. 71, 134 p., figs. 1-203.
—— 1911, A monograph of the Foraminifera of the north Pacific Ocean; Pt. 2, Textulariidae: U.S. Natl. Mus. Bull., no. 71, 156 p.
—— 1913, A monograph of the Foraminifera of the north Pacific Ocean; Pt. 3 Lagenidae: U.S. Natl. Mus. Bull. 71, p. 1-125, pls. 1-47.
—— 1914, A monograph of the Foraminifera of the north Pacific Ocean; Pt. 4, Chilostomellidae, Globigerinidae, Nummulitidae: U.S. Natl. Mus. Bull., no. 17, p. 1-46, pls. 1-19.
—— 1915, A monograph of the Foraminifera of the north Pacific Ocean; Pt. 5, Rotaliidae: U.S. Natl. Mus. Bull., no. 71, 87 p.
—— 1917, A monograph of the Foraminifera of the north Pacific Ocean; Pt. 6, Miliolidae: U.S. Natl. Mus. Bull., no. 71, 108 p.
—— 1918, Some Pliocene and Miocene Foraminifera of the coastal plain of the United States: U.S. Geol. Survey Bull. 676, 100 p., 31 pls.
—— 1923, The Foraminifera of the Atlantic Ocean; Pt. 4, Lagenidae: U.S. Natl. Mus. Bull. 104, 228 p., 42 pls.
—— 1925a, Recent Foraminifera from British Columbia: Cushman Lab. Foram. Research Contr., v. 1, p. 38-47, pls. 6-7.
—— 1925b, Miocene species of *Nonionina* from California: Cushman Lab. Foram. Research Contr., v. 1, p. 89-91, pl. 13.
—— 1925c, Some Textulariidae from the Miocene of California: Cushman Lab. Foram. Research Contr., v. 1, p. 29-35, pl. 5.
—— 1926a, Foraminifera of the typical Monterey of California: Cushman Lab. Foram. Research Contr., v. 2, p. 53-70, pls. 7-9.
—— 1926b, Some Pliocene bolivinas from California: Cushman Lab. Foram. Research Contr., v. 2, p. 40-47, pl. 6.
--- 1926c, Miocene species of *Nonion* from California: Cushman Lab. Foram. Research Contr., v. 2, p. 89-92.
—— 1927, Recent Foraminifera from off the west coast of America: Scripps Inst. Oceanography Tech. Ser. Bull., v. 1, p. 119-188, pls. 1-6.
—— 1929, Pliocene lagenas from California: Cushman Lab. Foram. Research Contr., v. 5, p. 67-72, pl. 11.
--- 1932, Foraminifera of the tropical Pacific collection of the *Albatross*; Pt. 1, Astrorhizidae to Trochamminidae: U. S. Natl. Mus. Bull. 161, 88 p.
--- 1933, Foraminifera of the tropical Pacific collection of the *Albatross*; Pt. 2, Lagenidae to Alveolinellindae: U. S. Natl. Mus. Bull. 161, 79 p.
—— 1937, A monograph of the subfamily Virgulininae of the foraminiferal family Bulminidae: Cushman Lab. Foram. Research Spec. Pub. 9, 228 p., 24 pls.
—— 1939a, A monograph of the foraminiferal family Nonionidae: U.S. Geol. Survey Prof. Paper 191, 100 p., 20 pls.
—— 1939b, Notes on some Foraminifera described by Schwager from the Pliocene of Kar Nicobar: Jour. Geological Soc. Japan, v. 46, no. 546, p. 149-154, pl. 10.
—— 1941, Some fossil Foraminifera from Alaska: Cushman Lab. Foram. Research Contr., v. 17,

p. 33-38, pls. 9-10.
—— 1942, The Foraminifera of the tropical Pacific collection of the *Albatross*, 1899-1900; Pt. 3, Heterohelicidae and Buliminidae: U.S. Natl. Mus. Bull. 161, 67 p. 15 pls.
--- 1950, Foraminifera, their classification and economic use: Cambridge, Mass., Harvard Univ. Press, 605 p., 55 pls.
—— and B. C. Adams, 1935, New late Tertiary bolivinas from California: Cushman Lab. Foram. Research Contr., v. 11, p. 16-20, pl. 3.
--- and U. S. Grant, IV, 1927, Late Tertiary and Quaternary elphidiums of the west coast of North America: San Diego Soc. Nat. History Trans., v. 5, no. 6, p. 69-82, pls. 7-8.
—— and H. B. Gray, 1946a, Some new species and varieties of Foraminifera from the Pliocene of Timms Point, California: Cushman Lab. Foram. Research Contr., v. 22, p. 65-69, pl. 12.
—— —— 1946b, A foraminiferal fauna from the Pliocene of Timms Point, California: Cushman Lab. Foram. Research Spec. Pub. 19, 46 p., 8 pls.
—— and D. H. Hughes, 1925, Some later Tertiary cassidulinas of California: Cushman Lab. Foram. Research Contr., v. 1, p. 11-17, pl. 2.
—— and R. M. Kleinpell, 1934, New and unrecorded Foraminifera from the California Miocene: Cushman Lab. Foram. Research Contr., v. 10, p. 1-24, pls. 1-3.
—— and Boris Laiming, 1931, Miocene Foraminifera from Los Sauces Creek, Ventura County, California: Jour. Paleontology, v. 5, p. 79-120, pls. 9-14, 5 figs.
—— and Irene McCulloch, 1939, Report on some arenaceous Foraminifera: Allan Hancock Pacific Exped., v. 6, p. 1-114, pls. 1-12.
—— —— 1942a, Some Nonionidae in the collections of the Allan Hancock Foundation: Allan Hancock Pacific Exped., v. 6, p. 145-178, pls. 17-20.
—— —— 1942b, Some Virgulininae in the collections of the Allan Hancock Foundation: Allan Hancock Pacific Exped., v. 6, p. 179-230, pls. 21-28.
—— —— 1948, The species of *Bulimina* and related genera in the collections of the Allan Hancock Foundation: Allan Hancock Pacific Exped., v. 6, p. 231-232, pls. 29-32.
—— —— 1950, Some Lagenidae in the collections of the Allan Hancock Foundation: Allan Hancock Pacific Exped., v. 6, p. 295-364, pls. 37-48.
—— and Y. Ozawa, 1930, A monograph of the foraminiferal family Polymorphinidae, Recent and fossil: U.S. Natl. Mus. Proc., v. 77, p. 1-195, pls. 1-40.
—— and F. L. Parker, 1938, Notes on some Pliocene and Pleistocene species of *Bulimina* and *Buliminella*: Cushman Lab. Foram. Research Contr., v. 14, p. 53-62, pls. 9-10.
—— —— 1946, *Bulimina* and related foraminiferal genera: U.S. Geol. Survey Prof. Paper 210-D, p. 55-176, pls. 15-30.
—— and R. E. Stewart, 1926, A new *Plectofrondicularia* from the Pliocene of California: Cushman Lab. Foram. Research Contr., v. 2, p. 39, pl. 6.
—— and Ruth Todd, 1947, Foraminifera from the coast of Washington: Cushman Lab. Foram. Research Spec. Pub. 21, 23 p., 4 pls.
—— R. E. Stewart, and K. C. Stewart, 1930, Tertiary Foraminifera from Humboldt County, California: San Diego Soc. Nat. History Trans., v. 6, no. 2, p. 41-94, pls. 1-8.
—— —— —— 1947a, Astoria Miocene Foraminifera from Agate Beach, Lincoln County, Oregon: Oregon Dept. Geology and Mineral Industries Bull., no. 36, pt. 2, p. 41-54, 1 fig., pls. 5-6.
—— —— —— 1947b, Astoria Miocene Foraminifera from the northwest corner of Tenth Street and Harrison Avenue, Astoria, Clatsop County, Oregon: Oregon Dept. Geology and Mineral Industries Bull., no. 36, pt. 1, p. 9-38, pls. 1-4.
--- --- --- 1949, Pliocene Foraminifera from Oregon; Pt. 7, Quinault Pliocene Foraminifera from western Washington: Oregon Dept. Geology and Mineral Industries Bull., v. 36, p. 148-162, pls. 17-18.
—— Ruth Todd, and R. J. Post, 1954, Recent Foraminifera of the Marshall Islands: U.S. Geol. Survey Prof. Paper 260-H, p. 319-384, pls. 82-93.
Diller, J. S., 1902, Topographic development of the Klamath Mountains: U.S. Geol. Survey Bull. 196, 69 p., 13 pls., 7 figs.
Dondi, L., and I. Papetti, 1968, Biostratigraphical zones of Po valley Pliocene: Gior. Geologia, ser. 2, v. 35, no. 3, p. 63-98, 3 figs., pls. 3-5.
D'Onofrio, Sara, 1964, Foraminiferi del neostratotipo del Messiniano: Gior. Geologia, ser. 2, v. 32, p. 409-461, 1 fig., 5 pls.
—— 1968, Biostratigrafia del Pliocene e Pleistocene inferiore nelle Marche: Gior. Geologia, ser. 2, v. 35, no. 3, p. 99-114, 3 figs.
d'Orbigny, A. D., 1826, Tableau methodique de la classe des Cephalopodes: Annales Sci. Natur., v. 7, p. 245-314, pls. 10-17.
—— 1839, Voyage dans l'Amérique Méridionale, Foraminiferes: Paris and Strasbourg, v. 5, no. 5, 86 p., 9 pls.
Dorf, E., 1930, Pliocene floras of California: Carnegie Inst. Washington Pub. 412, p. 1-112.
Driver, H. L., 1928, Foraminiferal section along Adams Canyon, Ventura County, California:

AAPG Bull., v. 12, p. 753-756, 1 fig.
Dunbar, C. O., 1960, Historical geology: New York, John Wiley & Sons (2d edition), 50 p., 406 figs.
Duncan, J. R., G. A. Fowler, and L. D. Kulm, 1970, Planktonic foraminifera-radiolarian ratios and Holocene-late Pleistocene deep sea stratigraphy off Oregon: Geol. Soc. America Bull., v. 81, p. 561-566, 3 figs.
Durham, J. W., 1950, Cenozoic marine climates of the Pacific Coast: Geol. Soc. America Bull., v. 61, p. 1243-1264, figs. 1-3.
—— 1955, Classification of clypeasteroid echinoids: California Univ. Pubs. Geol. Sci., v. 31, p. 73-198, pls. 3-4, 38 figs.
Ehrenberg, C. G., 1872, Mikrogeologische Studien über das kleinste Leben der Meeres-Tiefgründe aller Zonen und dessen geologischen Einfluss: Berlin, Abh. k. Ak. Wiss., v. 6, p. 131-397, pls. 1-12.
Ellis, B. V., and A. R. Messina, 1940ff; Catalogue of Foraminifera: N. Y., Amer. Mus. Natur. Hist., Spec. Publ.
Emiliani, Cesare, 1961, Cenozoic climatic changes as indicated by the stratigraphy and chronology of deep-sea cores in Globigerina-ooze facies: New York Acad. Sci. Annals, v. 95, art. 1, p. 521-536, figs. 1-10.
—— T. Mayeda, and R. Selli, 1961, Paleotemperature analysis of the Plio-Pleistocene section at La Castella, Calabria, southern Italy: Geol. Soc. America Bull., v. 72, p. 679-688.
Enbysk, B. J., 1960, Distribution of Foraminifera in the northeast Pacific: Univ. Washington, PhD thesis, 226 p., 21 pls., 23 figs.
Ericson, D. B., 1959, Coiling direction of *Globigerina pachyderma* as a climatic index: Science, v. 130, no. 3369, p. 219-220.
—— and G. Wollin, 1964, The deep and the past: New York, Alfred A. Knopf, Inc., 292 p.
—— M. Ewing, and G. Wollin, 1963, Pliocene-Pleistocene boundary in deep sea sediments: Science, v. 139, no. 3556, p. 727-737.
—— —— —— 1964, The Pleistocene epoch in deep sea sediments: Science, v. 146, no. 3645, p. 723-732, text-figs. 1-5.
Evenson, R. E., 1959, Geology and ground-water features of the Eureka area, Humboldt County, California: U.S. Geol. Survey Water-Supply Paper 1470, 80 p., 15 figs.
Everden, J. F., et al, 1964, Potassium-argon dates and the Cenozoic mammalian chronology of North America: Am. Jour. Sci., v. 262, p. 145-198.
Fairbridge, R. W., 1973, Friends of the Mediterranean Quaternary visit type sections: Geotimes, November, p. 24-26, 1 fig., 1 table.
Faustman, W. F., 1964, Paleontology of the Wildcat Group at Scotia and Centerville Beach, California: California Univ. Pubs. Geol. Sci., v. 41, no. 2, p. 97-160, 3 pls.
Flint, R. H., 1965, The Plio-Pleistocene boundary: Geol. Soc. America Spec. Paper 84, p. 497-533.
Follador, Umberto, 1967, Il Pliocene ed il Pleistocene dell'Italia centro-meridionale, versante Adriatico; biostratigrafia: Soc. Geol. Italiana Boll., v. 86, no. 3, p. 565-584, 2 figs.
Fowler, G. A., and L. D. Kulm, 1970, Foraminiferal and sedimentological evidence for uplift of the deep-sea floor, Gorda Rise, northeastern Pacific: Jour. Marine Research, v. 28, p. 321-329, 3 figs.
—— W. N. Orr, and L. D. Kulm, 1971, An upper Miocene diatomaceous rock unit on the Oregon continental shelf: Jour. Geology, v. 79, p. 603-608, 3 figs.
Frizzell, D. L., and Myra Keen, 1949, On the nomenclature and generic position of *Nautilus beccarii* Linne (Foraminifera, "Rotaliidae"): Jour. Paleontology, v. 23, p. 106-108.
Gabb, W. M., 1866-69, Cretaceous and Tertiary fossils: California Geol. Survey, Paleontology, v. 2, sec. 1, p. 1-124, pls. 1-36.
Galloway, J. J., and S. G. Wissler, 1927a, Correction of names of Foraminifera: Jour. Paleontology, v. 1, p. 193.
—— —— 1927b, Pleistocene Foraminifera from the Lomita quarry, Palos Verdes Hills, California: Jour. Paleontology, v. 1, p. 35-87, pls. 7-12.
Gianotti, Agostino, 1953, Microfaune della serie tortoniana del rio Mazzapiedi-Castellania (Tortona-Alessandria): Riv. Italiana Paleontologia e Stratigrafia, Mem. 6, p. 167-301, 10 pls.
Glaessner, M. F., 1970, Notes concerning a chronostratigraphic scale for the ECAFE region: United Nations, Mineral Resources Devel. Ser., no. 36, p. 25-28, 1 table.
Glen, W., 1959, Pliocene and lower Pleistocene of the western part of the San Francisco Peninsula: California Univ. Pubs. Geol. Sci., v. 36, p. 147-198, pls. 15-17, 5 figs.
Gordon, W. A., 1962, Problems of paleontological correlation with particular reference to the Tertiary: AAPG Bull., v. 46, p. 394-398.
Gradstein, F. M., 1970, Foraminifera from the type Sicilian at Ficarazzi, Sicily (lower Pleistocene): Koninkl. Nederlandse Akad. Wetensch. Proc., ser. B, v. 73, no. 4, p. 1-29, 4 figs., 1 pl.

—— 1973, Pliocene and Pleistocene planktonic Foraminifera from Santa Maria di Catanzaro, southern Italy: Newsletters Stratigraphy, v. 3, pt. 1, p. 45-58, 1 fig., 1 pl.

Grant, U. S., IV, 1935, Summary of the marine Pleistocene of California (abs.): Geol. Soc. America Proc. for 1935, p. 349-350.

—— and H. R. Gale, 1931, Catalogue of the marine Pliocene and Pleistocene Mollusca of California: San Diego Soc. Nat. History Mem. 1, 1036 p., 32 pls.

—— and L. G. Hertlein, 1943, Pliocene correlation chart, *in* Geologic formations and economic development of the oil and gas fields of California: California Div. Mines Bull. 118, p. 201-202.

Hanna, G. D., 1938, The Monterey Shale of California at its type locality with a summary of its fauna and flora: AAPG Bull., v. 12, p. 969-983.

Harmon, A. K. P., Jr., 1914, Eel River valley, Humboldt County, geology and oil possibilities: California State Mining Bur. Bull., v. 69, p. 455-459.

Harmon, R. A., 1964, Distribution of Foraminifera in the Santa Barbara basin, California: Micropaleontology, v. 10, p. 81-96, 12 text-figs., 4 tables.

Hawley, A. S., 1961, Tompkins Hill (Eureka) gas field, *in* California oil and gas fields. Pt. 1, San Joaquin-Sacramento valleys and northern coastal regions: California Div. Oil and Gas, p. 478-479.

Hay, O. P., 1927, The Pleistocene of the western region of North America and its vertebrate animals: Carnegie Inst. Washington Pub. 322B, 346 p., 12 pls.

Hays, J. D., 1970, Stratigraphy and evolutionary trends of radiolaria in North Pacific deep-sea sediments: Geol. Soc. America Mem. 126, p. 185-218, 10 figs., 1 pl., 2 tables.

—— and W. A. Berggren, 1971, Quaternary boundaries and correlations, *in* Micropaleontology of the oceans: Cambridge, England, Cambridge Univ. Press, p. 669-691, 9 figs.

Hofker, J., 1952, Zur Fassung der Foraminiferengattung *Bolivinoides* Cushman, 1927: Geol. Jahrb., v. 66, p. 377-382.

—— 1970, *Reprint of and introduction to* C. Schwager (1866, 1883) Fossile Foraminiferen von Kar Nicobar: Lochem, Holland, Antiquariaat Junk, 192 p., 10 pls.

Holman, W. H., 1958, Correlation of producing zones of Ventura basin oil fields, *in* Geology and oil fields of Los Angeles and Ventura regions: AAPG Pacific Sec. Ann. Mtg., Los Angeles, p. 190-199, 1 fig., 2 charts.

Hoskins, E. G., and J. R. Griffiths, 1971, Hydrocarbon potential of northern and central California offshore, *in* Future petroleum provinces of the United States—their geology and potential: AAPG Mem. 15, v. 1, p. 212-228, 18 figs.

Iaccarino, Silvia, 1967, Les foraminifères du stratotype du tabianien (Pliocène inférieur) de Tabiano bagni (Parme): Soc. Italiana Sci. Nat. e Museo Civico Storia Nat. Milano, Mem. 15, no. 3, p. 165-180, 3 figs., 1 pl.

Ingle, J. C., Jr., 1967, Foraminiferal biofacies variation and the Miocene-Pliocene boundary in southern California: Bulls. Am. Paleontology, v. 52, no. 236, p. 217-394.

—— 1968, Pliocene planktonic Foraminifera from northern California and paleo-oceanographic implications (abs.): Geol. Soc. America Spec. Paper 121, p. 147.

—— 1971, Neogene planktonic Foraminifera from the Pacific Coast of North America: biofacies, biostratigraphy, and paleo-oceanographic implications: Geol. Soc. America Abs., v. 3, no. 2, p. 139-140.

—— 1973a, Neogene Foraminifera from the northeastern Pacific Ocean, Leg 18, Deep Sea Drilling Project: Initial Repts. Deep Sea Drilling Project, v. 18, p. 517-567, 7 figs., 26 tables.

—— 1973b, Biostratigraphy and paleoecology of early Miocene through early Pleistocene benthonic and planktonic Foraminifera, San Joaquin Hills - Newport Bay - Dana Point area, Orange Co., California: SEPM Field Trip Guidebook No. 1, Ann. Mtg., Los Angeles, p. 18-38, figs. 1-12.

—— 1974, Oligocene to Holocene planktonic foraminiferal biofacies and paleooecanography of the California current system at lat. 39°N (abs.): Geol. Soc. America Abs. with Programs, v. 6, no. 3, p. 195.

International Geological Congress, 1950, Recommendations of the commision appointed to advise on the definition of the Pliocene-Pleistocene boundary: 18th Internat. Geol. Cong. Rept., Pt. IX, Proc. Sect. H, p. 6 (London).

Irwin, W. P., 1960, Geologic reconnaissance of the northern Coast Ranges and Klamath Mountains, California: Calif. Div. Mines and Geology Bull. 179, 80 p., 1 pl., 16 figs.

—— et al, 1960, Geology mineral resources, mineral industry, *in* Natural resources of northwestern California: U.S. Dept. Interior, Pacific Southwest Field Comm., 40 p., 1 pl., 10 figs.

Izett, G. A., C. W. Naeser, and J. D. Obradovich, 1974, Fission track age of zircons from an ash bed in the Pico Formation (Pliocene and Pleistocene) near Ventura, California (abs.): Geol. Soc. America Abs. with Programs, v. 6, no. 3, p. 197.

Jeletzky, J. A., 1973, Age and depositional environments of Tertiary rocks of Nootka Island, British Columbia (92-E): mollusks versus foraminifers: Canadian Jour. Earth Sci., v. 10, no. 3, p. 331-365, 3 pls.

Jenkins, D. G., 1966, Planktonic Foraminifera from the type Aquitanian-Burdigalian of France: Cushman Found. Foram. Research Contr., v. 17, p. 1-15, pls. 1-4.
—— 1971, New Zealand Cenozoic planktonic Foraminifera: New Zealand Geol. Survey Paleontology Bull. 42, 278 p., 23 pls., 2 figs., 57 tables.
Keen, A. M., 1939, The percentage method of stratigraphic dating: 6th Pacific Sci. Cong. Proc., v. 2, p. 659-663, 2 figs.
—— and H. Bentson, 1944, Checklist of California Tertiary marine Mollusca: Geol. Soc. America Spec. Paper 56, 300 p.
Kilkenny, J. E., 1970, Future petroleum potential of Region 2, Pacific coastal states and adjacent continental shelf and slope: AAPG Mem. 15, v. 1, p. 170-177.
Kleinpell, R. M., 1933, Miocene Foraminifera from Reliz Canyon, Monterey Co., California (abs.): Geol. Soc. America Bull., v. 44, p. 165.
—— 1934, Difficulty of using cartographic terminology in historical geology: AAPG Bull., v. 18, p. 374-379.
—— 1935, Proposed biostratigraphic classification of California Miocene (abs.): Geol. Soc. America Proc. for 1934, p. 390-391.
—— 1938, Miocene stratigraphy of California: Tulsa, Okla., AAPG, 450 p., 22 pls., 14 figs.
—— 1940, Foraminifera from Reef Ridge Shale, McClure Shale, and Gould Shale, *in* Geology of the Kettleman Hills oil field, California: U.S. Geol. Survey Prof. Paper 195, p. 121, 128-129, pls. 49-50.
—— 1946, Foraminiferal checklists from Altamira Shale, Valmonte Diatomite, and Malaga Mudstone, *in* Geology and paleontology of Palos Verdes Hills, California: U.S. Geol. Survey Prof. Paper 207, p. 19, 26, 32-33, 36, 39.
—— 1948, Miocene-Pliocene boundary in California as a typical example of series-epoch boundary problems in correlation: Geol. Soc. America Bull., v. 59, p. 1387.
—— 1967, Miocene: Chicago, Encyclopaedia Britannica, Inc., p. 558-561, 2 figs.
—— 1971, California's early "Oil Bug" profession: Jour. West, v. 10, no. 1, p. 72-101, 4 pls.
—— 1972, Some of the historical context in which a micropaleontological stage classification of the Pacific Coast middle Tertiary has developed, *in* E. H. Stinemeyer, ed., The Pacific Coast Miocene Biostratigraphic Symposium Proc.: Tulsa, Okla., SEPM, p. 89-110, 2 pls.
—— and D. W. Weaver, 1963, Oligocene biostratigraphy of the Santa Barbara Embayment, California: Calif. Univ. Pubs. Geol. Sci., v. 43, 250 p., 38 pls., 8 figs.
Lalicker, C. G., and Irene McCulloch, 1940, Some Textulariidae of the Pacific Ocean: Allan Hancock Pacific Exped., v. 6, p. 115-143, pls. 13-16.
Lamb, J. L., and J. H. Beard, 1972, Late Neogene planktonic foraminifers in the Caribbean, Gulf of Mexico, and Italian stratotypes: Kansas Univ. Paleont. Contr. 57, p. 1-67, 25 figs., 36 pls., 2 tables.
Lankford, R. R., 1962, Recent Foraminifera from the nearshore turbulent zone, western U.S. and northwest Mexico: Univ. California at San Diego, PhD thesis, 234 p., 6 pls., 9 figs.
—— and F. B. Phleger, 1973, Foraminifera from the nearshore turbulent zone, western North America: Jour. Foram. Research, v. 3, p. 101-132, 5 text-figs., 6 pls.
Lawson, A. C., 1893, The post-Pliocene diastrophism of the coast of southern California: California Univ. Dept. Geology Bull., v. 1, p. 115-160.
—— 1894, Geomorphogeny of coast of northern California: California Univ. Pubs. Dept. Geology Bull., v. 1, p. 241-272.
Leroy, L. W., 1943, Pleistocene and Pliocene Ostracoda of the coastal region of southern California: Jour. Paleontology, v. 17, p. 354-373, pls. 58-62.
Lidz, Barbara, 1972, *Globorotalia crassaformis* morphotype variations in Atlantic and Caribbean deep-sea cores: Micropaleontology, v. 18, p. 194-211, 7 pls.
Lipps, J. H., 1964, Miocene planktonic Foraminifera from Newport Bay, California: Tulane Studies Geology and Paleontology, v. 2, no. 4, p. 109-134, 4 pls.
—— 1965, Revision of the foraminiferal family Pseudoparrellidae Voloshinova: Tulane Studies Geology and Paleontology, v. 3, no. 2, p. 117-147, 3 pls.
—— 1972, Plankton biostratigraphy and paleoecology of the eastern north Pacific Ocean. Introduction: Palaeogeography, Palaeoclimatology, Palaeoecology, v. 12, p. 3-14, 4 figs.
Loeblich, A. R., Jr., and Helen Tappan, 1953, Studies of Arctic Foraminifera: Smithsonian Misc. Colln., v. 121, no. 7, 150 p., 24 pls., 1 fig.
—— —— 1957, Eleven new genera of Foraminifera, *in* Studies in Foraminifera: U.S. Natl. Mus. Bull. 215, p. 223-232, pls. 72-73.
—— —— 1961a, Remarks on the systematics of the Sarkodina (Protozoa), renamed homonyms and new and validated genera: Biol. Soc. Washington Proc., v. 74, p. 213-234.
—— —— 1961b, Suprageneric classification of the Rhizopodea: Jour. Paleontology, v. 35, p. 245-330.
—— et al, 1964, Sarcodina, chiefly "Thecamoebians" and "Foraminiferida," *in* Treatise on

invertebrate paleontology, pt. C, Protista 2 (1) and (2): Geol. Soc. America and Univ. Kansas Press, 900 p., 653 figs.

Lutz, G. C., 1951, The Sobrante Sandstone: California Univ. Pubs. Geol. Sci. Bull., v. 28, no. 13, p. 367-406, 3 figs., pls. 15-18.

Lutze, G. F., 1962, Variationsstatistik and Okologie bei rezenten Foraminiferen: Palaont. Zeitschr., v. 36, no. 3/4, p. 252-264, pl. 24.

Lyell, Charles, 1830-1833. Principles of geology: London, J. Murray (v. 1, 1830; v. 2, 1832; v. 3, 1833).

MacGinitie, H. D., 1943, Central and southern Humboldt County: California Div. Mines and Geology Bull. 118, p. 633-635.

Mandel, D. J., Jr., 1973, Latest Pliocene Foraminifera in the upper part of the San Diego Formation, California: San Diego Assoc. Geol., p. 33-36.

Mallory, V. S., 1959, Lower Tertiary biostratigraphy of the California Coast Ranges: Tulsa, Okla., AAPG, 416 p., 42 pls.

Martin, Bruce, 1914, Descriptions of new species of fossil Mollusca from the later marine Neocene of California: California Univ. Pubs. Dept. Geology Bull., v. 8, no. 7, p. 181-202, pls. 19-22.

—— 1916, The Pliocene of middle and northern California: California Univ. Pubs. Dept. Geology Bull., v. 9, no. 15, p. 215-259.

Martin, Lewis, 1952, Some Pliocene Foraminifera from a portion of the Los Angeles basin, California: Cushman Found. Foram. Research Contr., v. 3, p. 107-141, 3 figs., pls. 17-25.

McDonald, J. A., 1930, Summary and reproduction of plates of: fossil Foraminifera from Kar Nikobar, *in* Conrad Schwager, 1866, Reise der Osterreichischen Fregatte Novara. . . : Stanford Univ. Micropaleontology Bull., v. 2, no. 3, p. 67-70, 1 pl.

Miller, T. G., 1965, Time in stratigraphy: Paleontology, v. 8, pt. 1, p. 113-131.

Natland, M. L., 1933, Temperature and depth ranges of some Recent and fossil Foraminifera in the southern California region: Scripps Inst. Oceanography Tech. Ser. Bull., v. 3, no. 10, p. 225-230, 1 table.

—— 1938, New species of Foraminifera from off the west coast of North America and from the later Tertiary of the Los Angeles basin: Scripps Inst. Oceanography Tech. Ser. Bull., v. 4, p. 137-164, pls. 3-7.

—— 1950, Report on the Pleistocene and Pliocene Foraminifera, *in* 1940 E. W. Scripps cruise to the Gulf of California: Geol. Soc. America Mem. 43, pt. 4, p. 1-55, pls. 1-11.

—— 1952, Pleistocene and Pliocene stratigraphy of southern California: Univ. California at Los Angeles, PhD thesis, 165 p., 20 pls., 11 figs.

—— 1953, Pleistocene and Pliocene stratigraphy of southern California (abs.): Pacific Petroleum Geology Newsletters, v. 7, no. 2, p. 2 (correlation chart).

—— 1957, Paleoecology of West Coast Tertiary sediments, *in* Treatise on marine ecology and paleoecology; V. 2, paleoecology: Geol. Soc. America Mem. 67, p. 543-572, 2 figs., 6 pls.

—— 1963, Paleoecology and turbidites (presidential address, SEPM): Jour. Paleontology, v. 37, p. 946-951, 3 figs.

—— and Ph. H. Kuenen, 1951, Sedimentary history of the Ventura basin, California, and the action of turbidity currents: SEPM Spec. Pub. 2, p. 76-107, 25 figs.

—— and W. T. Rothwell, 1954, Fossil Foraminifera of the Los Angeles and Ventura regions, California: California Div. Mines Bull. 170, chap. 3, p. 33-42, 6 figs.

Norton, R. D., 1930, Ecologic relations of some Foraminifera: Scripps Inst. Oceanography Tech. Ser. Bull., v. 2, no. 9, p. 331-388, 6 tables.

Norvang, Aksel, 1958, *Islandiella* n.g. and *Cassidulina* d'Orbigny: Dansk Naturhistorisk Forening., Vidensk. Medd., v. 120, p. 25-41, pls. 6-9.

Oakeshott, G. B., 1964, Stratigraphic record of California: California Div. Mines and Geology Mineral Inf. Service, v. 17, no. 2, p. 17-28, 1 map.

Obradovich, J. D., 1965, Age of marine Pleistocene of California (abs.): AAPG Bull., v. 49, p. 1087.

Ogle, B. A., 1951, Geology of the Eel River valley area, Humboldt Co., California: Univ. California at Berkeley, PhD thesis, 392 p., 4 pls., 4 figs.

—— 1953, Geology of Eel River valley area, Humboldt County, California: California Div. Mines Bull. 164, 128 p., 6 pls., 14 figs.

—— 1960, The Eel River basin, *in* California–a traverse of the Klamath uplift, northern Coast Ranges and Eel River basin: Geol. Soc. Sacramento Ann. Field Trip Guidebook, p. 32-34.

Olsson, R. K., 1974, Pleistocene paleo-oceanography and *Globigerina pachyderma* (Ehrenberg) in site 36 DSDP, northeastern Pacific: Jour. Foram. Research, v. 4, p. 47-60, pls. 1-3.

Orr, W. N., and J. B. Zaitzeff, 1971, A new planktonic foraminiferal species from the California Pliocene: Jour. Foram. Research, v. 1, no. 1, p. 17-19, 1 pl.

Papani, Giovanni, and G. Pelosio, 1962, La serie Plio-Pleistocenica del T. stirone (Parmense occidentale): Soc. Geol. Italiana Boll., v. 81, pt. 4, p. 293-335, pl. 107, 10 figs. [1963].

Parker, F. L., 1962, Planktonic foraminiferal species in Pacific sediments: Micropaleontology, v. 8,

p. 219-254, pls. 1-10.
—— 1964, Foraminifera from the experimental Mohole Drilling near Guadalupe Island, Mexico: Jour. Paleontology, v. 38, p. 617-636.
Phillips, F. J., 1972, Age and correlation of the Eocene Ulatisian and Narizian Stages, California–discussion: Geol. Soc. America Bull., v. 83, p. 2217-2224, 2 figs.
Phleger, F. B., F. L. Parker, and J. F. Peirson, 1953, North Atlantic core Foraminifera: Swedish Deep-Sea Exped. Repts., v. 7, no. 1, 122 p.
Pierce, R. L., 1956, Upper Miocene Foraminifera and fish from the Los Angeles area, California: Jour. Paleontology, v. 30, p. 1288-1314, 6 figs.
Rau, W. W., 1948, Foraminifera from Miocene Astoria Formation in southwest Washington: Jour. Paleontology, v. 22, p. 774-782, 1 pl.
—— 1951, Tertiary Foraminifera from the Willapa River valley of southwest Washington: Jour. Paleontology, v. 25, p. 417-453, 3 pls.
—— 1964, Foraminifera from the Northern Olympic Peninsula, Washington: U.S. Geol. Survey Prof. Paper 374-G, 33 p., pls. 5-7.
—— 1970, Foraminifera, stratigraphy, and paleoecology of the Quinault Formation, Point Grenville - Raft River coastal area, Washington: Washington Div. Mines and Geology Bull. 62, 58 p.
—— 1973, Geology of the Washington Coast: Washington Geol. Earth Resources Div. Bull. no. 66, 58 p., 71 figs.
Reed, R. D., 1933, Geology of California: Tulsa, Okla., AAPG, 355 p., 58 figs.
—— 1937, California as a structural type: AAPG Bull., v. 21, p. 549-559, 5 figs.
Resig, J. M., 1958, Ecology of Foraminifera of Santa Cruz basin, California: Micropaleontology, v. 4, p. 287-308, 16 figs.
—— 1962, The morphological development of *Eponides repandus* (Fichtel & Moll), 1798: Cushman Found. Foram. Research Contr., v. 13, p. 55-57, pl. 14.
—— 1964, The southernmost occurrence of *Elphidiella hannai* (Cushman and Grant), 1927, off the west coast of North America: Jour. Paleontology, v. 38, p. 393-396, 3 figs.
Reuss, A. E., 1862, Die Foraminiferen-Familie der Lagenideen: Sitz. k. Akad. Wiss. Wien, Math-Naturw., v. 96, no. 1, p. 305-342, pls. 1-8.
Reyment, R. A., 1959, The foraminiferal genera *Afrobolivina* gen. nov. and *Bolivina* in the Upper Cretaceous and lower Tertiary of West Africa: Stockholm Contr. Geology, v. 3, no. 1, p. 1-57, 7 pls.
Rice, S. J., 1961, Geologic sketch of the Northern Coast Ranges: Calif. Div. Mines and Geology Mineral Inf. Service, v. 14, no. 1, p. 1-9, 4 pls., 1 fig.
Riedel, W. R., 1957, Radiolaria, a preliminary stratigraphy: Swedish Deep-Sea Expedition Repts., v. 6, pt. 3, p. 61-96, 4 pls.
—— 1973, Cenozoic planktonic micropaleontology and biostratigraphy: Ann. Rev. Earth and Planetary Sci., v. 1, p. 241-268, 3 tables.
—— M. N. Bramlette, and F. L. Parker, 1963, "Pliocene-Pleistocene" boundary in deep-sea sediments: Science, v. 140, no. 3572, p. 1238-1240, 1 fig.
Roda, Cesare, 1971, Lo stratotipo del piano Zancleano Sequenza, 1868: Accad. Gioenia Sci. Nat. Catania Boll., ser. 4, v. 10, no. 6-7, p. 443-453.
Ruggieri, Giuliano, 1965, A contribution to the stratigraphy of the marine lower Quaternary sequence in Italy: Geol. Soc. America Spec. Paper 84, p. 141-152, 2 figs., 1 pl.
Sars, M., 1868, Fortsatte Bemaerkninger over det dyriske Livs Udbredning: Havets Dybder: Vadensk.-Selsk. Christiania, Vorthandl., v. 1868, p. 246-275[1869].
Schenck, H. B., 1940, Applied paleontology: AAPG Bull., v. 40, p. 1752-1778, 5 figs., 3 pls.
—— 1945, Geologic application of biometrical analysis of molluscan assemblages: Jour. Paleontology, v. 19, p. 504-521, 3 figs., pls. 66-67.
Schenck, H. G., L. N. Waterfall, H. Hardy, and J. P. Fox, 1925, Some index fossils from Pacific Coast Tertiary horizons: privately printed, Univ. California and Stanford Univ., 27 p., 6 pls.
Schwager, Conrad, 1866, Fossile Foraminiferen von Kar-Nicobar: Reise der Österreichischen Fregatte Novara, Geol. Theil, v. 2, p. 187-268, pls. 4-7.
Sclater, J. G., et al, 1974, A comparison of the magnetic and biostratigraphic time scales since the Late Cretaceous, *in* Initial Reports of the Deep Sea Drilling Project, V. 22: Washington, D.C., U.S. Govt. Printing Office, p. 381-386, 2 figs., 2 tables.
Selli, Raimondo, 1967, The Pliocene-Pleistocene boundary in Italian marine sections and its relationship to continental stratigraphies: Prog. Oceanography, v. 4, p. 67-86, 2 tables.
Sherborn, O. D., 1955, An index to the genera and species of the Foraminifera: Smithsonian Misc. Colln., v. 132, 485 p.
Silver, E. A., 1971a, Transitional tectonics and late Cenozoic structure of the continental margin off northernmost California: Geol. Soc. America Bull., v. 82, p. 1-22, 19 figs.
—— 1971b, Small plate tectonics in the northeastern Pacific: Geol. Soc. America Bull., v. 82, p. 3491-3496, 5 figs.

—— 1971c, Tectonics of Mendocino Triple Junction: Geol. Soc. America Bull., v. 82, p. 2965-2978, 17 figs.
—— 1974a, Geometric principles of plate tectonics, *in* Geologic interpretations from global tectonics with applications for California geology and petroleum exploration: San Joaquin Geol. Soc., 3 p., 2 figs.
—— 1974b, Basin development along translational continental margins, *in* Geologic interpretations from global tectonics with applications for California geology and petroleum exploration: San Joaquin Geol. Soc., 5 p., 7 figs.
Sliter, W. V., 1970, Inner-neritic Bolivinitidae from the eastern Pacific margin: Micropaleontology, v. 16, p. 155-174, pls. 1-8.
Smith, J. P., 1919, Climatic relations of the Tertiary and Quaternary faunas of the California region: California Acad. Sci. Proc., ser. 4, v. 9, p. 123-173, 1 pl.
Smith, L. A., 1969, Pleistocene discoasters from the stratotype of Calabrian Stage (Santa Maria di Catanzaro) and the section at Le Castella, Italy: Gulf Coast Assoc. Geol. Socs. Trans., v. 19, p. 579-583, 3 figs.
Smith, P. B., 1960, Foraminifera of the Monterey Shale and Puente Formation, Santa Ana Mountains and San Juan Capistrano area, California: U.S. Geol. Survey Prof. Paper 294-M, p. 463-495, pls. 57-59.
—— 1964, Recent Foraminifera of the Pacific Ocean off Central America—ecology of benthonic species: U.S. Geol. Survey Prof. Paper 429-B, 55 p., 6 pls.
Smith, R. K., 1965, Foraminifera from glacio-marine deposits of the Pacific northwest coast of North America from Vancouver, British Columbia, north to Juneau, Alaska: Univ. British Columbia, PhD thesis, 179 p., 22 pls., 3 figs.
Snavely, P. D., Jr., W. W. Rau, and H. C. Wagner, 1964, Miocene stratigraphy of the Yaquina Bay area, Newport, Oregon: Ore Bin, v. 26, no. 8, p. 133-151, text-figs. 1-3.
Srinivason, M. S., and V. Sharma, 1969, The status of the late Tertiary Foraminifera of Kar Nicobar described by Schwager in 1866: Micropaleontology, v. 15, p. 107-110.
—— —— 1973, Stratigraphy and microfauna of Kar Nicobar Island, Bay of Bengal: Geol. Soc. India Jour., v. 19, p. 1-11, pl. 3.
—— and S. S. Srivastava, 1974, *Sphaeroidinella dehiscens* datum and Miocene-Pliocene boundary: AAPG Bull., v. 58, p. 304-323, 11 figs.
Stadler, W., 1914, Humboldt County, notes on geology and oil possibilities: California State Mining Bur. Bull. 69, p. 444-454, maps.
Stewart, R. E. and K. C. Stewart, 1930a, "Lower Pliocene" in eastern end of Puente Hills, San Bernardino County, California: AAPG Bull., v. 14, p. 1445-1450.
—— —— 1930b, Post Miocene Foraminifera from Ventura quadrangle, Ventura County, California: Jour. Paleontology, v. 4, p. 60-72, pls. 8-9.
—— —— 1933, Notes on the Foraminifera of the type Merced at Seven Mile Beach, San Mateo County, California: San Diego Soc. Nat. History Trans., v. 7, no. 21, p. 259-272, pls. 16-17.
—— —— 1949, Local relationships of the Mollusca of the Wildcat coast section, Humboldt County, California, with related data on the Foraminifera and Ostracoda: Oregon Dept. Geology and Mineral Industries Bull., no. 36, pt. 8, p. 165-208, figs. 5-7, pls. 19-22.
Story, J. A., V. E. Wessels, and J. A. Wolfe, 1966, Radiocarbon dating of recent sediments in San Francisco Bay: Calif. Div. Mines and Geology Mineral Inf. Service, v. 19, no. 3, p. 47-50.
Strand, R. G., compiler, 1962, Geologic map of California—Redding sheet: California Div. Mines and Geology Map, scale 1:250,000, and explanatory data sheet.
Sullivan, F. R., 1962, Foraminifera from the type section of the San Lorenzo Formation, Santa Cruz County, California: California Univ. Pubs. Geol. Sci., v. 37, no. 4, p. 233-352, 23 pls., 5 figs.
Tappan, Helen, 1951, Late Cenozoic Foraminifera from Alaska: Cushman Found. Foram. Research Contr., v. 2, no. 1, p. 4-18, pls.
Thalmann, H. E., 1950, *Uvigerina hollicki*, Nom. Nov.: Cushman Found. Foram. Research Contr., v. 1, no. 1, p. 45.
—— 1960, An index to the genera and species of the Foraminifera, 1890-1950: George Vanderbilt Foundation, Stanford Univ., 393 p.
Tipton, A., R. M. Kleinpell, and D. W. Weaver, 1973, Oligocene biostratigraphy, San Joaquin Valley, California: California Univ. Pubs. Geol. Sci., v. 105, 81 p., 12 figs., 14 pls.
Todd, Ruth, 1965, The Foraminifera of the tropical Pacific collections of the *Albatross* 1899-1900; Pt. 5, Rotaliform families and planktonic families: U.S. Natl. Mus. Bull. 161, 139 p., 28 pls.
Turner, D. L., 1968, Potassium-argon dates concerning the Tertiary time scale and San Andreas fault displacement: unpubl. PhD thesis, Univ. California at Berkeley, 99 p.
—— 1970, Paleontologic correlation and radiometric dating: Geol. Soc. America Spec. Paper 124, p. 91-129, 10 figs.
Uchio, T., 1960, Ecology of living benthonic Foraminifera from the San Diego, California area:

Cushman Found. Foram. Research Spec. Pub. 5, p. 1-72.
Ujiie, H., 1956, The internal structures of some Elphidiidae: Tokyo Kyoiku Daiguku, Sci. Repts., ser. C (Geol. Min., and Geog.), v. 4, no. 38, p. 267-282, pls. 14-15, figs. 1-2.
Valentine, J. W., 1961, Paleoecologic molluscan geography of the Californian Pleistocene: California Univ. Pubs. Geol. Sci., v. 34, no. 7, p. 309-442, 16 figs.
—— 1963, Biogeographic units as biostratigraphic units: AAPG Bull., v. 47, p. 457-466, 2 figs.
van der Vlerk, I. M., 1959, Problems and principles of Tertiary and Quaternary stratigraphy: Geol. Soc. London Quart. Jour., v. 115, no. 1, p. 49-63.
van Voorthuysen, J. H., 1952, *Elphidium oregonense* Cushman and Grant, a possible marker for the Amstelian (lower Pleistocene) in North America and northwestern Europe: Cushman Found. Foram. Research Contr., v. 3, p. 22-23, 1 fig.
—— 1953, Some remarks about the Plio-Pleistocene microbiostratigraphy in northwestern Europe and in North America: Jour. Paleontology, v. 27, p. 601-604.
—— 1957, The Plio-Pleistocene boundary in the North Sea basin: Geologie en Mijnbouw, no. 7, n.s., 19th Jaarg., p. 263-266.
Vaughan, T. W., 1924, Criteria and status of correlation and classification of Tertiary deposits: Geol. Soc. America Bull., v. 35, p. 677-742, 3 charts.
Vervloet, C. C., 1966, Stratigraphical and micropaleontological data on the Tertiary of southern Piedmont (northern Italy): Utrecht, Netherlands, Schotanus and Jens, 88 p.
Wade, M., 1957, Morphology and taxonomy of the foraminiferal family Elphidiidae: Washington Acad. Sci. Jour., v. 47, p. 330-339.
Walton, W. R., 1964, Recent foraminiferal ecology and paleoecology, *in* Approaches to paleoecology: New York, John Wiley & Sons, p. 151-237, 31 figs.
Weaver, C. E., et al, 1944, Correlation of the marine Cenozoic formations of western North America: Geol. Soc. America Bull., v. 55, p. 569-598, charts.
Wezel, F. C., 1968, Le cenozone del Pliocene superiore - Pleistocene inferiore in Sicilia e Lucania: Gior. Geologia, v. 35, no. 3, p. 437-448.
Wheeler, O. C., 1928, Zonal classification of the Pico Formation, Ventura County, California, on the basis of Foraminifera: Stanford Univ. Micropaleontology Bull., v. 1, no. 9, p. 1-4.
White, W. R., 1956, Pliocene and Miocene Foraminifera from Capistrano Formation, Orange County, California: Jour. Paleontology, v. 30, p. 237-260.
Williams, J. Steele, 1954, Problems of boundaries between geologic systems: AAPG Bull., v. 38, p. 1602-1606.
Wissler, S. G., 1937, Foraminiferal zones of the Domingues oil field, Los Angeles County, California (abs.): AAPG Ann. Mtg. Abs., p. 75.
—— 1943, Stratigraphic formations of the producing zones of the Los Angeles basin oil fields, *in* Geologic formations and economic development of the oil and gas fields of California: California Div. Mines and Geology Bull. 118, p. 209-234.
—— 1958, Correlation chart of producing zones of Los Angeles basin oil fields, *in* Guide to the geology and oil fields of Los Angeles and Ventura regions: AAPG Pacific Sec. Ann. Mtg., Los Angeles, p. 59-61.
—— and R. D. Crawford, 1948, Miocene-Pliocene boundary in the Los Angeles basin from the viewpoint of the microstratigrapher (abs.): Geol. Soc. America Bull., v. 59, p. 1390.
Wood, Alan, 1949, The structure of the wall of the test in the Foraminifera: Geol. Soc. London Quart. Jour., v. 104, pt. 2, p. 229-255, pls. 13-15.
—— J. Haynes, and T. D. Adams, 1963, The structure of *Ammonia beccarii* (Linné): Cushman Found. Foram. Research Contr., v. 14, p. 156-157, 1 pl.
Woodford, A. O., 1973, Johannes Walther's Law of the Correlation of Facies—discussion: Geol. Soc. America Bull., v. 84, p. 3737-3740.
Woodring, W. P., 1938, Lower Pliocene mollusks and echinoids from the Los Angeles basin: U.S. Geol. Survey Prof. Paper 190, p. 1-67, pls. 1-9.
—— 1951, Basic assumption underlying paleoecology (abs.): Science, v. 113, p. 482.
—— 1952, Pliocene-Pleistocene boundary in the California Coast Ranges: Am. Jour. Sci., v. 250, p. 401-410.
—— 1957a, Cenozoic mollusks of California: *in* Treatise on marine ecology and paleoecology; V. 2, paleoecology: Geol. Soc. America Mem. 67, p. 891-892.
—— 1957b, Marine Pleistocene of California, *in* Treatise on marine ecology and paleoecology, V. 2, paleoecology: Geol. Soc. America Mem. 67, p. 589-598, 1 fig.
—— and M. N. Bramlette, 1950, Geology and paleontology of the Santa Maria District, California: U.S. Geol. Survey Prof. Paper 222, 181 p.
—— —— and W. S. W. Kew, 1946, Geology and paleontology of Palos Verdes Hills: U.S. Geol. Survey Prof. Paper 207, 145 p.
—— —— and R. M. Kleinpell, 1936, Miocene stratigraphy and paleontology of Palos Verdes Hills, California: AAPG Bull., v. 20, p. 125-149.

—— D. D. Hughes, and S. G. Wissler, 1932, Sedimentary deposits in areas covered by excursions in coastal region of southern California: 16th Internat. Geol. Cong. Guidebook 15, Excursion C-1, pl. 2.

—— R. E. Stewart, and R. W. Richards, 1941, Geology of the Kettleman Hills oil field, California: stratigraphy, paleontology and structure: U.S. Geol. Survey Prof. Paper 195, 170 p., 57 pls., 15 figs.

Yeats, R. S., 1965, Pliocene seaknoll at South Mountain, Ventura basin, California: AAPG Bull., v. 49, p. 526-546.

—— and W. A. McLaughlin, 1970, Potassium-argon mineral age of an ash bed in the Pico Formation, Ventura basin, California: Geol. Soc. America Spec. Paper 124, p. 173-206, 4 figs., 3 tables.

—— —— and G. Edwards, 1967, K-Ar mineral age of ash bed in Pico Formation, Ventura basin, California: AAPG Bull., v. 51, p. 486.

Explanation of Indexing

A reference is indexed according to its important, or "key," words.

Three columns are to the left of a keyword entry. The first column, a letter entry, represents the AAPG book series from which the reference originated. In this case, S stands for Studies in Geology Series. Every five years, AAPG will merge all its indexes together, and the letter S will differentiate this reference from those of the AAPG Memoir Series (M) or from the AAPG Bulletin (B).

The following number is the series number. In this case, 11 represents a reference from Studies in Geology No. 11.

The last column entry is the page number in this volume where this reference will be found.

Note: This index is set up for single-line entry. Where entries exceed one line of type, the line is terminated. (This is especially evident with manuscript titles, which tend to be long and descriptive.) The reader must sometimes be able to realize keywords, although commonly taken out of context.